Pseudomonas syringae Pathovars
and Related Pathogens

Pseudomonas syringae Pathovars and Related Pathogens – Identification, Epidemiology and Genomics

Edited by

M'Barek Fatmi
Institut Agronomique et Vétérinaire Hassan II, Complexe Horticole, Agadir, Morocco

and

Alan Collmer
Cornell University, Ithaca, NY, U.S.A.

Nicola Sante Iacobellis
Università della Basilicata, Potenza, Italy

John W. Mansfield
Imperial College, Wye, Ashford, Kent, U.K.

Jesus Murillo
Universidad Pública de Navarra, Pamplona, Spain

Norman W. Schaad
USDA-ARS, Ft. Detrick, MD, U.S.A.

Matthias Ullrich
International University Bremen, Germany

Springer

M'Barek Fatmi
Institut Agronomique et
Vétérinaire Hassan II, Complexe Horticole
Agadir, Morocco

Nicola Sante Iacobellis
Università della Basilicata
Potenza
Italy

Jesus Murillo
Universidad Pública de
Navarra, Pamplona, Spain

Matthias Ullrich
International University
Bremen, Germany

Alan Collmer
Cornell University
Ithaca
NY, U.S.A.

John W. Mansfield
Imperial College, Wye
Ashford, Kent
U.K.

Norman W. Schaad
USDA-ARS, Ft. Detrick
MD, U.S.A.

ISBN 978-1-4020-6900-0   e-ISBN 978-1-4020-6901-7

Library of Congress Control Number: 2007940857

Printed on acid-free paper

9 8 7 6 5 4 3 2 1

springer.com

Preface

The Conference on *Pseudomonas syringae* which started in 1973 as an informal meeting of a group of scientists working on these bacteria in Angers, France, has become more and more important with time. Many meetings have been held since then: 1984, 1987, 1991, 1995, and 2002 in Cape Sounion, Greece; Lisbon, Portugal; Florence, Italy; Berlin, Germany; and Maratea, Italy; respectively. This Conference is considered as the most important scientific forum in which recent advances in different research aspects on *Pseudomonas syringae*, a plant pathogenic bacterial species that includes a high number of pathogens (referred as pathovars) and Related Pathogens such as *Acidovorax, Burkholderia, Ralstonia,* affecting several economically important crops. The proceedings resulting from these meetings are considered as valuable sources of information related to this group of pathogens.

The interest in organising this conference regularly is reflected by the attendance of more than 80 scientists from 20 countries worldwide, who participated at the 7th International Conference on *Pseudomonas syringae* pathovars and related pathogens organized by the Institut Agronomique et Vétérinaire Hassan II in Agadir, Morocco, from 13th to 16th November 2006.

Recent advances on:

- New methods and approaches for specific and sensitive detection and identification of *Pseudomonas syringae* and *Ralstonia solanacearum*
- Ecology and epidemiology bases of *Pseudomonas syringae* that enable the development of management strategies
- Pathogenesis and determinant of pathogenicity, and in particular, mechanisms involved in virulence and virulence gene expression
- Evolution and diversity of the pseudomonads through multilocus sequence typing (MLST) analysis
- Determination of pathogens associated with new and emerging diseases
- Effect of global warming on increase and emergence of new bacterial diseases; are reported in 43 papers written by leading scientists in the respective fields.

In this volume, manuscripts of the oral presentations and posters, presented at the 7th International Conference on *Pseudomonas syringae* pathovars and related pathogens, are combined under six section parts Identification and Detection; Epidemiology and Disease Management; Pathogenesis and Determinants of Pathogenicity; Genomics and Molecular Characterization; Taxonomy and Evolution; and New Emerging Pathogens. Each section part is introduced by a key review paper. All the papers presented in this volume have been reviewed by the editors.

I gratefully acknowledge the fruitful collaboration of the scientific and organizing committees, the financial support of several institutions and private companies and Professor M.C. Harrouni for his precious assistance during the organization of the Conference and the preparation of this book.

Professor M'Barek FATMI

Contents

Sponsors and Donors

The organizing committee is grateful to the sponsors and donors listed below. Their contribution was of great help for the success of this scientific event.

- Direction des Domaines Agricoles (DAA)
- Celtis France
- Comptoir Agricole du Souss (CAS)
- Coopérative Agricole Azro
- Omnium Agricole du Souss (SAOAS)
- Société de Développement Agricole (SODEA)
- Société Ezzouhour
- COPAG
- Domaines El Boura
- Groupe Delassus
- Société de Gestion des Exportations des Produits Agricoles (GPA)
- Syngenta, Maroc
- Agripharma
- Bureau d'Ingénierie en Horticulture et Agro-industrie (BIHA)
- Casem
- Coopérative Agricole Adrar
- Hortec

Part I
Identification and Detection

Current Technologies for *Pseudomonas* spp. and *Ralstonia solanacearum* Detection and Molecular Typing

M.M. López[1], J.M. Quesada[1], R. Penyalver[1], E.G. Biosca[2], P. Caruso[1], E. Bertolini[1], and P. Llop[1]

Abstract Standard protocols for detection of phytopathogenic *Pseudomonas* spp. and *Ralstonia solanacearum* in plant material, soil, water or other sources, often still rely on the isolation of bacterial colonies on appropriate media, and/or on the use of serological techniques. However, over the last several years, molecular techniques, mainly based on PCR methods after extraction of nucleic acids from samples, have improved enough to allow a more rapid, reliable detection of these bacteria. When maximum accuracy is required the use of multiple techniques in an integrated approach is advised. Other promising technologies like flow cytometry, electronic nose or microarrays are emerging due to the need for developing more rapid high throughput detection methods.

Molecular typing methods are very useful to study intra-specific diversity, to identify sources of inoculum and to track the spread of the infections, although phenotypic methods are also used. Several reliable molecular techniques including AFLP, BOX, ERIC and rep-PCR, IS-RFLP, MLST, and macrorestriction-PFGE are available. However there is no single technique that can be recommended and polyphasic analyses have provided to be most useful. The diversity of Spanish strains of *R. solanacearum* and *P. savastanoi* pv. *savastanoi* analysed by several techniques is discussed in detail.

Keywords PCR, diversity, AFLP, IS*53*-RFLP, PFGE, *P. savastanoi* pv. *savastanoi*

[1] Instituto Valenciano de Investigaciones Agrarias (IVIA). Carretera de Moncada a Náquera, Km 4.5. 46113 Moncada (Valencia), Spain

[2] Departamento de Microbiología y Ecología, Universidad de Valencia, Av. Dr. Moliner, 50. 46100 Burjassot (Valencia), Spain

Author for correspondence: M.M. López; e-mail: mlopez@ivia.es

M'B. Fatmi et al. (eds.), *Pseudomonas syringae Pathovars and Related Pathogens.*
© Springer Science + Business Media B.V. 2008

1 Introduction

Preventive measures for an integrated control of plant pathogenic bacteria include the use of pathogen-free propagative plant material, growth of plants in soil or substrate free of pathogens, and use of pathogen-free irrigation water. The implementation of such measures requires detection methods of high sensitivity, specificity and reliability (López et al., 2005). Furthermore, phytopathogenic bacteria may remain latent and can be present in low populations, making detection in plant material difficult. This required for rapid and accurate techniques is especially important for quarantine pathogens, because of the high risk they represent.

Efficient methods to detect phytopathogenic bacteria in soil, sediments, water, sewage, agricultural samples or air, are required to assess the role of the different inoculum sources on the ecology, epidemiology and life cycle of the target plant pathogen (López et al., 2006).

2 Techniques Commonly Used for Detection of Phytopathogenic *Pseudomonas* and *Ralstonia*

2.1 Standard Protocols

Detection of bacteria in seeds, fruits, plants, propagative material, or any other reservoir often relies on their isolation on common or semiselective media. This is followed by colony identification by their morphological, biochemical and serological characteristics and pathogenicity assays (López et al., 2003; Saettler et al., 1995; Schaad et al., 2001). However, when analysing samples in natural conditions, isolation on agar media will not detect stressed or injured bacteria or those in the viable but non culturable state (VBNC) (Grey & Steck, 2001; Ordax et al., 2006). In state the bacterial cells are unable to multiply sufficiently even on non-selective agar medium to yield visible growth as a colony.

For detection of bacteria commercial serological kits can be very useful, but they often fail to detect the pathogen in asymptomatic tissues. Furthermore, the number of useful specific antibodies available for detection is not great (Álvarez, 2004).

The current trend for detection of bacteria in the European Union (EU) is to use integrated approaches (Álvarez, 2004; López et al., 2005) including conventional, serological, and molecular techniques and to validate the protocol in ring tests (López et al., 2006). Diagnostic protocols for detection of 23 viruses, bacteria, fungi, nematodes, and insects considered as quarantine organisms in the EU, have been set up and validated in a project financed by the "Standard, Measurements and Testing" programme of the EU. The approved protocols are available through the web page of the Central Science Laboratory (www.csl.gov.uk/prodserv/know/diagpro) and have being published by the European and Mediterranean Plant Protection Organization (EPPO) (www.eppo.org). After a similar procedure, detection and protocols for *R. solanacearum* have been published as EU Directives

(Anonymous, 1998; Anonymous, 2006) and by the EPPO (Anonymous, 2004). The official protocol of the EU for *R. solanacearum* analysis is based on the combined use of several techniques (Fig. 1). It includes a detection scheme for symptomatic potato tubers and tomato or other host plants, a detection scheme for asymptomatic

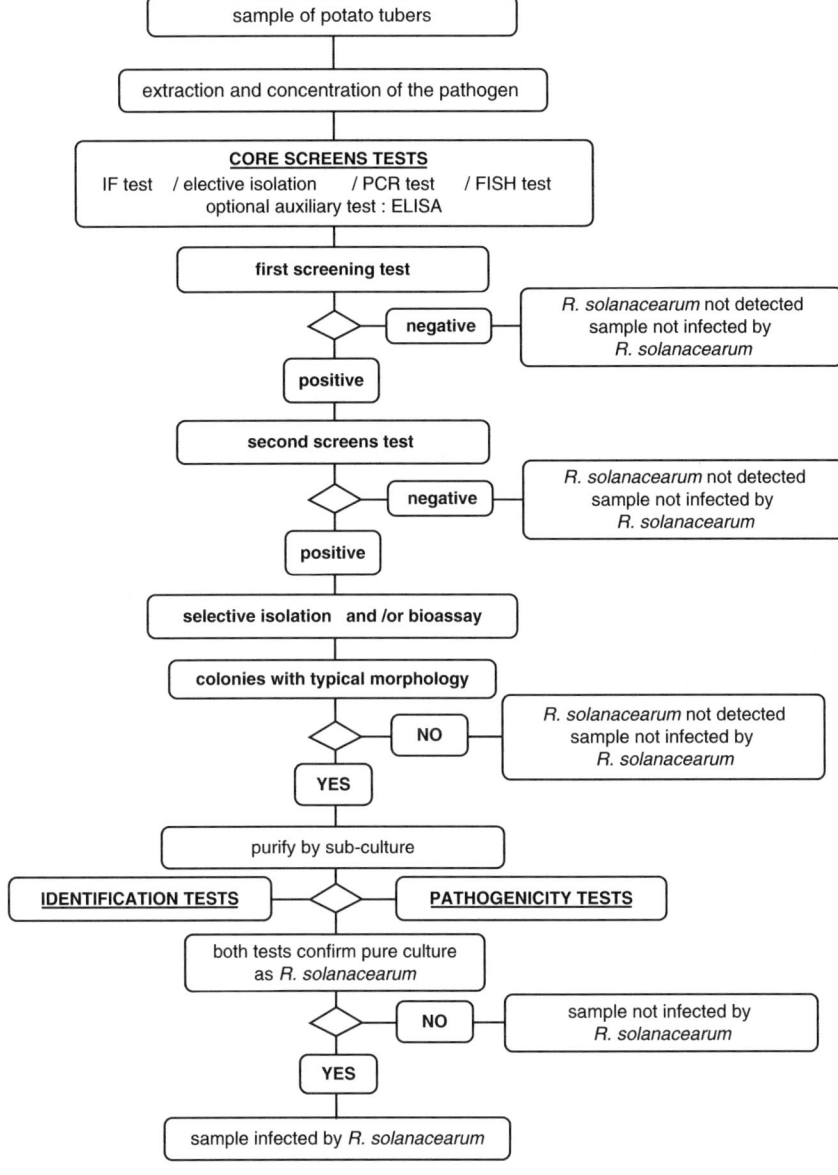

Fig. 1 Scheme for detection and identification of *Ralstonia solanacearum* in samples of asymptomatic potato tubers (Anonymous, 2006).

potato tubers and potato, tomato or other host plants, another for soil, water, potato, processing waste or sewage sludge, as well as techniques for identification of *R. solanacearum* strains. The only standard protocol recently published for the detection of phytopathogenic species of *Pseudomonas* is for *P. syringae* pv. *persicae* (Anonymous, 2005).

2.2 Molecular Methods

Nucleic-acids-based methods are stable, rapid and specific (López et al., 2003; Louws et al., 1999; Tenover et al., 1997; Vandamme et al., 1996) but have not yet completely replaced traditional isolation and phenotypic characterization in plant bacteriology. Most of the relatively new molecular tests are still being compared with conventional methods, validated and standardized before being accepted in official protocols (López et al., 2006).

The molecular methods most frequently used for detection and identification of plant pathogenic bacteria are amplification assays based on PCR. For their use in the analysis of plant or environmental samples, efficient DNA or RNA extraction or purification protocols are required. The preparation of samples is critical and target DNA or RNA should be purified to be available for the DNA polymerase. This aspect is crucial when detection methods are devised, but less important for bacterial identification, because the latter employs purified bacterial cells and a large amount of DNA is available. Depending on the material to be analysed the extraction method can be quite simple or more complex. The use of commercial kits has gained acceptance for detection due to the easiness of use and the avoidance of toxic reagents during the process (among others RNeasy and DNeasy Plant System, Qiagen, USA; Easy-DNA-Extraction kit, Invitrogen, USA; Nucleon plant tissue, Amersham, UK; EaZy Nucleic Acid Isolation Plant DNA kit, Omega Bio-tek, USA; Wizard Genomic DNA, Promega, USA; Extract-N-Amp Plant PCR kit, Sigma, USA) are used for different bacterial targets (López et al., 2006). Nevertheless, simple laboratory protocols have been developed reducing cost and time (Llop et al., 1999; Cubero et al., 1999; López et al., 2006). Several commercial automated systems, like ABI PRISM 6100 Nucleic Acid PrepStation (Applied Biosystems, USA) and QIAcube-Pure Efficiency (Qiagen, USA) allow the extraction and analysis of nucleic acids from plant samples and microorganisms, but they are not always efficient with all types of material.

In order to increase the population of the target and to avoid inhibitors, an enrichment step in the most appropriate general or semi-selective solid or liquid medium (Schaad et al., 1995, 2007; López et al., 1997, 2001) can be included before detection of *Pseudomonas* spp. and *Ralstonia solanacearum* from plant material, soil, sewage, water, etc. Afterwards, the target bacterium can be detected at a much lower level (López et al., 2001). Another alternative is to use immunomagnetic separation with magnetic beads coated with specific antibodies (De León et al., 2006; Walcott & Gitaitis, 2000).

2.2.1 Primer Design

The accuracy of nucleic acid technology for detection is based on the use of specific sequences (oligonucleotides/probes). As PCR is the most frequently used technique we will discuss here the design of primers for *Pseudomonas* spp. and *Ralstonia*, being used in the different formats of such technique. Different strategies have been developed to design PCR primers for specific pathogen detection but, in general, the DNA sequences from which the primers are designed for *Pseudomonas* and *Ralstonia* detection come from three main origins: pathogenicity/virulence genes, ribosomal genes, and plasmid genes (López et al., 2006). The pathogenicity genes used as targets can be involved in any of the several steps that lead to the development of symptoms. The ribosomal operon and the Internal Transcribed Spacer (ITS) region have been employed in several protocols (Louws et al., 1999). Genus-specific rDNA sequences of the phytobacteria are now available (Louws et al., 1999), and many diagnostic primers based on such sequences have been developed for detection of a number of plant pathogens (Louws et al., 1999). Then, this strategy employs primers from conserved regions of the 16S and 23S ribosomal genes to amplify the ITS region, that can include several tRNA genes and noncoding regions. Plasmid DNA is also widely employed in the design of primers: the plasmid genes amplified may be associated with pathogenicity, as indicated above, or be of unknown function. However a major problem with plasmids can be their possible lack of stability and their transmissibility (López et al., 2006; Louws et al., 1999).

Other sources of primers can be anonymous DNA, obtained through molecular analysis by different techniques, as random amplified polymorphic DNA (RAPDs) for *Pseudomonas corrugata* (Catara et al., 2000). Genomic subtraction, a powerful non-sequencing approach to find genetic differences between bacterial strains (Agron et al., 2002; Mills et al., 1997), can identify nearly all major sequence differences between two closely related bacteria and has been used to design specific probes to identify *R. solanacearum* (Seal et al., 1992).

The list of available primers for the detection of *Pseudomonas* species is increasing and often several sets of primers to several targets of the same pathogen are available (Table 1). At least 24 different primers pairs had been designed to detect *R. solanacearum* (Arahal et al., 2004). An important feature to take into account is the reliability of the information available in the sequence databases from which to perform the design of specific primers for detection. Unfortunately, often times published primers have discrepancies with the sequences to which they should match (Arahal et al., 2004).

2.2.2 Classical PCR, Nested PCR, Co-Operational PCR and Multiplex PCR

Classical PCR protocols have been developed for the many important plant pathogenic bacteria (López et al., 2003; Álvarez, 2004). They normally show a relatively good sensitivity for bacterial detection in plant material (10^3–10^4 bacterial cells/ml

Table 1 PCR protocols for detection or identification of several *Pseudomonas* species

Species	Target DNA	Number of sets of primers designed
Pseudomonas spp.	16S rRNA	1 (1998)
P. avellanae	*hrp W*	1(2002)
P. corrugata	Unknown	1 (2000)
P. syringae	IS*50*, Tabtoxin	1 (1997), 1 (1998)
pv. *actinidiae*	*arg K*	1 (1997), 1 (2002)
pv. *atropurpurea*	Coronatine, *cfl*	1 (1996), 1 (1995)
pv. *phaseolicola*	Phaseolotoxin, *Tox*-, arg K	5 (1991–1996), 1 (2006) 2 (1997), 1 (2006), 1 (2007)
pv. *papulans*	*hrp L*	1 (2002), 1 (2006)
pv. *pisi*	RAPD	1 (1996)
pv. *syringae*	*syr B, syr D*	1 (1998), 1 (2003)
pv. *tagetis*	Tagetitoxin	1 (2004)
pv. *tolaasi*	Tolaasin	1 (2004)
pv. *tomato*	*hrp 2*	1 (2005)
pv. *cannabina*	*efe*	1 (1997)

extract) and good specificity. For both reasons there is an increasing number of laboratories that include them in routine detection, after a DNA extraction step. Some sensitivity problems associated with classical PCR have been overcome by using nested PCR which can detect about $1–10^2$ bacterial cells/ml extract (Prosen et al., 1993). The process is based on two consecutive rounds of amplification but when it is performed in different tubes it increases the risk of contamination, especially when such method is used on large scale (López et al., 2006). To avoid it, the single tube nested PCR has been proposed for *P. savastanoi* pv. *savastanoi* (Bertolini et al., 2003b).

The co-operational amplification proposed for sensitive and specific detection of some viruses, has been applied to *R. solanacearum* (Caruso et al., 2003). The process can be performed easily in a simple reaction based on the simultaneous action of three or four primers. When coupled with colorimetric detection, the sensitivity is about 1–10 bacterial cells/ml extract. However, the low final volume of reagents can increase the susceptibility to inhibitors in the sample, requiring also an initial DNA extraction or an enrichment step.

Multiplex amplification is based on the use of a PCR mix with different compatible primers, specific to different targets. The use of a pair of common primers to amplify different targets is not advised because the reaction will be displaced to the most abundant target. Another technique, multiplex nested PCR in a single reaction, combines the advantages of the multiplex with the sensitivity, specificity and reliability of the nested, but it needs an accurate design of compatible primers. It can be used for the simultaneous detection of several pathogens (viral RNA and bacterial or fungal DNA targets) in a single analysis. Nevertheless, only in one case has this technology been carried out in a single reaction for including specific detection of four viruses (*Cucumber mosaic virus*, *Cherry leaf roll virus*, *Strawberry latent ring spot virus* and *Arabis mosaic virus*) and the bacterium *P. savastanoi* pv.

savastanoi in olive plant material, using 20 compatible primers (Bertolini et al., 2003a) and it is proposed as an analysis for certification purposes. Multiplex nested-PCR saves time and reagents because it can be performed in a single reaction. The sensitivity achieved for detection of *P. savastanoi* pv. *savastanoi* by multiplex nested RT-PCR was similar to the sensitivity reached by applying the monospecific nested PCR after an enrichment step, which demonstrated to be 100-fold more sensitive than conventional PCR (Penyalver et al., 2000; Bertolini et al., 2003b). This multiplex nested RT-PCR has been coupled with colorimetric detection increasing the sensitivity and facilitating the interpretation of results (Bertolini et al., 2003a).

2.2.3   Real-Time PCR

Classical PCR-based methods have good sensitivity and specificity but do not provide quantitative data and requires agarose gel electrophoresis and hybridisation as the confirming endpoint analysis. On the contrary, real-time (RT) PCR allows amplification, detection, monitoring and quantitation in a single step by employing several chemistries which are used to detect PCR products as they accumulate within a closed reaction vessel during the reaction. It has been proposed for detection of *R. solanacearum* (Weller et al., 2000; Ozakman & Schaad, 2003) and for *P. savastanoi* pv. *phaseolicola* (Schaad & Frederic, 2002; Schaad et al., 2007) among other plant pathogenic bacteria and it is expected that protocols for the most common *Pseudomonas* will be developed in the near future.

Moreover, the advantages this technology offers is leading to more quickly and accurate detection protocols. In addition, the identification of a pathogen in imported material may cause problems at the point of introduction, especially for perishable commodities because the time needed for sending the sample to a specialized laboratory represents a delay to take the appropriate measures. This can be solved with portable real-time PCR instruments (RAPID. system, Idaho Technology, Utah, USA; Smart Cycler, Cepheid, USA), that allow a rapid on-site diagnosis (Schaad et al., 2003).

A comparative evaluation of several described PCR protocols for their sensitivity in detection of *R. solanacearum performed in our laboratory* is shown in Table 2. The different protocols, based in the use of several sets of primers reached sensitivity levels from 10^{-1} to 10^3 cfu/ml in pure cultures and from 10 to 10^3 cfu/ml in plant material, after DNA extraction following Llop et al. (1999) or using commercial kits.

2.3   *Flow Cytometry*

Flow cytometry (Davey & Kell, 1996) is a technique for rapid identification of cells or other particles as they pass individually through a sensor in a liquid stream, providing quantitative and sensitive detection in few hours. Bacterial cells are

Table 2 Comparison of sensitivity of conventional and real-time PCR protocols for
R. solanacearum detection (López et al., unpublished data)

		Sensitivity	
Reference of the protocol	PCR type	Pure culture (cfu/ml)	Plant material (cfu/ml)
Caruso et al. (2003)	Co-PCR	10^{-1}	10
Seal et al. (1993)	OLI 1/Y2	10^3	10^4
Boudazin et al. (1999)	OLI 1/2	10^3	10^2
Weller et al. (2000)	Real-time	10^2	10^3
Ozakman & Schaad (2003)	Bio real-time	ND[a]	10

[a] ND Not determined

identified by fluorescent dyes conjugated to specific antibodies and detected
electronically using a fluorescence-activated cell sorter, which measures several cellular parameters based on light scatter and fluorescence. Multiparameter analysis
includes cell sizing, fluorescence imaging and gating out, or elimination of unwanted
background associated with dead cells and debris. Flow cytometry has excellent
potential as a research tool and possibility for routine use in seed health testing and
other fields (López et al., 2003). The cost for instrumentation is currently a major
disadvantage that will be solved when less expensive models become available.

2.4 Electronic Nose

Sensor systems for the easy detection of *R. solanacearum* potato tubers have been
developed recently (Stinton et al., 2006). The system operates through an electronic
nose containing a set of sensors selected for their sensitivity to marker volatile
organic compounds and the resulting data are captured, displayed and recorded by
computer. The electronic nose appears promising and this technology would be
applicable to the detection of statutory organisms by plant health and seed
inspectors (de Lacy Costello et al., 2006) but in addition it could be very useful for
accurate identification of pure cultures.

2.5 Microarray Technology

Although the potential of the microarray technology in the detection and diagnosis
of plant diseases is great, the practical development of these applications is still
under progress and few are available for diagnosis of plant pathogens (Schoen et al.,
2002, 2003; Fessehaie et al., 2003). Despite the slow development of the microarray
technology for detection of plant pathogenic bacteria, it shows some potential features that make it a very promising tool. The use of thousands of probes at the same
time allows the possibility of detection and differentiation of several pathogens in

only one analysis. Nevertheless, the need for a previous PCR reaction, the low level of sensitivity achieved, and the high cost of the equipment makes this technique still far from being used for routine analysis of plant pathogens (López et al., 2006) but it could be very useful for accurate identification of pure cultures.

3 Molecular Typing of Phytopathogenic *Pseudomonas* and *Ralstonia*

There are many techniques available for molecular typing of different species of plant pathogenic bacteria. Table 3 summarises the most frequently utilised and their efficiency for identification and diversity in species of both genus. Perhaps the most promising is Multilocus Sequence Typing (MLST). MLST studies are based on sequences from housekeeping genes located on the bacterial core genome, and the method has been proposed for sensitive and reproducible typing of bacterial strains. It allows the assessment of the relative contribution of mutations and recombinations in the evolution of a species and determines the clonal relationships between strains. MLST analysis of a wide range of *P. syringae* strains found that it was a clonal species (Sarkar & Guttman, 2004), and when applied to *R. solanacearum* 18 sequence types were described (Danial et al., 2006).

Other molecular typing methods have demonstrated their efficiency for intra-specific diversity studies in *Pseudomonas* spp. and *R. solanacearum*, but there is no universally advised technique neither for diversity studies nor for molecular typing. We briefly describe here, in two examples, the results obtained when analysing the molecular diversity of Spanish strains of *R. solanacearum* and a collection of *P. savastanoi* pv. *savastanoi* strains isolated from olive trees.

3.1 Diversity of R. solanacearum

Ralstonia solanacearum, responsible of bacterial wilt, is one of the most important bacterial diseases of crops in the world. The bacterium was first detected in potato in Spain in 1995 and, since then, several outbreaks have been identified in different Spanish regions but rapid eradication measures following European Directive 98/57/EC (Anonymous, 1998) have been effective to prevent the spread of the disease.

R. solanacearum represents a heterogeneous group of bacteria (Fegan & Prior, 2005). The species was subdivided into five races based on host range and six biovars based upon carbohydrate utilization patterns. The potato pathogen belongs to race 3, biovar 2. There is no relationship between races and biovars (Hayward et al., 1990).

PCR restriction fragment length polymorphism analysis (PCR-RFLP), amplified fragment length polymorphism (AFLP), sodium dodecyl sulphate polyacrylamide gel electrophoresis (SDS-PAGE), fatty acid methyl esters (FAME) analysis, 16S

Table 3 Some techniques employed for molecular characterization of *Pseudomonas* spp. and *R. solanacearum*

Technique[a]	Identification	Diversity	Remarks
16s rDNA	+++	+	Automated, discrimination at genus level
RFLP	–	++	Difficult, reliable
RAPDs	–	++	Simple, intermediate reliability
Rep-PCR	++	++	Simple, variable reliability
Ribotyping	++	+	Simple, variable reliability
PCR-RFLP	+++	++	Simple, low discrimination at strain level
SSR	–	++	Simple, variable reliability
AFLP	–	+++	Difficult, reliable, discrimination at strain level
PFGE	+	+++	Difficult, reliable, discrimination at strain level
MLST	++	+++	Simple, reliable, discrimination at species level
MLEE	–	++	Difficult
HMA-ITS	–	+++	Difficult, reliable, discrimination at infraspecific level

–: not appropriate, in general
+: sometimes useful
++: generally useful
+++: very useful
[a] Abbreviations: RFLP, restriction fragment-length polymorphism; RAPDs, random amplified polymorphic DNA analyses; Rep-PCR, repetitive PCR fingerprinting; PCR-RFLP, PCR restriction fragment length polymorphism analysis; SSR, short sequence repeats; AFLP, amplified fragment length polymorphism; PFGE, pulsed field gel electrophoresis; MLST, multilocus sequence typing; MLEE, multilocus enzyme profiles; HMA-ITS, heteroduplex mobility assay of internal transcribed spacer

rRNA, gene sequencing, RFLP, amplified ribosomal DNA restriction analysis (ARDRA), macrorestriction pulse field gel electrophoresis (PFGE) and sequence analysis of 16S–23S rRNA, indicate a possible selection of a *R. solanacearum* "European" variant (Timms-Wilson et al., 2001). To date, all potato strains isolated in Europe have been identified as race 3, biovar 2. However, only two studies have focused on the characterization of European strains (Van der Wolf, 1998; Timms-Wilson et al., 2001) showing that some of these techniques are able to discriminate the variability inside European strains of race 3, biovar 2.

The genetic diversity of 44 representative strains of *R. solanacearum* biovar 2 isolated from 1995 to 2002 from different sources in Spain, and reference strains from The Netherlands and USA, have been analysed by phenotypic and genotypic methods, including biochemical and serological characterization, repetitive PCR fingerprinting (Rep-PCR), macrorestriction followed by PFGE and AFLP (Caruso, 2005). This represents the first study of the characteristics of a large collection of Spanish strains of *R. solanacearum*.

Ten PFGE patterns were obtained after *Xba* I digestion and five patterns after using *Spe* I and, in general, the strains showing the same PFGE pattern shared their region of origin. Clustering analysis of the AFLP profiles yielded five clusters and in four of them, strains were also grouped according to their source and origin, but in the remaining one, strains from different geographical origins were also grouped. Figure 2 shows the combined analysis of all these results. It supports the hypothesis that several clones of the pathogen have been introduced into Spain, as previously was suggested by several authors for other European countries (Van der Wolf, 1998; Timms-Wilson et al., 2001).

3.2   *Diversity of* **P. savastanoi** *pv.* **savastanoi**

Studies on the genetic diversity of *P. savastanoi* (Psv) strains have focused in the relationships among strains isolated from different hosts. However, studies on the genetic diversity of a worldwide collection of *P. savastanoi* pv. *savastanoi* isolated from olive are lacking. Only Scortichini et al. (2004) studied the genetic structure of an Italian collection of Psv strains isolated from different provinces and olive cultivars by repetitive PCR, using short interspersed elements. They observed 20 patterns among the 360 Italian strains with an overall similarity of 81%, with no apparent grouping.

A collection of 62 strains of Psv isolated from olive knots (from which 44 were isolated in Spain) were examined for the distribution, variation in positions and copy number of the IS*53* insertion element, originally described in an oleander strain (Soby et al., 1993). Southern hybridization analysis revealed that the genetic

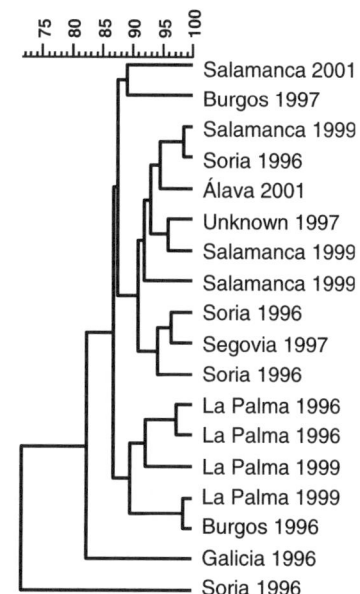

Fig. 2 Dendrogram based on combined results after macrorestriction with *Xba* I and *Spe* I followed by PFGE and AFLP, of *Ralstonia solanacearum* strains from different Spanish areas (Caruso, 2005). Average taken from experiments using Pearson correlation (PFGE), Dice coefficient (AFLP) and UPGMA clustering

Fig. 3 Dendrogram based on RFLP fingerprints of IS*53* element of *P. savastanoi* pv. *savastanoi* strains from olive. Spanish strains are those beginning by S and the following letter indicates the province. The others are reference strains from other countries (Quesada, 2007)

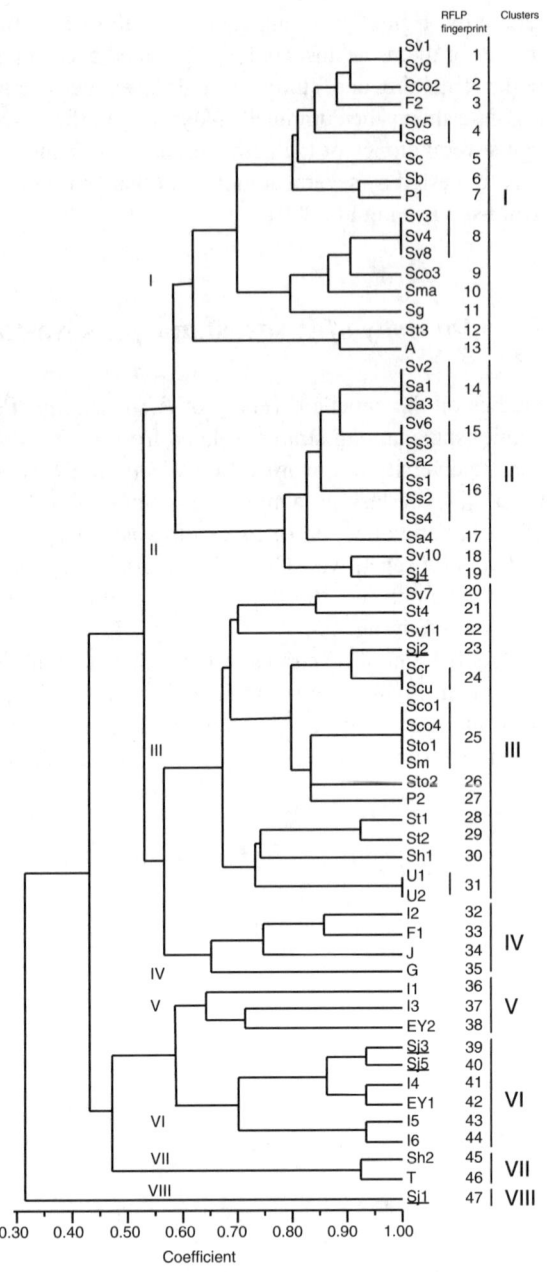

element IS*53* was present in multiple copies in all analysed strains isolated from olive knots. Copy number of IS*53* elements ranged from four up to ten. Southern hybridization analysis of plasmid DNA using IS*53* as a probe revealed that this genetic element is present in the chromosomal replicon in the six olive strains analysed, instead of in plasmids, as it was first described in the original strain. Although Psv strains displayed a remarkably high degree of IS*53* restriction fragment length polymorphism, transposition of this element was not detected in Psv olive strains grown either *in vitro* for up to 390 generations (Quesada, 2007). The genetic diversity of a worldwide collection of Psv strains based on 47 different IS*53*-RFLP fingerprints and UPGMA analysis allowed clustering all strains into eight groups with a similarity of 60% (Fig. 3).

Two groups (II and VIII) were composed by only Spanish strains. One of these groups (VIII), with only one strain from Jaén province, was separated from the rest of strains. Only nine out of 47 distinguishable RFLP fingerprints were shared by more than one strain (1, 4, 8, 14, 15, 16, 24, 25, 31) suggesting a common origin for all these strains. The genetic diversity within strains from Spanish provinces was quite similar to that between provinces, suggesting a lack of any geographical differentiation. Additionally, the Spanish strains were not distinguishable from strains from other sources. Due to the widespread distribution on Psv, its stability *in vitro* and *in vivo*, and the high degree of polymorphism generated, IS*53*-RFLP typing is considered a suitable marker for epidemiological and ecological studies.

4   Conclusions

Developing detection methods is a never-ending process and the detection of plant pathogenic *Pseudomonas* and *Ralstonia* is moving from conventional methods to the use of molecular techniques included in an integrated approach, that is required for maximum accuracy (Álvarez, 2004; López et al., 2005, 2006). Molecular detection is now widely used based on classical and/or PCR or Real-time PCR, but purification of nucleic acids from samples is generally required and should be optimised. PCR and especially real-time PCR are the methods of choice for rapid and accurate diagnosis of plant pathogenic bacteria, but conventional methods as IF are still widely used.

There is a lack of standard protocols for detection of plant pathogenic species and pathovars of most *Pseudomonas*, but there is an official EU protocol for detection of *R. solanacearum*.

The increased number of studies published on the diversity of some *Pseudomonas* species and *R. solanacearum* have shown that for each bacterial model, different techniques must be analysed in preliminary studies, to evaluate their respective value for the purpose of the study. It is expected that with the decrease in the costs of sequencing of total bacterial genomes, molecular techniques will replace the current analysis of diversity and will facilitate molecular typing.

The advances in genomics, proteomics and metabolomics have not had remarkable repercussions yet on developing more reliable detection and identification methods, but their impact should be important in the next few years. This could be due to the fact that developing diagnosis methods is considered of minor scientific interest. However, with the recent access to complete genome sequences, and the microarray possibilities, the function of gene products will soon be determined and their role in bacterial cells will be understood. This could lead to new research issues, especially in the field of diagnosis and hopefully innovative methods will facilitate early detection of bacterial diseases. However, the success of the practical use of these methods will depend on the heterogeneity of the target bacteria, the functional significance of the identified genes in the progression of the disease and the expression of the genes at the different steps of the interaction between the bacteria and the plant hosts.

The development of protocols of higher sensitivity and specificity for detection of plant pathogenic *Pseudomonas* and *Ralstonia* will have a positive effect in the sanitary status of their hosts, reducing the need for pesticide treatments, protecting ecosystems and enhancing the quality of the food and the environment.

Acknowledgements Authors wish to thank many colleagues for sending strains, E.A. Carbonell and J. Pérez Penadés for statistical analysis, I. Pérez-Martínez and C. Ramos for their advice and help and B. Álvarez and R. González for suggestions and typing. J.M. Quesada was the recipient of a predoctoral fellowships from I.F.A.P.A. (Andalucía). R. Penyalver has a contract from the Ministry of Education and Science of Spain (Programa Ramón y Cajal). P. Llop has a contract from IVIA-CCAA E. Bertolini is recipient of a contract Juan de la Cierva from MEC of Spain. This work was supported in part by grants CAO00-007 from INIA, AGL2002-0224T from MCYT and GRUPOS03/221 from Generalitat Valenciana of Spain.

References

Agron, P.G., Macht, M., Radnedge, L., Skowronski, E.W., Miller, W. & Andersen, G.L. (2002). Use of subtractive hybridisation for comprehensive surveys of prokaryotic genome differences. FEMS Microbiol. Lett. *211*, 175–182.

Álvarez, A.M. (2004). Integrated approaches for detection of plant pathogenic bacteria and diagnosis of bacterial diseases. Annu. Rev. Phytopathol. *42*, 339–366.

Anonymous (1998). Council Directive 98/57/EC of 20 July 1998 on the control of *Ralstonia solanacearum* (Smith) Yabuuchi et al. Official J. Eur. Communities, No. L235/1, 21.08.98.

Anonymous (2004). EPPO standards. Diagnostic protocols for regulated pests. *Ralstonia solanacearum*. Bull. OEPP-EPPO Bull. *34*, 173–178.

Anonymous (2005). EPPO standards. Diagnostic protocols for regulated pests. *Pseudomonas syringae* pv. *persicae*. Bull. OEPP-EPPO Bull. *35*, 285–287.

Anonymous (2006). Commission Directive 2006/63/EC of 14 July 2006 amending Annexes II to VII to Council.

Arahal, D.R., Llop, P., Alonso, M. & López, M.M. (2004). *In silico* evaluation of molecular probes for detection and identification of *Ralstonia solanacearum* and *Clavibacter michiganensis* subsp. *sepedonicus*. Syst. Appl. Microbiol. *27*, 581–591.

Bertolini, E., Olmos, A., López, M.M. & Cambra, M. (2003a). Multiplex nested reverse transcription-polymerase chain reaction in a single closed tube for sensitive and simultaneous detection of four RNA viruses and *Pseudomonas savastanoi* pv. *savastanoi* in olive trees. Phytopathology *93*, 286–292.

Bertolini, E., Peñalver, R., García, A., Olmos, A., Quesada, J.M., Cambra, M. & López, M.M. (2003b). Highly sensitive detection of *Pseudomonas savastanoi* pv. *savastanoi* in asymptomatic olive plants by nested-PCR in a single closed tube. J. Microbiol. Methods *52*, 261–266.

Boudazin, G., Le Roux, A.C., Josi, K., Labarre, P. & Jouan, B. (1999). Design of division-specific primers of *Ralstonia solanacearum* and application to the identification of European isolates. Eur. J. Plant Pathol. *105*, 373–380.

Caruso, P. (2005). Detección y caracterización serológica y molecular de *Ralstonia solanacearum* biovar 2, causante de la marchitez y podredumbre parda en patata. Tesis doctoral, Universidad de Valencia, Spain.

Caruso, P., Bertolini, E., Cambra, M. & López, M.M. (2003). A new and co-operational polymerase chain reaction (Co-PCR) for rapid detection of *Ralstonia solanacearum* in water. J. Microbiol. Methods *55*, 257–272.

Catara, V., Arnold, D., Cirvilleri, G. & Vivian, A. (2000). Specific oligonucleotide primers for the rapid identification and detection of the agent of tomato pith necrosis, *Pseudomonas corrugata*, by PCR amplification: evidence for two distinct genomic groups. Eur. J. Plant Pathol. *106*, 753–762.

Cubero, J., Martínez, M.C., Llop, P. & López, M.M. (1999). A simple and efficient PCR method for the detection of *Agrobacterium tumefaciens* in plant tumours. J. Appl. Microbiol. *86*, 591–602.

Danial, J., Mc Hugh, R.C. & Saddler, G.S. (2006). Molecular characterization of the potato brown rot pathogen *Ralstonia solanacearum*, race 3, biovar 2A. Proceedings of the 4th International Bacterial Wilt Symposium. CSL, York, UK, p. 28.

Davey, H.M. & Kell, D.B. (1996). Flow cytometry and cell sorting of heterogeneous microbial populations: the importance of single-cell analyses. Microbiol. Rev. *60*, 641–696.

de Lacy Costello, B.P.J., Ewen, R.J., Garner, K., Ratcliffe, N.M., Schleicher, T. & Spencer-Phillips, P.T.N. (2006). Sensors for detection of bacterial infections of potato tubers: from *Erwinia* to *Clavibacter*. Proceedings of the 11th International Conference on Plant Pathogenic Bacteria. Royal College of Physicians of Edinburgh, Scotland, pp. 27–28.

de León, L., Siverio, F. & Rodríguez, A. (2006). Detection of *Clavibacter michiganensis* subsp. *michiganensis* in tomato seeds using immunomagnetic separation. J. Microbiol. Methods *67*, 141–149.

Fegan, M. & Prior, P. (2005). How complex is the *Ralstonia solanacearum* species complex? In: Bacterial Wilt Disease, and the *Ralstonia solanacearum* Species Complex (Eds. Prior, P., Allen, J. & Hayward, A.C.). St. Paul, MN, APS Press, pp. 449–461.

Fessehaie, A.S.H., Boer, D. & Levesque, C.A. (2003). An oligonucleotide array for the identification and differentiation of bacteria pathogenic on potato. Phytopathology *93*, 262–269.

Grey, B.E. & Steck, T.R. (2001). The viable but nonculturable state of *Ralstonia solanacearum* may be involved in long-term survival and plant infection. Appl. Environ. Microbiol. *67*, 3866–3872.

Hayward, A.C., El-Nashaar, H.M., Nydegger, U. & De Lindo, L. (1990). Variation in nitrate metabolism in biovars of *Pseudomonas solanacearum*. J. Appl. Bacteriol. *69*, 269–280.

Llop, P., Caruso, P., Cubero, J., Morente, C. & López, M.M. (1999). A simple extraction procedure for efficient routine detection of pathogenic bacteria in plant material by polymerase chain reaction. J. Microbiol. Methods *37*, 23–31.

López, M.M., Bertolini, E., Caruso, P., Penyalver, R., Marco-Noales, E., Gorris, M.T., Morente, C., Salcedo, C., Cambra, M. & Llop, P. (2005). Advantages of an integrated approach for diagnosis of quarantine pathogenic bacteria in plant material. Phytopathol. Polonica *35*, 49–56.

López, M.M., Bertolini, E., Marco-Noales, E., Llop, P. & Cambra, M. (2006). Update on molecular tools for detection of plant pathogenic bacteria and viruses. In: Molecular Diagnostics. Current Technology and Applications (Eds. Rao, J.R., Fleming, C.C. & Moore, J.E.). Horizon Bioscience, Norfolk, UK, pp. 1–46.

López, M.M., Bertolini, E., Olmos, A., Caruso, P., Gorris, M.T., Llop, P., Penyalver, R. & Cambra, M. (2003). Innovative tools for detection of plant pathogenic viruses and bacteria. Int. Microbiol. *6*, 233–243.

López, M.M., Gorris, M.T., Llop, P., Cubero, J., Vicedo, B. & Cambra, M. (1997). Selective enrichment improves the isolation, serological and molecular detection of plant pathogenic bacteria. In: Diagnosis and Identification of Plant Pathogens (Eds. Dehne, H.W., Adam, G., Diekmann, M., Frahn, J. & Mauler, A.). Kluwer.

López, M.M., Llop, P., Cubero, J., Penyalver, R., Caruso, P., Bertolini, E., Peñalver, J., Gorris, M.T., Olmos, A. & Cambra, M. (2001). Strategies for improving serological and molecular detection of plant pathogenic bacteria. Proceedings of the 10th International Conference on Plant Pathogenic Bacteria. Charlottetown, Canada, Kluwer, pp. 83–85.

Louws, F.J., Rademaker, J.L.W. & Brujin, F.J. (1999). The three Ds of PCR-based genomic analysis of phytobacteria: diversity, detection, and disease diagnosis. Ann. Rev. Phytopathol. *37*, 81–125.

Mills, D., Russell, B.W. & Hanus, J.W. (1997). Specific detection of *Clavibacter michiganensis* subsp. *sepedonicus* by amplification of three unique DNA sequences isolated by subtraction hybridization. Phytopathology *87*, 853–861.

Ordax, M., Marco-Noales, E., López, M.M. & Biosca, E.G. (2006). Survival strategy of *Erwinia amylovora* against copper: induction of the viable-but-nonculturable state. Appl. Environ. Microbiol. *72*, 3482–3488.

Ozakman, M. & Schaad, N.W. (2003). A real-time BIO-PCR assay for detection of *Ralstonia solanacearum* race 3, biovar 2, in asymptomatic potato tubers. Can. J. Plant Pathol. *25*, 232–239.

Penyalver, R., García, A., Ferrer, A., Bertolini, E. & López, M.M. (2000). Detection of *Pseudomonas savastanoi* pv. *savastanoi* in olive plants by enrichment and PCR. Appl. Environ. Microbiol. *66*, 2673–2677.

Prosen, D., Hatziloukas, E., Schaad, N.W. & Panopoulas, N.J. (1993). Specific detection of *Pseudomonas syringae* pv. *phaseolicola* DNA in bean seed by polymerase chain reaction-based amplification of a phaseolotoxin gene region. Phytopathology *83*, 965–970.

Quesada, J.M. (2007). Epidemiología y control químico de *Pseudomonas savastanoi* pv. *savastanoi* causante de la tuberculosis del olivo. Tesis doctoral, Universidad Politécnica de Valencia, Spain.

Saettler, A.W., Schaad, N.W. & Roth, D.A. (1995). Detection of bacteria in seed and other planting material. St. Paul, MN, APS, 122 pp.

Sarkar, S.F. & Guttman, D.S. (2004) Evolution of the core genome of *Pseudomonas syringae*, a highly clonal, endemic plant pathogen. Appl. Environ. Microbiol. *70*, 1999–2012.

Schaad, N.W. & Frederick, R.D. (2002). Real-time PCR and its application for rapid plant disease diagnostics. Canadian J. Plant Pathol. *24*, 250–258.

Schaad, N.W., Berthier-Schaad, Y. & Knorr, D. (2007). A high throughput membrane BIO-PCR technique for ultra-sensitive detection of *Pseudomonas syringae* pv. *phaseolicola*. Plant Pathol. *56*, 1–8.

Schaad, N.W., Frederick, R.D., Shaw, J., Schneider, W.L., Hickson, R., Petrillo, M.D. & Luster, D.G. (2003). Advances in molecular-based diagnostics in meeting crop biosecurity and phytosanitary issues. Annu. Rev. Phytopathol. *41*, 305–324.

Schaad, N.W., Jones, J.B. & Chun, W. (2001). Laboratory guide for the identification of plant pathogenic bacteria, 3rd edn. St. Paul, MN, APS Press, 370 pp.

Schaad, N.W., Tamaki, S., Hatziloukas, E. & Panopoulos, N. (1995). A combined biological amplification (BIO-PCR) technique to detect *Pseudomonas syringae* pv. *phaseolicola* in bean seed extracts. Phytopathology *85*, 243–248.

Schoen, C., De Weerdt, M., Hillhorst, R., Boender, P., Szemes, M., & Bonants, P. (2003). Multiple detection of plant (quarantine) pathogens by micro-arrays: innovative tool for plant health management. In Abstracts of the 19th International Symposium on Virus and Virus-like Diseases of Temperate Fruit Crops, Valencia, Spain, p. 108.

Schoen, C., De Weerdt, M., Hillhorst, R., Chan, A., Boender, P., Zijlstra, C. & Bonants, P. (2002). Use of novel 3D microarray flow through system for plant pathogen multiplex detection. In Abstracts Agricultural Biomarkers for Array Technology, Management Committee Meeting, Wadenswil, Switzerland, p. 11.

Scortichini, M., Rossi, M.P. & Salerno, M. (2004). Relationship of genetic structure of *Pseudomonas savastanoi* pv. *savastanoi* populations from Italian olive trees and patterns of host genetic diversity. Plant Pathol. *53*, 491–497.

Seal, S., Jackson, L., Young, J.P. & Daniels, J. (1993). Differentiation of *Pseudomonas solanacearum*, *Pseudomonas syzygii*, *Pseudomonas pickettii* and the blood disease bacterium by partial 16S rRNA sequencing: construction of oligonucleotide primers for sensitive detection by polymerase chain reaction. J. Gen. Microbiol. *139*, 1587–1594.

Seal, S.E., Jackson, L.A., & Daniels, M.J. (1992). Isolation of a *Pseudomonas solanacearum*-specific DNA probe by subtraction hybridisation and construction of species-specific oligonucleotide primers for sensitive detection by the polymerase chain reaction. Appl. Environ. Microbiol. *58*, 3751–3758.

Soby, S., Kirkpatrick, B. & Kosuge, T. (1993). Characterization of an insertion sequence IS*53* located within IS*51* on the iaa-containing plasmid of *Pseudomonas syringae* pv. *savastanoi*. Plasmid *29*, 135–141.

Stinton, J.A., Persaud, K.C., Stead, D., Bryning, G. & Parkinson, N. (2006). Using an SPME-Enose to identify quarantine pathogens of potatoes. Proceedings of the 11th International Conference on Plant Pathogenic Bacteria. Royal College of Physicians of Edinburgh, Scotland, pp. 29–30.

Tenover, F.C., Arbiet, R.D. & Goering, R.V. (1997). How to select and interpret molecular typing strain methods for epidemiologic studies of bacterial infections: a review for health care epidemiologists. Infect. Control. Hosp. Epidemiol. *18*, 426–439.

Timms-Wilson, T.M., Bryant, K. & Bailey, M.J. (2001). Strain characterization and 16S-23S probe development for differentiating geographically dispersed isolates of the phytopathogen *Ralstonia solanacearum*. Environ. Microbiol. *3*, 785–797.

Van der Wolf, J.M., Bonants, P.J.M., Smith, J.J., Hagenaar, M., Nijhuis, E., van Beckhoven, J. R.C.M., Saddler, G.S., Trigalet, A. & Feuillade, R. (1998). Genetic diversity of *Ralstonia solanacearum* race 3 in Western Europe determined by AFLP, RC-PFGE and Rep-PCR. In: Bacterial Wilt Disease, Molecular and Ecological Aspects (Eds. Prior, P., Allen, J. & Elphinstone, J.). Berlin, Springer, pp. 44–49.

Vandamme, P., Pot, B., Gillis, M., de Vos, P., Kersters, K. & Swings, J. (1996). Polyphasic taxonomy, a consensus approach to bacterial systematics. Microbiol. Rev. *60*, 407–438.

Walcott, R.R. & Gitaitis, R.D. (2000). Detection of *Acidovorax avenae* subsp. *citrulli* in watermelon seed using immunomagnetic separation and the polymerase chain reaction. Plant Dis. *84*, 470–474.

Weller, S.A., Elphinstone, J.G., Smith, N.C., Boonham, N. & Stead, D.E. (2000). Detection of *Ralstonia solanacearum* strains with a quantitative, multiplex, real-time, fluorogenic PCR (TaqMan) assay. Appl. Environ. Microbiol. *66*, 2853–2858.

Siderophore Uses in *Pseudomonas syringae* Identification

A. Bultreys and I. Gheysen

Abstract The diversifying evolution observed in pyoverdin genes and the presence of numerous siderophore membrane receptors indicates that iron-supplying systems are important for fitness of the fluorescent pseudomonads. It is useful for a bacterium to produce a siderophore unusable by others and siderophores, or siderophore genes, that are specific to certain bacteria have been selected during evolution; they can therefore be used in identification. Pyoverdins are the principal siderophores of fluorescent pseudomonads and they contain a variable peptide chain. Because of the evolution in this peptide chain, the pyoverdins of *Pseudomonas cichorii* and *P. syringae*, and related species, are differentiable by visual and spectrophotometrical analysis from pyoverdins produced by species belonging to the saprophytic fluorescent *Pseudomonas* group. The common pyoverdin of *P. syringae* and related species *P. viridiflava* and *P. ficuserectae* has been shown to be specific. This characteristic can therefore be used for presumptive identification of these species by HPLC analysis of their pyoverdin. This significantly reduces the number of identification tests required. Additionally, these approaches worked well to rapidly discriminate among fluorescent strains in early stages of isolation, and to identify strains of *P. viridiflava* and of *P. syringae* pathogenic on fruit and horse-chestnut trees. *P. syringae* pathovars *antirrhini*, *apii*, *avii*, *berberidis*, *delphinii*, *lachrymans*, *passiflorae*, *persicae*, *tomato*, *maculicola*, *viburni*, *helianthi*, *tagetis*, *theae* and *morsprunorum* race 2 (genospecies 3, 7 or 8), as well as *P. phaseolicola* and *P. glycinea* (genospecies 2) possess an *irp1* gene involved in the production of the yersiniabactin siderophore. However, as yersiniabactin production is not systematically observed by HPLC, an *irp1*-based PCR test was used for these pathovars.

Keywords Siderophore, pyoverdin, yersiniabactin, *Pseudomonas syringae*, identification

---

Département Biotechnologie, Centre Wallon de Recherches Agronomiques, Chaussée de Charleroi 235, B-5030 Gembloux, Belgium

Author for correspondence: Alain Bultreys; e-mail: bultreys@cra.wallonie.be

1 Introduction

Iron is essential for life, but, in the environment, it is usually precipitated or chelated. Consequently, bacteria secrete siderophores to capture iron, and they incorporate the chelated siderophores via specific siderophore outer membrane receptors. It is useful for a bacterium to produce a siderophore unusable by others. Among fluorescent pseudomonads, two recent observations strengthen the hypothesis that the iron-supplying systems are important for fitness. First, the genomes of *P. aeruginosa* PAO1, *P. putida* KT2440, *P. fluorescens* Pf0 and *P. syringae* DC3000 contain 35, 29, 26 and 23 genes, respectively, that encode putative or confirmed siderophore outer-membrane receptors (Cornelis & Matthijs, 2002; Martins dos Santos et al., 2004). These receptors would have been conserved because of their utility to use the heterologous siderophores (siderophores produced by other organisms) that can be encountered in the environment. Three principal pyoverdin types can be encountered in *P. aeruginosa* and whole genome comparisons between strains from the three types showed that the central part of their pyoverdin locus was the most divergent locus between their genomes (Smith et al., 2005). This suggests that a diversifying evolution, or a positive selection, favored the apparition of modifications in pyoverdin genes encoding the pyoverdin peptide chain and receptor, which resulted in changes in the specificity of utilization of the secreted pyoverdin. Then, it is clear that siderophores, or siderophore genes, that are specific to certain bacteria, including *P. syringae*, have been selected during evolution; pyoverdin and yersiniabactin should therefore be useful in identification of *P. syringae* (reviewed in Bultreys [2007] for all the siderophores produced by fluorescent pseudomonads).

It is the evolution in the peptide chain of pyoverdins (Fig. 1) that renders them interesting for identification of *P. syringae*. A pyoverdin is made up of a constant quinoline chromophore, a variable peptide chain of 6–12 amino acids and a side chain consisting of a dicarboxylic acid (amide) (Budzikiewicz, 1993, 1997, 2004). The peptide chain is strain specific and variable among strains and species. About 50 peptide chains are known, but 106 are predicted (Meyer & Geoffroy, 2004). Several pyoverdins varying according to their peptide chain conformation and side chains can be found in the culture medium (Schäfer et al., 1991; Bultreys et al., 2004). Pyoverdin are synthesized in a non-ribosomal way by peptide synthetases (Merriman et al., 1995; Kleinkauf & von Döhren, 1996; von Döhren et al., 1999; Lehoux et al., 2000). Pyoverdin evolution results from replacements or deletions of amino acids, as well as from horizontal gene transfers (Bultreys et al., 2004; Smith et al., 2005). The catechol of the chromophore and two amino acids (Asp-based and/or Orn-based ligands) are involved in iron chelation (Fig. 1). In phytopathogenic fluorescent pseudomonads, an evolution is apparent in the peptide chains of pyoverdins (Bultreys et al., 2003, 2004): *P. syringae, P. viridiflava* and *P. ficuserectae* produce the same pyoverdin; the related species *P. cichorii* produces a pyoverdin differing in the replacement of one serine by glycine; and the distant species *P. fuscovaginae* and *P. asplenii* produce a clearly different, but related, pyoverdin.

Fig. 1 **a** The pyoverdin of *P. syringae* and its iron ligands: the catechol of the chromophore (I) and the two Asp-based ligands (II and III) (Bultreys et al., 2004). **b** A pyoverdin from a *P. putida* strain (Gwose & Taraz, 1992), having one *Asp*-based (*β*-OH-Asp) and one *Orn*-based ligand (N$^\delta$-OH-Orn), bound to the iron. The chelation of iron implies the loss of three protons by the siderophore, one by ligand. In *Orn*-based ligands, the proton from the hydroxyl is removed; in *Asp*-based ligands, two protons are candidates but only one will be lost. With *P. syringae*, at least one of the two carboxylic acids conserves its proton at acidic pH after iron chelation, and the removal of this proton, and the resulting apparition of one negative charge, occurring around pH 4.5 and at higher pH, is responsible for a net, visually detectable, conformational change at the iron binding site (Bultreys et al., 2001)

These pyoverdins are the only known pyoverdins that contain two Asp-based iron ligands. In other pseudomonads, the 4 pyoverdins of *P. aeruginosa* contain 2 Orn-based ligands, 19 pyoverdins of *P. fluorescens* contain either 2 Orn- or 1 Orn- and 1 Asp-based ligands, and 13 pyoverdins of *P. putida* always contain 1 Orn- and 1 Asp-based ligands; the rest of the peptide chains of these pyoverdins are highly variable. The presence of two Asp-based iron ligands in the peptide chain of the pyoverdins of *P. syringae*, *P. cichorii* and *P. fuscovaginae*, influences the color and spectral characteristics of the Fe(III)-chelates between pH 3 and 7; they are therefore named atypical pyoverdins (Bultreys et al., 2001, 2003). This behavior is presumably due to the presence of a carboxylic acid function (pKa near pH 4.5) in the chelation site (Fig. 1). In contrast, the color and spectral characteristics of the other pyoverdins (named typical pyoverdins) remain constant at these pH values. The change in atypical pyoverdins can easily be detected using visual and spectrophotometric tests which clearly differentiate phytopathogenic and saprophytic species (Bultreys et al., 2001). Additionally, HPLC enables distinction between the atypical pyoverdins because of their different general structure (Bultreys et al., 2003). The combination of HPLC with a photodiode array detector enables the confirmation of the type of the pyoverdin produced.

In the case of yersiniabactin $(C_{21}H_{27}N_3O_4S_3)$, it's heterogeneous distribution within *P. syringae* offers possibilities for identification. Yersiniabactin is a salicylic acid based siderophore initially characterized in *Yersinia pestis*, the causal agent of bubonic plague (Haag et al., 1993; Drechsel et al., 1995; Chambers et al., 1996). Yersiniabactin is widespread among human and animal pathogenic enterobacteria, such as *Yersinia* spp., *Escherichia coli*, *Citrobacter* spp., *Klebsiella* spp., *Salmonella enterica* and *Enterobacter* spp. (Bach et al., 2000; Schubert et al., 2000; Carniel, 2001; Oelschlaeger et al., 2003; Mokracka et al., 2004). The reason is that the complete yersiniabactin iron uptake system, called the yersiniabactin locus, is located in a genomic high-pathogenicity island transmissible by horizontal gene transfer (Carniel, 2001; Perry, 2004; Schmidt & Hensel, 2004; Antonenka et al., 2005). Genes that are similar to yersiniabactin have been detected in the plant pathogenic *P. syringae* (Buell et al., 2003; Joardar et al., 2005). In *P. syringae*. yersiniabactin genes appear to have been acquired by horizontal gene transfer, but no pathogenicity island has been identified and the genes appear to be stabilized in the chromosome in an apparently conserved location in the species (Bultreys et al., 2006b; Fig. 2). The locus was found in the genospecies 3, 7 and 8 and in the pathovars *glycinea* and *phaseolicola* of genospecies 2 (Bultreys et al., 2006b). This seems to indicate the yersiniabactin locus was acquired either by an ancestor of the producing pathovars followed by stabilization in the chromosome, or by an ancestor of *P. syringae* followed by a locus deletion in an ancestor of the non-producing pathovars. Yersiniabactin was detected by HPLC and a single PCR test using primers PSYE2/PSYE2R (Bultreys et al., 2006b). This enables the *irp1* detection in all the known yersiniabactin producing bacteria, including enterobacteria, but not in *P. syringae* pv. *glycinea* and pv. *phaseolicola*. However, the wide specificity of this test, and the negatives results with the yersiniabactin-positive pathovars of genospecies 2, restrict its use in the specific detection and identification of *P. syringae*. This required designing another PCR test.

Taken together, the pyoverdin and yersiniabactin identification tests proved to be very useful in field diagnostic assays to rapidly differentiate among fluorescent

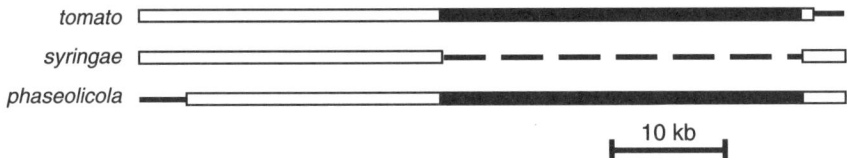

Fig. 2 Homologous regions around the yersiniabactin locus in *Pseudomonas syringae* pv. *tomato* DC3000, *P. syringae* pv. *syringae* B728a and *P. syringae* pv. *phaseolicola* 1448A (Bultreys et al., 2006b). Homologous regions are represented by rectangles; the yersiniabactin loci are black-coded; non-homologous regions are shown by a dark line; the dotted line represents a gap. Genes directly adjacent to the locus are conserved in the three pathovars

isolates in early stages of isolation, and to identify strains of *P. viridiflava* and of *P. syringae* pathogenic on fruit and horse chestnut trees and on legumes.

2   Materials and Methods

2.1   *Strains and Culture Conditions*

The following bacteria were included: 42 pathovars of *P. syringae*; the *Pseudomonas* species *P. viridiflava*, *P. meliae*, *P. ficuserectae*, *P. cannabina*, *P. cichorii*, *P. asplenii*, *P. fuscovaginae*, *P. agarici*, *P. marginalis*, *P. fluorescens*, *P. chlororaphis* and *P. putida*; *Escherichia coli*; several hundreds of field cultures of unknown fluorescent *Pseudomonas* species isolated over a five year period in Wallonia in Southern Belgium from cherry, pear, plum and horse chestnut, and occasionally from legumes such as corn salad (*Valerianella locusta*) or bean. The strains were obtained by isolation onto King et al.'s medium B agar (King et al., 1954) to detect the pyoverdin-related fluorescence. Siderophores were produced by growing cultures in/on GASN (glucose asparagines) solid/liquid medium in Petri dishes as described (Bultreys & Gheysen, 2000; Bultreys et al., 2006b).

2.2   *Pyoverdin Tests*

A modification of the visual, spectrophotometrical and HPLC tests for pyoverdin was used (Bultreys et al., 2001, 2003). Briefly, centrifugation and filtration of the culture medium was deleted to simplify the assay. The GASN solid and liquid media were prepared and stored at 4°C for several months. Each 8.5 cm diameter Petri dish contained 40 ml of solid medium and a grid of seven 1 × 3 cm rectangles drawn on the back; glass tubes contained 10 ml of GASN liquid medium. An occasional test was rapidly done by taking only one agar bloc and one tube from storage. After incubation for 3 days, the liquid fraction was collected, iron was added and the tube directly divided in two parts. The pH was adjusted to 3.0 and 7.0 using an NaOH or HCl solution. The pH was determined with an indicator paper. A strain was judged as an atypical pyoverdin producer by a color change. These colors could easily be observed by using one control. In rare cases, a spectrophotometric test at pH 7 was performed after filtration to confirm the identity. If further proof of identity of the pyoverdin was needed, HPLC was performed using our Waters Alliance 2695 HPLC system (Waters Belgium, Zellik) equiped with a photodiode array detector Waters 996 as described (Bultreys et al., 2003).

2.3 Yersiniabactin Tests

The yersiniabactin detection test using a HPLC assay has been described (Bultreys et al., 2006b). Both pyoverdin and yersiniabactin were produced and detected in the same way in a simple HPLC analysis of the GASN culture medium. The general PCR primers PSYE2 and PSYE2R allowed rapid detection of the *irp1* gene in all the yersiniabactin-producing bacteria, except *P. syringae* pv. *phaseolicola* and pv. *glycinea* (Bultreys et al., 2006b). The primers were used to verify the presence of the *irp1* gene in strains presumptively identified as *P. syringae* pv. *morsprunorum* race 2. To develop a PCR test specific to all the yersiniabactin producing *P. syringae* strains, the sequences of *irp1* of *P. syringae* pv. *tomato* DC3000, pv. *phaseolicola* 1448A and *Yersinia enterocolitica* WA-C were aligned to select primers specific to the two pathovars of *P. syringae*. After testing several primer pairs (data not shown) primers PT3 and PT3R were selected and the specificity to the following organisms determined: the pathotype strains of *P. syringae* pathovars *aceris, aptata, atrofaciens, dysoxyli, japonica, lapsa, panici, papulans, pisi, syringae, ciccaronei, eriobotryae, glycinea, mellea, mori, myricae, phaseolicola, savastanoi, sesami, tabaci, ulmi, anthirrhini, apii, berberidis, delphinii, lachrymans, maculicola, passiflorae, persicae, tomato, viburni, coronafaciens, garcae, oryzae, primulae, ribicola, helianthi, tagetis, theae, avii, morsprunorum* race 1 and 2; *Pseudomonas* species *P. cannabina, P. tremae, P. viridiflava, P. meliae, P. ficuserectae, P. cichorii, P. asplenii, P. fuscovaginae, P. agarici, P. marginalis, P. fluorescens, P. chlororaphis* and *P. putida*; and yersiniabactin-producing *E. coli* strain ECOR 10 (Bultreys et al., 2006b).

2.4 Additional Identification Tests

After the pyoverdin tests, a potato rot assay was used to discriminate between *P. syringae* and *P. viridiflava*. For strains for which more information was needed, assays for toxic lipodepsipeptide (Bultreys & Gheysen, 1999) and coronatine were included (Bereswill et al., 1994) The following biochemical tests oxidase, fluorescence on King's medium B, catalase, API 20NE, gelatin hydrolysis, aesculin hydrolysis, tyrosinase activity, use of lactic and tartric acid and Biolog assays were used to identify *P. syringae* pv. *syringae* and *P. syringae* pv. *morsprunorum* race 1 and 2 (data not shown).

3   Results and Discussion

3.1   Pyoverdin Tests

The same pyoverdin has always been detected in several hundred *P. syringae* strains belonging to more than 40 pathovars using the HPLC assay. The pyoverdin is also produced by strains of pathovars *savastanoi*, *phaseolicola* and *glycinea*, proposed to be grouped into a new species (Gardan et al., 1992), although the two nomenspecies remain valid (Young et al., 1996; Schaad et al., 2000). This pyoverdin is also common to strains of the species *P. viridiflava* and *P. ficuserectae*, which indicates a close relatedness of these species with *P. syringae*. *P. viridiflava* and *P. ficuserectae* belong to the *P. syringae* genospecies 6 and 2, respectively, which contain several recognized *P. syringae* pathovars (Gardan et al., 1999). According to the rules for differentiating new species, both genetic and phenotypic data are needed to define a new species (Wayne et al., 1987). But, according to the recent DNA/DNA hybridization studies (Gardan et al., 1999), genetic data are lacking to separate the formerly described species *P. viridiflava* and *P. ficuserectae* from certain *P. syringae* pathovars and, consequently, from *P. syringae*. The close proximity of these species is confirmed by pyoverdin analyses. This strengthens the information available from pyoverdins, as does the fact that strains from the different but closely related species *P. cichorii* produce a very similar pyoverdin varying only by one amino acid. The different pyoverdin tests commonly used in our laboratory were the visual, the spectrophotometric and the HPLC tests. These tests proved useful at different level of identification. They were used for research purposes but also in our phytopathological consultation service, which is accessible for professionals as well as private individuals.

The visual test is the least expensive and easiest to perform but it does not differentiate between the atypical pyoverdins of *P. syringae* and *P. cichorii*. Both atypical pyoverdins show a grayish-brown color near pH 3 and a yellow-orange color near pH 7. On the other hand, it allows differentiation from the atypical pyoverdin of *P. fuscovaginae* and *P. asplenii*, which produce the same pyoverdin and could be synonymous: this atypical pyoverdin appears similar at pH 7 but shows still a net orange color near pH 3. Also, the visual test enables distinction from all the other fluorescent species producing typical pyoverdins, which, most generally, show a constant brown color at both pH 3 and 7 (Fig. 3). However, with the continual augmentation of the field-collected bacteria investigated, strains were encountered that produced typical pyoverdins showing some color variation with the pH, but these changes were more limited, e.g. a more orange-brown color at pH

Atypical pyoverdin   **Typical pyoverdin**

pH ~ 7   pH ~ 3   pH ~ 7   pH ~ 3

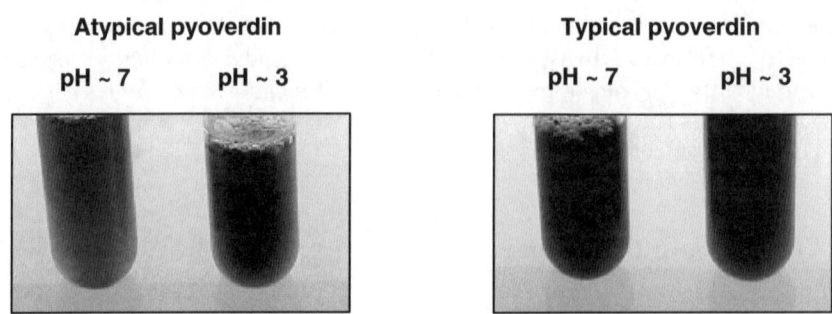

Fig. 3 Visual test for strains producing an atypical or a typical pyoverdin; the culture medium was not filtrated and the pH was measured with indicator paper. The clearer and different color of the chelated atypical pyoverdins near pH 7 is evident; on the other hand the color of the typical pyoverdin most generally remains constant

7, and no confusion occured when *P. syringae* controls were included at both pHs. Concerning the controls, it would be useful to those unfamiliar with this test to dilute solutions of a strong pyoverdin producer at both pH values to compare colors produced by weak pyoverdin producers. The initial test included a centrifugation of the culture medium and a filtration of the supernatant to discard the bacteria and precipitates (Bultreys et al., 2001), but these time-consuming steps proved unnecessary. Also, in order to avoid having routinely to disinfect the electrode of the pH meter, indicator paper can be used to adjust the pH to 3 and 7. It is not necessary to work at an exact pH as comparisons of solutions below pH 4 and over pH 6 are adequate. The visual test has been used in consultation with growers or private individuals who do not have the resources needed to confirm the cause of the disease was a known phytopathogenic bacterium. Also, it was often used to distinguish between *P. syringae* and the quarantine bacterium *Erwinia amylovora*, which can have similar and confusing symptoms in pear orchards. In this respect, it was useful for national and regional quarantine inspection services because it possibly enables to rapidly infirm the presence of the target quarantine pathogen. Additionally, the visual assay proved useful for presumptive identification of *P. syringae* as causing the horse chessnut disease (Bultreys et al., 2006a, 2006c).

    The spectrophotometric test includes a comparison of the spectral characteristics of the filtered culture medium at pH 7. No differentiation is possible with this test between the atypical pyoverdins of *P. syringae*, *P. cichorii* and *P. fuscovaginae*. On the other hand, at pH 7, typical and atypical pyoverdin can easily be discriminated by their spectrum between 300 and 700 nm. The absorbance maximum of a chelated typical pyoverdin is near 399 nm and two broad charge transfer bands near 470 and 550 nm are observed. On the other hand, the absorbance maximum of a chelated atypical pyoverdin is near 408 nm and the two broad charge transfer bands near 470 and 550 nm are not observed (Fig. 4A and B). In contrast, at pH 3, atypical pyoverdins show clear similarities with typical pyoverdins (Fig. 4A). The spectrophotometric test was performed when some doubt existed about the reading of a

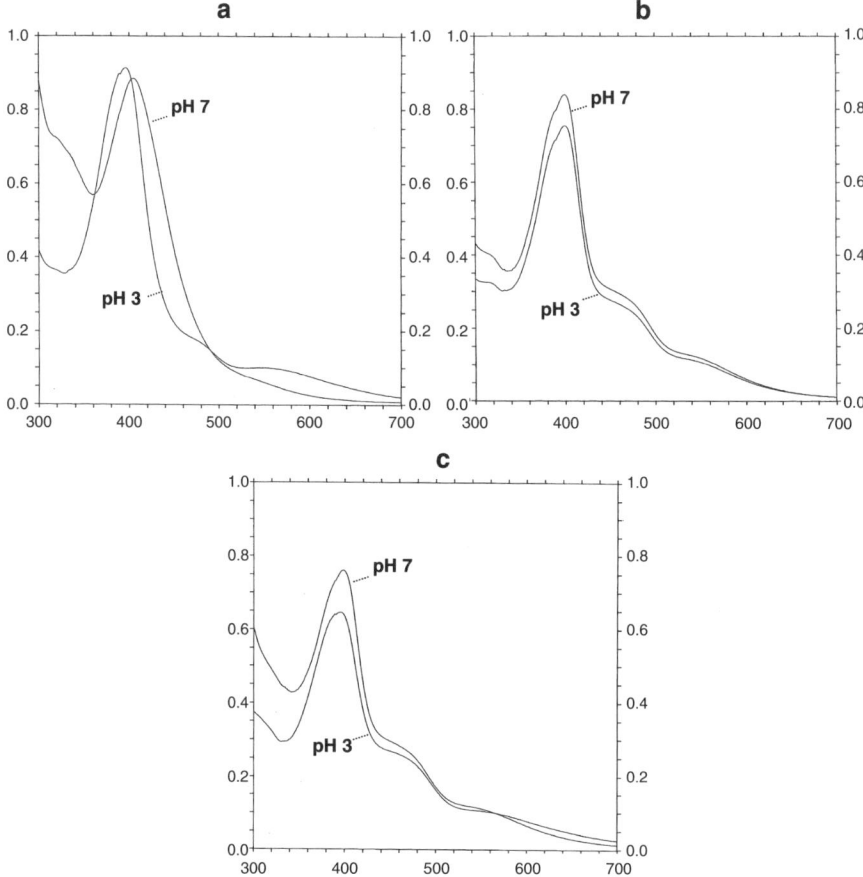

Fig. 4 Spectrophotometric analyses of the culture medium adjusted to pH 3 or 7 showing chelated pyoverdins produced by *Pseudomonas* strains; the y-axis is absorbance; the x-axis is wavelength. **a** A strain producing an atypical pyoverdin: the spectral changes between pH 3 and 7 are evident and no charge transfer bands are observed near 470 and 550 nm at pH 7. **b** A strain producing a typical pyoverdin whose color didn't change with pH: no spectral changes; the charge transfer bands near 470 and 550 nm are observable at both pH. **c** A strain producing a typical pyoverdin whose color slightly varied with pH: a conformational change is clear from the crossing near 570 nm of the spectra at different pH and from the faster extinction at pH 7, but the molecule at both pH has the spectrum of a typical pyoverdin. In the spectrophotometric test, the spectra at pH 7 are compared and strains B and C would then clearly be classified among typical pyoverdin producers, and strain A among atypical pyoverdin producers

visual test. This was the case when atypical color changes were for the first times observed with typical pyoverdins. As illustrated for a non conserved field isolate (Fig. 4C), the chromatographic test confirmed that the uncertain pyoverdins still had typical pyoverdin characteristics at pH 7, although the spectral characteristics were somewhat different at pH 3 and 7. Interestingly, as the chromatographic

characteristics of chelated pyoverdins in the visible range depend on the chromophore and on the way iron is bound to this chromophore, the analysis of the culture
medium at pH 7 and 3 clearly showed that the configuration changes had little
impact on the chelation site in the case of typical pyoverdins because the observed
spectral changes were inexistent or limited. Most probably, the configuration
changes resulted, in the cases of typical pyoverdins, from charge modifications in
the molecule, but not in the amino acid involved in the chelation, whereas for a
typical pyoverdin, they occured at the chelation site, in the direct proximity of iron
(Fig. 4A and C).

The HPLC analyses were performed to distinguish between the pyoverdin of
P. syringae and *P. cichorii* because the technique enables distinction between these
two closely related atypical pyoverdins. The use of a photodiode array detector
gives information about the spectrum of the HPLC peak and, consequently, on the
nature, typical or atypical, of the pyoverdins detected (Fig. 5A). Indeed, at pH 5.3

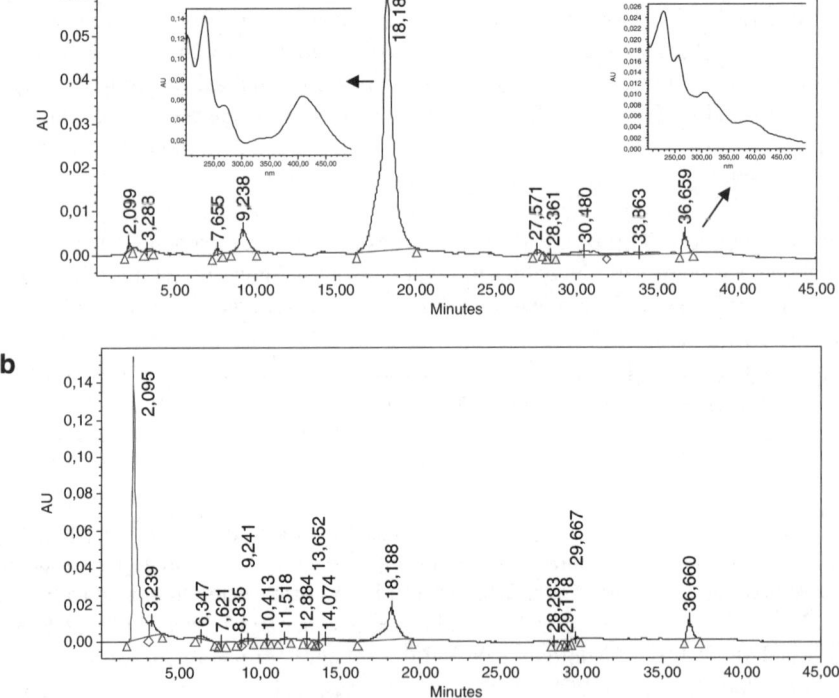

Fig. 2.5 HPLC detection in one HPLC analysis, at 403 nm **a** and 305 nm **b**, of chelated pyoverdin
and yersiniabactin produced by *P. syringae* strain PP441. The hydrophilic pyoverdin and the more
hydrophobic yersiniabactin are well separated and their absorbance at 403 nm **a** enables their
simultaneous detection. Their identity is confirmed by the spectra obtained in line from the photodiode array detector. At 305 nm **b**, the yersiniabactin induces a higher peak while the pyoverdin
peak is reduced

used in the HPLC assays, the distinction between the two types of spectra is evident. Without a photodiode array detector, a measure of absorbance near 400 nm enables pyoverdin peak detections without possible interferences from the very simple GASN culture medium constituents (Fig. 5A). The method needs to work with an automatic injector and to have a good precision in the 0–5% ratio of solvents. When the pyoverdin of *P. syringae* was detected for strains from fruit orchards, the potato rot test was used to determine if the strain belonged to *P. viridiflava*. For other strains, the coronatine and toxic lipodepsipeptide PCR or biological tests were used to identify *P. syringae* pv. *morsprunorum* race 1 and *P. syringae* pv. *syringae*, whereas *P. syringae* pv. *morsprunorum* race 2 was identified using different physiological tests.

3.2   Yersiniabactin Tests

The HPLC analyses carried out to identify the pyoverdins allowed rapid detection of yersiniabactin production (Fig. 5) among several strains from Belgian orchards, including 48 strains from sweet or sour cherry and 3 strains from plum orchards. The yersiniabactin peak was however sometimes very small.

The genetic tests using primers PSYE2/PSYE2R were previously reported to allow detection of the *irp1* gene in all the pathovars of the genospecies 3, 7 and 8 investigated, except the strains LMG 5295 and LMG 2276 reported to belong, respectively, to the pathovars *maculicola* and *ribicola* and to genospecies 3 (Bultreys et al., 2006b). However, the latter two strains do not belong to the geno-species 3. Although its correct genospecies was not specified, *P. syringae* pv. *maculicola* LMG 5295 (CFBP 1637) was previously reported to not belong to genospecies 3 (Clerc et al., 1998). Also, the negative pathotype strain *P. syringae* pv. *ribicola* LMG 2276 (CFBP 2348) actually belongs to genospecies 6 rather than genospecies 3 as reported for another strain misclassified in this pathovar (Gardan et al., 1999). A new PCR primer pair PT3/PT3R is under evaluation. These primers already allowed for the first time the detection of the *irp1* gene for strains of the pathovars *glycinea* and *phaseolicola* belonging to the genospecies 2. Also, they gave the expected positive results with pathovars *morsprunorum* race 2, *theae*, *tagetis*, *helianthi*, *viburni*, *tomato*, *persicae*, *passiflorae*, *lachrymans*, *delphinii*, *berberidis*, *apii* and *antirrhini* (Fig. 6). They also gave positive results with the pathotype strain *P. syringae* pv. *maculicola* CFBP 1657, which was used by Gardan et al. (1999) in their DNA hybridization study and was classified in the genospecies 3. Consequently, there is presently no more exceptions to the observation that all the strains belonging to the genospecies 3, 7 and 8 possess a yersiniabactin locus. On the other hand, the new PCR test gave negative responses with the strains of the other pathovars and *Pseudomonas* species investigated and with *E. coli* (Fig. 6). But the new PCR test utility and specificity have still to be evaluated for more pathovars of *Pseudomonas syringae* and within the different positive pathovars.

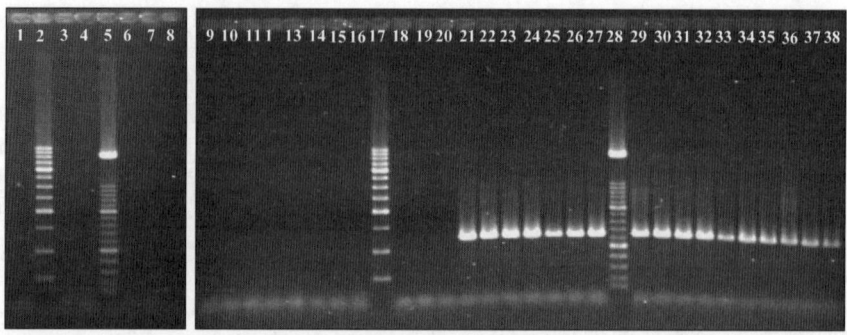

Fig. 2.6 PCR detection of the *irp1* gene using the primer pair PT3-PT3R. Lane 1, water; lane 2, 5, 17 and 28, MW markers; lane 3, *E. coli*; lanes 4, 6–16, 18 and 19, 14 strains belonging to 13 *Pseudomonas* species; lane 20, *P. syringae* pv. *syringae*; lanes 21–27 and 29–38, strains of *P. syringae* pv. *glycinea, phaseolicola, morsprunorum* race 2 (three strains), *theae, tagetis, helianthi, viburni, tomato, persicae, passiflorae, lachrymans, delphinii, berberidis, apii* and *antirrhini*

The use of classical identification procedures led us to the conclusion that certain strains of *P. syringae* pv. *morsprunorum* race 2 produced quantities of yersiniabactin detectable by HPLC, but others not. Consequently, the primers PSYE2/PSYE2R were used to test whether the strains of *P. syringae* pv. *morsprunorum* race 2 negative in HPLC possessed an *irp1* gene, which was always verified. Consequently, it is clear that a yersiniabactin PCR-based assay specific to *P. syringae* has real value in the rapid identification of this pathovar.

Acknowledgements We thank E. de Hoffmann, R. Rozenberg, B. Wathelet, M. Schäfer and H. Budzikiewicz for their help in the chemical characterization of the atypical pyoverdins and of yersiniabactin, and F. Legros for the collect of numerous Pseudomonas isolates from Walloon orchards. This work was supported by the Walloon Agricultural Research Centre subsided by the Ministry of Agriculture of the Walloon Region, and by an additional grant from the Ministry of Agriculture of the Walloon Region.

References

Antonenka U., Nölting C., Heesemann J. & Rakin A. (2005) Horizontal transfer of *Yersinia* high-pathogenicity island by the conjugative RP4 *attB* target-presenting shuttle plasmid. *Mol. Microbiol.* 57: 727–734.

Bach S., de Almeida A. & Carniel E. (2000) The *Yersinia* high-pathogenicity island is present in different members of the family *Enterobacteriaceae*. *FEMS Microbiol. Lett.* 183: 289–294.

Bereswill S., Bugert P., Völksch B., Ullrich M., Bender C.L. & Geider K. (1994) Identification and relatedness of coronatine-producing *Pseudomonas syringae* pathovars by PCR analysis and sequence determination of the amplification products. *Appl. Environ. Microbiol.* 60: 2924–2930.

Budzikiewicz H. (1993) Secondary metabolites from fluorescent pseudomonads. *FEMS Microbiol. Rev.* 104: 209–228.

Budzikiewicz H. (1997) Siderophores of fluorescent pseudomonads. *Z. Naturforsch.* 52c: 713–720.

Budzikiewicz H. (2004) Siderophores of the Pseudomonadaceae *sensu sticto* (fluorescent and non fluorescent *Pseudomonas* spp.), pp 81–237. Progress in the chemistry of organic natural products, vol. 87. Herz W., Falk H. & Kirby G.W. (Eds), Springer, Wien, New York.

Buell C.R., Joardar V., Lindeberg M., Selengut J., Paulsen I.T., Gwinn M.L., Dodson R.J., Deboy R.T., Durkin A.S., Kolonay J.F., Madupu R., Daugherty S., Brinkac L., Beanan M.J., Haft D.H., Nelson W.C., Davidson T., Zafar N., Zhou L., Liu J., Yuan Q., Khouri H., Fedorova N., Tran B., Russell D., Berry K., Utterback T., Van Aken S.E., Feldblyum T.V., D'Ascenzo M., Deng W.L., Ramos A.R., Alfano J.R., Cartinhour S., Chatterjee A.K., Delaney T.P., Lazarowitz S.G., Martin G.B., Schneider D.J., Tang X., Bender C.L., White O., Fraser C.M. & Collmer A. (2003) The complete genome sequence of the *Arabidopsis* and tomato pathogen *Pseudomonas syringae* pv. *tomato* DC3000. *Proc. Nat. Acad. Sci. USA* 100: 10181–10186.

Bultreys A. (2007) Siderotyping, a tool to characterize, classify and identify fluorescent pseudomonads, pp 67–89. Soil biology, volume 12: microbial siderophores. Varma A. & Chincholkar S.B. (Eds.), Springer, Berlin, Heidelberg, New York.

Bultreys A. & Gheysen I. (1999) Biological and molecular detection of toxic lipodepsipeptide-producing *Pseudomonas syringae* strains and PCR identification in plants. *Appl. Environ. Microbiol.* 65: 1904–1909.

Bultreys A. & Gheysen I. (2000) Production and comparison of peptide siderophores from strains of distantly related pathovars of *Pseudomonas syringae* and *Pseudomonas viridiflava* LMG 2352. *Appl. Environ. Microbiol.* 66: 325–331.

Bultreys A., Gheysen I., Maraite H. & de Hoffmann E. (2001) Characterization of fluorescent and nonfluorescent peptide siderophores produced by *Pseudomonas syringae* strains and their potential use in strain identification. *Appl. Environ. Microbiol.* 67: 1718–1727.

Bultreys A., Gheysen I., Wathelet B., Maraite H. & de Hoffmann E. (2003) High-performance liquid chromatography analyses of pyoverdin siderophores differentiate among phytopathogenic fluorescent *Pseudomonas* species. *Appl. Environ. Microbiol.* 69: 1143–1153.

Bultreys A., Gheysen I., Wathelet B., Schäfer M. & Budzikiewicz H. (2004) The pyoverdins of *Pseudomonas syringae* and *Pseudomonas cichorii*. *Z. Naturforsch.* 59c: 613–618.

Bultreys A., Gheysen I., Chandelier A., Zini J., Planchon V., Papart A.-T. & Decoux V. (2006a) *Pseudomonas syringae*, the possible responsible of horse-chestnut tree disease in Belgium. Proceedings of the 11th International Conference on Plant Pathogenic Bacteria, p. 139. Elphinstone J., Weller S., Thwaites R., Parkinson N., Stead D. & Saddler G. (Eds), Edinburgh.

Bultreys A., Gheysen I. & de Hoffmann E. (2006b) Yersiniabactin production by *Pseudomonas syringae* and *Escherichia coli*, and description of a second yersiniabactin locus evolutionary group. *Appl. Environ. Microbiol.* 72: 3814–3825.

Bultreys A., Gheysen I. & Planchon V. (2006c) Characterization of *Pseudomonas syringae* strains isolated from diseased horse-chestnut trees in Belgium. Book of abstracts of the 7th International Conference on *Pseudomonas syringae* Pathovars and Related Pathogens, p. 61. Fatmi M. et al. (Eds), Agadir.

Carniel E. (2001) The *Yersinia* high-pathogenicity island: an iron-uptake island. *Microbes. Infect.* 3: 561–569.

Chambers C.E., McIntyre D.D., Mouck M. & Sokol P.A. (1996) Physical and structural characterization of yersiniophore, a siderophore produced by clinical isolates of *Yersinia enterocolitica*. *Biometals.* 9: 157–167.

Clerc A., Manceau C. & Nesme X. (1998) Comparison of randomly amplified polymorphic DNA with amplified fragment length polymorphism to assess genetic diversity and genetic relatedness within genospecies III of *Pseudomonas syringae*. *Appl. Environ. Microbiol.* 64: 1180–1187.

Cornelis P. & Matthijs S. (2002) Diversity of siderophore-mediated iron uptake systems in fluorescent pseudomonads: not only pyoverdines. *Environ. Microbiol.* 4: 787–798.

Drechsel H., Stephan H., Lotz H., Haag H., Zähner H., Hantke K. & Jung G. (1995) Structure elucidation of yersiniabactin, a siderophore from highly virulent *Yersinia* strains. *Liebigs. Ann.* 1995: 1727–1733.

Gardan L., Bollet C., Abu Ghorrah M., Grimont F. & Grimont P.A.D. (1992) DNA relatedness among the pathovar strains of *Pseudomonas syringae* subsp. *savastanoi* Janse (1982) and proposal of *Pseudomonas savastanoi* sp. nov. *Int. J. Syst. Bacteriol.* 42: 606–612.

Gardan L., Shafik H., Belouin S., Broch R., Grimont F. & Grimont P.A.D. (1999) DNA relatedness among the pathovars of *Pseudomonas syringae* and description of *Pseudomonas tremae* sp. nov. and *Pseudomonas cannabina* sp. nov. (ex Sutic and Dowson 1959). *Int. J. Syst. Bacteriol.* 49: 469–478.

Gwose I. & Taraz K. (1992) Pyoverdins aus *Pseudomonas* putida. *Z. Naturforsch.* 47c: 487–502.

Haag H., Hankte K., Drechsel H., Stojiljkovic I., Jung G., Zähner H. (1993) Purification of yersiniabactin: a siderophore and possible virulence factor of *Yersinia enterocolitica. J. Gen. Microbiol.* 139: 2159–2165.

Joardar V., Lindeberg M., Jackson R.W., Selengut J., Dodson R.,. Brinkac L.M, Daugherty S.C., DeBoy R., Durkin A.S., Giglio M.G., Madupu R., Nelson W.C., Rosovitz M.J., Sullivan S., Crabtree J., Creasy T., Davidsen T., Haft D.H., Zafar N., Zhou L., Halpin R., Holley T., Khouri H., Feldblyum T., White O., Fraser C.M., Chatterjee A.K., Cartinhour S., Schneider D.J., Mansfield J., Collmer A. & Buell C.R. (2005) Whole-genome sequence analysis of *Pseudomonas syringae* pv. phaseolicola 1448A reveals divergence among pathovars in genes involved in virulence and transposition. *J. Bacteriol.* 187: 6488–6498.

King E.O., Ward M.K. & Raney D.E. (1954) Two simple media for the demonstration of pyocyanin and fluorescein. *J. Lab. Clin. Med.* 44: 301–307.

Kleinkauf H. & von Döhren H. (1996) A nonribosomal system of peptide biosynthesis. *Eur. J. Biochem.* 236: 335–351.

Lehoux D.E., Sanschagrin F. & Levesque R.C. (2000) Genomics of the 35-kb *pvd* locus and analysis of novel *pvdIJK* genes implicated in pyoverdine biosynthesis in *Pseudomonas aeruginosa. FEMS Microbiol. Lett.* 190: 141–146.

Martins dos Santos V.A.P., Heim S., Moore E.R.B., Strätz M. & Timmis K.N. (2004) Insights into the genomic basis of niche specificity of *Pseudomonas putida* KT2440. *Environ. Microbiol.* 6: 1264–1286

Merriman T.R., Merriman M.E. & Lamont I.L. (1995) Nucleotide sequence of *pvdD*, a pyoverdine biosynthetic gene from *Pseudomonas aeruginosa*: PvdD has similarity to peptide synthetases. *J. Bacteriol.* 177: 252–258.

Meyer J.M. & Geoffroy V.A. (2004) Environmental fluorescent *Pseudomonas* and pyoverdine diversity: how siderophores could help microbiologists in bacterial identification and taxonomy. Iron transport in bacteria, pp. 451–468. Crosa J.H., Mey A.R. & Payne S.M. (Eds), ASM, Washington, DC.

Mokracka J., Koczura R. & Kaznowski A. (2004) Yersiniabactin and other siderophores produced by clinical isolates of *Enterobacter* spp. and *Citrobacter* spp. *FEMS Immunol. Med. Micobiol.* 40: 51–55.

Oelschlaeger T.A., Zhang D., Schubert S., Carniel E., Rabsch W., Karch H. & Hacker J. (2003) The high-pathogenicity island is absent in human pathogens of *Salmonella enterica* subspecies I but present in isolates of subspecies III and VI. *J. Bacteriol.* 185: 1107–1111.

Perry R.D. (2004) *Yersinia.* Iron transport in bacteria, pp. 219–240. Crosa J.H., Mey A.R. & Payne S.M. (Eds), ASM, Washington, DC.

Schaad N.W., Vidaver A.K., Lacy G.H., Rudolph K. & Jones J.B. (2000) Evaluation of proposed amended names of several Pseudomomads and Xanthomonads and recommendations. *Phytopathology.* 90: 208–213.

Schäfer H., Taraz K. & Budzikiewicz H. (1991) Zur Genese der Amidisch an den Chromophor von Pyoverdinen gebundenen Dicarbonsäuren. *Z. Naturforsch.* 46c: 398–406.

Schmidt H. & Hensel M. (2004) Pathogenicity islands in bacterial pathogenesis. *Clin. Microbiol. Rev.* 17: 14–56.

Schubert S., Cuenca S., Fischer D. & Heesemann J. (2000) High-pathogenicity island of *Yersinia pestis* in enterobacteriaceae isolated from blood cultures and urine samples: prevalence and functional expression. *J. Infect. Dis.* 182: 1268–1271.

Smith E.E., Sims E.H., Spencer D.H., Kaul R. & Olson M.V. (2005) Evidence for diversifying selection at the pyoverdine locus of *Pseudomonas aeruginosa. J. Bacteriol.* 187: 2138–2147.

von Döhren H., Dieckmann R. & Pavela-Vrancic M. (1999) The nonribosomal code. *Chem. Biol.* 6: R273-R279.

Wayne L.G., Brenner D.J., Colwell R.R., Grimont P.A.D., Kandler O., Krichevsky M.I., Moore L.H, Moore W.E.C., Murray R.G.E., Stackebrandt E., Starr M.P. & Trüper H.G. (1987) Report of the ad hoc committee on the reconciliation of approaches to bacterial systematics. *Int. J. Syst. Bacteriol.* 37: 463–464.

Young J.M., Whenua M., Saddler G.S., Takikawa Y., DeBoer S.H., Vauterin L., Gardan L., Gvozdyak R.I. & Stead D.E (1996) Names of plant pathogenic bacteria 1864–1995. *Rev. Plant. Pathol.* 75: 722–763.

Chlorophyll Fluorescence Imaging for Detection of Bean Response to *Pseudomonas syringae* in Asymptomatic Leaf Areas

L. Rodríguez-Moreno[1], M. Pineda[2], J. Soukupová[3], A.P. Macho[1],
C. R. Beuzón[1], L. Nedbal[3], M. Barón[2], and C. Ramos[1]

Abstract The physiological condition of infected plants can be monitored prior to appearance of symptoms using non-destructive methods such as chlorophyll fluorescence imaging. A number of studies using different chlorophyll fluorescence imaging prototypes have shown that photosynthesis is severely impaired in symptomatic and asymptomatic leaves of fungal- or virus-infected plants. However, little information is available for bacterial infected plants. In the present study, chlorophyll fluorescence imaging was used to analyse the response elicited in *Phaseolus vulgaris* after inoculation with *Pseudomonas syringae* pv. *phaseolicola* 1448A (compatible interaction) and *P. syringae* pv. *tomato* DC300 (incompatible interaction). Quenching analysis was carried out and images of the different chlorophyll fluorescence parameters were obtained on infected plants as well as in mock controls at different post-inoculation times. Of the different parameters analysed, effective photosystem II quantum yield and non-photochemical quenching maximized the differences between plants infected with either of both pathogens. Significant changes in these parameters were observed before the appearance of visual symptoms. Images of effective photosystem II quantum yield did not reveal any differences between compatible and incompatible infected plants. In contrast, images of non-photochemical quenching showed considerable difference between such plants.

Keywords *Phaseolus vulgaris*, *Pseudomonas syringae*, chlorophyll fluorescence imaging, hypersensitive response

---

[1] Section of Genetics, Faculty of Sciences, University of Malaga, E-29071, Malaga, Spain

[2] Estación Experimental del Zaidín, CSIC, c/ Profesor Albareda 1, E-18008, Granada, Spain

[3] Institute of Landscape Ecology ASCR, CZ-37333, Zamek 136, Nové Hrady Czech Republic

Author for correspondence: C. Ramos; e-mail: crr@uma.es, M. Barón; e-mail: mbaron@eez.csic.es

M'B. Fatmi et al. (eds.), *Pseudomonas syringae Pathovars and Related Pathogens.*   37
© Springer Science+Business Media B.V. 2008

1 Introduction

The ability of *Pseudomonas syringae* strains to infect host plants (compatible interaction) and elicit the hypersensitive response (HR) in non-hosts (incompatible interaction) is dependent on the secretion of proteins encoded by avirulence genes (*avr*) into the plant cells (Klement, 1982). In host plants, the absence of interaction between bacterial effectors and plant resistance (R) proteins allows pathogen multiplication and systemic invasion of the plant. In contrast, during an incompatible interaction, recognition of specific effectors by R proteins leads to the induction of HR and, consequently disease was prevented (Staskawicz et al., 1995; Dangl and Jones, 2001).

It is known that disease development induces changes in the photosynthetic process and carbohydrate metabolism of the host plant (Balachandran et al., 1994; Esfeld et al., 1995; Peterson and Aylor, 1995; Rahoutei et al., 2000; Scharte et al., 2005). Pathogens manipulate the carbohydrate metabolism for their own benefit, to support replication and spreading within the plant; simultaneously, host plants reorganize their primary metabolism to mount defence mechanisms (Roitsch, 1999).

According to the heterogeneous distribution of the pathogen-induced symptomatology in the leaf, CO_2 assimilation, photosynthetic efficiency, and chlorophyll a (Chla) fluorescence (Chl-F) emission results in a heterogeneous spatial pattern in infected leaves (Chou et al., 2000). For this reason, the development of fluorescence imaging (FI) systems has been a valuable tool to study disease development in plants (Chaerle and Van Der Straeten, 2001; Nedbal and Whitmarsh, 2004). The technique allows pre-symptomatic analysis, in cases where changes on fluorescence parameters precede the symptom development (Peterson and Aylor, 1995; Soukupová et al., 2003).

Chlorophyll fluorescence imaging (Chl-FI) prototypes have showed photosynthesis to be severely impaired in symptomatic (Scholes and Rolfe, 1996; Osmond et al., 1998; Chou et al., 2000; Lohaus et al., 2000) and asymptomatic (Chaerle et al., 2006; Pérez-Bueno et al., 2006) plants infected by fungi and viruses. Berger and collaborators (2004) compared bacterial and fungal-infection in *Lycopersicum esculentum* plants and observed a co-regulation of defence, sink metabolism and gene expression of photosynthetic genes in the hosts. Based on this, they proposed the changes in non-photochemical quenching (NPQ) values in the tissue colonized by the pathogen to be and important step in plant pathogenesis. Both virulent and avirulent strains of *P. syringae* pv. *tomato* DC3000 resulted in reduced photochemical and non-photochemical quenching in the tissue around the inoculation point (Bonfig et al., 2006).

In the present study, Chl-FI was used to distinguish a compatible reaction from an incompatible reaction: *Phaseolus vulgaris* infected with either *P. syringae* pv. *phaseolicola* (Pph) or *P. syringae* pv. *tomato* (Pto) DC3000, respectively. Several Chl-FI parameters were analysed in leaves infected with either of both pathogens prior to the appearance of visible symptoms.

2   Materials and Methods

2.1   Bacterial and Plant Growth Conditions

Two bacterial strains were used in this study, *P. syringae* pv. *phaseolicola* 1448A (Fillingham et al., 1992) and *P. syringae* pv. *tomato* DC3000 (Cuppels, 1986). Seeds of *Phaseolus vulgaris* cultivar Canadian Wonder (Chase Organics Ltd, Hersham, Surrey, UK) were germinated and grown in a growth chamber with 16/8 h light/dark photoperiod at 24/18°C day/night, 70–75% of relative humidity and 130 μmol photons m^{-2} s^{-1} of white irradiance.

2.2   Plant Inoculation

Bacteria were grown for 48 h on Luria-Bertani (LB) plates. Cells were resuspended in 10 mM $MgCl_2$ and cell densities adjusted to 0.1 OD_{600nm} (corresponding to 5×10^7 colony forming units [cfu] per ml) and tenfold serial dilutions were carried out to reach the appropriate inoculation dose (approximately 10^7 or 10^4 cfu). Ten-day-old bean plants were inoculated by infiltrating 200 μl of bacterial suspension into the intracellular spaces of the abaxial side of one of the primary leaves using the blunt end of a 2 ml syringe. Negative control plants were inoculated with 200 μl of $MgCl_2$. In all cases, inoculations were carried out in the same half of the leaf allowing differentiation by the midrib of two areas within the inoculated leaf, infiltrated area (IF) and non-infiltrated area (NIF) (Fig. 1A).

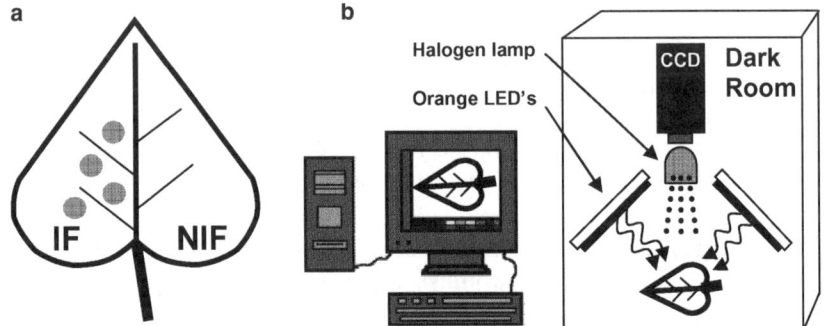

Fig. 1 a Differentiation of non-infiltrated (NIF) and infiltrated (IF) areas in leaves inoculated with *P. syringae* **b** FluorCam scheme. Fluorescence emission was induced in two sets of 325 super-bright orange light emitting diodes (LEDs) that provide either excitation flashes of a user-definable duration or a continuous actinic irradiance. In addition, a halogen lamp equipped with a shutter may also generate strong pulses of actinic light. The timing and duration of the excitation flashes and the duration of the actinic light exposure are controlled using the FluorCam software package (PSI, Ltd. Czech Republic; www.psi.cz)

2.3 Detection of Bacteria in Leaves

After 14 dpi bacteria were recovered from the infected leaves using a 10-mm-diameter cork-borer, homogenized by mechanical disruption into 1 ml of 10 mM $MgCl_2$ and enumerated by plating serial dilutions on to LB agar.

2.4 Chlorophyll Fluorescence Imaging

Red chlorophyll fluorescence of bean plants adapted to a 20 min dark period was measured by the kinetic imaging chlorophyll fluorometer FluorCam (PSI, Ltd. Czech Republic; www.psi.cz) as described by Nedbal et al. (2000) (Fig. 1B). Briefly, actinic light ($130 m^2 s^{-1}$) and measuring flashes are generated by two panels of orange light emitting diodes, whereas saturating flashes are generated by a 250 W halogen lamp. The chlorophyll fluorescence emission transients are captured by a CCD camera in a series of images with a resolution of 512×512 pixels. In this study, after measuring the minimum fluorescence in the dark-adapted state (F_0), samples were illuminated with a saturating pulse ($1500 \mu E m^{-2} s^{-1}$) to determine the maximal fluorescence in the dark-adapted state (F_M). For quenching analysis, after 20 s of darkness, leaves were illuminated with orange actinic light ($130 \mu E m^{-2} s^{-1}$) and saturating pulses at 2, 5, 10, 20, 30, 60, 90, 120, 180, 240 and 300 s. Fluorescence values registered at every saturating pulse are named F_M', whereas fluorescence values recorded just before them, are called F_t.

Numerical analysis of classical physiological parameters recorded as maximum photosystem II (PSII) quantum yield ($F_V/F_M = (F_M-F_0)/F_M$), effective PSII quantum yield ($\Phi_{PSII} = (F_M'-F_t)/F_M'$), and NPQ ($F_M-F_M'/F_M'$) were calculated from the IF area of the inoculated leaf (Fig. 1A) for plants inoculated with the lower number of cells (10^4 cfu). For plants inoculated with a higher number of bacteria (10^7 cfu), analysis was performed in a primary leaf located at the same level as the inoculated leaf (nearest neighbour leaf).

2.5 Chlorophyll Content Measurements

Chlorophyll content was measured (10 repeats per leaf) using the CL-01 Chlorophyll Content Meter (Hansatech Instruments, England; http://www.hansatech-instruments. com/cl01.htm), which determines relative Chl content using dual wavelength optical absorbance (620 and 940 nm) measurements from leaf samples.

3 Results and Discussion

Changes on the photosynthetic activity of *P. vulgaris* plants inoculated with Pph (compatible interaction) and Pto (non-compatible interaction) was followed by means of the FluorCam. Fluorescence induction and quenching analysis was carried out and pseudo-colour images of several Chl-F parameters were obtained for the infected plants and their corresponding negative controls at 5-, 7- and 9-day post-inoculation (dpi). Under these conditions, none of the leaf areas analysed showed visible symptoms. Infection of beans plants with large numbers of cells of Pph or Pto resulted in chlorosis of the tissue followed by the formation of water-soaked lesions or browning of the inoculated leaf area, respectively, after 3 days. However, no symptoms were visible on neighbouring leaves even after 14 days. In the case of inoculations with approximately 10^4 cfu, visible symptoms of the infected leaves did not appear in Pph-infiltrated plants during the first week after infection. Chlorotic lesions started became visible in the IF area after 11 days. In the incompatible reaction induced by Pto, as well as in the control plants, no HR was as observed at any time (data non shown).

No differences in maximum PSII quantum yield (F_v/F_M) were detected at any time among the leaves analysed, independently of the strain and cell concentration used to inoculate the plants. The values obtained for this parameter, between 0.79 and 0.82, for both mock controls and infected plants, are similar to those reported as optimal for plant photosynthetic performance (Björkman and Demmig, 1987). Bonfig et al. (2006) reported a pre-symptomatic decrease of F_v/F_M after infection of *Arabidopsis* plants with both virulent and avirulent strains of Pto DC3000. However, such changes were restricted to the vicinity of the point of inoculation and photosynthesis of the remaining leaf was unaffected.

In plants inoculated with 10^7 cfu, no significant differences in any of the photosynthetic parameters analysed were observed at any time in neighbouring leaves. NPQ at 20 s (NPQ_{20}) and Φ_{PSII} at 5 s (Φ_{PSII5}) varied in all plants from approximately 0.80 to 1.00 (Fig. 2A) and from about 0.25 to 0.40 (Fig. 2B), respectively, at 5, 7 and 9 dpi. In agreement with the absence of both visible symptoms and changes in photosynthesis efficiency, bacteria could not be detected at any time in such leaves after extraction and plating in LB medium.

With plants inoculated with the low numbers of bacteria, Φ_{PSII} and NPQ maximized the differences between infected and healthy plants. The most significant changes in these parameters were observed in the IF areas at 5 dpi. The kinetic points in which the control and Pph-infiltrated plants displayed clear differences were Φ_{PSII5} and NPQ_{20} (Fig. 2C and D). Values of Φ_{PSII5} and NPQ_{20} were similar for the controls and Pto-infiltrated plants; in contrast, Pph-infected plants showed significantly lower values of both parameters. When analysed at 7 and 9 dpi, Pph-infiltrated plants could not be differentiated from the rest of the treatments using NPQ_{20} values (Fig. 2C). However, at 7 dpi the Φ_{PSII5} resulted in lower values in Pph-infiltrated plants than in the controls and Pto-infiltrated plants. To our knowledge, this is the first time Chl-FI has been shown to differentiate a compatible from an incompatible interaction between plant and bacteria in asymptomatic tissues. Bonfig et al. (2006) reported a decrease in both parameters in *Arabidopsis* leaves infected with a virulent *P. syringae* strain, however, the same results were observed for these two parameters during incompatible interaction with an avirulent strain.

NPQ_{20} images showed a differential response of the leaf to Pph infection at 5 dpi, with lower values of this parameter than those observed in both Pto-infiltrated and

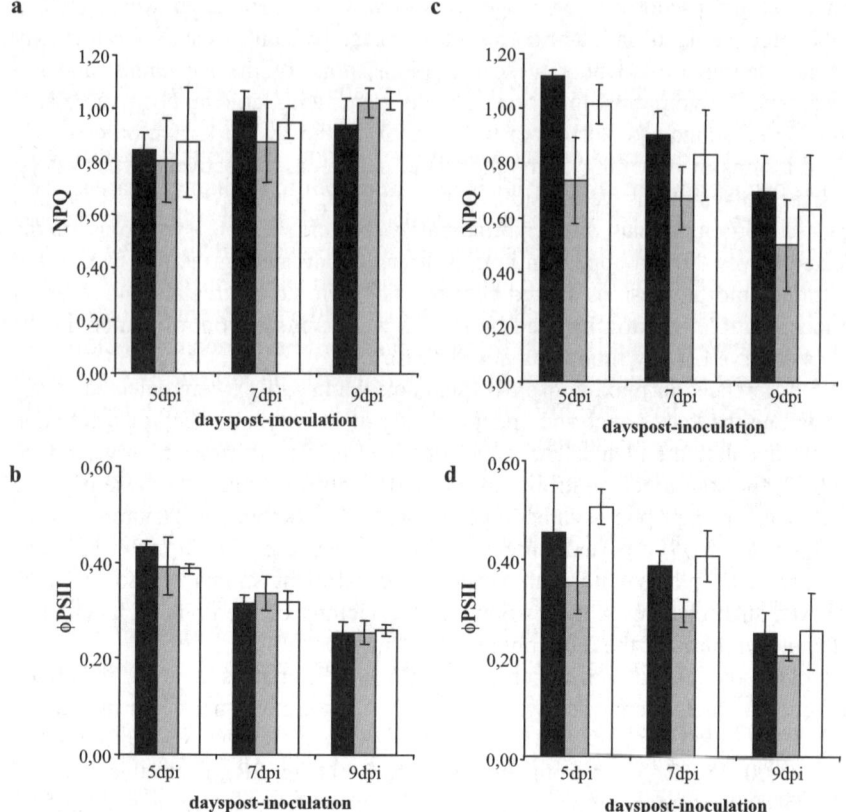

Fig. 2 Quantitative analysis of time-dependent changes induced by *P. syringae* in various chlorophyll fluorescence parameters. Data represent the mean of three independent plants ± standard deviation. Black, grey and white bars represents the values obtained for *P. syringae* pv. *tomato* DC3000, *P. syringae* pv. *phaseolicola* 1448a and control plants, respectively. **a** and **b** NPQ_{20} and Φ_{PSII} values, respectively, of the neighbour leaf in plants inoculated with approximately 10^7 cfu of bacteria. **c** and **d** NPQ_{20} and ΦP_{SII5} values, respectively, of the IF area of leaves inoculated with approximately 10^4 cfu of bacteria

control plants (Fig. 3). To investigate a possible correlation between Chl-F with those in Chl content, this parameter was also measured, however, none of the leaves analysed in any of the two experiments showed values significantly different from the control.

In summary, inoculation of bean leaves with a low dose of bacteria has allowed differentiation by Chl-FI quenching analysis, mainly reflected in NPQ, of a compatible from an incompatible bacterial interaction in asymptomatic leaf areas of *P. vulgaris*, encouraging the utilization of this parameter for the study of *P. syringae* infection of bean plants.

Mock   **Pto**   **Pph**

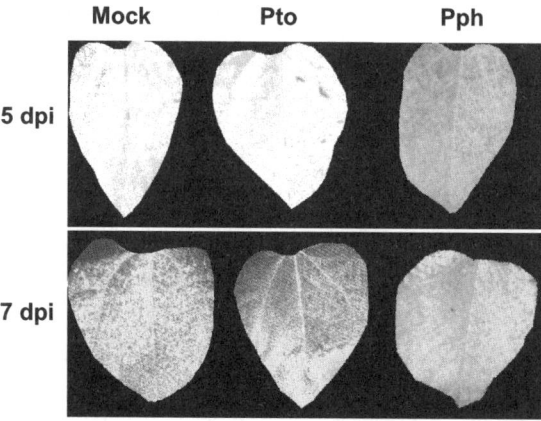

Fig. 3 Grey scale transformation of pseudo-colour images corresponding to non-photochemical quenching at 20 s (NPQ_{20}) of plants inoculated with *P.* syringae pv. *tomato* DC3000 or *P. syringae* pv. *phaseolicola* at 5- and 7-day post-inoculation (dpi). Dark and light tones correspond to low and high NPQ_{20} values, respectively

Acknowledgements This project was supported by Spanish MEC grants AGL2002-02214 and BIO2004-04968-C02-02 to C.R. and M.B., respectively.

References

Balachandran S., Osmond C.B., Daley P.F. (1994) Diagnosis of the earliest strain-specific interactions between tobacco mosaic virus and chloroplasts of tobacco leaves *in vivo* by means of chlorophyll fluorescence imaging. *Plant Physiol.* 104:1059–1065

Berger S., Papadopoulos M., Schreiber U., Kaiser W., Roitsch T. (2004) Complex regulation of gene expression, photosynthesis and sugar levels by pathogen infection in tomato. *Physiol Plantarum.* 122:419–428

Björkman O. Demmig B. (1987) Photom yield of O2 evolution and chlorophyll fluorescence at 77 K among vascular plants of diverse origins. *Planta* 170:489–504

Bonfig K.B., Schreiber U., Gabler A., Roitsch T., Berger S. (2006) Infection with virulent and avirulent *P. syringae* strains differentially affects photosynthesis and sink metabolism in *Arabidopsis* leaves. *Planta* 225:1–12

Chaerle L. Van Der Straeten D. (2001) Seeing is believing: imaging techniques to monitor plant health. *Biochim. Biophys. Acta.* 1519:153–166

Chaerle L., Pineda M., Romero-Aranda R., Van Der Straeten D., Barón M. (2006) Robotized thermal and chlorophyll fluorescence imaging of pepper mild mottle virus infection in *Nicotiana benthamiana. Plant Cell Physiol.* 47:1323–1336

Chou H., Bundock N., Rolfe S., Scholes J. (2000) Infection of *Arabidopsis thaliana* leaves with *Albugo candida* (white blister rust) causes a reprogramming of host metabolism. *Mol. Plant Pathol.* 1:99–113

Cuppels D.A. (1986) Generation and characterization of Tn*5* insertion mutations in *Pseudomonas syringae* pv. *tomato. Appl. Environ. Microbiol.* 51:323–327

Dangl J.L., Jones J.D.G. (2001) Plants pathogens and integrated deference responses to infection. *Nature* 411:826–833

Esfeld P., Siebke K., Wacker I., Weis E. (1995) Local defence-related shift in the carbon metabolism in chickpea leaves induced by a fungal pathogen. In Mathis P. (ed). Photosynthesis from Light to Biosphere, Vol. 5 (Dordrecht, The Netherlands: Kluwer), pp. 663–666

Fillingham A.J., Wood J., Bevan J.R., Crute I.R., Mansfield J.W., Taylor J.D., Vivian A. (1992) Avirulence genes from *Pseudomonas syringae* pathovars *phaseolicola* and *pisi* confer specificity towards both host and non-host species. *Physiol. Mol. Plant Pathol.* 40:1–15

Klement Z. (1982) Hypersensitivity. In Mount M.S. and Lacy G.H. (eds). Phytopathogenic Procaryotes (New York: Academic), pp. 149–177

Lohaus G., Heldt H.W., Osmond C.B. (2000) Infection with phloem limited abutilon mosaic virus causes localized carbohydrate accumulation in leaves of *Abutilon striatum*: Relationships to symptom development and effects on chlorophyll fluorescence quenching during photosynthetic induction. *Plant biol.* 2:161–167

Nedbal L. Whitmarsh J. (2004) Chlorophyll fluorescence imaging of leaves and fruits. In Papegeorgiou C.G. and Govindjee (eds). Chlorophyll a Fluorescence: A Signature of Photosynthesis (Dordrecht, The Netherlands: Springer) 14, pp. 389–407

Nedbal L., Soukupova J., Whitmarsh J., Trtilek M. (2000) Postharvest imaging of chlorophyll fluorescence from lemons can be used to predict fruit quality. *Photosynthetica* 38:571–579

Osmond C.B., Daley P.F., Badger M.R., Lüttge U. (1998) Chlorophyll fluorescence quenching during photosynthetic induction in leaves of *Abutilon striatum* Dicks infected with Abutilon mosaic virus, observed with a field portable imaging system. *Botanica Acta* 111:390–397

Pérez-Bueno M.L., Ciscato M., vandeVen M., García-Luque I., Valcke R., Barón M. (2006) Imaging viral infection. Studies on *Nicotiana benthamiana* plants infected with the pepper mild mottle tobamovirus. *Photosynth. Res.* 90:111–123

Peterson R.B. Aylor D.E. (1995) Chlorophyll fluorescence induction in leaves of *Phaseolus vulgaris* infected with bean rust (*Uromyces appendiculatus*). *Plant Physiol.* 108:163–171

Rahoutei J., García-Luque I., Barón M. (2000) Inhibition of photosynthesis by viral infection: Effect on PSII structure and function. *Physiol. Plantarum* 110:286–292

Roitsch T. (1999) Source–sink regulation by sugar and stress. *Curr. Opin. Plant Biol.* 3:198–206

Scharte J., Schon H., Weiss E. (2005) Photosynthesis and carbohydrate metabolism in tobacco leaves during an incompatible interaction with *Phytophthora nicotianae*. *Plant Cell Environ.* 28:1421–1435

Scholes J.D., Rolfe S. (1996) Photosynthesis in localised regions of oat leaves infected with crown rust (*Puccinia coronata*): quantitative imaging of chlorophyll fluorescence. *Planta* 196:573–582

Soukupová J., Smatanová S., Nedbal L., Jegorov A. (2003) Plant response to destruxins visualized by imaging of chlorophyll fluorescence. *Physiol. Plantarum* 118:1–8

Staskawicz B.J., Ausubel F.M., Baker B.J., Ellis J.G., Jones J.D.G. (1995) Molecular genetics of plant disease resistance. *Science* 268:661–667

Sensitive Detection of *Ralstonia solanacearum* Using Serological Methods and Biolog Automated System

A.E. Tawfik[1], A.M.M. Mahdy[2], and A.A. El Hafez Omnia[1]

Abstract Brown rot disease caused by *Ralstonia solanacearum* (Yabuuchi et al., 1995) is an important disease in tropical, subtropical and warm climates. Since the first authentic record on the occurrence of brown rot in Egypt (Sabet, 1961), it became a major problem affecting potato production and created a serious threat for potato export toward Europe. Thus plant quarantine services in several importing countries are quite alert to the Egyptian potatoes.

The present investigation was conducted in an attempt to evaluate different methods for rapid diagnosis of the disease. Accordingly the pathogen was isolated from imported potato seed tubers and exported potato and identified by serological and biochemical methods. Serological assays included Dot-immunobinding assay (DIA), immunofluorescent staining method (IFA). Biolog automated system was also used as biochemical assay.

Nineteen isolates of *R. solanacearum* were recovered from potatoes collected from two provinces. Cultural and physiological studies revealed that these isolates belonged to biovar II of *Ralstonia solanacearum*. Pathogenicity tests proved that 19 isolates were pathogenic with different degrees of virulence. The Biolog system showed that isolate 18F utilize 95 different carbon sources as compared to the other 10 bacterial species, which were grouped in 3 distinct clusters. The 18F isolate was found to have the closest metabolic fingerprint similarity to *R. solanacearum*. The results also revealed that DIA and IFA have a great potential as simple and rapid method for identification and detection of *R. solanacearum* in infected potato tubers.

Keywords Bacterial wilt, *Ralstonia solanacearum* DIA, IFA, Biolog system

[1] Plant Pathology Research Institute, Agricultural Research Center, Giza, Egypt

[2] Faculty of Agriculture, Benha University, Egypt

Author for correspondence: Ali Tawfik; e-mail: atawfiknasrcity@maktoob.com

1 Introduction

Ralstonia solanacearum (syn. *Pseudomonas solanacearum* and *Burkholderia solanacearum*) is a widespread and economically important bacterial plant pathogen (Sequeira, 1993). It causes bacterial wilt, a major disease that limits production of potato crop. *R. solanacearum* has an extended host range that includes hundreds of plant species belonging to 44 plant families (Hayward, 1991). Isolates of *R. solanacearum* are differentiated into five races according to host range (Buddenhagen et al., 1962) and five biovars according to utilization of various disaccharides and hexose alcohols (Hayward, 1964). The symptoms of the disease can be recognized by sudden wilting of the plant during sunny days, which results from blockage of the plant's water conducting vessels. The disease was first recorded in Egypt by Briton-Jones (1925) and the causal agent was identified as *P. solanacearum* by Sabet (1961). A significant reduction in crop loss due to bacterial wilt has been demonstrated through early detection of latent infected tubers (Elphinstone et al., 1998). Therefore, early detection of the bacterium in soil, water, tubers and plant residues could facilitate elimination and reduce the risk of crop loss. Several detection methods, including isolation, serological and biochemical assays were used in detection and identification of pathogenic bacteria. However, these assays were not satisfactory for routine diagnosis (Werres and Steffens, 1994). Immunofluorescent staining method has the advantage to be sensitive. Bochner (1989) introduced Biolog microplate system for the diagnosis of Gram negative bacteria. The system is based on the reaction to a series of 95 carbon sources which were indicated by color reactions.

The present investigation was conducted in an attempt to evaluate different methods for rapid diagnosis of brown rot of potato in imported and exported potato tubers.

2 Materials and Methods

2.1 Sampling and Isolation

Infected potato tubers and plants (*Solanum tuberosum* L.) showing brown rot symptoms were collected from fields at Minufia and Behera provinces during the period from 2002 to 2004. Infected tubers were surface sterilized by immersion in 0.5% sodium hypochlorite for 2 min then rinsed in sterile distilled water (SDW). The tubers were cut into two halves and a loopful of the bacterial ooze exuding from the vascular tissues suspended into SDW. A loopful from the resulting suspension was streaked on tetrazolium chloride medium TZC (Kelman, 1954). Single colonies were picked on NGA slants, pure cultures were maintained at 8°C for further utilizations.

2.2   Pathogenicity Tests

Tomato plants (*Lycopersicon esculentum*. Mill) were used. Plants grown in sterile potted soil and at the third leaf stage were inoculated by the stem puncture technique (Prior and Steva, 1990). Bacterial cell suspension was prepared from each isolate in SDW at a concentration of 10^9 CFU/ml. A sterile needle was forced into the stem through and 25 µl of bacterial suspension of each isolate placed in the axil of the third leaf below the stem apex. Uninoculated tomato seedlings served as control. Wilt severity was determined by calculating the proportion of wilted leaves in each plant as follows:

$$\frac{\text{No. of wilted leaves per plant}}{\text{Total No. of leaves per plant}} \times 100$$

Before using the bacterial cultures of *R. solanacearum*, the isolates were tested for their purity and were found to be typical in morophological, cultural and physiological characters to *R. solanacearum* according to Bergey's manual of systematic bacteriology

2.3   Biolog Microplate System

The isolate 18F, from Minufia province, was used in this test because it represents the virulent isolate of the pathogen. This isolates was grown on tryptic soy agar (TSA). The turbidity range of the bacterial cells was determined by Biolog turbidimeter at 590 nm. Cell density was adjusted to 3×10^8 CFU/ml. Wells of microplate were filled with 150 µl of the bacterial suspension. The microplate was incubated for 24 h and then the reading was done by a microplate reader. Results were analyzed by Biolog G-ve database to determine the identity of the bacterial isolate.

2.4   Serological Assays

1. Dot-immunobinding assay (DIA), according to Jahn et al. (1984), was carried out to identify the isolates from cultures or from diseased tubers by clearly defined color in the spot using fast stain.
2. Indirect immunofluorescent assay (IFA): This method was used according to method developed in the European community (Anonymous, 1987). It is based on indirect immunofluorescent antibody staining, then confirmed using pathogenicity test on tomato plants.

3   Results

R. solanacerum is found to be the important pathogen found in Egyptian soil which infects potato plants. This finding is in accordance to Sabet (1961). The biovar 2 was characterized based on the utilization of sorbitol (−), dulcitol (−) mannitol (−) and oxidation of lactose (+), maltose (+), cellobiose (+). Results of isolation from tubers showing typical symptoms of brown rot disease on medium containing triphenyletrazolium chloride (TZC), which differentiate between virulent and avirulent isolates, revealed that virulent colonies are fluidal, irregular in shape, slimy and appear creamy white or without pink centers. On the other hand, the avirulent colonies are truly round, deep red color with narrow bluish border (Fig. 1). Nineteen isolates obtained from diseased tubers and of which colonies were the virulent type gave wilted leaves of inoculated tomato plants (Table 1, Fig. 2). The obtained results showed that the isolates 1F, 7B and 19B were the most virulent and induced complete wilting after 25 days from inoculation. The isolates 2F, 3F, 4F, 6B, 9B, 10F, 12F, 13B, 14F, 15F and 17F induced 70–80% wilt of inoculated plants after 25 days from inoculation, while the isolate 5F was the least virulent inducing only 50% wilting after the same period.

3.1   Identification of Ralstonia solanacearum Using Biolog Automated System

R. solanacearum isolate 18F oxidizes sucrose, methyl pyruvate, aconitic acid, citric acid, D-galacturonic acid, L-aspargine, L-asparatic acid, L-glutamic acid, L-serine and L-threonine under the conditions used. Results of the Biolog system revealed

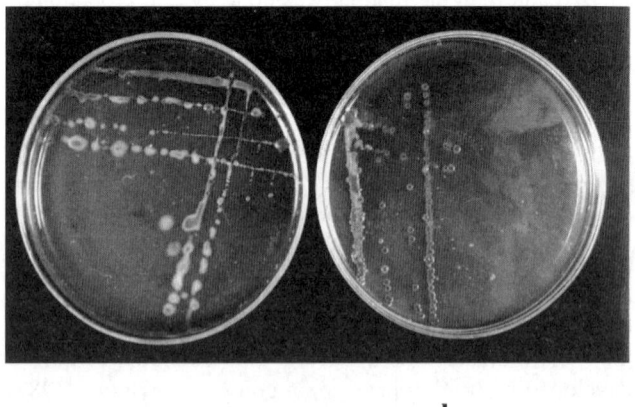

a   b

Fig. 1 Typical colonies of Virulent **a** and avirulent **b** of isolated bacteria from infected tubers with brown rot symptom on tetrazolium chloride (TZC) medium

that isolate 18F is identified as *R. solanacearum*. It utilizes 95 different carbon source as compared with other 10 pathogenic bacterial species which were grouped in 3 distinct clusters (Fig. 3). The isolate 18F was found to have the closest metabolic fingerprint similarity to *R. solanacearum* (*Pseudomonas*) *solanacearum* which was included with *Psychrobacter immobilis* in one cluster.

Isolate 4F   Isolate 19B

Fig. 2 Pathogenicity test on Tomato seedlings using two isolates of isolated bacterium from infected tubers with brown rot symptoms, under artificial inoculation conditions

Table 1 Disease severity on tomato seedlings inoculated by isolated bacteria from infected tubers with brown rot symptom, at different periods (7, 15 and 25 days)

Isolates	Disease severity (%)		
	7	15	25 days
1F[a]	33.3	62.0	100.0
2F	17.2	40.0	66.2
3F	22.1	48.2	70.5
4F	33.3	60.6	75.5
5F	16.6	37.5	50.6
6B[b]	33.3	62.6	77.5
7B	50.0	77.6	100.0
8B	37.4	50.2	83.2
9B	16.6	40.9	66.7
10F	16.6	35.2	61.2
11F	33.3	70.5	82.1
12F	17.2	40.0	66.2
13B	25.0	50.8	69.1
14F	16.6	38.2	65.5
15F	17.2	41.5	62.1
16B	37.4	50.2	80.2
17F	33.3	62.6	77.5
18F	16.6	48.6	82.5
19B	50.0	79.2	100.0
Control (uninoculated)	0.0	0.0	0.0

Source of isolates: [a]F: Minufia; [b]B: Behera

Fig. 3 Simplified dendogram showing the Biolog clusters of isolate number 18F of Ralstonia solanacearum, tested with the Biolog GN

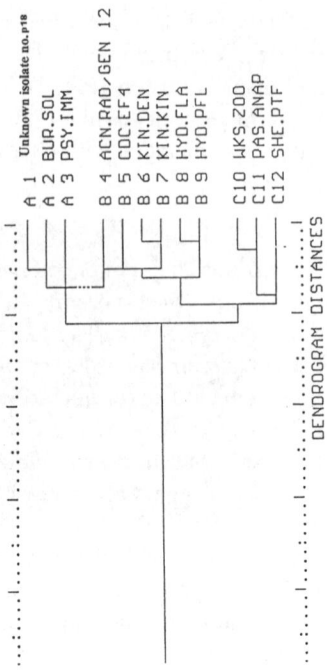

3.2   Immunological Studies

Normal serum was obtained from rabbit before immunization to serve as a negative control. Rabbit was intramuscularly injected every week for 2 months with 1 ml of 10^9 CFU/ml-suspension. Blood was collected from the rabbit 5 days after the last immunization. The antiserum was stored for further studies.

Dot-immunobinding assay (DIA) provided a better understanding of the interaction between virulent and avirulent isolates of *R. solanacearum*. It also provided a simple and rapid test for identification and detection of *R. solanacearum* in infected potato tubers.

The sensitivity of DIA was further evaluated with pure cultures of *R. solanacearum*. It appears, from the obtained data, that when using specific antiserum dilutions of *R. solanacearum* the color intensity was enhanced with all the isolates.

3.3   Detection of R. solanacearum in Pure Cultures by Immunofluorescence (IF)

The identity of the isolates of *R. solanacearum* and their pathogenicity were confirmed by IF and inoculation of tomato plants using isolate 18F as positive control (Table 2). The results showed that the identification of pure cultures could be confirmed by the IF test. Positive IF test with fluorescent bright colored short rod bacteria is indicative of *R. solanacearum*.

Table 2 Immunofluorescence assay of 19 isolates of *R. solanacearum* and their pathogenicity towards tomato plants

Isolates	Location	IFA	Pathogenicity[a]
1F	Minufia	+	5
2F	Minufia	+	3
3F	Minufia	+	4
4F	Minufia	+	4
5F	Minufia	+	3
6B	Behera	+	4
7B	Behera	+	5
8B	Behera	+	5
9B	Behera	+	3
10F	Minufia	+	4
11F	Minufia	+	3
12F	Minufia	+	4
13B	Behera	+	3
14F	Minufia	+	3
15F	Minufia	+	3
16B	Behera	+	5
17F	Minufia	+	4
18 F	Minufia	+	5
19 B	Behera	+	5

[a]Number of wilted tomato plants of five inoculated ones

4 Discussion

In this investigation 19 isolates of *R. solanacearum*, the causal agent of brown rot, were recovered from naturally infected potato tubers and plants grown in Behera and Minufia provinces. Results on the isolates have revealed that they belong to biotype or biovar II which is the only biotype in Egypt, based on the utilization of three disaccharides and three hexose alcohols confirming the results obtained by Abo-El-Dahab et al. (1978) and Gebreel et al. (2000). Pathogenicity test conducted with 19 isolates on tomato plants revealed that all isolates were significantly virulent.

Since identification of plant pathogenic bacteria by conventional procedures is time consuming and costly in terms of materials, the need for more rapid, reliable and inexpensive methods has led to the development of techniques that can be adopted for diagnostic purposes. These include the Biolog GN microplate system that can simply characterize and identify *R. solanacearum* isolates by examining their carbon source utilization profiles. Biolog has offered a directed systematic process for selecting sets of substrates that can be used to identify and characterize unknown microbes with a high degree of accuracy. Such profiles must include large sets of chemical substrates such as carbohydrates, carboxylic acids, amino acids, alcohols, amines, amid and aromatic chemicals.

Biolog has made this approach much more available by developing an oxidation–reduction reaction which detects the utilization of carbon source by color change. Therefore, the application of Biolog microplate system confirmed the identification of *R. solanacearum* isolates, as this method is more rapid and accurate than traditional procedures.

In this investigation, serological assays were applied for the identification and characterization of R. *solanacearum* and for immunodiagnosis of brown rot of potato in latently infected tubers. The detection of the latent infection of R. *solanacearum* based on the indirect immunofluorescence assay (IFA) and the pathogenicity on tomato plants was studied. Negative results indicated that the pathogen was not detected in symptomless tubers. In general IFA took less time than the direct isolation from diseased tubers and also it is a rapid and easy test for screening a large number of samples. Dot-immunobinding assay (DIA) procedure is potentially useful for the diagnosis of plant pathogens. Thus the chief advantage of this assay is in its specificity and sensitivity. In our study, the intensity of the DIA was roughly correlated with the number of bacteria in soaking solution, which required 10^5 cfu/ml for a positive reaction. The present study showed that sensitivity and specificity depend on the dilution of antiserum used. This can be achieved when the antiserum was in the highest dilution where color intensity would not be lost (sensitivity) and where cross-reaction with antigens of other bacteria would not be observed (specificity). It can be concluded of the importance of immunobinding assay for the detection of brown rot pathogen in naturally and latently infected tubers from cold stores or from field.

References

Abo-El-Dahab, M.K., El-Goorani, M.A. and Wagih, E.E.M., 1978, Race identification of *Pseudomonas solanacerarum* E.F. Smith in Egypt. Zenttralblatt fur Bak teriologie II Abt. 133: 211–216.

Anonymous, 1987, Scheme for the detection and diagnosis of the ring rot bacterium *Corynebacterium spedonicum* in batches of potato tubers. Commission on the European communities, Luxembourg, Report EUR 11288.

Bochner, B.R., 1989, Sleuting out bacterial identities. Nature 339: 157–158.

Briton-Jones, H.R., 1925, Mycological work in Egypt during the period 1920–1922. Egypt Min. Agr. Tech. and Sci. Serv. Bull. 49.

Buddenhagen, I., Sequeira, L. and Kelman, A., 1962, Designation of races in *Pseudomonas solanacearum* (Abst.) Phytopathology 52: 726.

Elphinstone, J.G., Stanford, H.M. and Stead, D.E., 1998, Detection of *Ralstonia solanacearum* in potato tubers, Solanum dulcamara and associated irrigation water. In: Prior, P. Allen, C. and Elphinstone, J. (eds.) Bacterial Wilt Disease: Molecular and Ecological Aspects. Reports of the Second International Bacterial Wilt Symposium, 22–27, 1997, Guadeloupe, France, pp. 133–139. Springer, Germany.

Gebreel, H.M., Madbouly, A.K., Tawfik, A.E. and Youssef, Y.A., 2000, Serological, phsiological and biochemical studies on *Burkholderia solanacearum* (*Pseudomonas solanacearum*) the bacterium causing potato brown rot and wilt disease. Al-Azhar J. Microb. 48(April): 41–60.

Hayward, A.C., 1964, Characteristics of *Pseudomonas solanacearum*. J. Apple. Bacterial. 27: 265–277.

Hayward, A.C., 1991, Biology and epidemiology of bacterial wilt caused by *Pseudomonas solanacearum*. Ann. Rev. Phytopath. 29: 65–87.

Jahn, R., Schieblen, W. and Greengard, P., 1984, A quantitative dot-immunobinding assay for proteins using nitrocellulose membrane filters. Proc. Nat. Acad. Sci. 81: 1684–1687.

Kelman, A., 1954, The relationship of pathogenicity of *Pseudomonas solanacearum* to colony appearance in a tetrazolium medium. Phytopathology 44: 693–695.

Prior, P. and Steva, H., 1990, Characteristics of strains of *Pseudomonas solanacearum* from the French West Indies. Plant Disease 74: 13–17.

Sabet, K.A., 1961, The occurrence of bacterial wilt of potatoes caused by *Pseudomonas solanacearum* (E.F. Smith) in Egypt. Min. Agr. Tech. Bul.

Sequeira, L., 1993, Bacterial wilt: past, present, and future. In: Hartman, G.L. and Hayward, A.C. (eds.) Bacterial Wilt. ACIAR Proceedings No. 45.

Werres, S. and Steffens, C., 1994, Immunological techniques used with fungal plant pathogens: aspects of antigens, antibodies and assay for diagnosis. Ann. Appl. Bio. 125: 615–643.

Yabuuchi, E., Kosako, Y., Yano, L., Hottas, H. and Nishiuchi, Y., 1995. Transfer of two Burkholder and Alkaligens species to Ralstonia Gen. Nov. Proposal of Ralstonia picketti (Ralston, palleronia solanacearum (smith 1986) Comb. Nov. Microbiology and Immunology 39: 879–904.

Part II
Epidemiology and Disease Management

Epidemiological Basis for an Efficient Control of *Pseudomonas savastanoi* pv. *savastanoi on* Olive Trees

J.M. Quesada, R. Penyalver, and M.M. López

Abstract *Pseudomonas savastanoi* pv. *savastanoi* (Psv) is the causal agent of olive knot disease and its epiphytic presence and measures for the disease control have been evaluated in Spanish olive orchards.

Seasonal dynamics of Psv populations on stem and leaf surfaces from symptomless shoots of naturally infected olive trees, was monitored by direct isolation and enrichment-PCR in five olive orchards, located in two olive-growing regions of southern Spain. No significant differences were found between leaf and stems in respect to the number of samples where Psv was isolated or detected by PCR. Average Psv populations varied from 2 to 10^2 cfu/g of tissue and were highly variable among field plots. Significant differences in Psv populations were found between summer and the other seasons in one field plot. Average high levels of Psv populations were observed with warm and raining months and, on the contrary, low levels were generally found in hot and dry months.

Plants of 'Picudo' and 'Arbequina' cultivars were inoculated once with Psv after the plantation and several treatments (copper oxychloride, cuprocalcic sulfate plus mancozeb or acibenzolar-S-methyl, Bion) were applied during four years. The effect of the copper treatments on Psv was more significant in the case of 'Picudo' where the average Psv population densities on untreated control and on treated samples reached 10^4 and 10^2 cfu/g of tissue, respectively. Both copper treatments also reduced the proportion of the samples where Psv was isolated. No resistance to copper in recovered isolates of Psv was detected. Furthermore, the average knot number was significantly lower in the plants treated with copper than in the untreated plants.

The treatments with Bion did not affect either the Psv population densities or the number of knots. No treatment significantly increased the olive yield in the four years of study, probably because very young trees were used. In conclusion, monitoring the populations of the olive knot pathogen in the orchards provides information necessary to design an integrated management of the disease and copper treatments should be regularly used for an efficient chemical control.

Instituto Valenciano de Investigaciones Agrarias (IVIA). Carretera de Moncada a Náquera, Km. 4.5. 46113 Moncada (Valencia), Spain

Author for correspondence: Maria M. López; e-mail: mlopez@ivia.es

M'B. Fatmi et al. (eds.), *Pseudomonas syringae Pathovars and Related Pathogens.*

© Springer Science + Business Media B.V. 2008

Keywords Phyllosphere, stem, leaf, *Olea europaea* L., PCR, copper compounds, acibenzolar-S-methyl, control

1   Introduction

Pseudomonas savastanoi pv. *savastanoi* (Psv), is the causal agent of olive knot, a disease characterized by knots (outgrowths) mainly on the trunk and branches and less frequently on leaves and fruits of the host plants (Wilson, 1935). The bacterium is widespread all over in most of the Mediterranean countries, where olive cultivation are growing for centuries. There are few available studies on the epidemiology of olive knot and most of the studies on Psv populations on the phyllosphere of olive trees were performed in Italy (Ercolani, 1978, 1985, 1991; Lavermicocca & Surico, 1987). Cupric compounds are the most frequently applied treatments for the control of olive knot (Schroth et al., 1973; Teviotdale & Krueger, 2004). However, there is a lack of information about their direct effect on the target bacterium as well as on their efficiency in the disease control under Mediterranean conditions (Quesada et al., 2006). As Psv populations on olive trees may serve as inoculum reservoir the monitoring of bacterial populations is necessary to forecast the disease development and to evaluate the control treatments efficiency. We report here: (1) a three-year study on the direct isolation and PCR detection of Psv on stem and leaf surfaces of symptomless shoots of naturally infected olive trees and (2) a four-year evaluation of the efficacy of several compounds on both the Psv population reduction and disease incidence.

2   Materials and Methods

2.1   *Epiphytic Psv on Naturally Infected Olive Trees*

Five olive orchards located in two olive-growing regions of Spain (Comunidad Valenciana and Andalucía) were selected. Plant samples were taken from summer 1999 to spring 2002 in two plots located in the Comunidad Valenciana and from spring 2001 to winter 2002 in three plots located in Andalucía. Five naturally infected olive trees with the same amount of knots (ca. 400 per tree) were selected in each plot. Seasonal dynamics of Psv populations on stem and leaf surfaces from symptomless shoots of naturally infected olive trees was monitored by washing and dilution plating using bulked samples on King's medium B and PVF-1 supplemented with cycloheximide at 250 and 125 mg/l, respectively and by enrichment-PCR according to Bertolini et al. (2003). A comparison between the pathogen densities on bulked leaves versus Psv distribution on individual leaves from the same shoot, by leaf printing, was realized. Total bacterial colonies and the number of yellow *Pantoea agglomerans*-like colonies were also evaluated on King's medium B. Psv populations were contrasted to the amount of total non-Psv remaining bacteria and to the populations of *P. agglomerans*. Covariance analysis using year as factor and cumulative rainfall, average temperature and average relative humidity as covariates, was used to assess the influence of climatic parameters on Psv populations only in the field where Psv was isolated more frequently.

2.2 Effect of Chemical Treatments on Psv Populations and Olive Knot Disease

Two-year-old olive trees of 'Picudo' and 'Arbequina' cultivars (both susceptible to Psv) established in an experimental orchard in Valencia, were inoculated once after the plantation with Psv strain IVIA 1628-3 by spotting 10 µl of 10^9 colony-forming units (cfu)/ml bacterial suspension on three V-shaped slits (about 2 mm deep by 5 mm ample) separated by 10 cm each other. They were made with a scalpel in three tagged branches per tree. The three chemical compounds assayed were: copper oxychloride (Curenox, 50% of copper active), cuprocalcic sulfate plus mancozeb (Cuppertine m, 20% of copper active) and acibenzolar-S-methyl registered in Europe as Bion (50WG). Control olive trees (non-treated) were sprayed with distilled water. Trees were treated twice per year on April and December, following the advice of the integrated control programs for olive knot disease. For the experimental design with the four treatments, two replicate groups were randomly selected for each combination of cultivar and treatment (20 replicate olive trees). Treatments were evaluated for: (1) their effect on Psv populations in asymptomatic samples, seasonally evaluated using the above described methodology, (2) their effect on olive knot incidence during the four years (2001–2005) of the study, and (3) their effect on olive cumulative yield.

3 Results

3.1 Epiphytic Psv on Naturally Infected Olive Trees

Psv population densities varied over several orders of magnitude among leaves sampled concurrently from the same shoot, as assessed by the comparison of leaf printing and washing isolation experiments. No significant differences were found between leaf and stems in respect to the number of samples where Psv was isolated or detected by PCR.

The number of samples where Psv was isolated was 89 out of 338 and the average Psv populations varied from 2 to 10^2 cfu/g of tissue. In the samples where Psv was isolated no correlation between Psv population density and that of non-Psv bacteria in any plant material or field plot was found. However, a correlation between Psv

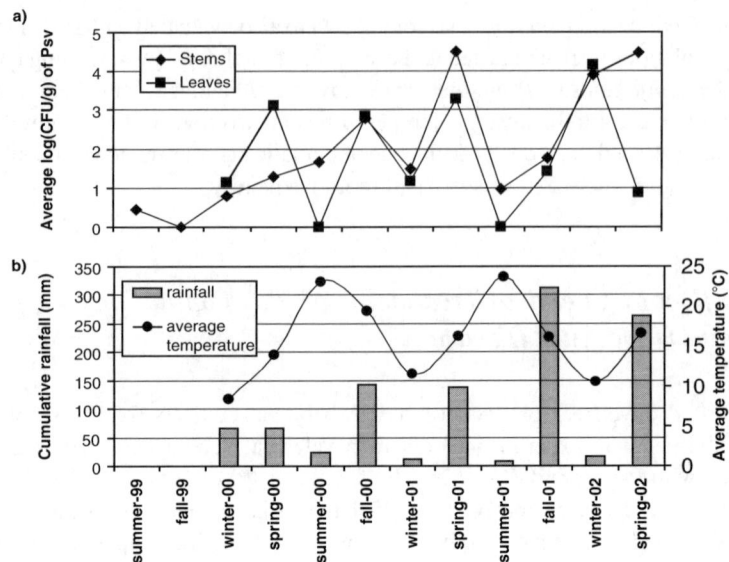

Fig. 1 (**a**) Average *Pseudomonas savastanoi* pv. *savastanoi* population densities on stem and leaf surfaces of naturally infected olive trees located in the field where Psv was more frequently isolated; (**b**) cumulative rainfall and average temperature during the sampling period

population sizes and those of *agglomerans*-like colonies was found. Frequency of Psv isolation and average population were highly variable among field plots. The seasonal fluctuations of Psv populations on stem and leaf surfaces and the influence of climatic parameters were analysed in a field plot where the pathogen was consistently isolated over the three-year study (Fig. 1). On stems and leaf surfaces, Psv populations reached the higher (ca. 10^3–10^4 cfu/g) densities mainly in warm and raining months (mainly spring season) and the lower (ca. 0–10 cfu/g) mainly in hot and dry months (summer season). Significant differences were observed between populations on summer and on other seasons. Interestingly, Psv population sizes in stems and leaf surfaces were correlated (with R^2 values of 0.7 and 0.43, respectively) to rainfall, temperature and relative humidity.

3.2   Effect of Chemical Treatments on Psv Populations and Olive Knot Disease

On 'Picudo' trees, over the four-year study both copper compounds significantly reduced the proportion of samples where Psv was isolated comparing to non-treated plants. The average Psv populations were significantly lower in copper-treated than in control plants since the first application of copper compounds. Average Psv population densities on control plants ranged from 6 to 10^4 cfu/g, while those on copper-treated plants ranged, (excluding the first sampling date where treatments were not yet applied), from 0 to 10^2 cfu/g. The population reduction was also observed on 'Arbequina' plants but no significant differences were found between control trees and plants treated with copper compounds, probably because Psv populations were always lower than 10^2 cfu/g.

Number of samples in which Psv was isolated from olive trees of 'Picudo' treated with acibenzolar-S-methyl was significantly lower than in non-treated plants in the first three years of study, but not in the fourth. In plants of 'Picudo' after the first treatment applications, average of Psv populations was only significantly lower in plants treated with acibenzolar-S-methyl than in control plants, in 5 out of 14 sampling dates. In 'Arbequina', the proportions of samples where Psv was isolated and average of Psv populations in non-treated plants and in plants treated with acibenzolar-S-methyl were quite similar over the four-year study, without significant differences.

Furthermore, in both cultivars the average number of knots was also significantly lower in plants treated with copper compounds than in control plants. In 'Picudo' cumulative average number of knots per plant at the end of the assay, was 251 ± 27 knots for control plants and 114 ± 31 or 76 ± 5 for plants treated with copper oxychloride or with cuprocalcic sulfate plus mancozeb, respectively (Fig. 2). In 'Arbequina' trees cumulative average number of knots was 108 ± 14 knots for control plants and 44 ± 4 or 41 ± 5 for plants treated with copper oxychloride or with cuprocalcic sulfate plus mancozeb, respectively. Treatments with acibenzolar-S-methyl did not affect the number of knots on olive plants.

Cumulative fresh olive yield in control plants of 'Picudo' was 72 ± 16 kg and in plants treated with copper oxychloride or cuprocalcic sulfate plus mancozeb was 97 ± 16 or 111 ± 16 kg, respectively. However, such values were not significantly different. Plants treated with acibenzolar-S-methyl showed a cumulative fresh olive yield of 98 ± 16 kg. Neither copper compounds nor acibenzolar-S-methyl treatments had a significant effect on yield of 'Arbequina' cultivar.

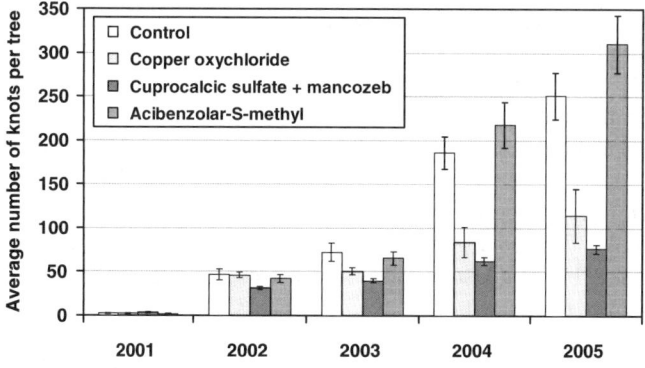

Fig. 2 Cumulative number of knots per year on cv. 'Picudo' trees after different treatments. Vertical bars represent the standard deviation of the mean

4 Discussion

Our study provides new data on the dynamics and plant distribution of *Pseudomonas savastanoi* pv. *savastanoi* (Psv) populations in non-irrigated olive orchards in dry Spanish areas. As described for other epiphytic bacteria and plant hosts, Psv populations were highly variable among leaves from a bulked sample, as inferred from the leaf printings assays. This suggests differences of several orders of magnitude among leaves and that Psv probably colonize a low numbers of leaves in a bulked sample.

Psv colonizes either stem and leaf surfaces, but in our conditions, at very low densities when considering the average populations of samples. However, as we have evaluated bacteria on stems after cutting them into pieces it is not excluded that also some endophytic Psv were counted (Penyalver et al., 2006). The fact that there was a high number of samples where Psv was only found in leaves but not in stems or the opposite, suggests that for monitoring Psv populations from olive trees both plant materials should be evaluated. This is relevant in terms of epidemiological studies as well as in setting up a certification scheme for olive plants.

In samples where Psv was isolated, we found a clear correlation between Psv population sizes and those of yellow *P. agglomerans*, either on stem or leaf surfaces. Both species have been previously described in association in knots and olive phyllosphere elsewhere (Ercolani, 1978; Marchi et al., 2006) but this is the first report showing the above correlation.

We found that seasonal fluctuations of Psv populations, in a field plot where Psv was consistently isolated from leaf and stem surfaces over the three-year study, fell into a recognizable pattern. The seasonal pattern displayed by Psv population densities, either on leaf or stem surfaces of naturally infected olive trees, showed their higher levels in warm and rainy months and the lower ones in hot and dry months, with the exception of winter 2002. Multiple regression analysis, between Psv population sizes on stem and leaf surfaces in this field plot and environmental parameters such as temperature, rainfall and relative humidity showed the high influence of these climatic parameters on the Psv population levels. In the case of stem surface up to the 70% of the variability in Psv population can be explained by such parameters which could allow to make a robust prediction model.

Epiphytic Psv on the host is considered an important source of inoculum and hence its elimination/reduction will be important for the development of knots in new wounds and the disease control. In fact, the efficiency of the copper treatments during the four year of this study in reducing the Psv populations and decreasing the olive knot incidence in 'Picudo' cultivar was demonstrated. Effectiveness of copper compounds against Psv was observed after the fifth treatment application in 'Picudo' trees. For the last two years of the study, average of Psv populations in 'Picudo' trees treated with copper compounds was very low compared to average population in non treated control plants (from 0 to 10 cfu/g of organ). The time between applications does not permit the recovery of Psv populations for reaching a level as high as in control plants. Besides, no resistance to copper in Psv isolates was detected during the study (data not shown) confirming the low accessibility to genetic variation reported for Psv strains including IVIA 1628-3 (Morea et al., 1989; Pérez-Martínez et al., 2007). In 'Arbequina' trees, copper compounds controlled the disease but it was not possible to demonstrate their effect on Psv population density because it was very low, even in control olive trees.

In our study, Psv populations in trees of 'Picudo' treated with acibenzolar-S-methyl were significantly lower than in non-treated plants, in some sampling dates, but Psv populations similar to those on non-treated trees were recovered the last year of the study. Unlike traditional pesticides, acibenzolar-S-methyl has no any direct activity on plant pathogens (Vallad & Goodman, 2004). Acibenzolar-S-methyl applications did not reduce the number of knots in the two evaluated cultivars. As only two treatment applications per year could be a too low number of sprays for a good control in a woody plant, we tested an extra intensive application schedule with five acibenzolar-S-methyl applications from April to July of 2003. However, also in this case no effect on either Psv populations or disease incidence were observed (data not shown).

The chemical compounds for control of olive knot disease seemed not significantly affect cumulative fresh olive yield when comparing control and treated plants, but the data refer to only two seasons in which fruit harvesting was possible because of the age of the trees and meteorological problems. We believe that further studies on older trees with a more stable production may demonstrate the role of copper treatments in yield increase.

The olive knot disease integrated control should combine appropriate cultural practices with the application, as preventive treatments, of chemical compounds. In such context, copper treatments should be regularly used for an effective control. The efficiency of these treatments was demonstrated during four-year on 'Picudo' cultivar, not only in reducing the Psv populations but also in decreasing olive knot incidence.

Acknowledgements Authors wish to thank F. Climent from Consellería de Agricultura from Generalitat Valenciana, Spain and E. Quesada (Andalucía) for kindly providing olive tree samples, C.I. Salcedo for technical work, E.A. Carbonell and J. Pérez Penadés for statistical analysis and I. Pérez-Martínez and C. Ramos for their helpful advice. J.M. Quesada was the recipient of predoctoral fellowship from I.F.A.P.A. (Andalucía). R. Penyalver had a contract from the Ministerio de Educación y Ciencia of Spain (Programa Ramón y Cajal). This work was supported in part by grants CAO00-007 from INIA, AGL2002-0224T from MCYT and by grant GRUPOS 03/221 from Generalitat Valenciana of Spain.

References

Bertolini, E., Penyalver, R., García, A., Olmos, A., Quesada, J.M., Cambra, M. & López, M.M. (2003). Highly sensitive detection of *Pseudomonas savastanoi* pv. *savastanoi* in asymptomatic olive plants by nested-PCR in a single closed tube. J. Microbiol. Methods *52*, 261–266.

Ercolani, G.L. (1978). *Pseudomonas savastanoi* and other bacteria colonizing the surface of olive leaves in the field. J. Gen. Microbiol. *109*, 254–257.

Ercolani, G.L. (1985). Factor analysis of fluctuation in populations of *Pseudomonas syringae* pv. *savastanoi* on the phylloplane of the olive. Microb. Ecol. *11*, 41–49.

Ercolani, G.L. (1991). Distribution of epiphytic bacteria on olive leaves and the influence of leaf age and sampling time. Microb. Ecol. *21*, 35–48.

Lavermicocca, P. & Surico, G. (1987). Presenza epifitica di *Pseudomonas syringae* pv. *savastanoi* e di alter batteri sull' olivo e sull' oleandro. Phytopathol. Mediterr. *26*, 136–141.

Marchi, G., Sisto, A., Cimmino, A., Andolfi, A., Cipriani, M.G., Evidente, A. & Surico, G. (2006). Interaction between *Pseudomonas savastanoi* pv. *savastanoi* and *Pantoea agglomerans* in olive knots. Plant Pathol. *55*, 614–624.

Morea M., Sisto, A., Iacobellis, N.S. & Surico, G. (1989). Introduction of cosmid pVK102 into *Pseudomonas syringae* pv. *savastanoi*. Phytopathol. Mediterr. *28*, 140–146.

Penyalver, R., García, A., Ferrer, A., Bertolini, E., Quesada, J.M., Salcedo, C.I., Piquer, J., Pérez-Panadés, J., Carbonell, E.A., del Río, C., Caballero, J.M. & López, M.M. (2006). Factors affecting *Pseudomonas savastanoi* pv. *savastanoi* plant inoculations and their use for evaluation of olive cultivar susceptibility. Phytopathology *96*, 313–319.

Pérez-Martínez, I., Rodríguez-Moreno, L., Matas, I.M. & Ramos, C. (2007). Strain selection and improvement of gene transfer for genetic manipulation of *Pseudomonas savastanoi* isolated from olive knots. Res. Microbiol. *158*, 60–69.

Quesada, J.M., Peñalver, R., García, A., Bertolini, E. & López, M.M. (2006). Bases para el control preventivo de la tuberculosis del olivo. Vida Rural. *228*, 50–54.

Schroth, M.N., Osgood, J.W. & Miller, T.D. (1973). Quantitative assessment of the effect of the olive knot disease on olive yield and quality. Phytopathology *63*, 1064–1065.

Teviotdale, B.L. & Krueger, B. (2004). Effects of timing of copper sprays, defoliation, rainfall, and inoculum concentration on incidence of olive knot disease. Plant Dis. *88*, 131–135.

Vallad, G.E. & Goodman, R.M. (2004). Systemic acquired resistance and induced systemic resistance in conventional agriculture. Crop Sci. *44*, 1920–1934.

Wilson, E.E. (1935). The olive knot disease: its inception, development, and control. Hilgardia *9*, 233–264.

Pseudomonas syringae pv. *syringae* on Kiwifruit Plants: Epidemiological Traits and Its Control

A. Rossetti and G.M. Balestra

Abstract *Pseudomonas syringae* pv. *syringae* (Pss) is one of the most dangerous bacterial pathogens of kiwifruit plants worldwide. In a trial period of four years (2003–2006) in a kiwifruit orchard in Northern Italy it was shown a remarkable Pss population on the different phylloplane organs (leaves, twigs and floral buds) and a relevant percentage of them showed ice nucleation activity (INA). A higher percentage of INA strains was recorded at the beginning of the vegetative kiwifruit seasons (59% of isolates collected in spring) whereas lower percentage of INA strains were observed in the other seasons. Among the different copper compounds utilised, the higher reduction of Pss population was observed after spray treatments with copper idroxide peptidate on twigs, copper sulphate on leaves and fruits, copper sulphate peptidate and copper oxychloride on buds. Natural extracts inhibited the *in vitro* growth of Pss strains.

Keywords Epiphytic survival, ice nucleation activity, phytopathogenic bacteria, copper compounds, natural extracts

1 Introduction

Pseudomonas syringae pv. *syringae* (Pss) is one of the most dangerous bacterial pathogens on kiwifruit plants worldwide. Pss causes remarkable damages in phyllosphere organs (canker on twigs, necrotic spots on leaves, browning and rots of flowers and buds) (Gaignard and Luisetti, 1992; Balestra and Varvaro, 1997). The pathogen is able to survive on kiwifruit organs as a typical epiphytic bacterium (Leben, 1965; Balestra, 2004). Furthermore, Pss is characterised by an ice nucleation

Dipartimento di Protezione delle Piante, Università degli Studi della Tuscia, Via S. Camillo de Lellis, 01100, Viterbo, Italy

Author for correspondence: Giorgio M. Balestra; e-mail: balestra@unitus.it

M'B. Fatmi et al. (eds.), *Pseudomonas syringae Pathovars and Related Pathogens.*   65
© Springer Science+Business Media B.V. 2008

activity (INA) which may cause severe frost damages especially when sudden decrease of temperature occurs (Young, 1987; Balestra, 2004). In a trial period of four years (2003–2006) in a kiwifruit orchard located in Northern Italy character-ised by a relevant presence of Pss, the bacterial dynamic population, the ice nuclea-tion activity (INA) of strains and control strategies have been evaluated.

2 Materials and Methods

In a kiwifruit orchard (cv. Hayward) located in Verona district, Northern Italy, five plots (100 plants each) were delimited. Four were treated in April, July and December with copper oxychloride (50%), copper sulphate (24%), copper sulphate peptidate (5%) and copper hydroxide peptidate (7%), respectively. One plot did not receive any treatment and was used as control. To evaluate the effects of different copper treatments on Pss population, samples from different kiwifruit organs (twigs, leaves, buds and fruits) were monthly collected and processed by washing them in sterile distilled water (SDW) for 2 h at 27°C at 150 rpm on orbital shaker. Serial dilutions of washing water were carried out and then plated on nutrient-agar with 5% of saccharose (NAS). After incubation at 27°C per 48 h the colonies, resembling those of Pss, were recorded (Balestra and Varvaro, 1997, 1998). Potential Pss isolates were evaluated for the morphological, nutritional, physiologi-cal, biochemical (LOPAT, acid from sucrose, pectate lyase production, growth on adonitol and lactate, API Biotype 100, Biolog) characters and INA tests.

The *in vitro* antibacterial activity of animal (albumen, 2.5%) and vegetal extracts from *Liliales* (*Allium sativum*) and *Urticales* (*Ficus carica*) plants was also evaluated following established protocols (Balestra et al., 1998). 10^4–10^8 cfu ml^{-1} bacterial suspensions of strains of Pss were used. *A. sativum* (fruits) and *F. carica* (fruits, leaves, bark) natural extracts were utilised at a concentration of 10 g l^{-1} and 300 g l^{-1}, respectively. Spot tests were performed on NSA medium. After for 72 h incubation at 25 ± 2°C inhibition zones were measured in mm. The assays were carried out under laboratory conditions and repeated three times, with two replicates per combination (bacterial pathogen/vegetal extract).

3 Results and Discussion

Pss population density on kiwifruit plants ranged from 1×10^4 to 6.6×10^5 cfu/cm^2. Among the four years study the average population was higher in 2006 with 5.8×10^5 cfu/cm^2 on twigs, 1.4×10^5 cfu/cm^2 on leaves, 6.6×10^5 cfu/gr on buds, 4.9×10^5 cfu/cm^2 on fruits (Table 1). This year was characterised by relative humidity (RH) >70%, low monthly rainfall values (52.7 mm) and mild temperatures (8°C and 18.7°C, min. and max, respectively).

Table 1 *Pseudomonas syringae* pv. *syringae* epiphytic year average population on kiwifruit organs

Year	Population density (cfu/cm^2 gr) on plant organs			
	Twigs	Leaves	Buds	Fruits
2003	1.8×10^4	3.7×10^4	1.7×10^4	1×10^5
2004	1×10^4	1.8×10^4	2.2×10^4	9.7×10^4
2005	3.1×10^5	3.8×10^4	4×10^5	1.8×10^5
2006	5.8×10^5	1.4×10^5	6.6×10^5	4.9×10^5

Table 2 Reduction (%) of *Pseudomonas syringae* pv. *syringae* epiphytic population on kiwifruit organs after treatment with different copper compounds

Copper compounds	Reduction (%) of epiphytic population density on plant organs			
	Twigs	Leaves	Buds	Fruits
Oxychloride	21.1	16.3	37.7	11.4
Sulphate	20.0	30.5	27.9	12.5
Idroxide peptidate	24.9	16.9	18.7	6.9
Sulphate peptidate	15.4	16.9	34.2	10.5

Among the Pss population collected in the different seasons a higher percentage of INA strains was recorded at the beginning of the vegetative kiwifruit seasons (59% of isolates collected in spring) whereas lower percentage of INA strains were observed in the other seasons (13% in summer, 20% in autumn, 8% in winter). Apparently cryogenic Pss strains were more able to survive on leaves than other plant organs. In fact, 50% of the INA strains were isolated from leaves whereas only 27% and 23% were isolated from twigs and floral buds, respectively. None of the Pss INA strains was isolated from fruits.

Amongst the different copper compounds tested, the highest reduction of Pss population was obtained after copper oxychloride and copper sulphate peptidate applications on twigs, copper sulphate on leaves and copper idroxide peptidate on twigs (Table 2). The better reduction of Pss population was obtained after copper idroxide peptidate application on twigs, copper sulphate on leaves and fruits, copper sulphate peptidate and copper oxychloride on buds (Table 2).

Copper compounds were effective in reducing Pss populations when applied at the beginning of kiwifruit vegetative season, in summer and in autumn but not in winter.

Copper peptidate compounds showed an efficacy in the pathogen population reduction which was similar to other copper compounds but the copper content was about 5–10 times lower.

The natural extracts resulted all effective in *in vitro* antibacterial tests. Albumen was able to inhibit the growth of 10^4 cfu/ml bacterial suspensions whereas *Liliales* (*A. sativum*) and *Urticales* (*F. carica*) fruit vegetal extracts were active on 10^4–10^8 cfu/ml suspensions (Table 3). Further studies are in progress on the epidemiology

Table 3 Antibacterial effects of albumen, *Liliales* and *Urticales* extracts on *Pseudomonas syringae* pv. *syringae* strains on NSA medium after 72 h of incubation at 25 ± 2°C

Vegetal extracts	Inhibition zones (mm)				
	Bacterial suspensions (cfu/ml)				
	10^4	10^5	10^6	10^7	10^8
Albumen	22 ± 1.8[a]	6 ± 0.3	4 ± 0.2	–	–
A. sativum	28 ± 1.6	26 ± 2.0	23 ± 1.9	20 ± 1.8	22 ± 1.8
F. carica	22 ± 1.2	19 ± 1.6	18 ± 1.4	14 ± 1.0	14 ± 1.0

[a] ± (standard deviation) of inhibition zones

traits of Pss on kiwifruit plants as well as on the improvement of copper compounds field doses and the *in vivo* effects of natural extracts.

Acknowledgements This research was supported by the Italian Ministry of Agricultural, Food and Forestry's Policies (M.I.P.A.A.F.), Project N° 893/2006.

References

Balestra G.M., 2004. Bacterial diseases on kiwifruit and their control. XVTH International Plant Protection Congress (IPPC 2004). Beijing, China, May, 11–16 2004, 457.

Balestra G.M., Varvaro L., 1997. *Pseudomonas syringae* pv. *syringae* causal agent of disease on floral buds of *Actinidia deliciosa* (A. Chev) Liang *et* Ferguson in Italy. Journal of Phytopathology 143: 375–378.

Balestra G.M., Varvaro L., 1998. Seasonal fluctuations in kiwifruit phyllosphere and ice nucleation activity of *Pseudomonas viridiflava*. Journal of Plant Pathology 80: 151–156.

Balestra G.M., Antonelli M., Varvaro L., 1998. Effectiveness of natural products for *in vitro* and *in vivo* control of epiphytic populations of *Pseudomonas syringae* pv. *tomato* on tomato plants. Journal of Plant Pathology 80: 251.

Gaignard J.L., Luisetti J., 1992. Role du pouvoir glacogene dans le procèssus infectieux de *Pseudomonas syringae* pv. *syringae* et de *Pseudomonas viridiflava* sur kiwi. Fruits 47: 495–501.

Leben C., 1965. Epiphytic microorganisms in relation to plant disease. Annual Review of Phytopathology 3: 209–230.

Young J.M., 1987. Ice-nucleation on kiwifruit. Annals Applied Biology 111: 697–704.

Head Rot of Cauliflower Caused
by *Pseudomonas fluorescens* in Southern Italy

P. Lo Cantore and N.S. Iacobellis

Abstract Bacterial isolation from water soaked and brown discoloured and rotted cauliflower heads in commercial fields in Apulia, Southern Italy, resulted, in all the specimens, in almost pure cultures with the LOPAT profile (++−+−) of the group Vb of fluorescent pseudomonads. A fluorescent pseudomonad with the LOPAT profile (−+++−) of the group IVb was also obtained from only one out of ten samples examined. The latter isolate, on the contrary of the former ones, produced pectolytic enzymes since it was able to rot potato in the potato disk assay and utilize the pectin. The nutritional analysis of the above isolates with the computerised system Biolog led to the identification of the isolates belonging to first and second groups as strains of *Pseudomonas fluorescens* and *Pseudomonas* spp., respectively.

Pathogenicity assays performed by spraying detached cauliflower florets with 10^8 cfu/ml bacterial suspensions showed that the above pseudomonads reproduced the floret brown discoloration and rotting observed in the field. A similar result was obtained when the florets were dipped in the above bacterial suspensions. Furthermore, the brown discoloration and rotting of internal tissues were obtained when the above suspensions were injected into the floret peduncle.

Keywords *Brassica oleracea*, pectolytic enzymes, romanesco

1 Introduction

During the winter of 2004 in some commercial cauliflower fields of "romanesco" type (cv. Navona) (*Brassica oleracea* L. convar. *botrytis* L. var. *italica*) in Apulia, Southern Italy, alterations of inflorescences, almost ready to harvesting, possibly

Dipartimento di Biologia, Difesa e Biotecnologie Agro-Forestali, Università degli Studi della Basilicata, Viale dell'Ateneo Lucano 10, 85100 Potenza, Italy

Author for correspondence: Nicola Sante Iacobellis; e-mail: iacobellis@unibas.it

M'B. Fatmi et al. (eds.), *Pseudomonas syringae Pathovars and Related Pathogens.*   69
© Springer Science+Business Media B.V. 2008

Fig. 1 Cauliflower corymbs showing
water soaked brown discoloured, and
soft rot superficial symptoms (**A**) and
brown discoloration and rotting of internal peduncles tissues (**B**)

caused by bacterial infections were observed. In particular, the corymbs showed
water soaked and brown discoloured areas which then rotted. The above alterations
interested the whole inflorescences or in some cases only a few florets (Fig. 1A).
Longitudinal sections of the symptomatic inflorescences or the single floret showed
the brown discoloration and rotting of the internal tissues (Fig. 1B). The alterations
caused severe crop losses because they also occurred in post harvest.

Given the economic importance of the cauliflower cultivation in the above area
and the remarkable crop loss caused by the disease it seemed useful to ascertain the
nature of the causal agent.

2   Materials and Methods

2.1   Bacterial Isolation

The bacterial isolation from infected tissue was performed following the usual procedures (Lelliott and Stead, 1987; Schaad et al., 2001). Pure cultures were maintained on nutrient agar plus 2% glycerol (NAG) slants at 4°C and, for long-term
storage, lyophilized.

2.2   Pathogenicity Assay

The bacterial isolates were assayed for pathogenicity either on cauliflower type "romanesco" or with a white corymb one. In particular, sub-corymbs, from fresh picked cauliflower heads, were washed under running tap water and then by sterile distilled water (SDW); the florets were then transferred on filter paper in sterile aluminium trays previously sterilised in autoclave at 121°C for 21 min. Bacterial suspensions containing about 10^8 cfu/ml were then sprayed on the sub-corymbs surface (three sub-corymbs/isolate). Furthermore, in others pathogenicity assays the florets were dipped in the above bacterial suspensions or small aliquots of them were injected with a sterile syringe into the peduncle of sub-corymbs. Sub-corymbs treated with SDW were used as control. After the inoculation, the sub-corymbs were maintained at 25°C and about 100% relative humidity for 48 h.

2.3   Biochemical Characterisation

Bacterial isolates were evaluated for Gram reaction, production of fluorescent pigments, LOPAT characters (Lelliott and Stead, 1987) and for pectolytic enzymes production (Schaad et al., 2001).

Furthermore, the nutritional profile of the isolates was evaluated with the computer-assisted system "Biolog" (Biolog, Inc., Hayward, California).

3   Results and Discussion

Bacterial isolation from water soaked and soft rotted cauliflower heads were always positive. In fact, the isolation plates showed colonies with similar morphology and producing, on KB agar (King et al., 1954), yellow–green fluorescent pigments. All the bacterial isolates, either applied on cauliflower sub-corymbs by suspension spray or dipping, were able to reproduce the disease symptoms (i.e. water soaking and soft rot) but the intensity of the alterations varied with the isolate and the inoculation method used. The injection of the above bacteria into the sub-corymb peduncle caused the water soaking and soft rot of cauliflower internal tissues.

Seventeen out of 18 isolates showed the LOPAT characters of group Vb (++–+–) of fluorescent pseudomonads (Lelliott and Stead, 1987). Only isolate USB1237 showed the LOPAT characters of group IVb (–+++–) of fluorescent pseudomonads (Lelliott and Stead, 1987). The pectolytic activity of the latter isolate was confirmed by the pectinase plate assay (Schaad et al., 2001).

The identity of representative isolates was confirmed by the nutritional profile obtained with the computer assisted system Biolog (Biolog, Inc., Hayward, California). In fact, strains USB1224, USB1226, USB1228, USB1231, USB1235,

USB1236, USB1238 and USB1239 (all with the LOPAT characters of group Vb) were identified as strains of *Pseudomonas fluorescens* with similitude values, in respect to the nutritional profile of bacteria deposited in the data base of Biolog, of 0.86, 0.52, 0.73, 0.81, 0.73, 0.74, 0.69 and 0.85, respectively. The pectolytic strain USB1237 was identified as a strain of *Pseudomonas* spp. with a similitude value, in respect to the nutritional profile of *P. putida*, of 0.45.

In conclusion, the above results indicate that *P. fluorescens*, is responsible for the head rot of cauliflower. A similar disease has been reported in the past on broccoli in different areas (Hildebrand, 1986; Canaday et al., 1987; Robertson et al., 1993) but to the authors knowledge this is the first report of this disease on cauliflower. The occurrence of the disease in Southern Italy has been probably favoured by the remarkable rain precipitations of November 2004.

Acknowledgements The "Assocodipuglia" (Apulia, Italy) was acknowledged for supplying meteorological data.

References

Canaday C.H., Mullins C.A., Wyatt J.E., Coffey D.L., Mullins J.A., Hall T. (1987). Bacterial soft rot of broccoli in Tennessee. Phytopathology 77, 1712.

Hildebrand P.D. (1986). Symptomatology and etiology of head rot of broccoli in Atlantic Canada. Canadian Journal of Plant Pathology 8, 350.

King E.O., Ward M.K., Raney D.E. (1954). Two simple media for the demonstration of pyocyanin and fluorescein. Journal of Laboratory and Clinical Medicine 44, 301–307.

Lelliott R.A., Stead D.E. (1987). Methods for the diagnosis of bacterial diseases of plants. In T.F. Preece (ed). Methods in Plant Pathology, vol 2. Blackwell, Oxford, 216 pp.

Robertson S., Brokenshire T., Kellock L.J., Sutton M., Chard J., Harling R. (1993). Bacterial spear (head) rot of calabrese in Scotland: Causal organisms, cultivar susceptibility and disease control. Proceedings Crop Protection in Northern Britain, 265–270.

Schaad N.W., Jones J.B., Chun W. (2001) Laboratory Guide for Identification of Plant Pathogenic Bacteria. Third edition, APS, St. Paul, MN, p. 373.

Internalization and Survival of *Pseudomonas corrugata* from Flowers to Fruits and Seeds of Tomato Plants

G. Cirvilleri[1], P. Bella[1,2], R. La Rosa[1], and V. Catara[1]

Abstract *Pseudomonas corrugata*, the causal agent of tomato pith necrosis, may cause severe losses on tomato worldwide. Control is based on the use of pathogen-free propagating materials and on the reduction of the inoculum in soil where the pathogen is apparently resident. Although *P. corrugata* has been isolated from soil, seeds and plantlets, little is known about its transmission, epidemiology and source of inoculum. Bioluminescent *P. corrugata* strain PVCT 4.3t lux 18 was spray-inoculated to flowers and immature tomato fruits of plants grown in greenhouse with the aim to monitor survival and transmission during fruit development and ripening. Bacterial population was evaluated on surface, pulp and seeds of tomato fruits harvested at the mature stage. Identity of *P. corrugata* PVCT 4.3t was confirmed by colony morphology on NDA, growth on selective medium, and PCR assay. All tomato fruits produced from inoculated flowers and green tomatoes contained at the mature stage the pathogen on the surface (100% in enriched samples) as well as in the pulp homogenates (100% before enrichment). *P. corrugata* was recovered from the surface and from the pulp of tomatoes after 20 days of fruit storage at 4°C. The pathogen was also recovered from the seeds of tomato fruits produced from inoculated flowers. Results suggests that *P. corrugata* can be internalized and transmitted from flowers to fruits and seeds, thus surviving in the tissues and seeds during fruit development and ripening.

Keywords Seed transmission, tomato plants, internalization

[1]Dipartimento di Scienze e Tecnologie Fitosanitarie, Via S. Sofia 100, Università degli Studi di Catania, Italy

[2]Parco Scientifico e Tecnologico della Sicilia, Z.I. Blocco Palma I, Catania, Italy

Author for correspondence: Gabriella Cirvilleri; e-mail: gcirvil@unict.it

M'B. Fatmi et al. (eds.), *Pseudomonas syringae Pathovars and Related Pathogens.*
© Springer Science+Business Media B.V. 2008

1 Introduction

Pseudomonas corrugata (Roberts and Scarlett, 1981 emended Sutra et al., 1997), is the main causal agent of tomato pith necrosis. The disease occurs worldwide in all tomato-growing areas, both in greenhouse and in open field (CABI, 2006). Characteristic symptoms of the disease are visible after sectioning tomato stem as necrosis and hollowing of the parenchymatic tissues of the stem. Frequently, the first visible symptom is chlorosis of the youngest leaves. This often happens on plants where the fruits of the first truss are fully grown. When the necrosis involves large portion of the stem the plant may losses collapses. *P. corrugata* has been isolated from the rhizosphere of host and non-host plants (Lukezic, 1979; Ryder and Borrett, 1991; Sutra et al., 1997), from soil of tomato fields and from bulk soil (Scortichini, 1989; Achouak et al., 2000), from irrigation water (Scarlett et al., 1978), seed lots (Zutra, 1989; Abdalla, 2000) and tomato seedlings in nursery (Sesto et al., 1996). *P. corrugata* was demonstrated to colonize tomato plant endophytically starting from the inoculum present in the soil (Bella et al., 2003) but also from inoculated seeds (Cirvilleri et al., 2000). Further spread of the pathogen can be achieved through irrigation water, rain splashes, handling, cultural practice and also by recirculating nutrient solution in soilless culture (Scarlett et al., 1978; Naumann et al., 1989; Sadowska-Rybak et al., 1997; Fiori, 2002).

Although its presence has been detected in seed lots, no information is available about the transmission of the pathogen to fruit and seeds. Since *P. corrugata* is not a vascular pathogen and symptoms on fruits have not been reported, we wondered about how the pathogen could reach the seeds. A process known as internalisation in which bacterial cells reach internal tissues of fruits through natural openings apertures in the early stages of fruit development has been demonstrated for other plant pathogens and for human opportunistic pathogens (Ercolani and Casolari, 1966; Bartz, 1982; Buchanan et al., 1999; Guo et al., 2001; Barak et al., 2002). Thus, the present study was carried out to clarify the role of infected flowers and green fruits in the epidemiology of *P. corrugata* by examining the transmission of the pathogen from inoculated flowers and fruits in greenhouse to the fruits and seeds of infected plants.

2 Materials and Methods

2.1 *Inoculum Production*

A bioluminescent mutant strain of *P. corrugata* PVCT4.3t lux 18, rifampicin and tetracycline resistant, was previously obtained (Cirvilleri et al., 2000). The lux-mutant was identical to its wild type, including colony morphology, pathogenicity, and growth *in vitro* and *in planta*. For the inoculum production, *P. corrugata* PVCT4.3t lux 18 was grown for 24h at 28°C on Nutrient Agar + 1% Dextrose (NDA) + Rif^{100} Tc^{10}. Cells were then suspended in phosphate buffer pH 7 at optical density of OD_{600} = ca. 0.1.

2.2 Plant Inoculation

Plant-to-fruit and seed transmission was tested in greenhouse experiments. Tomato seedlings (cv. Marmande) were purchased from a local market, planted in 4-l plastic pots and grown in greenhouse. Water was applied daily, and nutrient solution was applied weekly to maintain optimum soil moisture for plant growth, flowering and fruit development. Bacterial inoculation was carried out on tomato plants harbouring immature green fruits and fully opened flowers. Bacterial suspension was sprayed to run off on fruits and flowers. Ten plants for each replication were inoculated. Experiments were replicated twice. Ten plants were inoculated with buffer solution alone.

2.3 Microbiological Analysis

Tomato fruits were harvested when subjectively judged to be "red ripe" and ready for consumption. Each fruit was placed in a Stomacher bag (Seward, London, United Kingdom) containing 100 ml of 0.1% sterile peptone water and incubated in a rotary shaker for 1.30 h at 100 rpm. The peptone wash water was spiral plated (Eddy Jet IUL Instruments) on NDA with appropriate antibiotics. Peptone wash water was then enriched by adding 50 ml of Nutrient Broth (NB) in each Stomacher bag, incubated at 28°C in a rotary shaker at 100 rpm for 24 h, and spiral plated on NDA.

Tomato fruits were surface sterilized, peel removed and the pulp was placed in stomacher bag with 50 ml of sterile 0.1% peptone water and pummeled at medium speed for 1 min with a Stomacher 400 Instrument (Seward, London, United Kingdom). Portions of homogenate were surface plated on NDA added of appropriate antibiotics. Pulp homogenates were then enriched and plated as previously described.

Nine tomato fruits from inoculated flowers and 21 from inoculated immature green fruits were stored at 4°C. After 20 days of incubation, fruits were subjected to the previously described analyses and the presence of the pathogen on the surface and in the pulp homogenate of fruits was determined.

To estimate population of *P. corrugata* on seeds, approximately 20 seeds from each tomato fruit were separated from the pulp, washed in 20 ml of buffer solution for 30 min, and spiral plated on NDA $Rif^{100}Tc^{10}$.

2.4 Identification of Pathogen Isolates

Colonies formed on NDA + Rif^{100} + Tc^{10} were randomly picked for identification. As growth on plant could be a heterogeneous environment conducive to bacterial diversification, randomly picked colonies were plated onto tetrazolium chloride agar (TZCA) (Kelman, 1954) to look for morphology variants of 4.3t lux strain.

Colonies were analyzed by using *P. corrugata* specific primers (Catara et al., 2002). Amplifications were performed in 0.2-ml reaction tubes in a Perkin Elmer thermocycler and amplification products separated by electrophoresis in 1.2% agarose gels.

3   Results and Discussion

P. corrugata PVCT 4.3t lux 18 was recovered at population of approximately 10^3–10^4 cfu/ml from the surface and in the pulp of tomato fruits developed from inoculated flowers and immature green fruits. *P. corrugata* PVCT 4.3t lux 18 was detected on the surface of fruits developed from flower inoculation only after enrichment (Table 1). The bacterium was instead detected in the pulp at concentrations of approximately 10^4 cfu/fruit (40% of positive samples) and after enrichment in all samples. *P. corrugata* was detected on the surface of tomatoes that received fruit inoculation at concentration of approximately 4×10^3 cfu/fruit (53% of positive samples) and in all enriched samples. The pathogen was also detected in the pulp of the majority of the fruits at concentration of about 7×10^3 cfu/fruit (80% of positive samples) and in all enriched samples.

Although an higher percentage of fruits harboured *P. corrugata* PVCT 4.3t lux 18 in the pulp than on the surface, the enrichment experiments demonstrated that all the fruits contained the pathogen in both the surface and pulp tomato samples from flowers and immature fruits artificially inoculated.

Similar results were obtained by analysing tomato fruits stored for 20 days at 20°C and, in particular, no differences were observed between bacterial colonization before and after storage (Table 2).

P. corrugata was isolated from seeds of fruits obtained from inoculated flowers (2×10^3 cfu/g of seeds) but was not recovered from seeds of fruits produced from inoculated green immature tomatoes.

Table 1   Presence of *Pseudomonas corrugata* PVCT4.3t lux on and in tomato fruits produced on artificially inoculated plants

Treatment	No. of fruits	Surface % fruit	cfu/fruit	Enrichment % fruit	Pulp % fruit	cfu/fruit	Enrichment % fruit
Flower	15	0	0	100	40	9.4×10^4	100
Immature fruit	30	53	4.4×10^3	100	80	7.1×10^3	100

Table 2   Presence of Pseudomonas corrugata PVCT4.3t lux on and in tomato fruits after 20 days of storage

Treatment	No. of fruits	Surface % fruit	cfu/fruit	Enrichment % fruit	Pulp % fruit	cfu/fruit	Enrichment % fruit
Flower	9	0	0	60	66	5.0×10^2	100
Fruit	21	71	6.3×10^3	100	100	9.5×10^5	100

Identity of the recovered isolates was confirmed by morphology, resistance to Tc and Rif, and PCR. As growth on plant could be a heterogeneous environment conducive to bacterial diversification, randomly picked colonies were streaked onto TZCA to look for morphology variants. Colonies re-isolated from surface and pulp of tomato fruits maintained the same colony morphology on semiselective medium (TZCA) and were the only colony types detected. New phenotypes (morphology variants) with altered smooth colony morphology were detected in infected seeds from inoculated flowers (Fig. 1a). Bacterial colonies (typical and morphology variants) obtained from surface, pulp and seeds were identified as *P. corrugata* on the basis of an amplicon of 1.100 bp obtained with specific PCR primers (Fig. 1b). Spontaneous phenotype conversion from rough type to smooth type colonies has been reported in *P. corrugata in vitro* (Siverio et al., 1993) and after re-isolation from soil (Barnett et al., 1999). However, no information are available either on the extent of this phenomenon in other environment niches, the factors involved and the possible relationship of the morphology variation with the virulence.

Transmission of *P. corrugata* from artificially inoculated flowers and immature green fruits to seeds was demonstrated by the recovery of the pathogen from seeds of inoculated plants. Although the mechanism for transmission of the pathogen from plants to seeds was not elucidated, the transmission process appears to be relatively efficient. Several studies have reported the internalization, survival and growth in fruits and vegetables of plant and human pathogens such as *Xanthomonas campestris* pv. *vesicatoria* (Ercolani and Casolari, 1966) and *Erwinia carotovora* (Bartz, 1982) in tomato fruits, *X. c.* pv. *vitians* in lettuce (Barak et al., 2002), *Salmonella* spp. in tomatoes (Guo et al., 2001) or *Escherichia coli* in apples (Buchanan et al., 1999).

Some studies have suggested that bacteria might reach internal tissue of tomatoes through natural openings, such as stomata, lenticels, flowers, broken trichomes,

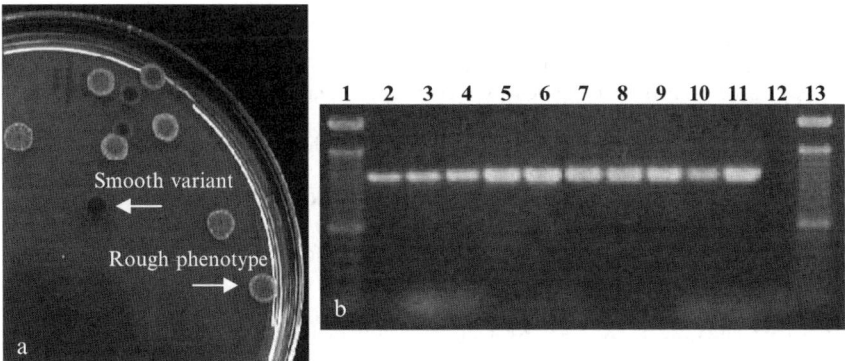

Fig. 1 Typical colonies (rough) and morphology variants (smooth) of *Pseudomonas corrugata* isolated from infected seeds originated from artificially inoculated flowers **a** Amplicon of 1.100 bp obtained with specific PCR primers from smooth and rough phenotypes isolated from seed, surface and pulp samples **b**

stem scar tissues, because of their small size and mobility. Samish et al. (1963) studied ten fruits and vegetables, and found that epiphytical bacteria, mostly *Pseudomonadaeae* and *Enterobacteriaceae,* were present and unevenly distributed within tomatoes tissues. This study suggested that epiphytical bacteria may readily enter fruit tissues in the early stages of fruit development when channels are not yet covered by corky or waxy materials. Tomato stem and flowers were more recently studied as possible route of entry of *Salmonella* into tomato tissues (Guo et al., 2001). As expected, the internalized bacteria can escape the traditional chemical control measures (i.e. copper compounds) and thus permitting the survival of pathogens in fruits and vegetables and, furthermore, providing the pathogen primary inoculum when associated to seeds.

Data from this study would support the hypothesis that *P. corrugata* can be transmitted from flowers and fruits to fruits and seeds. This is the first report on the occurrence of *P. corrugata* in tomato fruits and seeds after flowers inoculation.

References

Abdalla, M.E., 2000, Detection and identification of seed-borne pathogenic bacteria of imported tomato seeds in Egypt. *Bull. EPPO* 30: 327–331.

Achouak, W., Thie'ry, M., Roubaud, P., and Heulin, T., 2000, Impact of crop management on intraspecific diversity of *Pseudomonas corrugata* in bulk soil. *FEMS Microbiol. Ecol.* 31: 11–19.

Barak, J.D., Koikc, S.T., and Gilbertson R.L., 2002, Movement of *Xanthomonas campestris* pv. *vitians* in stems of lettuce and seed contamination. *Plant Pathol.* 51: 506–512.

Barnett, S., Alami, Y., Singleton, I., and Ryder, M., 1999, Diversification of *Pseudomonas corrugata* strain 2140 produces new phenotypes which cross taxonomic boundaries according to GC-Fame, BIOLOG and in vitro inhibition assays. *Can. J. Microbiol.* 45: 287–298.

Bartz, J.A., 1982, Infiltration of tomatoes immersed at different temperatures to different depths in suspensions of *Erwinia carotovora* subsp. *carotovora. Plant Dis.* 66: 302–306.

Bella, P., Greco, S., Polizzi, G., Cirvilleri, G., and Catara, V., 2003, Soil fitness and thermal sensitivity of *Pseudomonas corrugata* strains. *Acta Hortic. (ISHS)* 614: 831–836.

Buchanan, R.L., Edelson, S.G., Miller, R.L., and Sapers, G.M., 1999, Contamination of intact apples after immersion in an aqueous environment containing *Escherichia coli* 0157:H7. *J. Food Protec.* 62: 444–450.

CABI, 2006, Crop Protection Compendium. CABI Publishing, Nosworthy Way, Wallingford, Oxfordshire, OX10 8DE, UK.

Catara, V., Sutra, L., Morineau, A., Achouak, W., Christen, R., and Gardan, L., 2002, Phenotypic and genomic evidence for the revision of *Pseudomonas corrugata* and proposal of *Pseudomonas mediterranea* sp. nov. *Int. J. Syst. Evol. Microbiol.* 52: 1749–1758.

Cirvilleri, G., Bella, P., and Catara, V., 2000, Luciferase genes as a marker for *Pseudomonas corrugata. J. Plant Pathol.* 82: 237–241.

Ercolani, G. and Casolari, A., 1966, Ricerche di microflora in pomodori sani. *Industrie Conserve Parma* 41: 15–22.

Fiori, M., 2002, Severe damage caused by *Pseudomonas corrugata* Roberts et Scarlett on tomato soilless grown. *Inf. Fitopatol.* 52: 47–51.

Guo, X., Chen, J., Brackett, R.E., and Beuachat L.R., 2001, Survival of salmonellae on and in tomato plants from the time of inoculation at flowering and early stages of fruit development through fruit ripening. *Appl. Environ. Microbiol.* 67: 4760–4764.

Kelman A., 1954, The relationship of pathogenicity in *Pseudomonas solanacearum* to colony appearance on a tetrazolium medium. *Phytopathology* 44: 693–695.

Lukezic, F.L., 1979, *Pseudomonas corrugata*, a pathogen of tomato, isolated from symptomless alfalfa roots. *Phytopathology* 69: 27–31.

Naumann, K., Griesbach, E., and Lattauschke, E., 1989, Occurrence and importance of bacterial stem pith necrosis of tomato in the GDR. In: Proceedings of the 7th International Conference of Plant Pathogenic Bacteria, Budapest 1989 (Klement, Z., ed.). Akadémiai Kiadó, Budapest, pp. 473–478.

Ryder, M.H. and Borrett, M.A., 1991, Root colonization by non-fluorescent Pseudomonads used for the control of wheat take-all. *Bull. SROP* 14: 302–307.

Sadowska-Rybak, M., Pein, B., and Büttner, C., 1997, Transmission of *Pseudomonas corrugata* by watering and by nutrient solutions of low tide-high tide-irrigation-systems. *Gesunde Pflanzen* 49: 226–229.

Samish, Z., Etinger-Tulczynska, R., and Bick, M., 1963, The microflora within the tissues of fruits and vegetables. *J. Food Sci.* 28: 259–235.

Scarlett, C.M., Fletcher, J.T., Roberts, P., and Lelliott, R.A., 1978, Tomato pith necrosis caused by *Pseudomonas corrugata* n. sp. *Ann. Appl. Biol.* 88: 105–114.

Scortichini, M., 1989, Occurrence in soil and primary infections of *Pseudomonas corrugata* Roberts and Scarlett. *J. Phytopath.* 125: 33–40.

Sesto, F., Areddia, R., and Catara, V., 1996, Infezioni in vivaio di *Pseudomonas corrugata. Inf. Fitopatol.* 46(1): 62–64.

Siverio, F., Cambra, M., Gorris, M.T., Corzo, J., and Lopez, M.M., 1993, Lypopolysaccharides as determinant of serological variability in *Pseudomonas corrugata. Appl. Environ. Microbiol.* 59: 1805–1812.

Sutra, L., Siverio, F., Lopez, M.M., Hunault, G., Bollet, C., and Gardan, L., 1997, Taxonomy of *Pseudomonas* strains isolated from tomato pith necrosis: emended description of *Pseudomonas corrugata* and proposal of three unnamed fluorescent *Pseudomonas* genomospecies. *Int. J. Syst. Bacteriol.* 47: 1020–1033.

Zutra, D., 1989, Tomato pith necrosis in Israel. *Hassadeh* 69: 612–613.

Copper and Streptomycin Resistance in *Pseudomonas* Strains Isolated from Pipfruit and Stone Fruit Orchards in New Zealand

J.L. Vanneste[1], M.D. Voyle[1], J. Yu[1], D.A. Cornish[1], R.J. Boyd[1], and G.F. Mclaren[2]

Abstract Strains of plant pathogenic *Pseudomonas syringae* and some non pathogenic strains of Pseudomonas isolated from apple orchards in Hawke's Bay, New Zealand, were found to be resistant to streptomycin. Fifty of the 54 P. syringae strains tested carried the genes strA strB on a transposon identified by PCR and DNA sequencing as Tn5393. Other strains, including all strains of Erwinia amylovora, were streptomycin resistant due to a mutation in the rpsL gene. Strains of *P. syringae* able to grow on minimal medium containing 2 mM of copper sulphate were selected from the collection of streptomycin resistant strains. From one of these strains a 1.3 kb fragment of DNA was isolated by PCR using primers designed on two genes reported to be associated with copper resistance (the copA gene P. syringae pv. tomato and the pcoA from *Escherichia coli*). Sequencing of this fragment revealed a high level of similarity (95–98%) with a gene from *P. syringae* pv. *actinidiae* that codes for copper resistance. DNA homologous to this 1.3 kb fragment was detected in eight other strains of *P. syringae* isolated from Hawke's Bay. These strains were subsequently found to be resistant to copper. When the genes which conferred copper resistance were identified they were always carried by a plasmid. In five strains, the plasmid that carried genes coding for copper resistance also carried genes for streptomycin resistance. Therefore, the use of only one of those compounds could lead to selection of strains that carry resistance to both compounds. Strains of Pseudomonas resistant to streptomycin and to copper were also isolated from stone fruit orchards in Hawke's Bay (North Island) and Central Otago (South Island). Treatment of nectarines with copper seemed to affect within one day and for at least up to seven days, the percentage of pathogenic strains resistant to this compound.

[1] The Horticulture and Food Research Institute of New Zealand Limited, Ruakura Research Centre, East Street Private Bag 3123, Hamilton, New Zealand

[2] The Horticulture and Food Research Institute of New Zealand Limited, Clyde Research Centre, 990 Earnscleugh Road, RD1, Alexandra, New Zealand

Author for correspondence: Joel L. Vanneste; e-mail: jvanneste@hortresearch.co.nz

M'B. Fatmi et al. (eds.), *Pseudomonas syringae Pathovars and Related Pathogens.*
© Springer Science + Business Media B.V. 2008

Keywords *Pseudomonas syringae, Pseudomonas fluorescens, Erwinia amylovora, Pantoea agglomerans (Erwinia herbicola)*, genes *str*A *str*B, transposon Tn*5393*, *plasmid*

1   Introduction

Horticulture is a key component of the New Zealand economy; exports from the horticultural sector earned the country NZ$2.3 billion free on board (fob) in the year 2005. Apple exports contributed NZ$387 million of that total, while that same year stone fruit exports (apricots, cherries, nectarines peaches and plums) earned New Zealand NZ$15.9 million fob. Apples and stone fruit are susceptible to several bacterial diseases, including bacterial blast caused by *Pseudomonas syringae* and fire blight caused by *Erwinia amylovora* on apples, and bacterial blast, bacterial dieback or bacterial spot caused by *Pseudomonas syringae* pv. *syringae*, *P. syringae* pv. *persicae* and *Xanthomonas campestris* pv. *pruni* on stone fruit respectively. To control these bacterial diseases growers in New Zealand rely mostly on spraying either copper-based compounds or the antibiotic streptomycin. Streptomycin is registered in New Zealand for control of fire blight on pipfruit (apple, pear and nashi) and for the control of bacterial blast and bacterial spot on stone fruit (peach, nectarine, apricot, and plum).

Few countries in the world have allowed the use of antibiotics for the control of plant diseases, for fear that their use will lead to antibiotic resistance first in strains of plant pathogenic bacteria and then in animal or human pathogens. For example, outside New Zealand streptomycin is registered for the control of fire blight only in the USA and Israel (Psallidas and Tsiantos, 2000). In those two countries, strains of *E. amylovora* that became resistant to streptomycin have been isolated (Jones and Schnabel, 2000). In New Zealand the first record of streptomycin resistant strains of *E. amylovora* was published in 1993 (Thomson et al., 1993). All the streptomycin resistant strains of *E. amylovora* isolated in New Zealand are from Hawke's Bay, a pipfruit growing area in North Island. From the same apple orchards, some strains of *P. syringae* and some strains of epiphytic bacteria *Pantoea agglomerans* (ex-*Erwinia herbicola*), *Pseudomonas fluorescens* and some yellow *Pseudomonas* that were resistant to streptomycin, were also isolated (Vanneste and Yu, 1993). The first part of this paper will review the genetic basis of resistance to streptomycin in these different species of bacteria.

To determine whether the streptomycin resistant strains of *P. syringae* isolated from Hawke's Bay could be controlled with copper compounds, the ability of these strains to grow on a medium containing copper sulphate was determined. The second part of this paper will summarise the results of this investigation, including the genetic basis for copper resistance when this has been identified. Since copper-based compounds and streptomycin are also used in New Zealand for control of bacterial pathogens on stone fruit, a limited survey of the natural populations of streptomycin and/or copper resistant strains of *P. syringae* has been conducted in Hawke's Bay (North Island) and in Central Otago (South Island), the two major stone fruit growing areas in New Zealand. The direct impact of copper treatment on the percentage of strains of *P. syringae* resistant to copper was also determined in a research orchard located in Central Otago. All of these results are presented in the third part of the paper.

2 Streptomycin Resistance in Pipfruit Orchards in Hawke's Bay (North Island)

Streptomycin resistant strains of *E. amylovora* were isolated for the first time in New Zealand during the 1991–1992 seasons (Thomson et al., 1993). Later, strains of other bacterial species that were resistant to streptomycin were also isolated from those same pipfruit orchards. Some of those strains were identified as *P. syringae*, since they produced a fluorescent pigment on plates of King's B medium (King et al., 1954), were oxidase negative and were able to cause a hypersensitivity reaction (HR) when infiltrated onto tobacco plants. The other strains were identified as *P. agglomerans* or *P. fluorescens*. Plant pathogenic bacteria can become resistant to streptomycin because of a mutation in a chromosomal gene which alters the binding affinity of a ribosomal protein for streptomycin. This results in a complete or total resistance to streptomycin. Mutations of a chromosomal gene are transmitted, almost exclusively, by cell division, thus resistance to streptomycin resulting from such a mutation is usually not transmitted to other bacterial species. Alternatively, plant pathogenic bacteria can become resistant to streptomycin after acquiring genes that code for enzymes that modify streptomycin and render it inactive. Such genes are often carried by plasmids or transposons which are able to move from one bacterium to another even if they belong to different species or different genus. Thus this type of streptomycin resistance may spread rapidly to different bacterial species. However, this resistance can be overcome when the enzymatic system becomes overloaded.

All the strains of *E. amylovora* isolated from Hawke's Bay since 1991 have been found to be resistant to very high levels of streptomycin (1,000 ppm). The 67 strains of *E. amylovora* examined were all shown by ASA-PCR (Allele Specific Amplification-Polymerase Chain Reaction) using primers developed by Chiou and Jones (1995), to carry a mutation in the *rps*L gene that codes for the ribosomal protein S12. For all those strains the mutation occurred at codon 43 where a G was substituted for an A, leading to alteration of the binding affinity of the ribosome for streptomycin. This mutation was also found in 94% of the strains of *E. amylovora* resistant to high levels of streptomycin examined by Chiou and Jones (1995). It is not clear why no other mutations have been found in strains from New Zealand. Perhaps all streptomycin strains of *E. amylovora* in New Zealand originated from a few original mutants or perhaps not enough strains have been examined to detect less frequent mutations.

In contrast to strains of *E. amylovora*, 75% of the streptomycin resistant strains of the other bacterial species were resistant to only low concentrations of streptomycin (250 ppm), suggesting that in these strains resistance was not due to a mutation in the *rps*L gene and was therefore most probably due to the production of streptomycin modifying enzymes. Production of such enzymes was detected in the cell lysates of nine of those strains (two strains of *P. syringae*, two strains of *P. fluorescens* and five strains of *P. agglomerans*). The cell lysates obtained by treatment with lysozyme of an overnight culture followed by osmotic shock, prevented inhibition by streptomycin of a streptomycin sensitive indicator strain (Vanneste and Voyle, 1996). Furthermore, total DNA of 50 of the 54 fluorescent *Pseudomonas* and 31 of the 62 *P. agglomerans* or yellow *Pseudomonas* hybridised to a DNA probe containing the genes *str*A *str*B. Those two genes are the only genes conferring streptomycin resistance that have been detected in plant pathogenic bacteria; they are usually carried by a plasmid (Chiou and Jones, 1991; Minsavage et al., 1990; Norelli et al., 1991; Sundin and Bender, 1993). However, no plasmid common to these strains of *Pseudomonas* was found. Resistance conferred by *str*A *str*B has also often been associated with the presence of a Tn*3*-like transposon called Tn*5393* (Chiou and Jones, 1993). A DNA fragment from Tn*5393* containing the genes *tnp*A *tnp*R which code for the transposase and the resolvase respectively, was used as a DNA probe. The total DNA of the *Pseudomonas* strains which hybridised with the *str*A *str*B DNA probe also hybridised with this DNA probe, indicating the likely presence of a Tn*3*-like transposon. Based on the DNA sequence of Tn*5393* published by Chiou and Jones (Chiou and Jones, 1993) a primer which binds to the inverted repeats that border Tn*5393* (JVTn3) was designed (Vanneste and Voyle, 2002). Using this primer, a c. 5.5 kb product that hybridised with the *str*A, *str*B and *tnp*A, *tnp*R DNA probes was amplified from 24 strains of *P. agglomerans* and 39 strains of fluorescent *Pseudomonas*. The product amplified from pEa34, the Tn*5393*-carrying plasmid isolated from some strains of *E. amylovora* in the USA (Chiou and Jones, 1993), was c. 6.7 kb. The difference in size between the products obtained from the New Zealand strains and that obtained from pEa34 is 1.2 kb and it corresponds to the size of the insertion sequence IS1133 which is present in the Tn*5393* from pEa34 (Fig. 1). To confirm the presence of Tn*5393* and the absence of IS1133 in the New Zealand strains, two primers (JVTn1 and JVTn2) which allowed sequencing out from each end of *str*A, *str*B were designed (Vanneste and Voyle, 2002). The DNA obtained after PCR with these two primers from two strains of *Pseudomonas* was sequenced. The sequences showed 96% and 97% identity respectively with that of Tn5393. No sequence similar to the insertion sequence IS1133 was found.

Presence of Tn*5393* was detected in all strains of *Pseudomonas* which have the genes *str*A, *str*B. However, total DNA of 19 strains of *P. agglomerans* hybridised with the *str*A, *str*B probe but not with the *tnp*A, *tnp*R probe. For 16 of these strains we showed that the *str*A, *str*B genes were located on a plasmid that has the same size and the same restriction pattern as that of RSF1010 (Vanneste and Voyle, 1999).

The total DNA of three strains of *Pseudomonas* did not hybridise with the *str*A, *str*B probe. To determine the genetic basis of streptomycin resistance in these

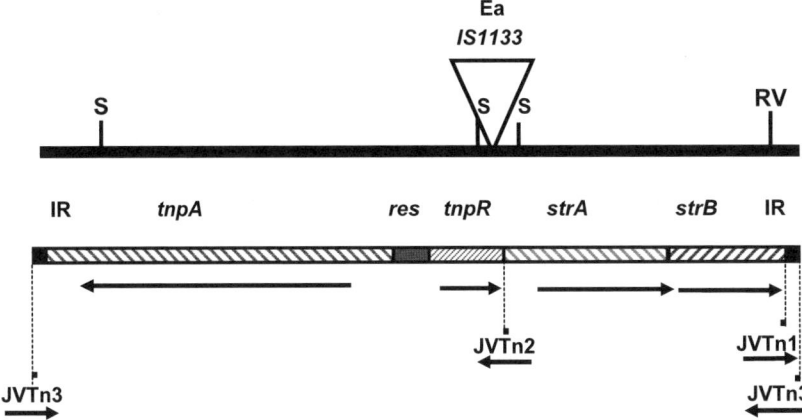

Fig. 1 Physical and functional maps of Tn*5393*. The locations of the primers developed to confirm the presence of Tn*5393* are noted on the map, as well as the location of the IS1133 found in some of the strains of *E. amylovora* isolated from the USA

strains, two primers, PsrpsLF and PsrpsLR (Vanneste and Voyle, 2002), were designed based on the *rps*L gene sequence from *Pseudomonas aeruginosa* PA01 (genebank AE4082.1). These primers were used to amplify and sequence a portion of the *rps*L gene from these three strains of *Pseudomonas* and from two strains which carried a transposon and which were suspected not to have a mutation in the *rps*L gene. When the DNA sequences of these five strains were compared, there appeared to be a consistent difference between the strains which carried the transposon and those which did not. However, the DNA sequence of the *rps*L gene of four additional streptomycin sensitive strains of *Pseudomonas* was similar to that of the *rps*L gene sequence of the streptomycin resistant strains, indicating that the differences previously noted are not linked to streptomycin resistance (Vanneste and Voyle, 2002). The genetic basis of streptomycin resistance in these strains is not yet determined.

3 Copper Resistance in Streptomycin Resistant Strains of *Pseudomonas syringae* Isolated from Hawke's Bay

Since copper-based compounds are in most cases the only compounds available to control streptomycin resistant strains of plant pathogenic bacteria, it was important to determine whether the streptomycin resistant strains of plant pathogenic bacteria isolated in Hawke's Bay were susceptible or resistant to copper. Sixty-five streptomycin resistant strains, including 21 strains of *P. syringae*, isolated between 1991 and

1994 in the Hawke's Bay area from pipfruit orchards, were evaluated for resistance to copper by determining their ability to grow on media supplemented with copper sulphate (Vanneste and Voyle, 2003). The medium used was the minimal medium, called Ceria 132 (Vanneste et al., 1992); it was supplemented with 500 mg/l of copper sulphate (i.e. 2 mM of copper). For all these experiments the strains of *P. syringae* FF5 (Sundin and Bender, 1993) and *P. syringae* 7B40 (Sundin and Bender, 1996) were used as a negative and as a positive control respectively.

Eight of the 21 strains of *P. syringae* initially tested were able to grow on minimal medium supplemented with copper sulphate. To determine whether resistance in any of these strains was coded by genes similar to genes which code for copper resistance in other bacterial strains, two primers were designed based on the DNA sequence of two genes involved in resistance to copper (Vanneste and Voyle, 2003): the *cop*A gene isolated from strain PT23 of *P. syringae* pv. *tomato* (Bender and Cooksey, 1986) and the *pco*A gene carried in *Escherichia coli* by the plasmid pRJ1004 (Brown et al., 1995). Using these primers, only one of the eight copper resistant strains identified yielded by PCR a 1.3 kb long DNA fragment, which was the expected size. Analysis of the DNA sequence of this 1.3 kb DNA fragment revealed that a 572 bp fragment was 98% similar to a portion of ORF A of pPaCu1, a plasmid which codes for copper resistance identified in a strain of *P. syringae* pv. *actinidiae* isolated in Japan (Nakajima et al., 2004). A 275 bp fragment was 95% similar to another portion of ORF A of pPaCu1. Smaller fragments ranging in size from 86 to 20 bp were highly similar (89–100%) to portions of *cop*A from *P. syringae* pv. *tomato*.

When this 1.3 kb DNA fragment was used as a probe against the total DNA extracted from 44 streptomycin resistant strains of *P. syringae* isolated from Hawke's Bay between 1992 and 1994, eight strains were found to carry DNA homologous to this fragment. These eight strains were able to grow on minimal medium supplemented with 2 mM of copper sulphate. When the hybridisation was repeated using as template the plasmid DNA isolated from those eight strains, in every case a plasmid was found carrying the gene homologous to this fragment of DNA. When the same membrane was hybridised with a fragment of DNA corresponding to the genes *str*A and *str*B, in five cases the plasmid which hybridised with the 1.3 kb probe also hybridised with the *str*A *str*B probe. In those five strains the same plasmid carries the genes for streptomycin resistance and the genes for copper resistance.

Of the eight strains of *P. syringae* for which total DNA hybridised with the 1.3 kb DNA fragment used as a probe, seven yielded a fragment of DNA of c. 1.3 kb after PCR using the previously designed primers. Analysis of the DNA sequence of these PCR products obtained from two of those strains revealed that fragments of a total length of 1.1 kb were 98% similar to DNA sequence of ORF A of pPaCu1 from *P. syringae* pv. *actinidiae*. On that basis, of the 16 strains of streptomycin resistant *P. syringae* that were able to grow on a medium supplemented with copper sulphate, eight might have acquired genes homologous to *cop*A and *pco*A. The genetic basis for resistance in the other strains has yet to be determined.

4 Copper and Streptomycin Resistance in *Pseudomonas syringae* Isolated from Stone Fruit in Central Otago (South Island) and Hawke's Bay (North Island)

The discovery of strains that were streptomycin and copper resistant from pipfruit orchards in Hawke's Bay prompted the search for similar strains in other orchards (stone fruit orchards) and in other fruit growing areas of New Zealand (Central Otago). Samples consisting of twigs with leaves, flowers or fruits, depending on the time of the year, were couriered from Central Otago or Hawke's Bay to the laboratory in Hamilton, where surface bacteria were washed with 10 mM MgSO4, and plated on King's medium B (Vanneste et al., 2005). When strains were characterised, it was based on their ability to produce or not a pigment on such a medium, and the colour of pigment produced. Bacterial strains representative of the epiphytic population present on each of the samples were tested for resistance to copper and/or to streptomycin. Resistance to copper was determined by the ability of the strain to grow on minimal medium supplemented with either 250 mg of copper sulphate/litre, corresponding to 1 mM of copper or supplemented with 500 mg of copper sulphate/litre corresponding to 2 mM of copper. While only one sample from a commercial orchard in Hawke's Bay was received in the 2004/2005 season and one in the 2005/2006 season, a total of 47 and 46 samples from commercial orchards of nectarines, apricot and peaches in Central Otago were received during the 2004/2005 season and the 2005/2006 season respectively.

Of the 120 bacteria selected from the Hawke's Bay orchard sample received during the first season, 67% of the strains were resistant to 1 mM of copper, 46% were resistant to 2 mM of copper and 97% were resistant to streptomycin. From the sample received the following season, all the strains of *Pseudomonas* tested were resistant to 2 mM of copper and 50% were resistant to streptomycin. In 42 of the 47 samples received from Central Otago orchards, more than 50% of the bacteria tested were resistant to 1 mM of copper, and in 35 of the 47 samples tested more than 50% of the bacteria were resistant to 2 mM of copper. In 33 samples more than half the bacteria tested were streptomycin resistant. From the 46 samples received the following season, a total of 1061 isolates of *Pseudomonas* were examined; 38% were resistant to streptomycin, 84% were resistant to 1 mM of copper, and 72% were resistant to 2 mM of copper.

From a block of 'Fantasia' nectarines at Clyde Research Centre, Central Otago, about 50% of the bacteria isolated from twigs with open flowers were found to be resistant to 2 mM of copper in 2004–2005. Among the strains tested, there were 42 strains of *P. syringae*. Thirty-eight of these strains (90%) were resistant to 1 mM of copper and five (12%) were resistant to streptomycin. Interestingly, only some parts of this research orchard had received some copper-based compounds during the late dormant season. Some trees were treated twice during the month of August with Kocide 2000, while other trees were not treated. There tended to be fewer copper-resistant strains on trees that had received no copper (7/24, i.e. 29%) than on trees that were treated with copper (51/96, i.e. 53%) (Vanneste et al., 2005). Overall, only

10% of the strains tested were resistant to streptomycin, and 4% were resistant to both streptomycin and copper.

To determine the impact of copper spray on the percentage of copper resistant bacteria, the percentage of copper resistant bacteria present on twigs of 'Fantasia' nectarines at the Clyde Research Centre, Central Otago, was determined before and after treatment with copper (Kocide 2000 at 30 g/100 l). The percentage of strains of *P. syringae* resistant to 1 mM of copper before treatment was 89%; it rose to 99% the day following the treatment and stayed at that level up to seven days after treatment. Before treatment, the percentage of *P. syringae* strains that were resistant to 2 mM of copper was 58%. This percentage rose to 76% one day after treatment and reached 92% seven days after treating the trees with copper. Only 4% of the total population were resistant to both copper and streptomycin before copper treatment. This percentage was 2% seven days after treatment.

5   Conclusions

Copper-based compounds and antibiotics such as streptomycin are the two most important types of compounds available for control of bacterial diseases. Plant pathogenic and plant epiphytic bacteria resistant to either or both of these compounds have been isolated from pipfruit orchards in the Hawke's Bay area, and from stone fruit orchards in Hawke's Bay and Central Otago. Although the results from the commercial orchards in Central Otago do not represent a survey of copper resistance in stone fruit orchards in New Zealand, they indicate that the presence of strains of *P. syringae, P. fluorescens* and *P. agglomerans* resistant to copper and/or streptomycin is not limited to pipfruit orchards in Hawke's Bay.

In contrast to the strains of *E. amylovora* which became resistant to streptomycin because of a mutation in the *rps*L gene, most of the strains of *Pseudomonas* (including *P. syringae*) and *P. agglomerans* isolated from pipfruit orchards were found to carry genes which confer streptomycin resistance. These genes, *str*A, *str*B, were carried either on a transposon identified as Tn*5393* or on a plasmid. It is significant that this resistance is carried by plasmids, as some plasmids have the ability to transfer from one bacterium to another, even if these bacteria are from different species or even different genera. One of the plasmids identified in this study is similar to RSF1010, which has a very wide host range (Grinter and Barth, 1976).

In the case of *P. syringae*, all the strains which acquired *str*A, *str*B did so by acquiring the transposon Tn*5393*. Eight of these streptomycin resistant strains were also resistant to copper. Genes that conferred copper resistance to some of those strains are similar to genes that code for copper resistance in plant pathogenic strains isolated overseas, such as ORF A of pPaCu1 from *P. syingae* pv. *actinidiae* (Nakajima et al., 2004) and *cop*A of *P. syringae* pv. *tomato* (Bender and Cooksey, 1986). However, other mechanisms that lead to copper resistance seem to be present in some New Zealand strains, since some copper resistant strains did not hybridise with the DNA fragment which contains the genes similar to those of the *cop* operon.

All but two of the eight strains that carry genes similar to ORF A of pPaCu1 also carry *str*A, *str*B on the same plasmid. The presence of streptomycin and copper resistance determinants on the same plasmid suggests that the use of either streptomycin or copper derivatives could result in the selection of strains which carry resistance to both of these compounds. This is made worse by the ability of some plasmids to transfer from one bacterium to another. This raises the possibility of selecting strains resistant to one of these compounds when using the other one. The use of copper, for example, to control bacterial or fungal pathogens could help the spread of streptomycin-resistant plasmids in epiphytic and pathogenic bacteria. The spread of copper and streptomycin resistance in plant pathogenic bacteria would leave very few options for the control of bacterial diseases.

Acknowledgements This project was funded by the Foundation for Research, Science and Technology, Summer fruit New Zealand, MAF Policy and by The Horticulture and Food Research Institute of New Zealand Limited.

References

Bender C.L. and Cooksey D.A., 1986, Indigenous plasmids in *Pseudomonas syringae* pv. *tomato*: conjugative transfer and role in copper-resistance. *J. Bacteriol.* 165: 534–541.

Brown N.L., Barrett S.R., Camakaris J., Lee B.T.O. and Rouch D.A., 1995, Molecular genetics and transport analysis of the copper-resistance determinant (pco) from *Escherichia coli* plasmid pRJ1004. *Mol. Microbiol.* 17: 1153–1166.

Chiou C.S. and Jones A.L., 1991, The analysis of plasmid-mediated streptomycin resistance in *Erwinia amylovora*. *Phytopathology* 81: 710–714.

Chiou C.S. and Jones A.L., 1993, Nucleotide sequence analysis of a transposon (Tn*5393*) carrying streptomycin resistance genes in *Erwinia amylovora* and other gram-negative bacteria. *J. Bacteriol.* 175: 732–740.

Chiou C.S. and Jones A.L., 1995, Molecular analysis of high-level streptomycin resistance in *Erwinia amylovora*. *Phytopathology* 85: 324–328.

Grinter N.J. and Barth P.T., 1976, Characterisation of SmSu plasmids by restriction endonuclease cleavage and compatibility testing. *J. Bacteriol.* 128: 394–400.

Jones A.L. and Schnabel E.L., 2000, The development of streptomycin resistant strains of *Erwinia amylovora*. p. 235–251. In: J.L. Vanneste (ed.), Fire blight the disease and its causative agent *Erwinia amylovora*. CAB International, Wallingford, UK.

King E.O., Ward M.K. and Raney D.E., 1954, Two simple media for the demonstration of pyocyanin and fluorescin. *J. Lab. Clin. Med.* 44: 301–307.

Minsavage G.V., Canteros B.I. and Stall R.E., 1990, Plasmid mediated resistance to streptomycin in *Xanthomonas campestris* pv. *vesicatoria*. *Phytopathology* 80: 719–723.

Nakajima M., Goto M., Akutsu A. and Hibi T., 2004, Nucleotide sequence and organisation of copper resistance genes from *Pseudomonas syringae* pv. *actinidiae*. *Eur. J. Plant Pathol.* 110: 223–226.

Norelli J.L., Burr T.J., Lo Cicero A.M., Gilbert M.T. and Katz B.H., 1991, Homologous streptomycin resistance gene present among diverse gram-negative bacteria in New York State apple orchards. *Appl. Environ. Microbiol.* 57: 486–491.

Psallidas P.G. and Tsiantos J., 2000, Chemical control of fire blight. p. 199–234. In: J.L. Vanneste
    (ed.), Fire blight the disease and its causative agent *Erwinia amylovora*. CABI, Wallingford,
    UK.
Sundin G.W. and Bender C.L., 1993, Ecological and genetic analysis of copper and streptomycin
    resistance in *Pseudomonas syringae* pv. *syringae*. *Appl. Environ. Microbiol.* 59: 1018–1024.
Sundin G.W. and Bender C.L., 1996, Molecular analysis of closely related copper- and streptomy-
    cin-resistance plasmids in *Pseudomonas syringae* pv. *syringae*. *Plasmid* 35: 98–107.
Thomson S.V., Gouk S.C., Vanneste J.L., Hale C.N. and Clark R.G., 1993, The presence of
    streptomycin resistance in pathogenic and epiphytic bacteria isolated in apple orchards in
    New Zealand. *Acta Hort.* 489: 671–672.
Vanneste J.L. and Voyle M.D., 1996, Production of streptomycin degrading enzyme by bacterial
    strains isolated from pipfruit orchards in New Zealand. *Phytopathology* 86: S79.
Vanneste J.L. and Voyle M.D., 1999, Genetic basis of streptomycin resistance in pathogenic and
    epiphytic bacteria isolated in apple orchards in New Zealand. *Acta Hort.* 489: 671–672.
Vanneste J.L. and Voyle M.D., 2002, Characterisation of transposon, genes, and mutations which
    confer streptomycin resistance in bacterial strains isolated from New Zealand orchards. *Acta
    Hort.* 590: 493–495.
Vanneste J.L. and Voyle M.D., 2003, Genetic basis of copper resistance in New Zealand strains of
    Pseudomonas syringae. *N.Z. Plant Prot.* 56: 109–112.
Vanneste J.L. and Yu J., 1993, Significance of the presence of streptomycin resistant plant patho-
    genic and epiphytic bacteria in New Zealand. *Proc. 46th N.Z. Plant Prot. Conf.* 46: 171–173.
Vanneste J.L., Yu J. and Beer S.V., 1992, Role of antibiotic production by *Erwinia herbicola*
    Eh252 in biological control of *Erwinia amylovora*. *J. Bacteriol.* 174: 2785–2796.
Vanneste J.L., McLaren G.F., Yu, J., Cornish D.A. and R. Boyd R., 2005, Copper and streptomycin
    resistance in bacterial strains isolated from stone fruit orchards in New Zealand. *N. Z. Plant
    Prot.* 58.

Basal Defence in Arabidopsis Against *Pseudomonas syringae* pv. *phaseolicola:* Beyond FLS2?

A. Forsyth[1], N. Grabov[1], M. de Torres[1,2], V. Kaitell[1], S. Robatzek[3], and J. Mansfield[1]

Abstract We have developed a system to use natural variation between Arabidopsis ecotypes to study the basis of non-host/basal resistance. Our experiments suggest that the leucine rich repeat receptor kinase FLS2 and the flagellin perception system are at the core of this resistance mechanism. However, several genes/loci work in concert with FLS2 to specify non-host resistance in Arabidopsis to *P.s.* pv. *phaseolicola* race 6 1448A.

Keywords Innate immunity, genetic analysis, basal resistance

1   Introduction

Innate immunity in plants is based on the perception of elicitors that induce basal or *R* gene-mediated defences (de Torres et al., 2006; Jones and Dangl, 2006; Arnold et al., 2007). Basal defences typically involve modifications to the plant cell wall, whereas most *R* genes control activation of the hypersensitive reaction (HR). Common molecules found in microbes, described as microbe associated molecular patterns (MAMPs, Ausubel, 2005), such as subunits of flagellin, elongation factor proteins and lipopolysaccharides trigger basal defence (Gomez-Gomez and Boller, 2002). Perception of MAMPs involves receptor proteins with extracellular domains; for example the leucine rich repeat receptor kinases FLS2 and EFR1 detect peptides from flagellin (flg22, Felix et al., 1999) and elongation factor (elf26, Zipfel et al., 2006) respectively. We are investigating the role of MAMP perception in non-host resistance using the bean pathogen *Pseudomonas syringae* pv. *phaseolicola* race 6 strain 1448A to probe Arabidopsis accessions for natural variation in basal defence.

[1] Imperial College, London, Wye Campus, Ashford, Kent TN25 5AH, UK

[2] School of Biosciences, University of Exeter, Exeter, Devon EX4 4QD, UK

[3] Max-Planck-Institute for Plant Breeding Research, Carl-von-Linne Weg 10, 50829 Cologne, Germany

Author for correspondence: Alec Forsyth; e-mail: a.forsyth@imperial.ac.uk

M'B. Fatmi et al. (eds.), *Pseudomonas syringae Pathovars and Related Pathogens.*
© Springer Science + Business Media B.V. 2008

2 Basal Defence: Symptom Phenotypes and Bacterial Multiplication

Arabidopsis is a non-host for *P.s.* pv. *phaseolicola* race 6 which will multiply to limited amounts within challenged leaves but does not activate a rapid HR. Bacterial multiplication is associated with the development of differential symptom phenotypes, ranging from no visible reaction, to yellowing, and collapse of the infiltrated area over the course of several days. A survey of different accessions revealed that the progression of symptom development was directly associated with different levels of bacterial multiplication. This is clearly illustrated by the accessions Niedersenz (Nd-1), Columbia (Col-5 and Col-0), and Wassilewskija (Ws-3). Nd-1 shows no (few) symptom post inoculation (Fig. 1A-1), whereas Col-0 (and Col-5) shows a clear yellowing of the inoculated area by 3 days (Fig. 1A-2). By contrast Ws-3 inoculated leaves collapse by 3 days after inoculation (Fig.1A-3). The different symptoms produced correlate with different amounts of bacterial growth (Fig.1B), where the accumulation of bacteria is a log fold more for each increase in symptoms phenotype. Although *P.s.* pv. *phaseolicola* race 6 does multiply in Ws-3, the population levels reached by 3 days is similar to levels reached by the Arabidopsis pathogen *P.s.* pv. *tomato* DC3000 by 12h after inoculation (data not shown).

3 The Role of FLS2 in Non-host Defence

The presence or absence of the flagellin perception system coordinated by FLS2 has a major role in regulating the success or failure of non-pathogenic bacteria to multiply *in planta* (Gomez-Gomez and Boller, 2002; Zipfel et al., 2004). Both Nd-1 and Col-0 possess an active flagellin perception system and respond to treatment with the peptide flg22 by producing callose, which can be easily observed by staining treated tissue with aniline blue and examination with epi-fluorescence microscopy (Fig. 1C-1, 2). By contrast, Ws-3 does not produce callose when challenged with the flg22 peptide (Fig. 1C-3). All three accessions respond to the elf26 peptide from *P.s.* pv. *phaseolicola* by callose deposition (data not shown). de Torres et al. (2006) demonstrated that FLS2 was responsible for resistance to the attenuated strain *P.s.* pv. *phaseolicola* RW60, where RW60 was used to deliver different effector proteins. Sequence analysis of FLS2 from Col-0 and Nd-1 revealed coding differences (data not shown), which may suggest a quantitative difference between the effectiveness of the two alleles and explain the increase in bacterial multiplication in Col-0 compared to Nd-1. To test this hypothesis, we have instigated genetic experiments to determine if FLS2 is responsible for the different levels of bacterial multiplication and symptom development observed between the three accessions.

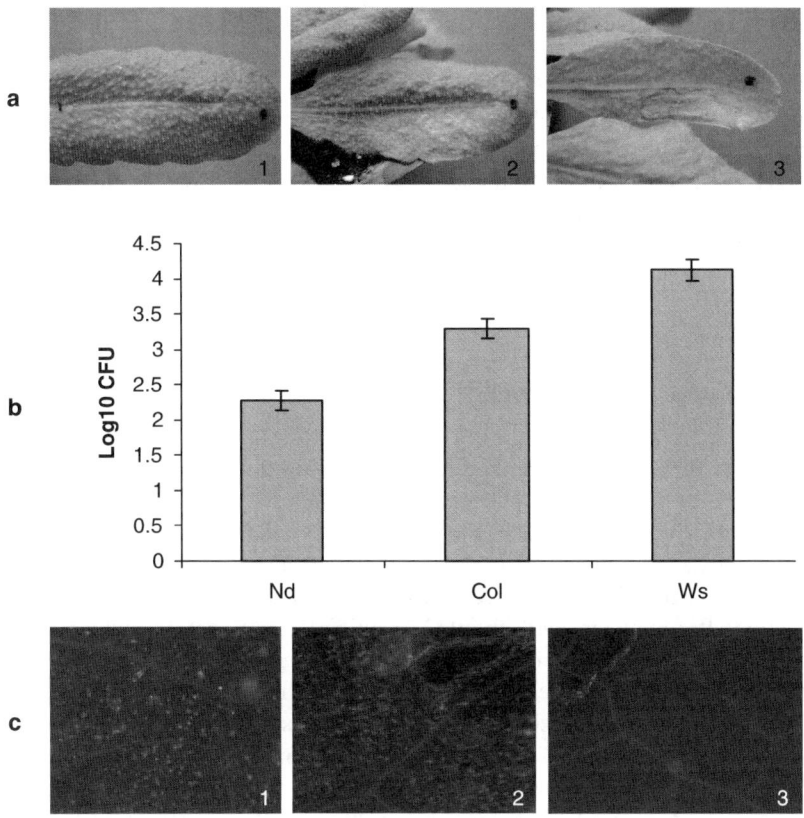

Fig. 1 **a** Disease symptoms 3 days after inoculation of Arabidopsis leaves with a bacterial suspension (OD_{600} = 0.05 approx 5 × 10^7 cfu/ml, 10 mM $MgCl_2$) of *P.s.* pv. *phaseolicola* race 6 1448A. (1) Nd-1 no obvious symptoms, (2) Col-0 a clear yellowing of the inoculated area, (3) Ws-3 collapse of the infiltrated area. **b** Bacterial populations in leaf tissue recovered at 3 days post inoculation from plants inoculated with *P.s.* pv. *phaseolicola* race 6 1448A suspensions (OD_{600} = 0.05 approx 5 × 10^7 cfu/ml, 10 mM $MgCl_2$). Mean of 18 replicates shown with standard errors. **c** Leaves treated with the flg22 peptide (100 mM) stained for callose with 0.05% aniline blue in 50 mM phosphate buffer pH 8.0. (1) Nd-1 callose deposits seen as fluorescent spots, (2) Col-0 1 callose deposits seen as fluorescent spots, (3) Ws-3 no callose observed

4 Genetics of Differential Non-host Resistance

4.1 Comparing Nd-1 and Col-0

The segregation of resistance was studied in an F2 population derived from Nd-1 × Col-0. Plants were scored on the basis of symptoms at daily intervals using a numerical system (1 = slight yellow up to 12 = total collapse), however the resultant scores gave rise to continuous data with segregants giving scores from 1 to 9. Nd-1 control plants had a score of 1 or below, whereas variation was observed with Col-0 plants with scores which ranged from 1 to 9 with an average of 8. This result suggested that several quantitative trait loci (QTLs) may be segregating within the cross. To map QTLs believed to be involved in specifying basal resistance we used 90 well-characterised Nd-1 × Col recombinant inbred lines (RILs, Werner et al., 2005). As in the F2 population, resistance segregated as a continuous trait. Analysis with QTL Café (1998) predicted several loci throughout the genome specifying resistance and susceptibility, one of which is in the region of FLS2.

4.2   Comparing Nd-1 and Ws-3

Analysis of F2 populations derived form Nd-1 × Ws-3 suggested that two loci are responsible for the resistance phenotype observed in Nd-1. Segregants were scored susceptible if three or more (of 4) inoculated leaves collapsed by 3 days after inoculation. Resistance segregated in a ratio of 9:7 (resistance:susceptible) indicating that resistance in Nd-1 is specified by two loci. Plants were also scored for callose induction after treatment with the flg22 peptide (FLS2 function), which revealed that *FLS2* segregated as a single dominant trait (as predicted). No resistant plants were found that failed to respond to the flg22 peptide, indicating that one of the resistance loci is *FLS2* which acts in concert with a second locus to give the symptomless resistance observed in Nd-1. That FLS2 alone is insufficient to give complete resistance was supported by the detection of segregants that respond to the flg22 peptide (possess FLS2) but are as susceptible as Ws-3. Experiments are being performed to map this second locus.

4.3   Isolation of Mutants of Nd-1 Compromised
in Basal Resistance

Further support for the requirement of a second locus comes from screening an ethyl methanesulfonate (EMS) mutagenised Nd-1 population. Mutant populations have been screened by inoculation to isolate plants with compromised resistance that show increased symptoms. Several mutants have been isolated and are being characterised. One such mutant, while allowing increased bacterial growth and displaying disease symptoms, still retains a functioning flagellin perception system as assayed by flg22 callose induction.

Acknowledgements We wish to acknowledge financial support from the UK Biotechnology and Biological Research Council.

References

Arnold, D., Jackson, R.J., Waterfield, N. and Mansfield, J.W., 2007, Evolution of microbial virulence – the benefits of stress. *Trends Genet.* 23, 293–300.

Ausubel, F., 2005, Are innate immune signalling pathways in plants and animals conserved? *Nature Immunol.* 6, 973–979.

de Torres, M., Mansfield, J., Grabov, N., Brown, I., Ammouneh, H., Tsiamis, G., Forsyth, A., Robatzek, S., Grant, M. and Boch, J., 2006, *Pseudomonas syringae* effector AvrPtoB suppresses basal defence in Arabidopsis. *Plant J.* 47(3), 368–382.

Felix, G., Duran, J.D., Volko, S. and Boller, T., 1999, Plants have a sensitive perception system for the most conserved domain of bacterial flagellin. *Plant J.* 18, 265–276.

Gomez-Gomez, L. and Boller, T., 2000, FLS2: an LRR receptor-like kinase involved in the perception of the bacterial elicitor flagellin in Arabidopsis. *Mol. Cell* 5, 1003–1011.

Gomez-Gomez, L. and Boller, T., 2002, Flagellin perception: a paradigm for innate immunity. *Trends Plant Sci.* 7, 251–256.

Jones, J.D. and Dangl, J.L. (2006) The plant immune system. *Nature* 444, 323–329.

QTL Café, 1998, http://www.biosciences.bham.ac.uk/labs/kearsey/applet.html.

Werner, J., Borevitz, J., Warthmann, N., Trainer, G., Ecker, J., Chory, J. and Weigel, D., 2005, *PNAS* 102(7), 2460–2465.

Zipfel, C., Robatzek, S., Navarro, L., Oakeley, E.J., Jones, J.D.G., Felix, G. and Boller, T., 2004, Bacterial disease resistance in *Arabidopsis* through flagellin perception. *Nature* 428, 764–767.

Zipfel, C. et al., 2006, Perception of the bacterial PAMP EF-Tu by the receptor EFR restricts *Agrobacterium*-mediated transformation. *Cell* 19, 749–760.

Agrobacterium Suppresses *P. syringae*-Elicited Salicylate Production in *Nicotiana tabacum* Leaves

A. Rico and G.M. Preston

Abstract *Agrobacterium tumefaciens* strain GV3101 is widely used in transient gene expression assays in plants, particularly *Nicotiana* spp. Transient expression assays have previously been used to study the role of pathogen effectors and elicitors in modulating plant defences, and to investigate the signal transduction pathways involved in expression of defence responses. We have used *Agrobacterium*-mediated transient expression to monitor the fate of fluorescent proteins targeted to specific cellular compartments in healthy and *P. syringae*-infected tobacco leaves. However, inoculation of *A. tumefaciens* into tobacco leaves followed by co-infiltration with the tobacco pathogen *P. syringae* pv. *tabaci* resulted in delayed macroscopic symptoms when compared with leaves infiltrated with a procedural $MgCl_2$ control followed by *P. s.* pv. *tabaci* infiltration. We have explored the mechanistic basis of this phenomenon by monitoring production of the defence signal salicylic acid (SA), which is induced during compatible and incompatible *P. syringae*–plant interactions. We monitored SA production in tobacco leaves inoculated with *A. tumefaciens* or with $10\,mM$ $MgCl_2$, followed by inoculation with *P. s.* pv. *tabaci*, *P. s.* pv. *tomato*, Flg22 or $MgCl_2$. Both *Pseudomonas* pathogens elicited SA production in tobacco leaves but SA levels were significantly reduced by pre-treatment with *A. tumefaciens*. This clearly shows that *Agrobacterium* transient expression is not a phenotypically neutral background for studying plant defence responses and suggests that researchers should be cautious when interpreting results obtained using this method.

Keywords Photon-counting, bioluminescence, hypersensitive response, flagellin, MAMPs

Department of Plant Sciences, University of Oxford, South Parks Road, OX1 3RB, Oxford, UK

Author for correspondence: Gail Preston; e-mail: gail.preston@plants.ox.ac.uk

1 Introduction

Agrobacterium tumefaciens is a soil-borne pathogen that causes crown galls in a wide range of dicotyledonous plants (DeCleene & DeLey, 1976) by injecting a region of DNA (T-DNA) from its tumor-inducing (Ti) plasmid into plant cells, which then is integrated in the plant nuclear genome. T-DNA regions are flanked by two highly homologous 25 bp border sequences, which are the target for VirD2 protein. Vir proteins are involved in nicking the T-DNA from the Ti plasmid and in assembling a type IV secretion system, thereby transferring the T-DNA into the plant cells as a single-stranded molecule (T-strand). However, the exact molecular mechanisms of how the T-strands are transmitted and integrated into the plant genomes are not well known (Ditt et al., 2001; Gelvin, 2003; Veena et al., 2003). Once integrated in the plant chromosome, it elicits a hyperplastic response that results in a tumour, which produces opines, unusual amino acids that act as a nutrient source for the bacterium.

   A. tumefaciens is the only known organism capable of interkingdom DNA transfer, and is able to transfer DNA to plants (Chilton et al., 1977), fungi (Piers et al., 1996) and human cells (Kunik et al., 2001). *A. tumefaciens* has been widely used as a biotechnological tool to engineer plants and express, disrupt or silence genes for studies on plant functional genomics (Ditt et al., 2001). *Agrobacterium*-mediated transformation is typically carried out using a binary vector system, which consists of two plasmids, the binary vector and the *vir* vector. The binary vector is a small broad host range plasmid disarmed of the *vir* and T-DNA functions but containing the region that will be transferred to plants flanked by the 25 bp border sequences. The large *vir* plasmid is disarmed of the T-region but carries all the *vir* functions required for T-DNA transfer (Hajdukiewicz et al., 1994; Hellens et al., 2000). Recently, GATEWAY technology has been incorporated into the *Agrobacterium* binary vectors system, facilitating rapid cloning of any gene or sequence for its transformation in plants (Karimi et al., 2002).

   While the creation of stably transformed plants has many biotechnological advantages, it also has some limitations for plant functional genomic studies. The process of obtaining transgenic plants is time-consuming even with species for which the procedure has been simplified (Wroblewski et al., 2005). Other issues include transgene silencing, and the need to express transgenes non-constitutively (Gelvin, 2003). An alternative to the analysis of stable transformants is the use of *Agrobacterium*-mediated transient expression assays, where transgenes can be assayed within a few days of infiltration. In the context of plant–pathogen interactions, *Agrobacterium*-mediated transient expression assays have been used to study plant defence responses to bacterial elicitors and effectors, including responses to microbe associated molecular patterns (MAMPs; Hann & Rathjen, 2007; Shao et al., 2003), type III-secreted effectors known to suppress the plant immune response in susceptible plants (Jamir et al., 2004), or to elicit programmed cell death in resistant or non-host plants (Van der Hoorn et al., 2000; Vinatzer et al., 2006).

   However, *Agrobacterium*-mediated transient expression assays have some limitations. Transient assays are restricted to species and tissues that are biologically compatible with *A. tumefaciens* and are limited by environmental factors that affect *A. tumefaciens* virulence and plant physiology (Wroblewski et al., 2005). It is well known that *A. tumefaciens* elicits necrosis in some plants, which has been described as an apoptotic response (Gelvin, 2003). There is increasing interest in identifying plant genes or products that are involved in *A. tumefaciens* transformation, which could help to improve plants for *A. tumefaciens* infection or conversely, increase resistance to the pathogen. Recently, it was demonstrated that the expression of plant transcripts involved in signal transduction and defence response was altered upon infection with *A. tumefaciens* (Ditt et al., 2001). Veena and co-authors (2003) also identified plant genes that were differentially regulated during various stages of *Agrobacterium*-mediated transformation. They concluded that *Agrobacterium* suppressed host defence responses in order to facilitate the integration process.

We have used *Agrobacterium*-mediated transient expression assays in tobacco to monitor the cellular trafficking of green fluorescent protein (GFP) between different compartments before and after infection with *P. syringae*. However, macroscopic observations of plant symptoms produced by *P. s.* pv. *tabaci* 11528 (Pstab 11528) showed that symptoms were severely reduced when *A. tumefaciens* GV3101(pMP90) (At GV3101) was pre-infiltrated into tobacco leaves. This observation led us to investigate the mechanistic basis of the interaction between At GV3101, Pstab 11528 and the tobacco plant. The results outlined here suggest that changes in plant or *P. syringae* physiology due to the use of *Agrobacterium*-mediated transient expression assays may mask or alter the phenotypic results observed in further experiments regarding pathogen-induced plant responses.

2 *A. tumefaciens* Interferes with the Growth of *P. syringae* in Tobacco Plants

When a virulent *P. syringae* strain is infiltrated into tobacco leaves at a high density (e.g. *P. s.* pv. *tabaci* at ~10^7 cfu/ml) it elicits a rapid necrotic response, whose macroscopic symptoms are visible between 12 and 24 h after inoculation. At high inoculum densities, necrotic lessions caused by pathogenic *P. syringae* can be distinguished by necrosis associated with non-host resistance because the lesions caused by the pathogenic strains are often surrounded by a chlorotic halo and may increase in size with time. The observation that pre-treatment with *A. tumefaciens* reduced necrosis caused by Pstab 11528 prompted us to investigate whether the presence of *A. tumefaciens* was interfering with the ability of *P. syringae* to multiply in tobacco leaves. We infiltrated At GV3101 in tobacco leaves at 10^8 cfu/ml and challenged leaves 2 days later with Pstab 11528. The growth of Pstab 11528 was significantly reduced in tobacco leaves that had been pre-infiltrated with At GV3101, which suggests that At GV3101 has a direct or indirect effect on the growth of *P. syringae* strains.

3 *A. tumefaciens* Suppresses *P. syringae*-Elicited SA Production

Salicylic acid (SA) is a plant signalling molecule that has been widely studied in the context of plant defence. SA is produced locally in incompatible interactions when programmed cell death occurs (Mur et al., 1997) and is known to be involved in signalling processes during local and systemic acquired resistance (SAR), where SA is one of the systemic signal molecules that is translocated to and synthesized in uninfected areas of the plant in order to promote increased resistance to pathogen attack (Gaffney et al., 1993). Recently, Huang et al. (2006) used the bacterium *Acinetobacter* ADP1, to develop a bacterial biosensor that can be used to detect apoplastic SA. *Acinetobacter* ADP1 contains the salicylate-responsive *sal* operon that allows bacteria to use salicylate as a sole carbon source. *Acinetobacter* ADPWH-*lux* contains a chromosomally located fusion of the *salA* (SA hydroxylase) gene with the promotorless *luxCDABE* operon from *Photorhabdus luminescens* (Winson et al., 1998). In this biosensor, SA-induced bioluminescence is proportional to SA from as low as 5–200 nM (Huang et al., 2006). The method is non-destructive and has been used to show local changes in apoplastic SA in tobacco leaves infected with viral and bacterial pathogens (Huang et al., 2006).

We used *Acinetobacter* ADPWH-*lux* to assess the effect of *A. tumefaciens* on SA production in tobacco leaves infected with *P. syringae* strains. As expected, both the non-host pathogen *P. s.* pv. *tomato* DC3000 and Pstab 11528 elicited SA production 24 h after inoculation. However, SA levels were significantly lower in leaves that had been pre-infiltrated with At GV3101. Further experiments showed that (i) heat-killed At GV3101 did not affect SA; (ii) At GV3101 did not degrade SA *in vitro* when incubated in synthetic media and apoplast extracts and (iii) the effect of At GV3101 on *P. syringae*-elicited SA production was time and dose-dependent.

4 Concluding Remarks

Our experiments have shown that pre-treatment with *A. tumefaciens* at inoculum densities used in transient expression experiments actively suppresses *P. syringae* growth and results in lower levels of SA production in leaves inoculated with *P. syringae*. We could find no evidence that *Agrobacterium* acts downstream of SA biosynthesis, degrading SA, which suggests that *Agrobacterium* interferes with the signal transduction events that induce SA synthesis. There are several hypotheses that could explain how *A. tumefaciens* interferes with *P. syringae* pathogenesis and plant defence signalling. Growth suppression could be due to direct interactions between At GV3101 and *P. syringae* cells involving nutrient competition, antagonism or alteration of physiological conditions in the intercellular space. This would reduce *P. syringae* growth and possibly suppress the production of *P. syringae* pathogenicity factors such as toxins or the type III secretion system, which in turn would attenuate *P. syringae*-dependent elicitation of SA. Alternatively *A. tumefaciens* could have an indirect effect on *P. syringae* growth by increasing plant resistance. However, the observation that *A. tumefaciens* suppresses SA production indicates that *A. tumefaciens* promotes some defence pathways while suppressing others. Collectively, these results confirm that *A. tumefaciens* is not invisible to plant surveillance and defence mechanisms and show that stably transformed plants are likely to provide more accurate information on plant defence responses than transiently transformed plants.

References

Chilton, M. D., Drummond, M. H., Merio, D. J., Sciaky, D., Montoya, A. L., Gordon, M. P., & Nester, E. W. (1977). Stable incorporation of plasmid DNA into higher plant cells: the molecular basis of crown gall tumorigenesis. *Cell* 11, 263–271.

DeCleene, M. & DeLey, J. (1976). The host range of crown gall. *Bot Rev* 42, 389–466.

Ditt, R. F., Nester, E. W., & Comai, L. (2001). Plant gene expression response to *Agrobacterium tumefaciens*. *Proc Natl Acad Sci USA* 98, 10954–10959.

Gaffney, T., Friedrich, L., Vernooij, B., Negrotto, D., Nye, G., Uknes, S., Ward, E., Kessmann, H., & Ryals, J. (1993). Requirement of salicylic acid for the induction of systemic acquired resistance. *Science* 261, 754–756.

Gelvin, S. B. (2003). *Agrobacterium*-mediated plant transformation: the biology behind the "gene-jockeying" tool. *Microbiol Mol Biol Rev* 67, 16–37

Hajdukiewicz, P., Svab, Z., & Maliga, P. (1994). The small, versatile pPZP family of *Agrobacterium* binary vectors for plant transformation. *Plant Mol Biol* 25, 989–994.

Hann, D. R. & Rathjen, J. P. (2007). Early events in the pathogenicity of *Pseudomonas syringae* on *Nicotiana benthamiana*. *Plant J* 49, 607–618.

Hellens, R., Mullineaux, P., & Klee, H. (2000). Technical focus: a guide to *Agrobacterium* binary Ti vectors. *Trends Plant Sci* 5, 446–451.

Huang, W. E., Huang, L., Preston, G., Naylor, M., Carr, J. P., Li, Y., Singer, A. C., Whiteley, A. S., & Wang, H. (2006). Quantitative *in situ* assay of salicylic acid in tobacco leaves using a genetically modified biosensor strain of *Acinetobacter* sp. ADP1. *Plant J* 46, 1073–1083.

Jamir, Y., Guo, M., Oh, H. S., Petnicki-Ocwieja, T., Chen, S., Tang, X., Dickman, M. B., Collmer, A., & Alfano, J. R. (2004). Identification of *Pseudomonas syringae* type III effectors that can suppress programmed cell death in plants and yeast. *Plant J* 37, 554–565.

Karimi, M., Inze, D., & Depicker, A. (2002). GATEWAY vectors for *Agrobacterium*-mediated plant transformation. *Trends Plant Sci* 7, 193–195.

Kunik, T., Tzfira, T., Kapulnik, Y., Gafni, Y., Dingwall, C., & Citovsky, V. (2001). Genetic transformation of HeLa cells by *Agrobacterium*. *Proc Natl Acad Sci USA* 98, 1871–1876.

Mur, L. A., Bi, Y. M., Darby, R. M., Firek, S., & Draper, J. (1997). Compromising early salicylic acid accumulation delays the hypersensitive response and increases viral dispersal during lesion establishment in TMV-infected tobacco. *Plant J* 12, 1113–1126.

Piers, K. L., Heath, J. D., Liang, X., Stephens, K. M., & Nester, E. W. (1996). *Agrobacterium tumefaciens*-mediated transformation of yeast. *Proc Natl Acad Sci USA* 93, 1613–1618.

Shao, F., Golstein, C., Ade, J., Stoutemyer, M., Dixon, J. E., & Innes, R. W. (2003). Cleavage of *Arabidopsis* PBS1 by a bacterial type III effector. *Science* 301, 1230–1233.

Van der Hoorn, R. A., Laurent, F., Roth, R., & De Wit, P. J. (2000). Agroinfiltration is a versatile tool that facilitates comparative analyses of Avr9/Cf-9-induced and Avr4/Cf-4-induced necrosis. *Mol Plant Microbe Interact* 13, 439–446.

Veena, J. H., Doerge, R. W., & Gelvin, S. B. (2003). Transfer of T-DNA and Vir proteins to plant cells by *Agrobacterium tumefaciens* induces expression of host genes involved in mediating transformation and suppresses host defense gene expression. *Plant J* 35, 219–236.

Vinatzer, B. A., Teitzel, G. M., Lee, M. W., Jelenska, J., Hotton, S., Fairfax, K., Jenrette, J., & Greenberg, J. T. (2006). The type III effector repertoire of *Pseudomonas syringae* pv. syringae B728a and its role in survival and disease on host and non-host plants. *Mol Microbiol* 62, 26–44.

Winson, M. K., Swift, S., Hill, P. J., Sims, C. M., Griesmayr, G., Bycroft, B. W., Williams, P., & Stewart, G. S. (1998). Engineering the *luxCDABE* genes from *Photorhabdus luminescens* to provide a bioluminescent reporter for constitutive and promoter probe plasmids and mini-Tn5 constructs. *FEMS Microbiol Lett* 163, 193–202.

Wroblewski, T., Tomczak, A., & Michelmore, R. (2005). Optimization of *Agrobacterium*-mediated transient assays of gene expression in lettuce, tomato and *Arabidopsis*. *Plant Biotechnol J* 3, 259–273.

Characterisation of an Inhibitory Strain of *Pseudomonas syringae* pv. *syringae* with Potential as a Biocontrol Agent for Bacterial Blight on Soybean

S.D. Braun and B. Völksch

Abstract The soybean epiphyte *Pseudomonas syringae* pv. *syringae* 22d/93 (Pss22d) exhibits strong potential to control *Pseudomonas syringae* pv. *glycinea* (Psg), the causal agent of bacterial blight. Under greenhouse and field conditions the antagonism has been proven, but the underlying mechanisms are unclear, up to now. The secondary metabolites are of particular interest for this antagonism. We focus our investigation on the influence of produced toxins of the biological control system.

The antagonist produces the well-known syringomycin, syringopeptin and one unknown toxin, which inhibits the pathogen in vitro. The unknown toxin was detected by an agar diffusion bioassay using Psg as indicator strain. This growth inhibition was reversed by arginine and not by argininosuccinate, suggesting that the unknown toxin could interfere with this step of citrulline/arginine biosynthesis.

Via marker exchange mutagenesis in genes coding for nonribosomal protein synthesis responsible for initial steps in biosynthesis of syringomycin (*syrE*) and syringopeptin (*sypA*), toxin negative mutants were constructed. Subsequently, a double mutant in which both genes were deleted was constructed, as well. In planta coinoculation experiments with the double mutant of Pss22d and the pathogen indicated that syringomycin and syringopeptin did not affect the antagonism.

Keywords Pseudomonas syringae, biocontrol, bacterial blight, soybean, novel toxin

1 Introduction

The ability of naturally occurring microorganisms to inhibit growth or metabolic activity of pathogens can be used to control plant diseases in agriculture and horticulture (Ji and Wilson, 2003; Mascher et al., 2003). Such biological control agents offer an alternative to conventional pesticides.

University of Jena, Institute of Microbiology, Microbial Phytopathology,

Neugasse 25, D-07743 Jena, Germany

Author for correspondence: Sascha Braun; e-mail: sascha.braun@uni-jena.de

M'B. Fatmi et al. (eds.), *Pseudomonas syringae Pathovars and Related Pathogens.* 103
© Springer Science+Business Media B.V. 2008

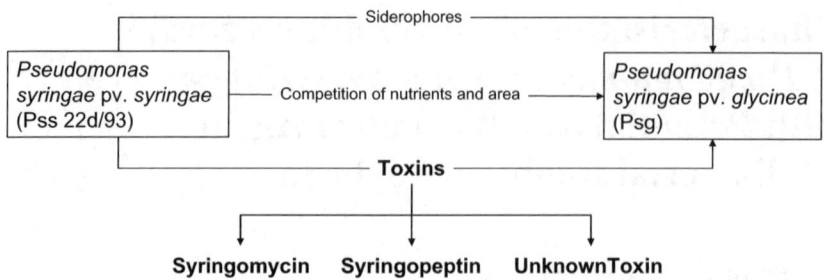

Fig. 1 Three possible mechanisms in the antagonism between antagonist and pathogen

The object of our research is the biological control of bacterial leaf spot diseases using resident bacterial epiphytes of the target plants. We investigate the antagonism between *Pseudomonas syringae* pv. *glycinea* (Psg) the pathogen of bacterial blight on soybean and the epiphytic bacteria *Pseudomonas syringae* pv. *syringae* 22d/93 (Pss22d). Pss22d has a strong potential to control bacterial blight. It has been demonstrated successfully in vitro, in planta, and under field conditions (Völksch and May, 2001). The mechanisms of antagonistic action are unknown. Three widely known mechanisms of microbial control mediated by biocontrol organisms are production of siderophores (Bultreys and Gheysen, 2000; Wensing et al., in press), production of toxins (Cirvilleri et al., 2005), and competition of nutrients and area on the leaf surface (Fig. 1).

Here, we study the toxin production as a potential mechanism in the antagonism between Pss22d and Psg. Pss22d produces three different active substances: syringopeptin, syringomycin and an unknown toxin. Syringomycin and syringopeptin are known to be produced by *Ps. syringae* pv. *syringae* strains and are well described in the literature (Bender, 1998; Scholz-Schroeder et al., 2001). To determine the potential role of each toxin in the antagonism we produced toxin negative mutants. These mutants were tested on soybean plants under greenhouse condition. Furthermore, we are planning to elucidate the chemical structure and biosynthetic pathway of the unknown toxin.

2   Materials and Methods

The bacterial strains and plasmids used in this study are listed in Table 1.

To determine syringomycin, syringopeptin and unknown toxin production of Pss22d agar diffusion bioassays with *Geotrichum candidum*, *Bacillus megaterium* and *Pseudomonas syringae* pv. *glycinea* as indicator strains were used, respectively.

The genes *syrE* and *sypA* were disrupted by marker exchange mutagenesis (Fig. 2) with pGEM*e12*SP and pGEM*a12*Km (Table 1).

Table 1 Bacterial strains and plasmids in this study

Strains or plasmid	Relevant characteristics[a]	Source
Pseudomonas syringae pv. *Syringae*		
PssB301D	Wild type from pear	Cody and Gross, 1987
PssB728a	Wild type	Willis and Hirano, 1987
PssJ59	Wild type	Völksch and Weingart, 1998
Pss22d	Wild type from soybean	Ji and Wilson, 2003; Völksch et al., 1996
Pss22dΔAE	Syringomycin, syringopeptin deficient mutant	This study
Pseudomonas syringae pv. *Glycinea*		
Psg1a/96	Wild type from soybean	Völksch et al., 1996
Eschericha coli		
DH5α	*supE44 ΔlacU169* (φ80 *lacZΔM15*) *hsdR17 recA1 endA1 gyrA96 thi-1 relA1*	Hanahan, 1983
Plasmids		
pGEM-TEasy	cloning vector for T/A cloning, Amp[r]	Promega
pBBR1MSC	Cloning vector, broad host range, *IncP IncQ*, Cm[R]	Kovach et al., 1994
pGEM*a12*Km	For homologues fragmants primer derived from PssB728a, DNA template Pss22d, Km[r]-cassette	This study
pGEM*e12*Sp	For homologues fragments primer derived from PssB728a, DNA template Pss22d, Sp[r]-cassette	This study

[a]Cm[R], Sp[r], Km[r], Amp[r] = resistance to chloramphinicol, spectinomycin, kanamycin, ampicillin respectively

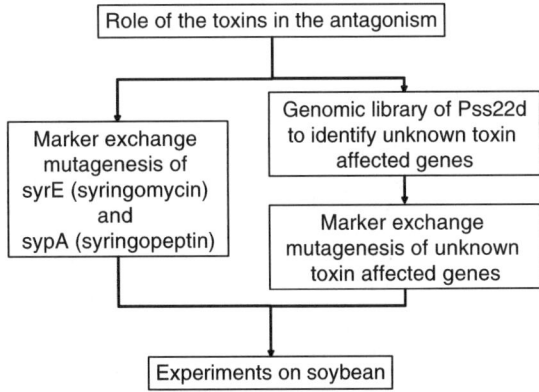

Fig. 2 To determine the role of each toxin in the antagonism between Pss22d (antagonist) and Psg (pathogen) toxin negative mutants were generated using different methods. In addition, a genomic library was created to get genomic information about the biosynthesis cluster from unknown toxin

To identify the biosynthesis genes of the unknown toxin a genomic library was carried out. For this library the cloning vector pBBR1MSC (Table 1) was used with partial digested genomic DNA of Pss22d (10U PstI, 10 min) in ligation overnight at 16°C. The ligated vectors were transformed in *E. coli* DH5α by electroporation. All resistant colonies of transformed *E.coli* were pooled in a "pBBR1MSC_Pss22d" vector library. A conjugation was carried out between vector library and PssJ59 (Table 2). All resistant PssJ59 colonies were screened in agar diffusion bioassays with Psg searching for activity. All unknown toxin positive PssJ59 conjugates were used to get information about the genes of unknown toxin production. This information will help us to carry out a marker exchange mutagenesis (Fig. 1).

For plant experiments the double mutant Pss22dΔAE, the wild type Pss22d, and the pathogen Psg were coinoculated via prick technique on soybean under greenhouse conditions. Following ratios were used: Pss22dΔAE–Psg (2:1); Pss22d–Psg (2:1). The pathogen, the antagonist, and the double mutant were single inoculated in the same ratio with water as control. Five microlitres of each inoculum with approximately 10^4 cfu was placed on the wounds. Three parallels on different plants each with five wounds were carried out (Fig. 3). After inoculation the samples were taken on day 1, 7, 14 and 21. To determine the development of bacterial populations in the wounds 15 leaf spots (diameter 7 mm) of each inoculation were homogenized in 15 ml isotonic NaCl and serially diluted for plating onto agar media.

Table 2 Toxin phenotypes of used strains and mutants

	Syringomycin[a]	Syringopeptin[b]	Unknown toxin[c]
Pss22d	+	+	+
PssJ59	+	+	−
PssB301D	+	+	−
PssB728a	+	+	−
Pss22dΔAE	−	−	+

Tested by agar diffusion bioassays with [a]*Geotrichum candidum*; [b]*Bacillus megaterium*; [c]*Pseudomonas syringae* pv. *glycinea*

Fig. 3 Plant experiments on soybean under greenhouse conditions: inoculation of leaves with cell suspension of Pss22d (antagonist), Psg1a (pathogen), Pss22dΔAE (double mutant) as control, and coinoculation of Pss22d–Psg1a (2:1) and Pss22dΔAE–Psg1a (2:1) by prick technique (three parallels on different plants)

3   Results and Discussion

Pss22d was screened in agar diffusion bioassays and compared to the well-characterized related strains PssB728a and PssB301D. The fungus *Geotrichum candidum* and the bacterial species *Bacillus megaterium* are well known as indicator strains for syringo-mycin and syringopeptin, respectively. Pss22d, PssB728a, and PssB301D showed growth inhibitions in these bioassays (Table 2). The unknown toxin was identified in vitro by bioassays with Psg as indicator strain. Of the three strains tested only Pss22d inhibited Psg in vitro (Table 2). We, therefore, assume that the unknown toxin is pro-duced only by Pss22d. The growth inhibition of Psg caused by Pss22d was compensated only by arginine (Fig. 4c). Neither the direct precursor of arginine argininosuccinate nor another precursor in the arginine biosynthesis compensated the inhibition. These results suggest that the unknown toxin affects in the biosynthesis of arginine. We suppose that argininosuccinase, that converts argininosuccinate to arginine, is the possible target enzyme. A similar mechanism is well described by phaseolotoxin (Bender, 1999).

Experiments on soybean showed that the antagonist Pss22d, the double mutant Pss22dΔAE, and the pathogen (Psg) built a stable population in single inoculation (Fig. 5A). Both the antagonist and the double mutant reached a cell density between

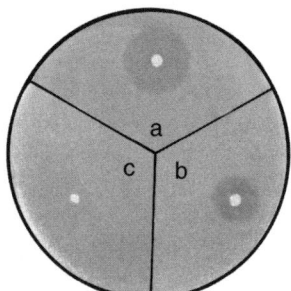

Fig. 4 Agar diffusion bioassays with Psg as indicator strain, growth inhibition caused by Pss22d; (a) without amino acid, (b) with 1mM argininosuccinate, (c) with 1 mM arginine

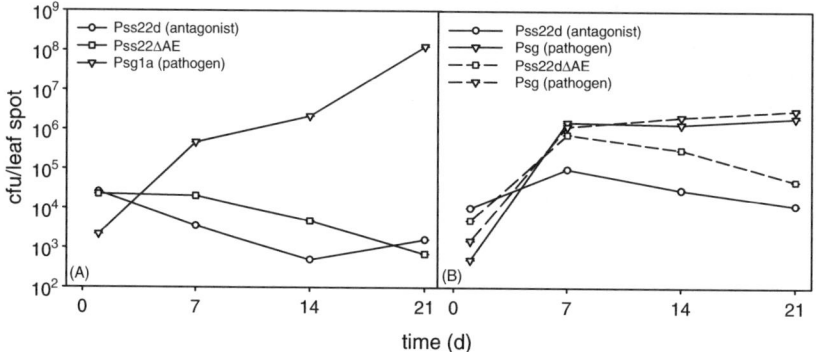

Fig. 5 The growth of (A) Pss22d (antagonist), Pss22ΔAE, Psg1a in single inoculation (B) Coinoculation of Psg1a with Pss22d or Pss22ΔAE

10^3 and 10^4 cfu/wound during the whole sampling time. This density is typically for epiphytic bacteria (Völksch et al., 1996). However, the pathogen grew fast in single inoculation and reached the threshold level at approximately 5×10^6 cfu/wound after 10 days. Yellow chlorosis was visible in surrounding of the wounds at this time (data not shown). After 21 days the population size of the pathogen was around 10^8 cfu/leaf spot and at this time the surrounding tissue of the wounds was clearly necrotic (Fig. 6C). The antagonist and the double mutant did not show symptom development after this time (Fig. 6A and B).

If the pathogen grew in coinoculation with the antagonist or the double mutant, the pathogen did not reach the threshold level at 5×10^6 cfu/leaf spot (Fig. 5B). Both the antagonist and the double mutant grew better in coinoculation than in single inoculation (Fig. 5A and B). These increases could be explaining by the higher availability of nutrients through the pathogen. In both coinoculation experiments no symptoms were visible in surrounding of the leaf spots (Fig. 6D and E). These results showed that syringomycin and syringopeptin were irrelevant to the mechanism of antagonism. Additionally, in planta experiments with mutants which deleted in the unknown toxin production or deleted in all toxin productions will carry out.

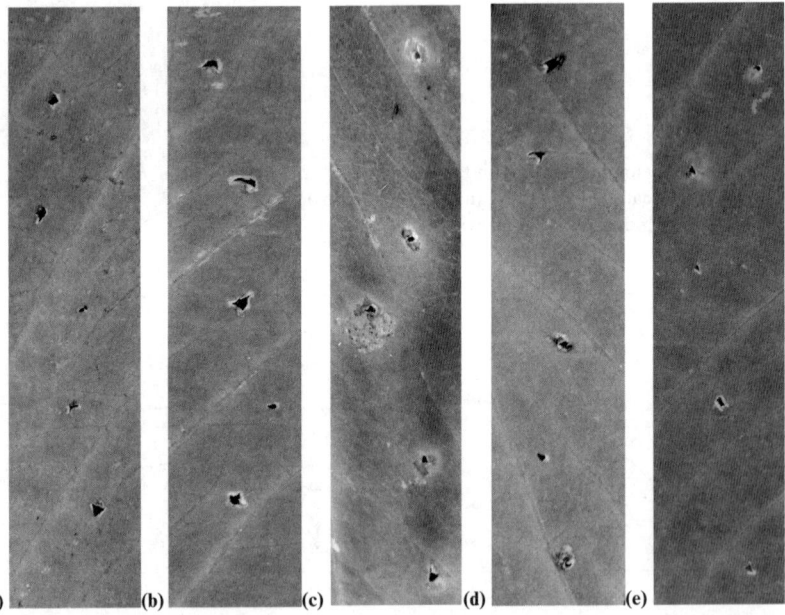

(a)   (b)   (c)   (d)   (e)

Fig. 6 Symptom development on leaves of soybean under greenhouse conditions after 21 days. **a** Pss22d (antagonist), **b** Pss22dΔAE, **c** Psg1a (pathogen), **d** Pss22d–Psg1a (2:1), **e** Pss22dΔAE–Psg1a (2:1)

Acknowledgements We thank the Deutsche Forschungsgemeinschaft (DFG) for the grant and the members of the group of Professor M. Ullrich (IU Bremen) for experimental support.

References

Bender, C.L. 1998. Bacterial phytotoxins. *In Methods in Microbiology.* Vol. 27, pp. 169–175.

Bender, C.L. 1999. Chlorosis-inducing phytotoxins produced by *Pseudomonas syringae.* Eur. J. Plant Pathol. 105: 1–12.

Bultreys, A., Gheysen, I. 2000. Production and comparison of peptide siderophores from strains of distantly related pathovars of *Pseudomonas syringae* and *Pseudomonas viridiflava* LMG 2352. Appl. Environ. Microbiol. 66: 325–331.

Cirvilleri, G., Bonaccorsi, A., Scuderi, G., Scortichini, M. 2005. Potential biological control activity and genetic diversity of Pseudomonas syringae pv. syringae strains. J. Phytopathol. 153: 654–666.

Cody, Y.S., Gross, D.C. 1987. Outer membrane protein mediating iron uptake via pyoverdinpss, the fluorescent siderophore produced by *Pseudomonas syringae* pv. *syringae.* J. Bacteriol. 169: 2207–2214.

Hanahan, D. 1983. Studies on Transformation of Escherichia-Coli with Plasmids. J. Mol. Biol. 166: 557–580.

Ji, P.S., Wilson, M. 2003. Enhancement of population size of a biological control agent and efficacy in control of bacterial speck of tomato through salicylate and ammonium sulfate amendments. Appl. Environ. Microbiol. 69: 1290–1294.

Kovach, M.E., Phillips, R.W., Elzer, P.H., Roop, R.M., Peterson, K.M. 1994. Pbbr1mcs – a broad-host-range cloning vector. Biotechniques 16: 800–802.

Mascher, F., Schnider-Keel, U., Haas, D., Defago, G., Moenne-Loccoz, Y. 2003. Persistence and cell culturability of biocontrol strain Pseudomonas fluorescens CHA0 under plough pan conditions in soil and influence of the anaerobic regulator gene anr. Environ. Microbiol. 5: 103–115.

Scholz-Schroeder, B.K., Soule, J.D., Lu, S.E., Grgurina, I., Gross, D.C. 2001. A physical map of the syringomycin and syringopeptin gene clusters localized to an approximately 145-kb DNA region of *Pseudomonas syringae* pv. *syringae* strain B301D. Mol. Plant Microbe Interact. 14: 1426–1435.

Völksch, B., May, R. 2001. Biological control of *Pseudomonas syringae* pv. *glycinea* by epiphytic bacteria under field conditions. Microb. Ecol. 41: 132–139.

Völksch, B., Weingart, H. 1998. Toxin production by pathovars of *Pseudomonas syringae* and their antagonistic activities against epiphytic microorganisms. J. Basic Microbiol. 38: 135–145.

Völksch, B., Nüske, J., May, R. 1996. Characterization of two epiphytic bacteria from soybean leaves with antagonistic activities against *Pseudomonas syringae* pv. *glycinea.* J. Basic Microbiol. 36: 453–462.

Wensing, A. 2006. Siderophore production of *Pseudomonas syringae* and its implication on the biological control of bacterial blight of soybean. PhD Thesis, International University of Bremen.

Willis, D.K., Hirano, S.S. 1987. Molecular diversity of a pathogenicity-specific gene within populations and pathovars of *Pseudomonas syringae.* Phytopathology 77: 1709–1709.

Characterization of the Inhibitory Strain *Pantoea* sp. 48b/90 with Potential as a Biocontrol Agent for Bacterial Plant Pathogens

B. Völksch and U. Sammer

Abstract Biological control by using epiphytic bacteria against plant pathogens has been considered as an alternative method for disease controlling. *Pantoea* sp. strain 48b/90 isolated from soybean leaf (synonym: *Erwinia herbicola*) produced a stabile antibiotic in chemically defined medium which is active against a variety of bacterial plant pathogens. The characteristics of this antibiotic substance are not identical with the well-known antibiotics produced by *Pantoea agglomerans*. Additionally, 48/90 produced two different siderophores (the known ferrioxamine E and a non-identified catechole siderophore) and quorum sensing signal molecules. These characteristics of strain 48b/90 justify further investigation.

Keywords *Pantoea*, biological control, antibiotic production

1 Introduction

Aerial leaf surfaces are habitats for epiphytic bacteria which are able to grow and survive on plants. Many of these saprophytic bacteria can antagonize phytopathogenic bacteria by inhibition of their growth or metabolic activity. Such natural occurring antagonists can be used to control plant diseases in agri- and horticulture and offer an interesting alternative to conventional pesticides.

The *Pantoea* sp. 48b/90 (formerly *Erwinia herbicola*) is a naturally occurring epiphyte which was isolated from a soybean leaf. It inhibited the growth of a broad spectrum of bacterial species of distinct genera including plantpathogenic bacteria (e.g. various *Pseudomonas syringae* and *Xanthomonas campestris* pathovars, *Agrobacterium tumefaciens*) *in vitro* on minimal medium (Völksch, 2006). It also affected *Pseudomonas syringae* pv. *glycinea*, the pathogen of bacterial blight of

Friedrich-Schiller-Universität Jena, Biologisch-Pharmazeutische Fakultät, Mikrobielle Phytopathologie, Neugasse 25, D-07743 Jena

Author for correspondence: Beate Völksch; e-mail: beate.voelksch@uni-jena.de

soybean, by reducing the population size of the pathogen *in planta*. Consequently, disease symptoms were suppressed (May et al., 1997). The economically important plant pathogen *Erwinia amylovora* which causes fire blight is inhibited by strain 48b/90, too. Therefore, the strain 48b/90 is interesting as a suitable candidate for biocontrol.

The aim of this study is to characterize the biologically active compounds of 48b/90 which could be responsible for this antagonism.

2   Results and Discussion

2.1   Possible Mechanisms for the Antagonistic Activities

2.1.1   Toxin

The toxin produced by 48b/90 was investigated by agar plate diffusion assays on minimal medium (MM 5b) with *Erwinia amylovora* Ea7 as indicator strain. Sterile cell free culture filtrates were used in these tests.

When 5b MM agar was used zones of inhibition around colonies of the strain 48b/90 were consistently observed in lawns of a variety of indicator bacteria (Fig. 1A; Völksch, 2006). This antibiotic activity could only be found on minimal medium, but not on complex medium (e.g. ST1-agar), suggesting that one or more constituents of the natural products blocked antibacterial activity (Fig. 1B). Essential amino acids and casein hydrolysate were checked to determine if the inhibition could be compensated. However, this activity is neither inhibited by essential amino acids nor by casein hydrolysate (Fig. 1C). In contrast, both the strain and its culture filtrate did not show any activity against fungi tested.

Figure 2 shows the time course of a typical batch culture with 48b/90. Strain 48b/90 synthesized the toxin during the exponential growth phase of the culture. The highest toxin value was detected when the strain reached the stationary growth phase. The toxin production is strictly growth associated (Fig. 2). This is unusual for the production of most secondary metabolites (classical antibiotics) but typically for toxins produced by phytopathogenic and epiphytic bacteria.

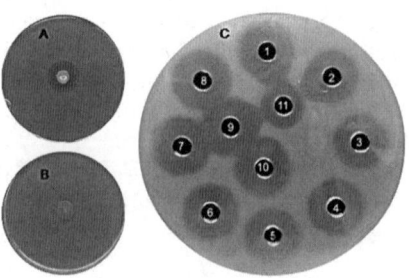

Fig. 1 Activity of 48b/90 grown on minimal (**A**) and complex agar (**B**). Influence of different amino acids on toxic activity of culture filtrate produced by 48b/90 (**C**). **1**: L-arginine (5 mM), **2**: L-aspartic acid (5 mM), **3**: L-asparagine (5 mM), **4**: L-leucine (5 mM), **5**: L-histidine (5 mM), **6**: L-glutamic acid (5 mM), **7**: L-lysine (5 mM), **8**: L-tryptophan (5 mM), **9**: untreated sample, **10**: casein hydrolysate (0.15%), **11**: spectinomycin (5 mg/ml) as control

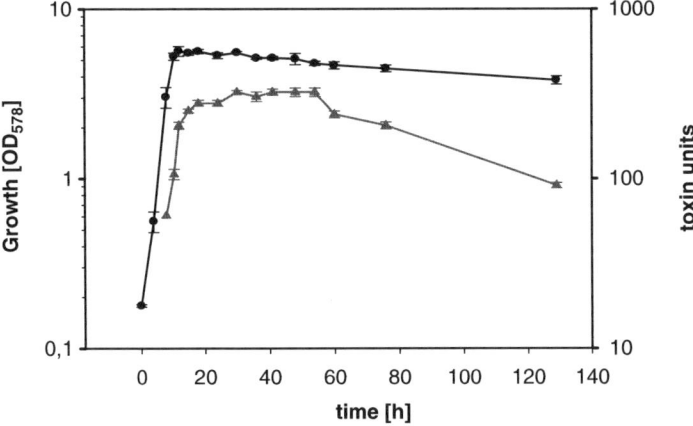

Fig. 2 Growth kinetics (• black line) and toxin production (▲ grey line) of *Pantoea* sp. 48b/90 in Pipes medium at 28°C

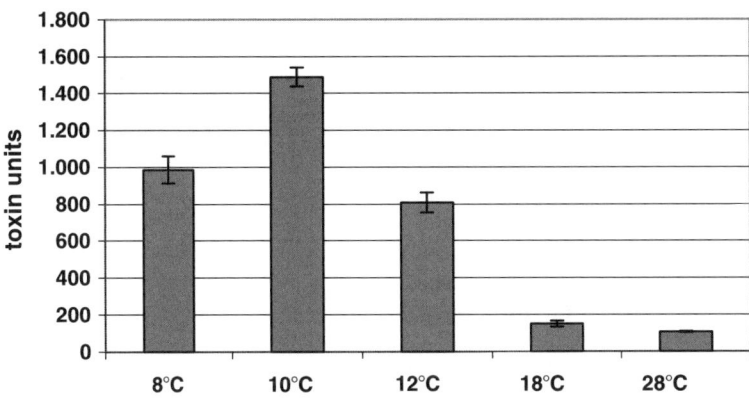

Fig. 3 Toxin production at different temperatures in Pipes medium (sampling at $OD_{578nm} = 5$)

The toxin production depends on temperature. Surprisingly, the highest production was reached at 10°C and was about 15-fold of the production at 28°C (Fig. 3). However, 48b/90 grew to the same high populations independent of the cultivation temperatures used (10°C, 18°C, 28°C).

The toxic activity of the culture filtrates showed a clear resistance to high temperature and to extreme pH values. Heat treatment for 10 min at 80°C did not have an effect on the activity, and at 100°C the activity was half of it. However, after treatment at 100°C for 60 min the activity was markedly reduced, but it was still detectable. After extreme pH treatments (pH 3 and 10) for 10 min, with subsequent reestablishing of the pH value at 7, the toxin remained stable.

2.1.2 Siderophore

Strain 48b/90 produced at least two different siderophores – the hydroxamate siderophore ferrioxamine E a principle siderophore of *E. herbicola* (Berner et al., 1988) and an unidentified catechol siderophore during growth under iron limitation (Fig. 4).The level of siderophores at 18°C was about the 3.0-fold in the catechol assay (Rioux et al., 1983) and 1.5-fold in the CAS assay (Schwyn and Neilands, 1987) of that at 28°C after 48 h (Fig. 5).

2.1.3 Quorum Sensing

48b/90 induced weakly the production of violacein in the *Chromobacterium violaceum* CV026 (Fig. 6) but not the β-galactosidase biosynthesis in *Agrobacterium tumefaciens* NTL4(pCF218)(pCF372). Probably, it produced an acyl-homoserine lactone (AHL) with a short acyl side chain or another class of quorum sensing signal molecules.

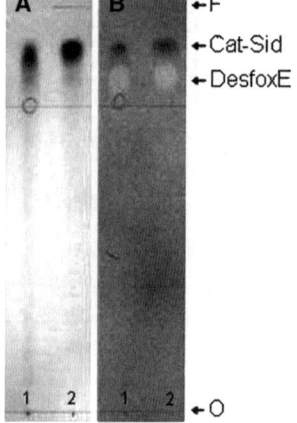

Fig. 4 Thin layer chromatograms of SP-extracts (LiChrolut-RP18, Merck) from 48b/90 grown in low-iron 5b medium for 48 h at 28°C (**1**) and 18°C (**2**). Chromatograms were developed in methanol to chloroform to acetic acid (90:5:5) and analyzed by spraying with 0,1 M FeCl$_3$ in 0,1 M HCl (**A**) and CAS-assay-solution (**B**) (F: solvent front, O: origin of migration, DesfoxE: Desferrioxamine E, Cat-Sid: catecholate siderophore)

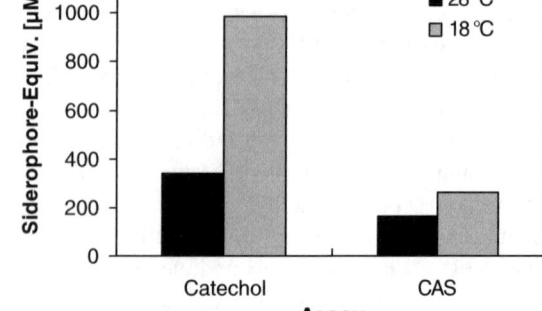

Fig. 5 Siderophore production of 48b/90 in low-iron Pipes medium after 48 h growth at 18°C and 28°C (DFOM-equivalents = CAS-assay; DHBA-equivalents = Catechol-assay)

Fig. 13.6 Different strains were screened on solid medium in a streak plate assay using AHL biosensor strain *Chromobacterium violaceum* CV026. AHL production (*N-acyl* side chains C4-C8) was judged by its ability to induce violacein production (purple pigment: arrows)

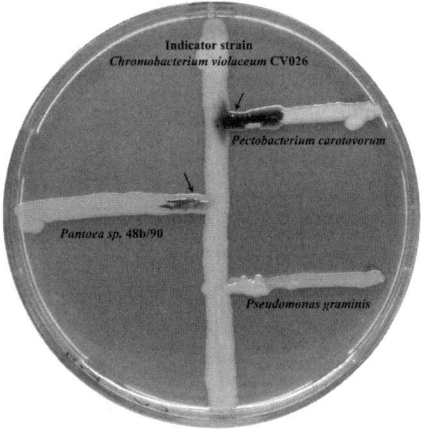

Pantoea agglomerans or *Pantoea dispersa*, also known as *Erwinia herbicola* (Löhnis) Dye are ubiquitous in nature. Strains of this species produce a family of antibiotics which have different patterns of biological activity and chemical structures in a poorly understood strain dependent fashion (Ishimaru et al., 1988; Wodzinski and Paulin, 1994). The antibiotics pantocin A, pantocin B and the phenazine antibiotic D-alanyl-griseoluteic acid (AGA) are well-characterized. Pantocin A is inactive in the presence of histidine and pantocin B in the presence of arginine (Sutton and Clardy 2001); whereas AGA is not inhibited by amino acids (Kearns and Hale, 1996). *Pantoea* sp. 48b/90 produced a broad-spectrum antibiotic activity in vitro. This antibiotic activity was not inhibited by essential amino acids (neither by histidine nor by arginine). It was stable to acidic and basic pH and was also stable at high temperature. The characteristics of the 48b/90 antibiotic are similar to AGA. The genes required for the production of AGA are described (Giddens et al., 2002). With PCR technique we could not identify biosynthetic genes responsible for AGA in strain 48b/90. Therefore, we suppose that the toxin produced by 48b/90 is not identical to AGA.

3 Conclusion

A biological control organism reduces the population size of the pathogen, thus preventing it to reach a critical population density necessary for disease formation (Wilson, 1997). Competition for limited resources, induction of host plant defense, antibiosis, and interference with quorum sensing system of the pathogen have been determined as mechanisms involved in biological control. Quite often, multiple mechanisms are participated.

In conclusion, it was demonstrated here that the strain *Pantoea* sp. 48b/90 produces toxin(s) with a wide antibiotic spectrum against bacteria, siderophores (the known ferrioxamine E and an unknown chatechol-siderophore), and a not yet more detailed analyzed quorum sensing substance. Efforts are now being focused on detailed characterization of these metabolites and on investigating whether these substances play any role in the antagonism.

References

Berner, I., Konetschny-Rapp, S., Jung, G., Winkelmann, G., 1988, Characterization of ferrioxamine E as the principal siderophore of *Erwinia herbicola* (*Enterobacter agglomerans*). Biol. Metals 1: 51–56.
Giddens, R.S., Feng, Y., Mahanty, H.K., 2002, Characterization of a novel phenazine antibiotic gene cluster in *Erwinia herbicola* Eh1087. Mol. Microbiol. 45: 769–783.
Ishimaru, C.A., Klos, E.J., Brubaker, R.R., 1988, Multiple antibiotic production by *Erwinia herbicola*. Phytopathololgy 78: 746–750.
Kearns, L.P., Hale, C.N., 1996, Partial characterization of an inhibitory strain of *Erwinia herbicola* with potential as a biocontrol agent for *Erwinia amylovora*, the fire blight pathogen. J. Appl. Bacteriol. 81: 369–374.
May, R., Völksch, B., Kampmann, G., 1997, Antagonistic activities of epiphytic bacteria from soybean leaves against *Pseudomonas syringae* pv. *glycinea* in vitro and in plant. Microb. Ecol. 32: 118–124.
Rioux, C., Jordan, D.C., Rattray, J.B.M., 1983, Colorimetric determination of catechol siderophores in microbial cultures. Anal. Biochem. 133: 163–169.
Schwyn, B., Neilands, J.B., 1987, Universal chemical assay for the detection and determination of siderophores. Anal. Biochem. 160: 47–56.
Sutton, A., Clardy, J., 2001, Synthesis and biological evaluation of analogues of the antibiotic pantocin B. J. Am. Chem. Soc. 123: 9935–9946.
Völksch, B., 2006, Characterization of the inhibitory strain *Pantoea* sp. 48b/90 with potential as a biocontrol agent for bacterial plant pathogens. Mitt. Biol. Bundesanst. Land-Forstwirtsch. 408: 216–219.
Wilson, C.L., 1997, Biological control and plant diseases – a new paradigm. J. Ind. Microbiol. Biotechnol. 19: 158–159.
Wodzinski, R., Paulin, J., 1994, Frequence and diversity of antibiotic production by putative *Erwinia herbicola* strains. J. Appl. Bacteriol. 76: 603–607.

Pseudomonas syringae: Prospects for Its Use as a Weed Biocontrol Agent

B.M. Thompson, M.M. Kirkpatrick, D.C. Sands, and A.L. Pilgeram

Abstract *Pseudomonas syringae* is found on numerous weeds, occasionally causing serious disease. The challenge lies in developing this ubiquitous bacterium into a safe biocontrol agent with consistent knockdown, host-specificity, and cost efficacy for weed control in environments where chemical herbicides are not an option. We have demonstrated that an overabundance of certain amino acids will have a negative impact on certain plants presumably due to feedback inhibition of branched pathways of amino acid biosynthesis. Strains of *P. syringae* that overproduce a specific amino acid can be selected using appropriate amino acid analogs. Such amino acid overproducing strains of *P. syringae* retain the ability to colonize plants and, importantly, are inhibitory to plants sensitive to the overproduced amino acid. The host-specificity of the amino acid excreting pathogens might be narrowed by selection for one or more amino acids that are abundant in the target weed.

Keywords *Pseudomonas syringae*, frenching disease, plant pathogen, amino acid, virulence, biological control, weeds

1 Introduction

It is rare in the natural environment for a pathogen to eliminate all of its host population. Plant disease epidemics are limited by the genetic diversity of the host plants, spatial distribution of the host plant, energy expenditure to pathogen, and ease of pathogen dispersal. The success of a biocontrol agent for elimination or suppression of a weed population is dependent upon the same criteria. There have been marginally successful classical biocontrol efforts using obligate rust fungi (Hasan and Wapshere, 1973; Morris et al., 1999). These successes are somewhat limited by the inability to culture the pathogen in vitro or natural genetic diversity of the target weed.

Plant Sciences and Plant Pathology, Montana State University, Bozeman, MT, 59717, USA

Author for correspondence: Dave Sands; e-mail: dsands@montana.edu

M'B. Fatmi et al. (eds.), *Pseudomonas syringae Pathovars and Related Pathogens.*   117
© Springer Science+Business Media B.V. 2008

Another way to overcome natural barriers to epidemics is utilization of mycoherbicides (Miller et al., 1989). Mycoherbicides are weed pathogens that are cultured in the laboratory and applied directly to existing weed infestations. The efficacy of mycoherbicides is limited by production and application costs (reference), lack of host specificity (Miller et al., 1989), and non-conducive environmental conditions (primarily free moisture or humidity) (reference).

Bacteria, such as pseudomonads, have potential as agents for biocontrol of weeds (Johnson et al., 1996). *Pseudomonas syringae* epiphytically and saprophytically colonizes the leaf surfaces of a large number of plant species. In most cases, this association is not detrimental to the host plant. However, in certain specific pairings, the bacterium causes severe disease and plant death (Zidack and Backman, 1996). Of particular interest is Frenching disease, a naturally occurring disease of tobacco described by Steinberg (1946). Frenching disease of tobacco is caused by saprophytic bacteria growing on the roots of plants that overproduce the amino acid isoleucine (Steinberg et al., 1950). Isoleucine is synthesized in plants via the branched chain amino acid pathway. The end products of this pathway (valine, leucine, and isoleucine) allosterically regulate the activity of acetolactate synthase (ALS). In Frenching disease, the isoleucine produced by the saprophytic bacteria inhibited ALS activity in the tobacco plant, not only halting synthesis of isoleucine but also valine and leucine. The lack of valine and leucine disrupted essential protein metabolism. Interestingly, several modern chemical herbicides mimic this strategy. Some modern herbicides inhibit a single biosynthetic enzyme in plants, rendering treated plants incapable of producing a metabolite essential for plant growth.

Subsequently, we have demonstrated that amino acid producing plant pathogenic fungi and bacteria have increased virulence to their host plant (Tiourebaev et al., 2001). We hypothesize that saprophytic bacterium such as *Pseudomonas syringae* could also provide effective biocontrol of weeds. First, saprophytic bacteria are easily cultured and disseminated. Second, bacteria have relatively high mutation frequencies facilitating isolation of variants that overproduce high concentrations of inhibitory amino acids. Such variants pose minimal risk to the environment. While over-excretion of an amino acid will increase virulence, it will also decrease the environmental fitness of the bacterium. Amino acid production requires energy and overproducers may not compete with wild-type organisms in the environment.

2   Amino Acid Inhibition of Plants, an Exploitable Weakness

Every weed, in fact every plant that we have so far examined is inhibited by at least one amino acid, leading to the conclusion that weeds have a readily exploitable weakness (Fig. 1). The fitness of a weed, and therefore its resistance to plant pests, can be reduced by direct application of inhibitory amino acids. In our studies with

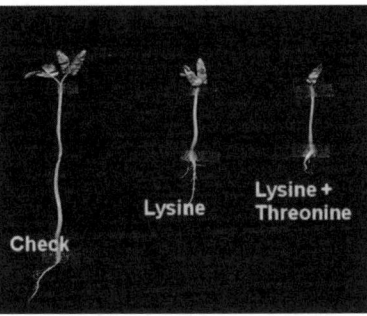

Fig. 1 Inhibition of field bindweed with amino acids

Fusarium oxysporum (Tiourebaev, 1999) amino acid excretion did not alter the host range of a plant pathogen. The implementation of this biocontrol strategy thus offers a novel tool to increase the efficacy of biocontrol pathogens or in this case, to turn a common saprophyte into a safe and effective biocontrol agent.

3   Enhancement of Bioherbicides

3.1   Criteria for Selection of Biocontrol Agents

Classical biocontrol has proven successful in select situations, including biocontrol of rush skeletonweed with *Puccinia chondrilla* in Australia (Hasan and Wapshere, 1973) and *Acacia saligna* by the rust fungus *Uromycladium tepperianum* in South Africa Morris et al. (1999). These successes utilized obligate pathogens that are highly host-specific, highly virulent, and capable of naturally spreading from a focal inoculation point. Prime candidates for biological control among bacterial plant pathogens are *Ralstonia solanacearum*, *Pseudomonas syringae* and various pathovars of *Xanthomonas campestris*. These agents offer the added advantages of rapid culture growth and swift dissemination. The ideal biocontrol agent probably would be host-specific, soil-borne and disseminated by water. Preferably it has good knockdown, is easy to culture, is prototrophic for all amino acids, and does not produce any mammalian toxins. Rhizosphere competence on numerous non-host species is a highly desirable trait for dissemination, and therefore represents an important biocontrol selection factor. The ideal agent should also possess, as do most bacteria, a natural mutation frequency that permits mutant selection in the absence of mutagenic agents. More commonly studied bacteria are preferable due to knowledge base and DNA characterization.

3.2   Selection of Biocontrol Agents that Excrete Target Amino Acids

The virulence and efficacy of bioherbicides is enhanced by selection of variants of the pathogen that overproduce and excrete amino acids that are inhibitory to the target plant. Our approach is modeled after "Frenching disease", a naturally occurring disease of tobacco (Steinberg, 1946). Steinberg et al. (1950) discovered that saprophytic bacteria growing on the roots of symptomatic plants overproduced a single amino acid, isoleucine. Isoleucine is synthesized in plants via the branched chain amino acid pathway. The end products of this pathway (valine, leucine, and isoleucine) allosterically regulate the activity of acetolactate synthase (ALS). In "Frenching disease", overproduction of isoleucine by saprophytic bacteria inhibited ALS activity in tobacco plants, thus halting valine and leucine synthesis, which in turn disrupted essential protein metabolism. Interestingly enough, several well-studied modern chemical herbicides mimic this strategy inhibiting a key biosynthetic enzyme in plants, rendering treated plants incapable of producing a metabolite essential for plant growth. In Tiourebaev's work (1999) the growth of *Cannabis sativa* is inhibited by the amino acid valine. Therefore, a search for mutant pathogens resistant to valine analogs (penicillamine or norvaline), yielded possible useful candidates.. For example, he isolated variants of *Fusarium oxysporum* f. sp. *cannabis* that were resistant to valine analogs (Tiourebaev, 1999). When analyzed these variants excreted 10–55 times more valine than their wild-type parent strains. These strains proved to be more virulent to *C. sativa* than the wild-type parent. The wild-type strain resulted in 25% mortality of the target plant, while the valine mutants increased mortality to 70–90%. In addition, the development of wilt disease was more rapid in the plants infested with the valine overproducers. Limited studies on host range did not reveal any changes. Thus, overproduction of an essential amino acid provided a highly effective means of enhancing the virulence of a biocontrol agent (Tiourebaev, 1999) and has been used to enhance the virulence of *Fusarium oxysporum* f. sp. *cannabis*, and *F. oxysporum* f. sp. *papaveris*, Nina Zidack (personal communication) found similar results with *Pseudomonas syringae* pv. *tagetis*. It appears that enhanced strains of *Fusarium oxysporum* for control of *Orobanche* (Vurro et al., 2006) and *Xanthomonas campestris* pv. *poae* (A. Pilgeram, personal communication) can be used for this same biocontrol purpose.

3.3   Inhibition of Weeds by Amino Acids

Amino acids, when applied to plants or seeds, have a definite effect on plant health in all cases examined (Vurro et al., 2006; A. Pilgeram, personal communication). This includes weeds with effects varying from necrosis to wilting and stunting of growth.

For example, when lysine is applied to leaves of *Cirsium arvense*, necrosis is observed within hours of application. Application of methionine to *Cirsium arvense* resulted in yellowing on new leaf buds as well as necrosis. Other amino acids had little or no effect on the plants. Conversely certain amino acids also selectively enhance the growth and vigor of plants.

4   General Methodology

4.1   *Determination of Amino Acids or Combinations of Amino Acids That are Most Inhibitory to the Growth and Development of the Target Weed*

Surface sterilized seeds (10% commercial household sodium hypochlorite for 10 min) are placed on water agar (1.5% agar, 1 mM Tris, pH 6.8) supplemented with a single amino acid (2–5 mM l-form). The inhibitory effects of amino acids in the branched chain pathway (valine, leucine, isoleucine), the aspartate pathway (lysine, threonine, and methionine) and the aromatic pathway (tyrosine, tryptophan, phenylalanine) are evaluated by decreasing seed germination, inhibition of shoot growth or necrosis.

4.2   *Selection of Variants of the Bioherbicide Resistant to Analogs of the Selected Amino Acid*

Amino acid overproducing strains of each fungus or bacterium are selected via exposure to specific amino acid analogs (Sands and Hankin, 1974; Tiourebaev, 1999). Resistant colonies can be selected using a well zone-diffusion assay on CUTS minimal medium (Czapek-Dox Agar (Difco) (35 g/l) supplemented with ammonium sulfate (0.5 g/l), uracil (20 mg/l), thiamine (4 mg/l) and a vitamin mixture (100 mg/l of crushed Sesame Street Complete Vitamins). The zone diffusion plates are prepared by cutting a plug from the center of the CUTS plate with a sterile cork borer. The plates are then inoculated with 10^6–10^7 fungal spores, a suspension of 10^3–10^5 mycelial fragments, or a suspension of 10^7–10^8 bacteria. A sterile solution of the amino acid analog (0.1 ml of a 100 mM solution) is then added to the well. The plates are then incubated at 28°C and monitored daily for zones of inhibition and resistant colonies. Resistant colonies are isolated and analyzed for amino acid excretion. This selection may need to be repeated several times using increasing concentrations of analog and/or different analogs.

4.3   Assay for Amino Acid Excretion

In our laboratory amino acid excretion is assayed by growth of a bacterial auxotroph seeded into media lacking the amino acid required for growth (Tiourebaev et al., 2001). Subsequent growth of the auxotroph in the media is dependent upon and proportional to the quantity of added amino acid. A standard dose–response can be determined by placing discs containing various levels of amino acid onto the auxotroph-seeded agar.

4.4   Testing Virulence and Host Range of the Amino Acid Overproducing Variants in Growth Chamber Studies

The virulence (rate of kill and % mortality) of amino acid producing variants of each pathogen should be first evaluated in environmental growth chambers in order to eliminate most external factors. During initial studies, target weed plants are inoculated with each amino acid excreting variant and its respective wild-type parent. Amino acid excreting variants that are more virulent than the parent are further evaluated in host range and scale-up experiments.

5   Improving Dissemination

A soil-applied pathogen will not be an efficacious mycoherbicide, even if it has specificity, sufficient lethality, and long-term soil survival, unless it can be delivered in a cost effective manner. Fungi grown in liquid or solid-phase fermentation are inherently expensive when applied to large acreages at 10^4 spores per gram of soil. Conventional formulation methods with spore suspensions and food-based formulations did not provide enough spores in the root zone of the target weed (Eparvier and Alabouvette, 1994; Fravel et al., 1996; Connick et al., 1998; Ciotola et al., 2000). However, plant pathogenic fungi such as *Fusarium oxysporum* saprophytically colonize the roots of many non-host plants and thus, *F. oxysporum* mycoherbicides could be delivered to farmer's fields on non-host seed such as crop or grass seed. This offers the added advantage of positioning the mycoherbicide directly in the rhizosphere of target weed (Tiourebaev, 1999). The multiplication of fungal biomass in the rhizosphere of the carrier seedling allows for application of low levels of the mycoherbicide, greatly reducing the cost of inoculum production.

6   Conclusions

Over the last 30 years, numerous pathogens have been investigated as potential bioherbicides. Despite this intensive research effort, few pathogens have been successful as biocontrol agents. The inherent constraints associated with biological species are largely responsible for this failure, yet our preconceived ideas about these agents are also at fault. The authors believe that a paradigm shift must occur if bioherbicides are to enjoy wider success as a weed control method. In the past, researchers have focused on lethality and host specificity as initial screening requirements for successful agents. However, many pathogens that do not meet these criteria could be enhanced by synergistic additions or genetic modification. Embracing new methodologies may allow many "unsuitable" pathogens and likewise saprophytes to be developed into successful biocontrol agents. Success in controlling a serious agricultural parasitic weed such as Striga would do wonders for farmers, farm-based economies, and biocontrol researchers.

References

amino acids, *Recent Adv. Phytochem.* 20, 83–117.

Ciotola, M., DiTommaso, A., and A. K. Watson. 2000. Chlamydospore production, inoculation methods and pathogenicity of *Fusarium oxysporum* M12-4A, a biocontrol for *Striga hermonthica*, *Biocontrol Sci. Tech.* 10, 129–145.

Connick, Jr., W. J., Daigle, D. J., Pepperman, A. B., Hebbar, K. P., Lumsden, R. D., Anderson, T. W., and D. C. Sands. 1998. Preparation of stable granular formulations containing *Fusarium oxysporum* pathogenic to narcotic plants, *Biol. Control* 13, 79–84.

Eparvier, A. and C. Alabouvette. 1994. Use of ELISA and GUS-transformed strains to study competition between pathogenic and non-pathogenic *Fusarium oxysporum* for root colonization, *Biocontrol Sci. Tech.* 4, 35–47.

Fravel, D. R., Stosz, S. K., and R. P. Larkin. 1996. Effect of temperature, soil type, and matrix potential on proliferation and survival of *Fusarium oxysporum* f. sp. *erythroxyli* from *Erythroxylum coca*, *Phytopathology* 86, 236–240.

Hasan, S. and A. J. Wapshere. 1973. The biology of *Puccinia chondrillina* a potential biological control agent of skeleton weed. *Ann. Appl. Bio.* 74, 325–332.

Johnson, D. R., Wyse, D. L., and K. J. Jones. 1996. Controlling weeds with phytopathogenic bacteria, Weed Technol. 10(3), 621–624.

Miller, R. V., Ford, E. J., Zidack, N. J., and D. C. Sands. 1989. A pyrimidine auxotroph of Sclerotinia sclerotiorum for use in biological weed control, *J. Gen. Microbiol.* 135, 2085–2091.

Morris, M. J., Wood, A. R., and A. den Breeÿen. 1999. Plant pathogens and biological control of weeds in South Africa: a review of projects and progress during the last decade. In: T. Olckers and M. P. Hill (eds.), African Entomology Memoir No. 1. Entomological Society of South Africa, Hatfield, PA, pp.125–128.

Sands D. C. and L. Hankin. 1974. Selecting lysine-excreting mutants of lactobacilli for use in food and feed enrichment, *Appl. Microbiol.* 28, 523–524.

Steinberg, R. A. 1946. A "Frenching" response of tobacco seedlings to isoleucine, *Science* 103, 329–330.

Steinberg, R. A., Bowling, J. D., and J. E. McMurtrey, Jr. 1950. Accumulation of free amino acids as a chemical basis for morphological symptoms in tobacco manifesting frenching and mineral deficiency symptoms, *Plant Physiol.* 25, 279–288.

Tiourebaev, K. 1999. Virulence and dissemination enhancement of a mycoherbicide. Ph.D. thesis. Montana State University, Bozeman, MT.

Tiourebaev, K. S., Semenchenko, G. V., Dolgovskaya, M., McCarthy, M. K., Anderson, T. W., Carsten, L. D., Pilgeram, A. L., and D. C. Sands. 2001. Biological control of infestations of ditchweed (*Cannabis sativa*) with *Fusarium oxysporum* f. sp. *cannabis* in Kazakhstan. *Biocontrol Sci. Techn.* 11, 535–540.

Vurro, M., Boari, A., Pilgeram, A. L., and D. C. Sands. 2006. Exogenous amino acids inhibit seed germination and tubercle formation by *Orobanche ramosa* (Broomrape): potential application for management of parasitic weeds, *Biol. Control* 36, 258–265.

Zidack, N. K. and P.A. Backman. 1996. Biological control of Kudzu (*Pueraria lobata*) with the plant pathogen *Pseudomonas syringae* pv *phaseolicola*. Weed Sci. 44 (3), 645–649.

Analysis of *Pseudomonas syringae* Populations and Identification of Strains as Potential Biocontrol Agents Against Postharvest Rot of Different Fruits

G. Cirvilleri[1], G. Scuderi[1], A. Bonaccorsi[1], and M. Scortichini[2]

Abstract A collection of Pseudomonas syringae strains was evaluated for in vitro and in vivo growth inhibition of a broad spectrum of bacteria and fungi, for production of hydrolytic enzymes, presence of syrB gene and for pathogenicity on detached fruits and then they were genetically characterized by ARDRA, ERIC-PCR and fAFLP.

P. syringae strains and their culture filtrates inhibited a wide range of phytopathogenic bacteria and fungi *in vitro* and the inhibition depended on the bacterial and fungal strain. The cell-wall degradation enzymes N-acetyl-β-D-glucosaminidase (NAGase), β-glucosidase, cellobiohydrolase, cellulase and protease were detected in culture filtrates of the majority of strains while only a few strains showed chitinase and glucanase activities. The antagonistic activity *in vivo* of *P. syringae* strains – evaluated on wounded lemon, orange and mandarin fruits against *Penicillium digitatum*, on apples against *P. expansum*, on grape berries against Botrytis cinerea – was apparent for some of the strains evaluated. *P. syringae* strains isolated from *Citrus* spp., pear and strelitzia were moderately to highly pathogenic on tested detached citrus fruits.

All *P. syringae* strains showed the presence of *syr*B gene and their molecular characterization with ARDRA, ERIC-PCR and fAFLP allowed their clustering in distinct groups. A partial correlation between groups delineated on the basis of antagonistic activity *in vitro* and genomic fingerprints was also apparent. Fluorescent AFLP analysis also produced characteristic profiles for each strains useful for risk assessment, monitoring and identification of released antagonistic strains.

Keywords *Pseudomonas syringae*, antagonistic activity, ARDRA, ERIC-PCR, fAFLP

[1] Dipartimento di Scienze e Tecnologie Fitosanitarie- Sezione Patologia Vegetale, Università degli Studi di Catania, Via S. Sofia 100, 95100 Catania, Italy

[2] C.R.A.-Istituto Sperimentale per la Frutticoltura, Via di Fioranello, 52, I-00040 Ciampino aeroporto, Roma, Italy

Author for correspondence: Gabriella Cirvilleri; e-mail: gcirvil@unict.it

M'B. Fatmi et al. (eds.), *Pseudomonas syringae Pathovars and Related Pathogens.*   125
© Springer Science + Business Media B.V. 2008

1   Introduction

Fruits and vegetables suffer significant losses after harvest due to fungal diseases (Snowdon, 1990, 1991). Traditionally, postharvest diseases are controlled by post-harvest application of synthetic fungicides (Eckert and Ogawa, 1988). However, the development of resistance to fungicides and the concern for public safety have led to increase interest in alternative and effective natural methods of postharvest diseases control.

Biological control of postharvest pathogens using microbial antagonist has been considered an emerging alternative to the use of synthetic chemicals (Janisiewicz, 1988; Wilson et al., 1991; Janisiewicz and Korsten, 2002). Among bacteria, a number of Gram positive and Gram negative bacteria have been studied as Biological Control Agents (BCAs). *Bacillus subtilis* (Obagwu and Korsten, 2003), *Pantoea agglomerans* (Nunes et al., 2001, 2002), *Serratia plymuthica* (Meziane et al., 2006), *Pseudomonas cepacia* (Janisiewicz and Roitman, 1988) and *P. syringae* (Janisiewicz and Jeffers, 1997) were reported as effective BCAs against postharvest diseases of different fruit crops.

The bacterial species *Pseudomonas syringae* includes both strains responsible of diseases, for which the mechanism of action is actively investigated, and biocontrol agents already applied in agriculture at a commercial level. Strains of *P. syringae* have been reported to be effective as biocontrol agents against *Penicillium expansum* and *Botrytis cinerea* on pear fruits (Janisiewicz and Marchi, 1992; Sugar and Spotts, 1999; Northover and Zhou, 2002), *Monilinia fructicola* and *Rhizopus stolonifer* on peaches (Zhou et al., 1999), *P. digitatum* on *Citrus* spp. (Bull et al., 1998; Cirvilleri et al., 2000, 2005), *P. expansum* on apple fruits (Janisiewicz and Bors, 1995; Conway et al., 1999; Zhou et al., 2001; Vinale et al., 2005), *E. coli* O157:H7 on apple wounds (Janisiewicz et al., 1999). *P. syringae* strains ESC-10 and ESC-11 were registered in 1995 and developed by EcoScience Corporation in the Bio-Save product line for the suppression of postharvest diseases of *Citrus* spp., pome, stone fruit as well as tubers.

Pseudomonas spp. strains produce a wide spectrum of lipodepsipeptides (LDPs) (Bender et al., 1999) showing antimicrobial activity against a broad range of fungi, including yeast and human pathogens (Sorensen et al., 1996; Lavermicocca et al., 1997). Production of the lipodepsinonapeptide syringomycin E by *P. syringae* strains ESC-10 and ESC-11 has been correlated with the ability of these strains to control postharvest diseases of citrus and pome fruits caused by *P. digitatum*, *P. italicum*, *P. expansum*, *B. cinerea* and *Geotricum candidum* (Bull et al., 1998).

Molecular typing of *P. syringae* strains is an important prerequisite in risk assessment analysis for tracking their dispersal and fate in the environment when used as BCAs and various molecular techniques have been used to characterize them. They include Pulsed-Field Gel Electrophoresis (PFGE) (Grothues and Rudolph, 1991), Restriction Fragment Length Polymorphism (RFLP) (Scholz et al., 1994), Random Amplified Polymorphic DNA (RAPD) (Clerc et al., 1998), repetitive-sequence PCR (rep-PCR) (Little et al., 1998; Scortichini et al., 2003), Amplified 16S Ribosomal DNA Restriction Analysis (ARDRA) (Scortichini et al., 2001) and Amplified Fragment Length Polymorphism (AFLP) (Clerc et al., 1998; Manceau and Brin, 2003; Cirvilleri et al., 2006). Presently, AFLP analysis can be considered one of the most discriminating genomic methods to distinguish among bacterial strains (Vos et al., 1995).

This study was conducted to explore criteria for selecting *P. syringae* antagonistic strains using both biological and molecular methods. Strains of *P. syringae* and their culture filtrates were evaluated for the antagonistic activities either *in vitro* against a broad spectrum of bacteria and fungi, for pathogenicity on fruit, or *in vivo* on citrus, apple and grapes, for production of hydrolytic enzymes, for presence of *syr*B gene, and were genetically characterized by ARDRA, ERIC-PCR and fAFLP.

2   Materials and Methods

2.1   Biochemical Characterization

P. syringae strains used in this study were isolated from different woody and herbaceous host plants whereas some other were obtained from international culture collections (Table 1). All the strains were evaluated for LOPAT characters following the procedures described by Schaad et al. (2001).

2.2   In Vitro Antagonistic Activity

All *Pseudomonas syringae* strains were tested for the growth inhibition activity on Potato Dextrose Agar plate (PDA, Oxoid) of microorganisms listed in Table 2 by following the technique described elsewhere (Cirvilleri et al., 2005). The presence and size of a clear zone around *P. syringae* colonies, indicating the growth inhibitory effect, were scored after 2–4-day incubation.

2.3   Antimicrobial and Phytotoxic Activity of Culture Filtrates

Cell-free culture filtrates of *P. syringae* strains in IMM (Surico et al., 1988) were evaluated for the antimicrobial activity on PDA as previously described (Cirvilleri et al., 2005). The phytotoxic activity of cell-free culture filtrates was evaluated on tobacco leaves and bean pods and the necrosis scored after 2 and 6 days, respectively.

Table 1 Original source and main characteristic of *Pseudomonas syringae* strains

Strains[a]	Host	Geographic origin	Year	Antagonistic group[b]	syrB	Pathogenicity on lemon
PVCT 10.2	*Citrus sinensis*	Italy	1990	A	+	8.0[b-d]
PVCT 40_2	"	Italy	1990	A	+	n.t.
PVCT 41_2	"	Italy	1990	A	+	8.0[b-d]
PVCT 119_2	"	Italy	1990	A	+	0.0[a]
PVCT 130_1	"	Italy	1990	A	+	10.0[cd]
PVCT 147_1	"	Italy	1990	A	+	8.0[b-d]
PVCT 280_2	"	Italy	1990	A	−	10.0[cd]
PVCT 281_1	"	Italy	1990	A	+	5.0[b]
PVCT 282_1	"	Italy	1990	A	+	5.0[b]
PVCT 285_1	"	Italy	1990	A	+	10.0[cd]
PVCT 287_1	"	Italy	1990	A	+	n.t.
PVCT 290_2	"	Italy	1990	A	+	10.0[cd]
PVCT 291_1	"	Italy	1990	A	+	9.0[b-d]
PVCT 293_1	"	Italy	1990	A	+	5.0[b]
PVCT 295_1	"	Italy	1990	A	−	8.0[b-d]
PVCT 310	"	Italy	1990	A	+	10.0[cd]
PVCT 334	"	Italy	2000	A	+	8.0[b-d]
PVCT 335_2	"	Italy	2000	A	+	8.0[b-d]
PVCT 337_1	"	Italy	2000	A	−	10.0[cd]
PVCT 337_2	"	Italy	2000	A	+	7.5[b-d]
PVCT 339_1	"	Italy	2000	A	+	7.5[b-d]
PVCT 342_1	"	Italy	2000	A	+	n.t.
PVCT 48SR1	"	Italy	1990	A	+	10.0[cd]
PVCT 48SR2	"	Italy	1990	A	+	10.0[cd]
PVCT 40SR4	"	Italy	1990	A	+	n.t.
PVCT Al513		USA (S.E. Lindow)	1986	A	+	0.0[a]
ISF 242	*Citrus lemon*	Italy	1996	A	+	10.0[cd]
ISF 243	*Citrus reticulata*	Italy	1996	A	+	12.0[d]
PVCT 23P	*Pyrus communis*	Italy	1998	A	+	8.0[b-d]
PVCT 26P	"	Italy	1998	A	+	10.0[cd]

PVCT 46P	"	Italy	1998	A	+	10.0[cd]
PVCT 76P	"	Italy	1998	A	+	10.0[cd]
ISF 280	"	Italy	1996	A	+	9.0[b-d]
ISF 281	"	Italy	1996	A	+	5.0[b]
ISF 288	"	Italy	1996	A	+	6.5[bc]
ISF 347	"	Italy	1996	A	+	10.0[cd]
PVCT 1.1S	*Strelitzia reginae*	Italy	2000	A	+	n.t.
PVCT 1.2S	"	Italy	2000	A	+	5.0[b]
PVCT 1.3S	"	Italy	2000	A	+	8.5[b-d]
PVCT 1.4S	"	Italy	2000	A	+	5.5[bc]
PVCT B.I.	"	Italy	2004	A	−	0.0[a]
ISF 106 = NCPPB 2426	*Prunus avium*	Switzerland		A	+	0.0[a]
ISF 107 = NCPPB 1093	*Prunus armeniaca*	New Zealand		A	+	0.0[a]
ISF 231	"	Italy	1996	A	+	0.0[a]
ISF 290 = B3A	*Prunus persicae*	USA (J.E. De Vay)	1992	A	+	12.0[d]
ISF 015 = NCPPB3869	*Laurus nobilis*	Italy	1996	A	+	8.0[b-d]
ISF 282	*Castanea sativa*	Italy	1988	A	+	12.0[d]
AID 48	*Fragaria x ananassa*	Italy		A	+	7.0[bc]
HRI 1480A	*Pisum sativum*	UK		A	+	0.0[a]
ISF 292	*Triticum aestivum*	USA (J.E. De Vay)		A	+	7.0[bc]
ISF 300	"	Italy		A	+	0.0[a]
ISF 304	"	Italy	1996	A	+	8.0[b-d]
ISF 309	"	Italy	1996	A	+	5.5[bc]
ISF 310 = NCPPB2612	"	New Zealand		A	+	5.0[b]
ISF 355	*Hordeum vulgare*	Italy	1996	A	+	5.0[b]
ISF 356	"	Italy	1996	A	+	5.0[b]
ISF 359 = 475A	"	(J.E. De Vay)		A	+	12.0[d]
ISF 293 = B359	*Setaria italica*	Australia (J.E. DeVay)		A	+	8.0[b-d]
PVCT 7NC	*Corylus avellana*	Italy	2005	A	n.t.	n.t.
PVCT 44NC	"	Italy	2005	A	n.t.	n.t.
AID 122A	*Prunus amygdali*	Japan (D.C. Gross)	1988	B	+	0.0[a]
ISF 291 = SY12	*Syringa vulgaris*	Italy	1987	B	+	0.0[a]
AID 24	*Fragaria x ananassa*	Italy		B	+	0.0[a]
AID 33	"	Italy		B	+	0.0[a]

(continued)

Table 1 (continued)

AID 76	"	Italy	1988	B	+	0.0[a]
AID 88	"	Italy	1988	B	+	n.t.
PVCT B728a	Phaseolus vulgaris	USA (S.E. Lindow)	1986	B	+	4.0[b]
ISF 286 = Y37	"	USA (D.C. Gross)		B	+	0.0[a]
ISF 332		Italy	1996	B	+	0.0[a]
PVCT 4	Cynara scolimus	Italy	1992	B	+	0.0[a]
PVCT 14	"	Italy	1992	B	+	n.t.
PVCT 29	"	Italy	1992	B	+	n.t.
PVCT 40	"	Italy	1992	B	+	0.0[a]
PVCT 74	"	Italy	1992	B	+	n.t.
PVCT 96	"	Italy	1992	B	+	n.t.
PVCT 98	"	Italy	1992	B	+	n.t.
PVCT 106	"	Italy	1992	B	+	0.0[a]
PVCT 113	"	Italy	1992	B	+	n.t.
PVCT 120	"	Italy	1992	B	+	0.0[a]
PVCT 133	"	Italy	1992	B	+	n.t.
PVCT 141	"	Italy	1992	B	+	n.t.
PVCT 152	"	Italy	1992	B	+	0.0[a]
PVCT 169	"	Italy	1992	B	+	n.t.
ISF 353 = ISF-PP2	Capsicum annuum	Italy	1997	B	+	0.0[a]
ISF 284 = PSS61	Triticum aestivum	USA (D.C. Gross)		B	+	0.0[a]
ISF 294 = W451	"	USA (D.C. Gross)		B	+	0.0[a]
ISF 295 = SD202	"	USA (D.C. Gross)		B	+	0.0[a]
ISF 357	Hordeum vulgare	Italy	1996	B	+	0.0[a]
PVCT 38NC	Corylus avellana	Italy	2005	B	n.t.	n.t.

[a] AID: Agricultural Industrial Development, Catania, Italy; HRI: Horticulture Research International, Wellesbourne, United Kingdom; ISF: C.R.A.: Istituto Sperimentale per la Frutticoltura, Roma, Italy; NCPPB: National Collection of Plant Pathogenic Bacteria, York, United Kingdom; PVCT: Plant Pathology, University of Catania, Italy

[b] Antagonistic groups (see also Cirvilleri et al., 2005)

[c] Pathogenicity tests performed with *P. syringae* strains. The number refers to the mean diameter of lesion recorded seven days after artificial inoculation of lemon cv. Femminello fruits. Mean are based on four replications per experiment. Each experiment was repeated twice. Number in column followed by the same letter are not significantly different using the Student–Newman–Keul's mean separation test at P ≤ 0.05 (see also Cirvilleri et al., 2005).

Table 2 Antagonistic activity *in vitro* of *Pseudomonas syringae* strains recorded as mean inhibition zones (mm)

Target microorganism	Inhibition zone (mm) ± SD[a]	
	Antagonistic group A	Antagonistic group B
Rhodotorula pilimanae ATCC26423	17.5 ± 7.3	–
Bacillus megaterium ITM100	12.0 ± 7.4	23.8 ± 10.9
Salmonella spp[b]	1.0 ± 2.7	1.1 ± 4.3
Listeria monocytogenes[b]	14.2 ± 10.5	23.3± 10.9
Pseudomonas syringae PVCT48SR2	11.4 ± 6.3	21.0 ± 7.7
Botrytis cinerea[c]	13.4 ± 6.7	–
Penicillium digitatum[c]	4.8 ± 2.9	–
Alternaria alternata[c]	6.5 ± 4.2	–
Aspergillus ochraceus ATCC18641	6.0 ± 4.0	–
Fusarium solani 1A[c]	3.6 ± 3.5	–

[a] Inhibition zones ± standard deviation (SD) of the mean of four replications for each strain
[b] Kindly provided by the Istituto Zooprofilattico Regionale, Catania, Italy
[c] From mycologial laboratory of Plant Pathology, University of Catania, Italy

2.4 Pathogenicity Tests

Pathogenicity tests were performed at 20°C on lemon (*Citrus lemon* Burm) cv. Femminello, orange (*Citrus sinensis* Osbeck) cv. Tarocco, mandarin (*Citrus reticulata* L.) cv. Fortuna, apple (*Malus domestica* Borkh.) cv. Golden Delicious fruits, grape (*Vitis vinifera* L.) cv. Italia berries and bean pods (*Phaseolus vulgaris* L.) as elsewhere described (Cirvilleri et al., 2005). Fruits were observed repeatedly and the width of necrosis was measured 7 days after inoculation.

2.5 In Vivo Antagonistic Activity

P. syringae strains were tested for the ability to inhibit the growth of *P. digitatum* on lemon, orange and mandarin fruits, of *P. expansum* on apple fruits and of *B. cinerea* on grape berries. Fruits were co-inoculated with *P. syringae* suspensions (1×10^9 cfu/ml) and fungal spore suspension (1×10^6 cfu/ml). Incidence and severity of disease were calculated one week after inoculation as previously described (Cirvilleri et al., 2005).

2.6 Enzymatic Activity

N-acetyl-β-D-glucosaminidase (NAGase), β-glucosidase, cellobiohydrolase, cellulase, protease, chitinase and glucanase activity were detected in cell free culture filtrates as previously described (Nielsen and Sorensen, 1997; Madesen and de Neergard, 1999; Campisano et al., 2001).

2.7   Molecular Characterization

The presence of *syr*B gene, Amplified 16S Ribosomal DNA Restriction Analysis (ARDRA), repetitive-sequence PCR (rep-PCR) using ERIC-PCR primers, and fluorescent Amplified Fragment Length Polymorphism (fAFLP) were performed as described elsewhere (Cirvilleri et al., 2005, 2006; Manceau and Brin, 2003; Scortichini et al., 2001). Cluster analysis was performed according to the unweighted pair-group method with average linkages (UPGMA). Similarity coefficients were determined using the Dice's coefficient (Dice, 1945) and the robustness of the tree was assessed by bootstrap analysis (1000 repeated samplings) (Felsenstein, 2004).

3   Results

3.1   In Vitro Antagonistic Activity

Pseudomonas syringae strains evaluated in this study inhibited *in vitro* the growth of a wide range of phytopathogenic bacteria and fungi and the inhibition rate depended on bacterial and fungal strains. *P. syringae* strains were divided into two distinct groups A and B on the basis of *in vitro* inhibitory activity (Table 2). In particular, *P. syringae* strains inhibiting to a different extent the growth of *R. pilimanae* and of all the others target microorganisms were included in the antagonistic group A whereas strains unable to inhibit *R. pilimanae* and the majority of the target microorganisms were included in the antagonistic group B. Strains of the latter group showed the widest inhibition zones against *B. megaterium*, *L. monocytogenes* and *P. syringae*.

3.2   Antimicrobial and Phytotoxic Activity of Culture Filtrates

Culture filtrates of strains belonging to group A were phytotoxic on tobacco and caused bean pods necrosis with an intensity sometimes similar to those caused by the pathogen strain inoculation. Cell-free culture filtrates of strains of group B did not cause necrosis on tobacco and on bean pods (data not shown).

3.3   Pathogenicity Tests

P. syringae strains belonging to the antagonistic group A and isolated from *Citrus* spp., pear and strelitzia were moderately to highly virulent on lemon, orange and mandarin fruits (Table 3). Mean size of lesions ranged between 5 and 12 mm on

Table 3 Percentage of *Pseudomonas syringae* strains from different hosts and belonging to antagonistic groups A and B causing lesions (mm) on inoculated fruits

Host of origin	Antagonistc groups	Fruits											
		Lemon		Orange		Mandarin		Apple		Grape		Bean pods	
		Strains (%)	Lesions (mm)[a]	Strains (%)	Lesions (mm)[a]	Strains (%)	Lesions (mm)[a]	Strains (%)	Lesions (mm)[a]	Strains (%)	Lesions (mm)[a]	Strains (%)	Pathogenicity[b]
Citrus spp.	A	100	5–12	47	0.5–4	94	0.6–3	100	0	100	0	100	+
Pyrus communis		100	5–10	71	2–4	43	2.6–3.6	100	0	100	0	100	++
Strelitzia reginae		100	5–8.5	33	1–2	66	0.6–3	100	0	100	0	100	++
Prunus armeniaca		100	0	0	0	0	0	100	0	100	0	100	++
Laurus nobilis		100	12	0	0	0	0	100	0	100	0	100	++
Castanea sativa		100	8	100	2	0	0	100	0	100	0	100	+++
Fragaria x ananassa		100	12	0	0	100	2.3	100	0	100	0	100	++
Pisum sativum		100	7	0	0	0	0	100	0	100	0	100	++
Triticum aestivum		50	5–8	25	0.5–2	0	0	100	0	100	0	100	++
Hordeum vulgare		100	5–12	100	2	0	0	100	0	100	0	100	++
Setaria italica		100	8	100	2	0	0	100	0	100	0	100	+++
Corylus avellana		100	0	100	0	0	0	100	0	100	0	100	+++
Fragaria x ananassa	B	100	0	100	0	100	0	100	0	100	0	100	+
Triticum aestivum		100	0	100	0	100	0	100	0	100	0	100	++
Hordeum vulgare		100	0	100	0	100	0	100	0	100	0	100	++
Corylus avellana		100	0	100	0	100	0	100	0	100	0	100	++
Syringa vulgaris		100	0	100	0	100	0	100	0	100	0	100	++
Phaseolus vulgaris		100	4	100	0	100	0	100	0	100	0	100	++
Cynara scolymus		100	0	100	0	100	0	100	0	100	0	100	++

[a] Diameter of lesions (mm) recorded 7 days after artificial inoculation with 1×10^9 CFU ml^{-1} and incubated at 20°C. Mean values are based on four replication per experiment and each experiment was repeated twice.

[b] +, lesion associated with inoculated sites; ++, progressive necrotic lesion; +++, progressive necrotic lesion and red halo on bean.

Table 4 Disease severity on lemon, orange, mandarin, apple fruits and grape berries 7 days after artificial inoculation with *Penicillium digitatum*, *P. expansum* and *Botrytis cinerea* and subsequently treated with *Pseudomonas syringae* strains

	P. digitatum/lemon[a]		P. digitatum/orange[a]		P. digitatum/mandarin[a]		P. expansum/apple[a]		B. cinerea grape[a]	
	Strains (%)	Disease severity (%)	Strains (%)	Disease severity (%)	Strains (%)	Disease severity (%)	Strains (%)	Disease severity (%)	Strains (%)	Disease severity (%)
Antagonistic group										
A	79%	0	21%	0	0%	0	30%	0	0%	0
(60 strains)	11%	22–45	56%	17–45	93%	30–60	51%	4–45	0%	45
	10%	77–90	23%	53–90	7%	90	19%	47–90	100%	57–90
B	14%	0	7%	0	0%	0	57%	0	0%	0
(29 strains)	50%	17–45	14%	21–38	0%	30–60	29%	13–45	0%	45
	36%	68–90	79%	49–90	100%	90	14%	52–90	100%	49–90

[a] Disease severity was evaluated with an empiric scale and ratings was converted to percentage midpoint values, where 0% = no visible symptoms, 35% = initial soft rot, 65% = presence of mycelium, 90% = sporulation. Mean values are based on four replications and four fruits per replication

lemon, 0.5 and 4 mm on orange and 0.6 and 3 mm on mandarin. By contrast, group B strains did not induce necrosis on lemon, orange and mandarin fruits. None of the strains of either group was virulent on apple fruits and grape berries whereas almost all strains induced necrotic lesions on bean pods.

3.4   In Vivo Antagonistic Activity

Several strains belonging to group A prevented (disease severity 0%) the growth of *P. digitatum* on lemon and orange fruits (respectively 79% and 21% of strains) and of *P. expansum* on apple fruit (30% of strains) whereas only a few strains of group B prevented the growth of *P. digitatum* on lemon and orange (14% and 7% of strains, respectively) and *P. expansum* on apple fruits (57% of strains) (Table 4). None of the strains of groups A and B totally controlled *P. digitatum* on mandarin fruits and *B. cinerea* on grapes. One strain of group A (PVCT 119$_2$) and two strains of group B (ISF284 and ISF294), that did not induce symptoms on artificially inoculated fruits, totally inhibited the growth of *P. digitatum* and *P. expansum* on lemon and apple fruits.

3.5   Enzymatic Activity

The cell-wall degradation enzymes NAGase, b-glucosidase, cellobiohydrolase, cellulase and protease were detected in culture filtrates of the majority of strains while only a few strains showed chitinase and glucanase activities (Table 5). At least one of the seven tested enzymatic activities was detected in the cell-free culture filtrates of *P. syringae* strains, irrespective of the antagonistic group and of the host of origin. NAGase, β-glucosidase and cellobiohydrolase (up to 10, 30 and 7 mM *p*-nitrophenol ml^{-1} min^{-1} respectively) and cellulase and protease activity were detected in culture filtrates of the most part of the strains. Chitinase and glucanase activities were almost undetectable in the majority of the strains.

Table 5 Enzymatic activity of *Pseudomonas syringae* strains

		Production of[a]						
	% of strains	NAGase	β-gluc	cellbio	pro	cell	chi	glu
Group								
A	36	–	–	–	+	+	–	–
	21	+	+	+	+	+	–	–
	29	+	+	–	+	+	–	–
	14	+	–	–	+	+	–	–
B	19	+	+	+	+	–	–	–
	56	+	–	+	+	+	–	–
	25	–	–	+	–	–	–	–

[a]NAGase = *N*-acetyl-β-D-glucosaminidase; β-gluc = β-glucosidase; cell-bio = cellobiohydrolase; pro = protease; cell = endocellulase; gluc = β-glucanase; chi = chitinase; glu = glucanase

3.6   Molecular Characterization

All *P. syringae* strains belonging to both group A and B had the *syr*B gene. ARDRA analysis was performed with many representative *P. syringae* strains. P0 and P6 primers amplified an approximately 1600 bp fragments (Fig. 1), and restriction analysis with *Alu*I (Fig. 2a) revealed two different patterns: pattern I, comprising all strains belonging to the antagonistic group A, and pattern II, comprising all strains belonging to the group B. The same analysis with *Hae*III revealed a single pattern with all *P. syringae* strains (Fig. 2b). ARDRA analysis did not permit to distinguish *P. syringae* strains within the two groups A and B.

PCR amplification with ERIC primers yielded six to ten distinct PCR products, ranging in size from approximately 150 to over 3000 bp. A representative gel is illustrated in Fig. 3. Fingerprints showed high degree of genetic diversity among the strains and UPGMA analysis revealed seven main clusters (Fig. 4). The clusters included strains obtained from different host plants species and belonging to either the antagonistic groups A or B, with the exception of two clusters (2 and 7) which included strains all belonging to antagonistic group A, and two clusters (3 and 7) which included the majority of strains isolated from *Citrus* spp.

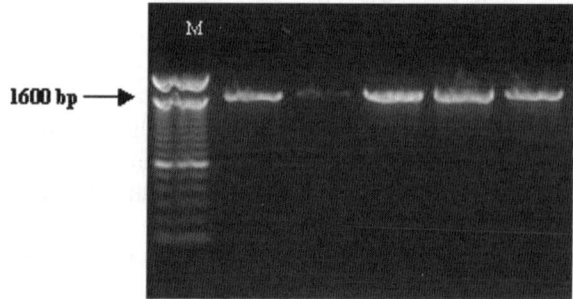

Fig. 1 ARDRA-16S PCR with P0 and P6 primer. All Pseudomonas syringae strains showed a fragment of about 1600 bp. Ladder used (M) was 100 bp (Invitrogen-Life Technologies, Paisley, UK)

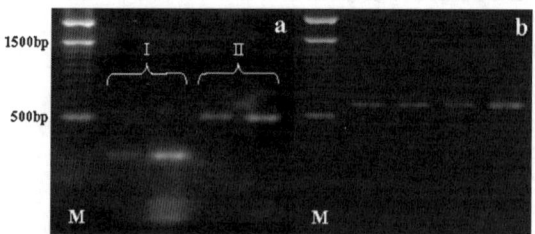

Fig. 2 ARDRA analysis with restriction enzymes *Alu*I and *Hae*III. Restriction performed with *Alu*I (a) showed two different patterns (I, II) corresponding to the groups (A, B) previously identified. Restriction permormed with *Hae*III (b) showed single pattern group. Ladder used (M) was 100 bp (Invitrogen-Life Technologies, Paisley, UK)

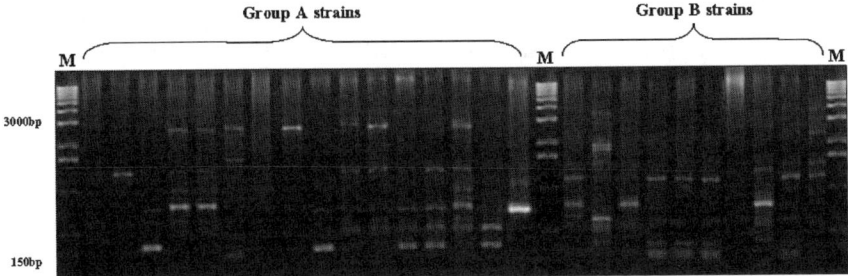

Fig. 3 Representative ERIC-PCR patterns of *Pseudomonas syringae* strains used in this study. Amplification with ERIC primers (ERIC 1R and ERIC 2) yielded six to ten distinct PCR products, ranging in size from approximately 150 bp to over 3000 bp and allowed differentiation of strains. Ladder used (M) was 1 Kb Invitrogen-Life Technologies, Paisley, UK (From Cirvilleri et al., 2005 partially modified)

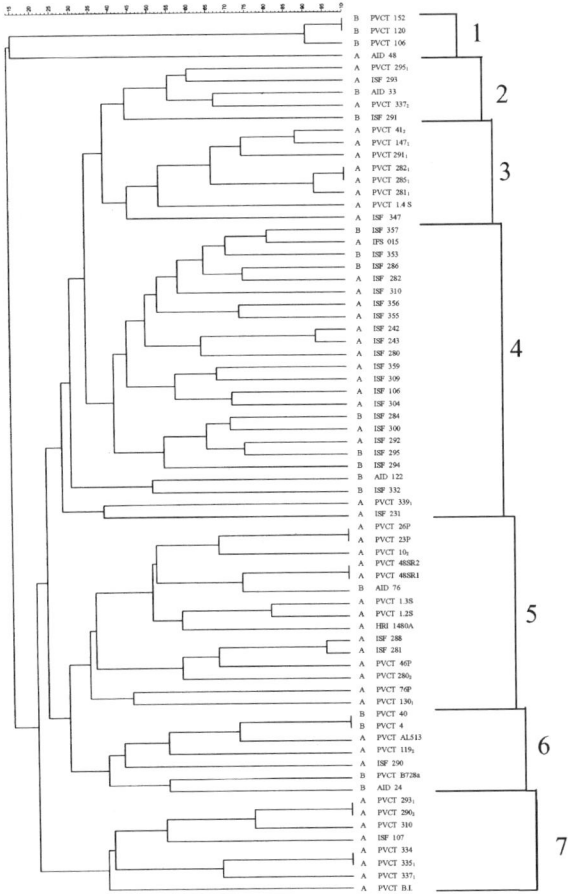

Fig. 4 Dendrogram of relationships among *Pseudomonas syringae* strains obtained using repetitive-sequence PCR and ERIC primer sets. PCR products in the range of 150–3000 bp were compared by numerical analysis using GelCompar II software. Similarity between fingerprints was calculated with the Dice's coefficients using the unweighted pair-group method with average linkages (UPGMA) (From Cirvilleri et al., 2005)

fAFLP analysis generated 29–43 fragments sized within 1 bp upon amplification with *MseI/EcoRI* primer set. fAFLP analysis allowed to distinguish between groups of strains isolated from different hosts and between strains isolated from the same host. Eight distinct fAFLP clusters were identified on the basis of host origin (Fig. 5).

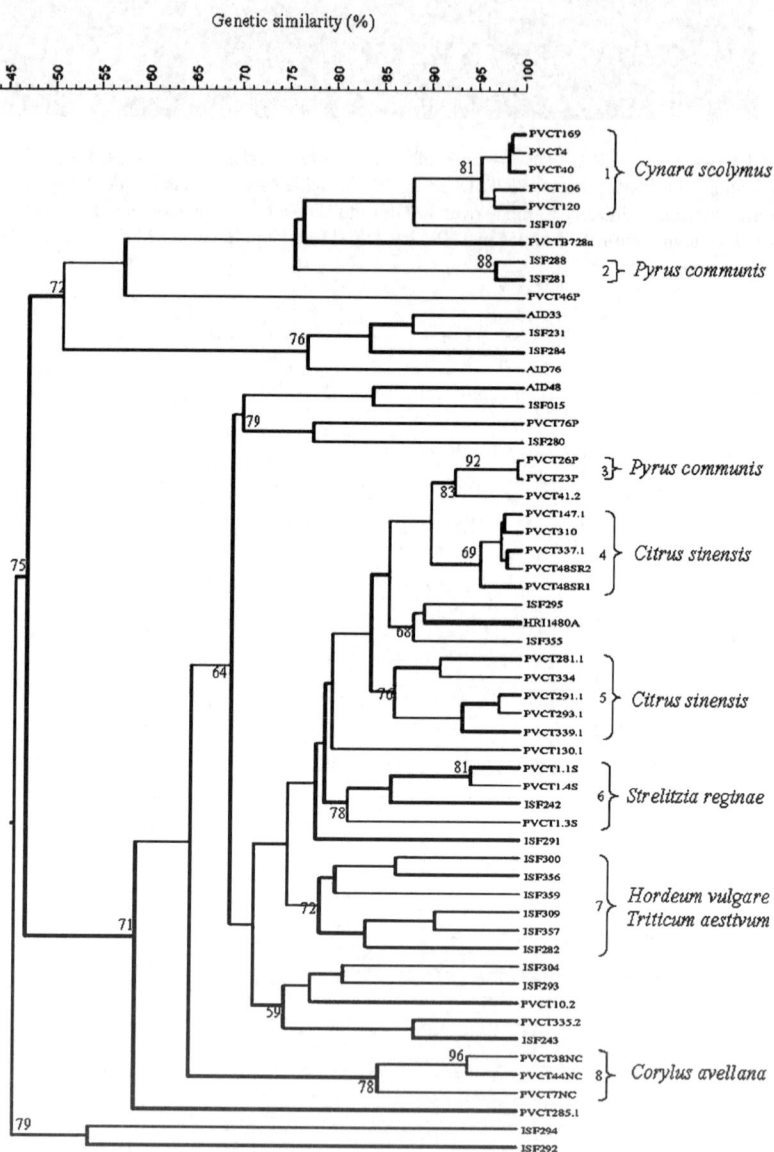

Fig. 5 Dendrogram of *Pseudomonas syringae* strains analysed by fluorecent AFLP. Profiles in the range of 100–700bp were compared by numerical analysis using Phylip 3.6 software. Similarity between fingerprints was calculated with unweighted pair-group method with average linkages (UPGMA) and the robustness of the tree was assessed by bootstrap analysis (1000 repeated samplings)

Table 6 fAFLPs characteristic fragments in selected *Pseudomonas syringae* strains

		Fragments (bp)[a]							
Host	Strains	113	217	242	246	317	415	555	583
Cynara scolymus	PVCT4, PVCT40, PVCT106, PVCT120, PVCT169	–	–	+	–	–	+	+	–
Corylus avellana	PVCT7NC, PVCT38NC, PVCT44NC	–	–	+	–	–	–	–	+
Strelitzia reginae	PVCT1.1S, PVCT1.3S, PVCT1.4S	–	–	+	–	+	–	–	–
Citrus sinensis	PVCT10.2, PVCT 337_1, PVCT48SR1, PVCT48SR2	–	+	+	–	–	–	–	–
Hordeum vulgare	ISF356, ISF357, ISF359	+	–	+	–	–	–	–	–
Fragaria x ananassa	AID 33, AID 76	–	–	+	+	–	–	–	–

[a] The presence or absence of differential fragments is indicated: +, a fragment characteristically present in fAFLP profiles of that strain; –, characteristic absence of fAFLP fragment from that profiles

The eight clusters included strains belonging to either the antagonistic groups A or B. Bootstrap values depicted the robustness of the dendrogram and almost all nodes showed moderate consistency in clustering.

Analysis of electropherograms showed that fifteen fragments were shared by most of strains and a 242 bp fragment was common to 90% of strains (Table 6). Strains isolated from *Cynara scolymus* showed two specific fragments of 415 and 555 bp, strains isolated from *Corylus avellana* showed a specific fragment of 583 bp, strains isolated from *Strelitzia reginae* showed a specific fragment of 317 bp and some strains isolated from *Citrus* spp. (PVCT 10.2, PVCT 337_1, PVCT $48SR_1$ and PVCT $48SR_2$) showed a specific fragment of 217 bp. In addition, three strains isolated from *Hordeum vulgare* (ISF356, ISF357, ISF359) and two from *Fragaria x ananassa* (AID33, AID76) showed specific fragments of 113 and 246 bp, respectively.

4 Discussion

Pseudomonas syringae strains and the relative culture filtrates evaluated in this study showed an antagonistic activity *in vitro* towards a variety of microorganisms. Strains showing a broader *in vitro* activity and those with reduced antagonistic activity were placed in the group A and group B, respectively.

Screening tests useful for minimizing the risk of introducing noxious organisms into the environment should be an important part of any effort to develop biological control antagonists and this is particularly true with pseudomonads which contain plant pathogens. Like other plant pathogenic pseudomonads, *P. syringae* strains isolated from lesions are virulent on either on the original host or occasionally on other non-host plants (Schaad et al., 2001).

In the present study, some *P. syringae* strains isolated from *Citrus* spp., *Pyrus communis*, *Strelitzia reginae* and belonging to antagonist group A produced necrotic lesions of different size on lemon, orange and mandarin fruits but not on apple and grape ones. On the contrary, strains isolated from *Prunus amygdali*, *P. armeniaca*, *Corylus avellana*, *Fragaria* x *ananassa*, *Phaseolus vulgaris* and *Cynara scolimus* (antagonistic groups A and B) did not produce any lesions on all fruits tested.

The consequences of necrotic lesions caused by antagonistic *P. syringae* strains evaluated in this study on the quality of the fruit might be of minor importance under commercial conditions. Moreover, the relative large and fresh wounds employed in laboratory may not simulate many of those inflicted during typical harvest and subsequent handling.

As observed in previous studies (Smilanick et al., 1996), the ability of antagonistic *P. syringae* strains ESC-10 to control postharvest green mold was not associated with virulence to citrus fruits when 10^6–10^7 CFU ml^{-1} was used whereas 10^8 CFU ml^{-1} caused small necrotic lesions on lime fruits. In the same studies, *P. syringae* strain 485–10 able to control green mould better that ESC-10 (Smilanick et al., 1996), caused cosmetically unacceptable black pit symptoms on fruit and blast symptoms on shoots and foliage. Light brown and dark brown lesions have been also reported as reactions associated, respectively, with virulent and non virulent *P. syringae* strains (Burkowicz and Rudolph, 1994), and dark brown reactions were supposed to represent a more vigorous plant defence reaction involving oxidation of phenolic compounds.

In this study we showed the ability of some *P. syringae* strains to completely or strongly control green and blue mould and this was not associated with significant alteration of citrus and apple fruits. One strain of group A (PVCT 119_2) and two strains of group B (ISF284 and ISF294), that did not induce symptoms on artificially inoculated citrus and apple fruits, totally inhibited *P. digitatum* and *P. expansum* on lemon and apple fruits even if high inoculum doses were used (10^9 CFU ml^{-1}) in both tests. Moreover, several strains belonging to group A controlled green and blue mould significantly better than PVCT 119_2, but produced large lesions (5–12 mm) on citrus fruits, thus precluding their potential use as biological control agents.

Strains of *P. syringae* groups A and B were effective to a different extent in reducing the incidence and severity of disease on lemon, orange, mandarin and apple fruits artificially inoculated with *P. digitatum*, *P. expansum* and *B. cinerea*. However, the antagonistic action of the A and B strains appeared different. Thus, strains of group A showed antagonistic activity against the above pathogens either *in vitro* and *in vivo*, whereas strains of group B were active only in *in vivo* assays. Two strains of group B showed interesting prerequisite for use as biocontrol agents towards *P. digitatum* and *P. expansum* on orange and apple fruits, respectively.

The mechanism of biocontrol ability of *P. syringae* strains ESC-10 and ESC-11 has been proposed to be due to competition for sites and nutrients in wounds since their application to wounds before or shortly after the introduction of the pathogen (Smilanick and Denis-Arrue, 1992; Smilanick et al., 1999), and large and metabolically active populations are required for optimum efficacy (Bull et al., 1997; Janisiewicz and Marchi, 1992). Bull et al. (1998) also found that ESC-10 and ESC-11 produced syringomycin E in culture at quantities sufficient to inhibit growth of *P. digitatum in vitro*, and that purified syringomycin E controlled green mold of lemons to levels similar to application of the producing strains but since syringomycin E or other antibiotics have not been detected in fruits treated with ESC-10 and ESC-11 it may not be important in the biocontrol activity. Thus the role of syringomycin E as well as of other LDPs produced *in vitro* by the majority of *P. syringae* strains is still uncertain and remain to be clarified. *P. syringae* strains are also able to produce enzymes such as chitinases and glucanases which degrade fungal cell walls, and recently Fogliano et al. (2002) demonstrated that LDPs and cell-wall degrading enzymes can act synergistically in the biocontrol of different pathogens. Two *P. syringae* strains, HRI 1480A and PVCT 46P, isolated respectively from pea and pear and with the behaviour of the strains belonging to group A, were able to control *P. expansum* on apple fruits also at 10^7 and 10^6 CFU ml^{-1} (Vinale et al., 2005) and showed considerable level of glucanase and chitinase activity They were not able to produce LDPs antibiotics and this suggest that ability to produce LDPs *in vitro* is not related to biocontrol activity of these strains.

The genetic characterization of *P. syringae* by ERIC-PCR and fAFLP pointed out an high variability among strains from the same host plant, as well as among strains within the same antagonistic group or with similar pathogenic feature. This variability was in agreement with other investigations performed on the same strains using BOX analysis (Scortichini et al., 2003), or on other strains of *P. syringae* using ERIC (Little et al., 1998) or AFLP analysis (Clerc et al., 1998; Manceau and Brin, 2003). The latter analyses revealed that there was no strict correlation between the ERIC-PCR and fAFLP profiles and the host of origin. Clusters of *P. syringae* strains defined with ERIC-PCR, with antagonistic groups and pathogenicity, were partially revealed with fAFLP clustering.

fAFLP fingerprints seems useful for tracking the biocontrol agents in the environment. Such a capability would be very important to verify the presence of the biocontrol agent in the long-term period under commercial conditions. Moreover, strain-specific discriminative amplicons, typically detected in strains isolated from the same host, may be useful for rapid identification of biocontrol agents strains and for epidemiological studies.

Although our tests showed that some strains are effective biological control agents under laboratory conditions, full-scale commercial evaluation is needed to demonstrate the possible value of these agents to the citrus and apple industry. Further testing is needed under commercial conditions, including assessment of biological control efficacy, population dynamics of these antagonists on fruit and observation of the quality of treated fruit.

Acknowledgements This work was supported by PRIN 2002 prot. 2002075322 "*Pseudomonas syringae* in lotta biologica: meccanismi d'azione e valutazione del rischio". We thank "Parco Scientifico e Tecnologico della Sicilia" for providing us with CEQ 8000 Beckman & Coulter automated DNA sequencer.

References

Bender C.L., Alarcon-Chaidez F., and Gross D.C., 1999. *Pseudomonas syringae* phytotoxins: mode of action, regulation and biosynthesis by peptide and polyketide synthetases. *Microbiol. Mol. Biol. Rev.* **63**:266–292.

Bull C.T., Stack J.P., and Smilanick J.L., 1997. *Pseudomonas syringae* strains ESC-10 and ESC-11 survive in wounds on citrus and control green and blue molds of citrus. *Biol. Control* **8**:81–88.

Bull C.T., Wadsworth M.L., Sorensen K.N., Takemoto J.Y., Austin R.K., and Smilanick J.L., 1998. Syringomyvcin E produced by biological control agents controls green mold on lemons. *Biol. Control* **12**:89–95.

Burkowicz A. and Rudolph K., 1994. Evaluation of pathogenicity and of cultural and biochemical tests for identification of *Pseudomonas syringae* pathovars *syringae, morsprunorum* and *persicae* from fruit trees. *J. Phytopathol.* **141**:59–76.

Campisano A., Di Silvestro S., Coco V., and Cirvilleri G., 2001. Influence of culture media on enzymatic activity of interesting yeasts in post-harvest pest management. *J. Plant Pathol.* **83**:232.

Cirvilleri G., Bella P., and Catara V., 2000. Biocontrol activity in plant pathogenic *Pseudomonas* species. Proceedings of the 5th Congress of the European Foundation of Plant pathology, Taormina, pp. 534–538.

Cirvilleri G., Bonaccorsi A., Scuderi G., and Scortichini M., 2005. Potential biological control activity and genetic diversity of *Pseudomonas syringae* pv. *syringae* strains. *J. Phytopathol.* **153**:654–666.

Cirvilleri G., Scuderi G., Bonaccorsi A., and Scortichini M., 2006. Molecular characterization of *Pseudomonas syringae* pv. *syringae* strains from different host plants using fluorescent amplified fragment length polymorphism. *J. Plant Pathol.* **88**:327–331.

Clerc A., Manceau C., and Nesme X., 1998. Comparison of randomly amplified polymorphic DNA with amplified fragment length polymorphism to assess genetic diversity and genetic relatedness within genospecies III of *Pseudomonas syringae. Appl. Environ. Microbiol.* **64**:1180–1187.

Conway W.S., Janisiewicz W.J., Klein J.D., and Sams C.E., 1999. Strategy for combining heat treatment, calcium infiltration, and biological control to reduce postharvest decay of Gala apples. *HortScience* **34**:700–704.

Dice L.R., 1945. Measures of the amount of ecologic association between species. *Ecology* **26**:297–302.

Eckert J.W. and Ogawa J.M., 1988. The chemical control of postharvest diseases: deciduous fruits, berries, vegetables and root/tuber crops. *Annu. Rev. Phytopathol.* **26**:433–469.

Felsenstein J., 2004. PHYLIP (Phylogeny Inference Package) Version 3.6 Distributed by the Author. Department of Genome Sciences, University of Washington, Seattle, WA.

Fogliano V., Ballio A., Gallo M., Woo S., Scala F., and Lorito M., 2002. *Pseudomonas* Lipodepsipeptides and fungal cell wall-degrading enzymes act synergistically in biological control. *Mol. Plant Microbe Interact.* **15**:323–333.

Grothues D. and Rudolph K., 1991 Macrorestriction analysis of plant pathogenic *Pseudomonas* species and pathovars. *FEMS Microbiol.* **79**:83–88.

Janisiewicz W. and Bors B., 1995. Development of a microbial community of bacterial and yeast antagonists to control wound-invading postharvest pathogens of fruits. *Appl. Envir. Microbiol.* **61**:3261–3267.

Janisiewicz W., Conway W.S., and Leverentz B., 1999. Biological control of postharvest decays of apples can prevent growth of *Escherichia coli* 0157:H7 in apple wounds. *J. Food Prot.* **62**:1372–1375.

Janisiewicz W. and Korsten L., 2002. Biological control of postharvest diseases of fruits. *Annu. Rev. Plant Pathol.* **40**:411–441.

Janisiewicz W.J., 1988. Biological control of diseases of fruits. In: Mukerji K.G. and Garg K.L. (eds), Biocontrol of Plant Diseases, Vol. 2, CRC, Boca Raton, FL, pp. 153–165.

Janisiewicz W.J. and Jeffers S.N., 1997. Efficacy of commercial formulation of two biofungicides for control of blue mold and gray mold of apples in cold storage. *Crop Prot.* **16**:629–633.

Janisiewicz W.J. and Marchi A., 1992. Control of storage rots on various pear cultivars with a saprophytic strain of *Pseudomonas syringae*. *Plant Dis.* **76**:555–560.

Janisiewicz W.J. and Roitman J., 1988. Biological control of blue-mold and gray-mold on apple and pear with *Pseudomonas cepacia*. *Phytopathology*. **78**:1697–1700.

Lavermicocca P., Iacobellis N.S., Simmaco M., and Graniti A., 1997. Biological properties and spectrum of activity of *Pseudomonas syringae* pv. *syringae* toxins. *Physiol. Mol. Plant Pathol.* **50**:29–140.

Little E.L., Bostock R.M., and Kirkpatrick B.C., 1998. Genetic characterization of *Pseudomonas syringae* pv. *syringae* strains from stone fruits in California. *Appl. Environ. Microbiol.* **64**:3818–3823.

Madesen A.M. and de Neergard E., 1999. Interaction between the mycoparrasite *Pythium oligandrum* and sclerotia of the plant pathogen *Sclerotinai sclerotiorum*. *Eur. J. Plant Pathol.* **105**:761–768.

Manceau C. and Brin C., 2003. Pathovars of *Pseudomonas syringae* are structured in genetic populations allowing the selection of specific markers for their detection in plant samples. In: Iacobellis N.S., Collmer A., Hutcheson S.W. et al. (eds), *Pseudomonas syringae* and Related Pathogens, Kluwer, Dordrecht, The Netherlands, pp. 503–512.

Meziane H., Gavriel S., Ismailov Z., Chet I., Chenin L., and Hofte M., 2006. Control of green and blue mould on orange fruit by *Serratia plymutyca* strains IC14 and IC1270 and putative modes of action. *Postharvest Biol. Technol.* **8**:191–198.

Nielsen P. and Sorensen J., 1997. Multi-target and medium-independent fungal antagonism by hydrolytic enzymes in *Paenibacillus polymyxa* and *Bacillus pumilus* strains from barley rhizosphere. *FEMS Microbiol. Ecol.* **22**:183–192.

Northover J. and Zhou T. 2002. Control of rhizopus rot of peaches with postharvest treatments of tebuconazole, fludioxonil, and *Pseudomonas syringae*. *Can. J. Plant Pathol.* **24**:144–153.

Nunes C., Usall J., Teixido N., and Viñas I., 2001. Biological control of postharvest pear diseases using a bacterium, *Pantoea agglomerans* CPA-2. *Int. J. Food Microbiol.* **70**:53–61.

Nunes C., Usall J., Teixido N., Fons E., and Viñas I., 2002. Post-harvest biological control by *Pantoea agglomerans* (CPA-2) on Golden Delicious apples. *J. Appl. Microbiol.* **92**:247–255.

Obagwu J. and Korsten L., 2003. Integrated control of citrus green and blue molds using *Bacillus subtilis* in combination with sodium bicarbonate or hot water. *Postharvest Biol. Technol.* **28**:187–194.

Schaad N.W., Jones J.B., and Chun W., 2001. *Pseudomonas*. In: Laboratory Guide for Identification of Plant Pathogenic Bacteria, APS, St. Paul, MN, pp. 84–120.

Scholz B.K., Jakobek J.L., and Lindgren P.B., 1994. Restriction fragment length polymorphism evidence for genetic homology within a pathovar of *Pseudomonas syringae*. *Appl. Environ. Microbiol.* **60**:1093–1100.

Scortichini M., Marchesi U., Rossi M.P., and Di Prospero P., 2001. Bacteria associated with hazelnut (*Corylus avellana* L.) decline are of two groups: *Pseudomonas avellanae* and strains resembling *P. syringae* pv. *syringae*. *Appl. Environ. Microbiol.* **68**:476–484.

Scortichini M., Marchesi U., Dettori M.T., and Rossi M.P., 2003. Genetic diversity, presence of the syrB gene, host preference and virulence of *Pseudomonas syringae* pv. *syringae* strains from woody and herbaceous host plants. *Plant Pathol.* **52**:277–286.

Smilanick J.L. and Denis-Arrue R., 1992. Control of green mold of lemon with *Pseudomonas* species. *Plant Dis.* **76**:481–485.

Smilanick J.L., Gouin-Behe C.C., Margosan D.A., Bull C.T., and Mackey, B.E., 1996. Virulence on citrus of *Pseudomonas syringae* strains that control postharvest green mold of citrus fruit. *Plant Dis.* **80**:1123–1128.

Smilanick, J.L., Margosan D.A., Milkota F., Usall J., and Michael I., 1999. Control of citrus green mold by carbonate and biocarbonate salts and influence of commercial postharvest practices on their efficacy. *Plant Dis.* **83**:139–145.

Snowdon A.L., 1990. A colour atlas of post-harvest diseases and disorders of fruit and vegetables. In: Snowdon A.L. (ed), General Introduction and Fruits, Vol. 1, Wolfe Publishing, London..

Snowdon A.L., 1991. A colour atlas of post-harvest diseases and disorders of fruit and vegetables. In: Snowdon A.L. (ed), vegetables Vol. 2, Wolfe Publishing, London.

Sorensen K.N., Kim K.H., and Takemoto J. Y., 1996. *In vitro* antifungal and fungicidal activities and erythrocyte toxicities of cyclic lipodepsipeptides produced by *Pseudomonas syringae* pv. *syringae*. *Antimicrob. Agents Chemother.* **40**:2710–2713.

Sugar D. and Spotts R.A., 1999. Control of postharvest decay in pear by four laboratoty-grown yeasts and two registered biocontrol products. *Plant Dis.* **83**:155–158.

Surico G., Lavermicocca P., and Iacobellis N.S., 1988. Produzione di siringomicina e siringotossina in colture di *Pseudomonas syringae* pv. *syringae*. *Phytopathol. Mediter.* **27**:163–168.

Vinale F., Fiore A., Foglaino V., Scala F., Gallo M., Gigante S., Cirvilleri G., Catara A., and Lorito M., 2005. Biocontrol Pseudomonas strains againts postharvest pathogens of apple. *J. Plant Pathol.* **87**:307.

Vos P., Hogers R., Bleeker M., Reijans M., Van De Lee T., Hornes M., Frijters A., Pot J., Pelman J., Kuiper M., and Zabeau M., 1995. AFLP: a new technique for DNA fingerprinting. *Nucleic Acids Res.* **23**:4407–4414.

Wilson C.L., Wisniewski M.E., Biles C.L., Mclaughlin R., Chalutz E., and Droby S., 1991. Biological control of post-harvest diseases of fruits and vegetables: alternatives to synthetic fungicides. *Crop Prot.* **10**:172–177.

Zhou T., Northover J., and Schneider K.E., 1999. Biological control of postharvest diseases of peach with phyllosphere isolates of *Pseudomonas syringae*. *Can J. Plant Pathol.* **21**:375–381.

Zhou T., Chu C.-L., Liu W.T., and Schaneider K.E., 2001. Postharvest control of blue mold and gray mold on apples using isolates of *Pseudomonas syringae*. *Can. J. Plant Pathol.* **23**:246–252.

Part III
Pathogenesis and Determinants
of Pathogenicity

The Distribution of Multiple Exopolysaccharides in *Pseudomonas syringae* Biofilms

H. Laue, A. Schenk, H. Li, and M. Ullrich

Abstract Exopolysaccharides play important roles in attachment of bacterial cells to a surface and/or in building and maintaining the three-dimensional, complex structure of bacterial biofilms. To elucidate the spatial distribution and the impact of the exopolysaccharides, levan and alginate, on biofilm formation, we compared biofilms of *Pseudomonas syringae* PG4180 strains with different exopolysaccharide patterns. The mucoid strain PG4180 (PG4180.muc), which produced levan and alginate, and its levan- and/or alginate-deficient derivatives all formed biofilms in the wells of microtitre plates and in flow-chambers. We applied confocal laser scanning microscopy (CLSM) with fluorescently labelled lectins to investigate the spatial distribution of levan and an additional yet-unknown exopolysaccharide in flow chamber biofilms. The lectin Concanavalin A (ConA) bound specifically to levan and accumulated in cell-depleted voids in the centers of microcolonies and in blebs. No binding of ConA was observed in biofilms of the levan-deficient mutants or in wild-type biofilms grown in the absence of sucrose. Production of a yet-to-be characterized additional exopolysaccharide might be more important for biofilm formation than the syntheses of levan and alginate.

Keywords Exopolysaccharides, Pseudomonas syringae, gene expression, biofilm

1 Introduction

Biofilms are defined as communities of microbial cells growing on a surface and embedded in a self-synthesized matrix composed of extracellular polymeric substances (EPS). The major components of EPS are exopolysaccharides (Sutherland, 2001), but DNA (Whitchurch et al., 2002), proteins and lipids (Wingender et al., 1999) can make up a significant proportion of EPS in biofilms. EPS is considered

School of Engineering and Sciences, Jacobs University Bremen, D-28759 Bremen, Germany

Author for correspondence: Matthias Ullrich; e-mail: m.ullrich@jacobs-university.de

M'B. Fatmi et al. (eds.), *Pseudomonas syringae Pathovars and Related Pathogens.*   147
© Springer Science + Business Media B.V. 2008

to be the key component that determines the physicochemical and biological properties of biofilms (Wolfaardt et al., 1999). Exopolysaccharides have been shown to be required for the initial attachment to a surface of *Vibrio cholerae* (Watnick and Kolter, 1999) and *Staphylococcus epidermidis* (McKenney et al., 1998) or for the structural development of mature biofilms like colanic acid in *Escherichia coli* K-12 biofilms (Danese et al., 2000). *Pseudomonas syringae* pv. glycinea is the causal agent of bacterial blight on soybean plants. The bacterium is known to produces two different exopolysaccharides, alginate and levan, which may function as virulence factors (Osman et al., 1986). Levan is a high-molecular-weight (2,6)-polyfructan with extensive branching through (2,1)-linkages. Its synthesis from sucrose is catalyzed by one single, extracellular enzyme, levan-sucrase (EC 2.4.1.10). Many Gram-negative (e.g. *Acetobacter diazotrophicus*, *P. syringae*, *Erwinia amylovora*) (Arrieta et al., 1996; Bogs and Geider, 2000; Hettwer et al., 1995) and Gram-positive bacteria (e.g. *Bacillus subtilis*, *Streptococcus mutans*) (Dedonder, 1966; Sato et al., 1984) produce levansucrases. Recently, three *lsc* genes encoding levansucrase (of which only two are expressed *in vitro*) were identified in *P. syringae* pv. glycinea PG4180 and mutants defective in levan formation were constructed (Li and Ullrich, 2001). Like production of the phytotoxin coronatine (Ullrich et al., 2000), levan synthesis is regulated by temperature and is maximal at 18°C, whereas only minor amounts of levan are formed at 28°C (Li and Ullrich, 2001). With the exception of dental biofilms, the role of levan in biofilm formation has not been studied so far. Fructans produced by *S. mutans* are believed to primarily function as extra-cellular storage compounds that are metabolized during periods of nutrient deprivation (Burne et al., 1996). *P. syringae* mainly produces levan when sucrose is in the medium. In contrast, alginate was shown to be the only exopol-ysaccharide produced by this bacterium when it infects plant leaves (Fett and Dunn, 1989; Osman et al., 1986). Alginate is a copolymer of 1,4-linked D-mannuronic and L-guluronic acid, which is normally *O*-acetylated at the second and/or third position(s) of the D-mannuronate residues (Osman et al., 1986). The arrange-ment of the alginate structural gene cluster of *P. syringae* is virtually identical to that previously described for *Pseudomonas aeruginosa* (Penaloza-Vazquez et al., 1997). Alginate production by *P. syringae* has been associated with increased epiphytic fitness, resistance to desiccation and toxic molecules (Yu et al., 1999), and the induction of water-soaked lesions on infected leaves (Fett and Dunn, 1989). The *algA* gene is one of the genes induced during *P. syringae* pv. tomato infection of *Arabidopsis thaliana* as shown by IVET technology (Boch et al., 2002). Likewise, the *algD* gene is expressed when *P. syringae* attempts to colonize both susceptible and resistant plant hosts (Keith et al., 2003). Due to their specific binding to carbohydrates, fluorescently labeled lectins combined with confocal laser scanning microscopy (CLSM) (Lawrence et al., 1998; Mathee et al., 1999; Neu and Lawrence, 1999; Strathmann et al., 2002) or with enzyme-linked lectinsorbent assays (ELLA) have been applied for analysis of exopolysaccharides in biofilms (Leriche et al., 2000; Strathmann et al., 2002; Thomas et al., 1997).

2 Results

2.1 Levan and Alginate Are Not Required for Formation of Static Biofilms

Exopolysaccharides had previously been shown to play a role in attachment and biofilm formation. Consequently, we compared exopolysaccharide-positive and levan- and/or alginate-deficient derivatives of *P. syringae* for their ability to form biofilms on polystyrene surfaces in wells of microtitre plates using the crystal violet staining method. PG4180.muc, PG4180, PG4180.M6, and PG4180.M6 (pBBR3-AXSalgT) all formed surface-associated biofilms under static conditions (Fig. 1). No substantial differences in the amount of biomass could be observed as indicated by crystal violet staining of any of the PG4180 biofilms. When sucrose was added to the flow chambers, PG4180 (levan+ alginate–), PG4180.M6 (levan+ alginate–), and PG4180.M6 (pBBR3-AXSalgT) (levan– alginate+) showed the same phenotype as observed above. However, a considerable decrease in biofilm formation was observed when PG4180.muc (levan+ alginate+) was incubated in the presence of sucrose. Our results suggest that neither levan nor alginate is required for biofilm formation of *P. syringae* in a static system. Interestingly, the presence of sucrose seemed even to inhibit biofilm formation when both, alginate and levan were produced.

2.2 Biofilm Development by the Levan and Alginate Producing Strain PG4180.Muc

To investigate the individual impact of the exopolysaccharides, levan and alginate, on *P. syringae* biofilm structure, PG4180 derivatives either deficient in levan or alginate synthesis or both were analyzed and compared with PG4180.muc (alginate

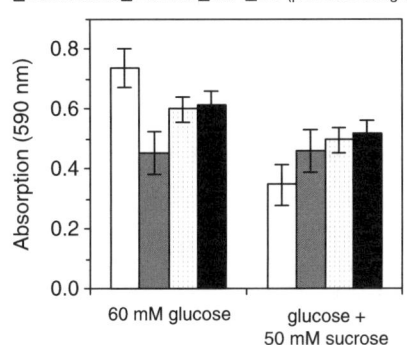

Fig. 1 Biofilm formation of *P. syringae* PG4180.muc, PG4180, PG4180.M6 and PG4180.M6 (pBBR-AXSalgT) in microtitre plates. The strains were grown as surface attached biofilms in polyvinylchloride microtitre plates using minimal medium supplemented with glucose (60 mM) or glucose (10 mM) plus sucrose (50 mM) for 2 days at 20°C. After removing the planktonic cells by washing, the biofilm was stained with 0.1% crystal violet. Bars represent the standard deviation between quadruplicates of a representative experiment

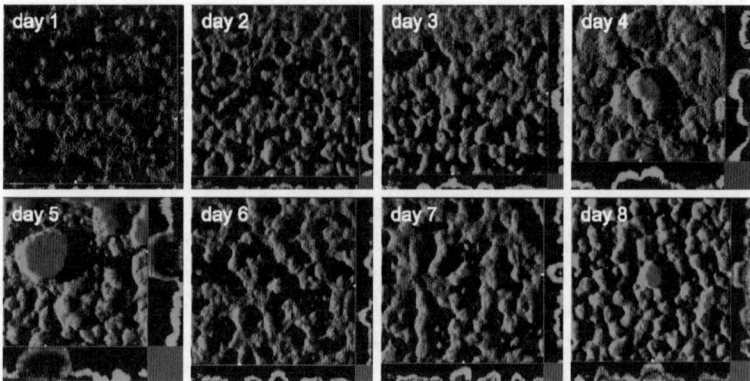

Fig. 2 Biofilm development of *P. syringae* PG4180.muc. A diluted overnight culture of the GFP tagged PG4180.muc-GFP was inoculated into a flow chamber irrigated with minimal medium containing 1 mM citrate and 1 mM sucrose. The development of the flow chamber biofilm was observed by CSLM for 8 days

+ levan+). Experimental conditions were optimized for maximum levan production by an incubation temperature of 20°C (Li and Ullrich, 2001) and by supplementing the cultures with sucrose. Biofilms were stained after 6 and 9 days with SYTO 62 prior to CSLM analysis (Fig. 2). All four derivatives formed heterogeneous, dynamic biofilms with similar structures, i.e. dense towers and ball-shaped micro-colonies. In general, PG4180.muc and PG4180 biofilms were thicker than those of the levan-deficient derivatives, PG4180.M6 and PG4180.M6 (pBBR3-AXSalgT), with about 100 μm vs. 63 μm maximum height. Dissolution of the biofilms by either cell motility or physical sloughing appeared to proceed regardless whether or not levan or alginate were formed. To check for the integrity of the exopolysaccharide phenotypes of the four *P. syringae* derivatives, biofilms were recovered from the flow chambers after 10 days and grown on KB or minimal agar plates containing sucrose. Levan synthesis was confirmed by formation of mucoid colonies on agar plates containing sucrose and alginate formation by the presence of mucoid colonies on KB agar. Under any condition, the bacteria isolated from the flow channels continued to exhibit the expected phenotypes.

2.3   Specific Binding of Fluorescently Labeled Lectins to P. syringae Flow Chamber Biofilms

To determine the specificity of diverse fluorescently labeled lectins for levan or alginate produced in *P. syringae* biofilms, we tested 64 commercially available lectins for binding to PG4180.muc biofilms. The lectins were applied to flow chamber biofilms, which had been fixed and embedded in polyacrylamide. This procedure allowed individual staining of several samples from the same biofilm using each individual lectin. A total of 11 out of 64 tested fluorescently labeled lectins bound to the PG4180.muc biofilm. Furthermore, the specificity of the 11 positive lectins was screened in flow chamber biofilms from all four PG4180 derivatives using CSLM analysis. Surprisingly, none of the 11 lectins was found to bind specifically to the levan-deficient and alginate producing transconjugant, PG4180.M6 (pBBR3-AXSalgT). In contrast, all but one of the lectins showed no significant differences in dependence of the genotype of the other three tested PG4180 derivates. Interestingly, FITC-labeled ConA did not bind to biofilms of the levan-deficient mutant PG4180.M6 suggesting that ConA might specifically interact with levan (Fig. 3). As reported above, ConA did not exhibit binding to biofilms from PG4180. M6 (pBBR-AXSalgT), which produces alginate but not levan suggesting that alginate is not bound by ConA. Binding of this lectin to biofilms of *P. syringae* was studied in more detail (see below).

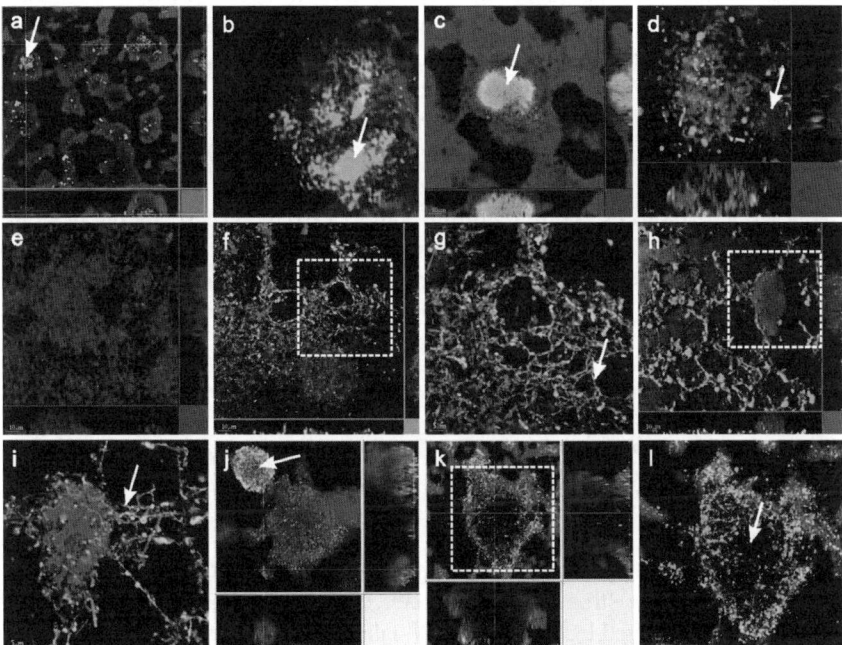

Fig. 3 Binding of lectins to *P. syringae* biofilms. CLSM optical sections of 9- to 10-day-old biofilms of *P. syringae* PG4180.muc, stained using the fluorescently labelled lectins (green) and SYTO 62 (red). Except for e (diluted LB without sucrose) and (**j–l**) (minimal medium with 1 mM citrate plus 1 mM glucose) all biofilms were grown in minimal medium supplemented with 1 mM citrate plus 1 mM sucrose. (**a–d**) Binding of the FITC-labelled ConA (green) to biofilms of PG4180.muc grown in the presence of sucrose (pictures were taken at different positions of the biofilm close to the inlet), arrows point out the presence of levan indicated by ConA-binding. ConA did not bind to *P. syringae* PG4180.muc biofilms grown without sucrose (**e**). Binding of the Alexa 488-labelled lectin from *Naja mossambica* to *P. syringae* PG4180.muc biofilms (**f–i**) and to PG4180 biofilms (**j–l**). Boxes in f, h and k indicate positions which were magnified in g, i and l, respectively. Arrows point out to the presence of a third exopolysaccharide indicated by binding of the Naja lectin

2.4   Spatial Distribution of Levan in P. syringae Biofilms

Biofilms of the levan-producing derivatives, PG4180.muc and PG4180, were grown in minimal or rich medium supplemented with sucrose as described above, incubated with FITC-labeled ConA, and analyzed by CLSM. The fluorescently labeled ConA specifically bound to the center of microcolonies in a condensed form. Interestingly, ConA mainly accumulated in the internal voids of mature microcolonies and in the blebs characteristic for levan production. However, neither all voids nor all microcolonies were stained evenly by ConA (Fig. 3). Moreover, ConA did not significantly bind to biofilm regions consisting of a network of flat microcolonies. Control experiments with SYTO 62 alone gave no signals under the CLSM settings used for FITC-linked ConA. This lectin also bound to microcolonies of non-fixed biofilms suggesting that the fixation step had no influence on the specificity of the binding. ConA did not bind to biofilms when PG4180.muc and PG4180 were grown in rich medium lacking sucrose. Additionally, ConA did not bind to biofilms of the levan-deficient mutant PG4180.M6 grown either in presence or absence of sucrose (data not shown). In summary, these findings suggested that ConA binds specifically to levan of P. syringae and that levan mainly accumulates in the internal voids of mature microcolonies of this bacterium under biofilm conditions. To investigate the dynamics of levan production in biofilms, we stained channels of PG4180.muc biofilms at different time points (3–10 days). No binding of ConA to early stage biofilms was observed. In contrast, the lectin bound to mature biofilms after about 7 days of growth. As describe above, the preferred sites of binding were the internal voids of microcolonies and differently sized blebs (Fig. 3a–d). Consequently, the blebs might either consist of or partially contain levan.

2.5   Binding of Naja Lectin Reveals Presence of a Potential
New Exopolysaccharide

Since ConA showed specificity to levan and neither alginate nor levan were significantly influencing the structure of the produced P. syringae biofilms, one of the other positively screened lectins, Alexa 488-labelled lectin from Naja mossambica, was subsequently tested in more detail. Binding of Naja lectin to mature P. syringae biofilms revealed a different binding pattern than ConA. Interestingly, this lectin bound in a web structure-like manner to the analyzed biofilms. In contrast to ConA, binding of Naja lectin was not detected in condensed form in the voids of large microcolonies or blebs, but was found specifically surrounding dense microcolonies of non-fixed biofilms. Additionally, fibers penetrating the dense microcolonies were stained by Naja lectin (Fig. 3, panels j–l). Naja lectin bound to biofilms of levan-deficient PG4180 derivatives and to those of levansucrase-positive PG4180 strains when grown without sucrose suggesting that this lectin was not specific to levan.

3   Discussion

Exopolysaccharides have been shown to play important roles in attachment and structural development of mature biofilms (Sutherland, 2001; Watnick and Kolter, 1999). We investigated the impact and distribution of multiple polysaccharides in *P. syringae* biofilms using PG4180 derivatives deficient in the synthesis of levan and/or alginate. Surprisingly, both investigated exopolysaccharides had no major impact on biofilm formation in either static microtitre plate assays or in continuous flow chamber experiments. Fluorescently labelled lectins were applied to native *P. syringae* flow chamber biofilms to study the spatial localization of exopolysaccharides. Herein, we demonstrate for the first time that the lectin, ConA, specifically binds to levan of *P. syringae*. The levan-specific binding of ConA to mature *P. syringae* biofilms suggested that levan might accumulate within cell-depleted voids in the centers of microcolonies and in blebs. However, at present it can not be ruled out that the intense ConA labeling of the cell-free voids is due to a better accessibility of this lectin within these spaces. Consequently, it remains hypothetical that ConA-binding to the levan-associated blebs is specific to levan. In support of this hypothesis, ConA did only bind to biofilms of cells which produced the enzyme, levansucrase, and were supplemented with sucrose. These results were furthermore confirmed by applying ELLA on *P. syringae* biofilms. Pretreatment of biofilms of levansucrase-expressing cells grown in presence of sucrose with levanase reduced the binding of peroxidase-linked ConA to background levels.

Li and Ullrich (2001) have shown that *P. syringae* PG4180 produces two levan-sucrases, which accumulate in different compartments, i.e. LscC in the periplasm and LscB in supernatants of liquid cultures. Since the tested levan-producing derivatives, PG4180 and PG4180.muc, of this study produce both, LscB and LscC, future experiments with mutants defective in either LscB or LscC might give valuable insights into potential differences between the structural nature of levans produced by either isoenzymes or their respective interaction with ConA. ConA did not bind to flow chamber or microtitre plate biofilms of the levan-deficient, alginate-producing transconjugant, PG4180.M6 (pBBR3-AXSalgT). This finding clearly ruled out that *P. syringae* alginate is bound by ConA. Thus, our results are contradictory to those for alginate-containing *P. aeruginosa* biofilms, in which ConA was demonstrated to bind to alginate (Strathmann et al., 2002). This discrepancy may be due to different chemical properties of the alginate(s) produced by either pseudomonad as reflected in different proportions of mannuronic and guluronic acid monomers and different degrees of *O*-acetylation.

Besides the structural differences between the exopolysaccharides involved in ConA binding for *P. syringae* and *P. aeruginosa* biofilms, respectively, the topological binding pattern was distinct for either organism as well. In mature *P. syringae* biofilms, ConA bound in condensed form preferably within voids of hollow microcolonies and to blebs whereas the same lectin yielded cloud-like cell-containing regions that were heterogeneously distributed throughout *P. aeruginosa* biofilms (Strathmann et al., 2002). When comparing biofilms of both species, it should be noted that the experimental set-up of the earlier authors (Wingender et al., 1999) and our study described herein, i.e. a static system on agar plates vs. a flow chamber system, differed significantly. When analyzing the growth stage dependency of levan production during development of *P. syringae* biofilms, significant ConA binding was only observed in mature 7- to 10-day-old biofilms. In contrast to our findings, time-resolved studies on exopolysaccharide production in *Sphingomonas* sp. flow chamber biofilms using ConA revealed a particular role of those exopoly-saccharides for the initial establishment of the biofilm (Kuehn et al., 2001). In the same line, we had used a ConA-specific ELLA with static microtitre plates and found an earlier appearance of levan build-up under those conditions. However, mature biofilms are formed much faster in static systems as opposed to the flow chamber set-up suggesting that the experimental design has rather a major influ-ence on the speed by which biofilms mature but not on levan production within this developmental process in *P. syringae*.

There are three potential scenarios for the late occurrence of ConA binding to levan in *P. syringae* biofilms: (i) levan is produced constitutively and accumulates but ConA binding signals are too weak in initial stages; (ii) levan is produced con-stitutively but it does not have a structural impact possibly because it is flushed out or accumulates in the observed blebs; (iii) levan is only produced in mature bio-films, when sufficiently high cell numbers are present which lead to inducing con-ditions for levan production (e.g. changes in pH, osmolarity, nutrient concentrations, autoinducer molecules). Since we observed the formation of large blebs in *P. syringae* PG4180 biofilms only in the presence of sucrose and because ConA bound to these blebs, the second explanation seems to be the most likely; yet a low staining sensi-tivity of ConA cannot be ruled out. Known ConA-specific sugar residues are glu-cose and mannose, from which mannose is not a structural component of levan. However, glucose is present in a single terminal copy for each polyfructosyl chain of levan (Rapoport et al., 1966). Consequently, it is tempting to speculate that such terminal glucose residues may be involved in ConA binding to levan, thus also explaining the relative low sensitivity of ConA staining.

A recent report showed that extra-cellular DNA made up the major part in the extra-cellular matrix and is required for the initial establishment of biofilms in *P. aeruginosa* PAO1 (Whitchurch et al., 2002). Currently, we can not rule out the possibility that *P. syringae* biofilms also partially consist of DNA. Thus, this question needs to be addressed in future experiments. Binding of a *Naja mossambica* lectin indicated the presence of an additional exopolysaccharide in biofilms of *P. syringae* regardless of the investigated genotypes. Thus, our model organism seems to produce at least one additional polysaccharide aside of levan and alginate. This yet-to-be identified polysaccharide might stabilize the biofilm structure by forming a dense web of fibers. Although this study revealed a unique spatial distribution of levan, which seems to accumulate in cell-depleted voids in centers of microcolonies and in blebs, our data give no conclusive information on any particular structure-determining role for alginate. Most importantly, results of this study suggested the presence of at least one additional exopolysaccharide forming a web-like structure. It has long been assumed that alginate is the primary secreted exopolysaccharide in

P. aeruginosa biofilms (Evans and Linker, 1973), but recent data indicated that alginate is not a significant component of the primary structural matrix of non-mucoid *P. aeruginosa* biofilms (Friedman and Kolter, 2004a; Hentzer et al., 2001; Jackson et al., 2004; Matsukawa and Greenberg, 2004; Nivens et al., 2001; Wozniak et al., 2003).

In conclusion it can be stated that the combination of genetic mutant analysis and flow chamber biofilm experiments give valuable insights into the role and particular function of individual exopolysaccharides in *P. syringae* biofilm development. Consequently, our future studies will focus on an in-depth analysis of environmental parameters and particular gene products for levan and alginate synthesis, respectively. Additionally, our future experiments will aim at the missing exopolysaccharide(s), which might be fundamental to the formation of *P. syringae* biofilms.

Acknowledgements We thank Soeren Molin and Thomas Neu for tremendous support and fruitful discussions. This study was supported by the Deutsche Forschungsgemeinschaft. Part of this work has been published in Microbiology (2006) Volume 152, pages 2909–2918 (Society of General Microbiology, United Kingdom).

References

Arrieta, J., Hernandez, L., Coego, A., Suarez, V., Balmori, E., Menendez, C., Petit-Glatron, M. F., Chambert, R. and Selman-Housein, G., 1996, Molecular characterization of the levansucrase gene from the endophytic sugarcane bacterium *Acetobacter diazotrophicus* SRT4. *Microbiology* 142, 1077–1085.

Boch, J., Joardar, V., Gao, L., Robertson, T. L., Lim, M. and Kunkel, B. N., 2002, Identification of *Pseudomonas syringae* pv. tomato genes induced during infection of *Arabidopsis thaliana*. *Mol Microbiol* 44, 73–88.

Bogs, J. and Geider, K., 2000, Molecular analysis of sucrose metabolism of *Erwinia amylovora* and influence on bacterial virulence. *J Bacteriol* 182, 5351–5358.

Burne, R. A., Chen, Y. Y., Wexler, D. L., Kuramitsu, H. and Bowen, W. H., 1996, Cariogenicity of Streptococcus mutans strains with defects in fructan metabolism assessed in a program-fed specific-pathogen-free rat model. *J Dent Res* 75, 1572–1577.

Danese, P. N., Pratt, L. A. and Kolter, R., 2000, Exopolysaccharide production is required for development of *Escherichia coli* K-12 biofilm architecture. *J Bacteriol* 182, 3593–3596.

Dedonder, R., 1966, Levansucrase from *Bacillus subtilis*. *Methods Enzymol* 8, 500–506.

Evans, L. R. and Linker, A., 1973, Production and characterization of the slime polysaccharide of Pseudomonas aeruginosa. *J Bacteriol* 116, 915–924.

Fett, W. F. and Dunn, M. F., 1989, Exopolysaccharides produced by phytopathogenic *Pseudomonas syringae* pathovars in infected leaves of susceptible hosts. *Plant Physiol* 89, 5–9.

Friedman, L. and Kolter, R., 2004a, Two genetic loci produce distinct carbohydrate-rich structural components of the *Pseudomonas aeruginosa* biofilm matrix. *J Bacteriol* 186, 4457–4465.

Hentzer, M., Teitzel, G. M., Balzer, G. J., Heydorn, A., Molin, S., Givskov, M. and Parsek, M. R., 2001, Alginate overproduction affects Pseudomonas aeruginosa biofilm structure and function. *J Bacteriol* 183, 5395–5401.

Hettwer, U., Gross, M. and Rudolph, K., 1995, Purification and characterization of an extracellular levansucrase from *Pseudomonas syringae* pv. phaseolicola. *J Bacteriol* 177, 2834–2839.

Jackson, K. D., Starkey, M., Kremer, S., Parsek, M. R. and Wozniak, D. J., 2004, Identification of psl, a locus encoding a potential exopolysaccharide that is essential for Pseudomonas aeruginosa PAO1 biofilm formation. *J Bacteriol* 186, 4466–4475.

Keith, R. C., Keith, L. M., Hernández-Guzmán, G., Uppalapati, S. R. and Bender, C. L., 2003, Alginate gene expression by *Pseudomonas syringae* pv. tomato DC3000 in host and non-host plants. *Microbiology* 149, 1127–1138.

Kuehn, M., Mehl, M., Hausner, M., Bungartz, H. J. and Wuertz, S., 2001, Time-resolved study of biofilm architecture and transport processes using experimental and simulation techniques: the role of EPS. *Water Sci Technol* 43, 143–150.

Lawrence, J. R., Neu, T. R. and Swerhone, G. D. W., 1998, Application of multiple parameter imaging for the quantification of algal, bacterial and exopolymer components of microbial biofilms. *J Microbiol Methods* 32, 253–261.

Leriche, V., Sibille, P. and Carpentier, B., 2000, Use of an enzyme-linked lectinsorbent assay to monitor the shift in polysaccharide composition in bacterial biofilms. *Appl Environ Microbiol* 66, 1851–1856.

Li, H. and Ullrich, M. S., 2001, Characterization and mutational analysis of three allelic lsc genes encoding levansucrase in *Pseudomonas syringae*. *J Bacteriol* 183, 3282–3292.

Mathee, K., Ciofu, O., Sternberg, C. et al., 1999, Mucoid conversion of *Pseudomonas aeruginosa* by hydrogen peroxide: a mechanism for virulence activation in the cystic fibrosis lung. *Microbiology* 145, 1349–1357.

Matsukawa, M. and Greenberg, E. P., 2004, Putative exopolysaccharide synthesis genes influence *Pseudomonas aeruginosa* biofilm development. *J Bacteriol* 186, 4449–4456.

McKenney, D., Hubner, J., Muller, E., Wang, Y., Goldmann, D. A. and Pier, G. B., 1998, The ica locus of Staphylococcus epidermidis encodes production of the capsular polysaccharide/adhesin. *Infect Immun* 66, 4711–4720.

Neu, T. R. and Lawrence, J. R., 1999, Lectin-binding analysis in biofilm systems. *Methods Enzymol* 310, 145–152.

Nivens, D. E., Ohman, D. E., Williams, J. and Franklin, M. J., 2001, Role of alginate and its O acetylation in formation of Pseudomonas aeruginosa microcolonies and biofilms. *J Bacteriol* 183, 1047–1057.

Osman, S. F., Fett, W. F. and Fishman, M. L., 1986, Exopolysaccharides of the phytopathogen *Pseudomonas syringae* pv. glycinea. *J Bacteriol* 166, 66–71.

Penaloza-Vazquez, A., Kidambi, S. P., Chakrabarty, A. M. and Bender, C. L., 1997, Characterization of the alginate biosynthetic gene cluster in *Pseudomonas syringae* pv. syringae. *J Bacteriol* 179, 4464–4472.

Rapoport, G., Dionne, R., Toulouse, E. and Dedonder, R., 1966, Initiation of levan chains in *Bacillus subtilis*. *Bull Soc Chim Biol (Paris)* 48, 1323–1348.

Sato, S., Koga, T. and Inoue, M., 1984, Isolation and some properties of extracellular D-glucosyltransferases and D-fructosyltransferases from *Streptococcus mutans* serotypes c, e, and f. *Carbohydr Res* 134, 293–304.

Strathmann, M., Wingender, J. and Flemming, H. C., 2002, Application of fluorescently labelled lectins for the visualization and biochemical characterization of polysaccharides in biofilms of Pseudomonas aeruginosa. *J Microbiol Methods* 50, 237–248.

Sutherland, I., 2001, Biofilm exopolysaccharides: a strong and sticky framework. *Microbiology* 147, 3–9.

Thomas, V. L., Sanford, B. A., Moreno, R. and Ramsay, M. A., 1997, Enzyme-linked lectinsorbent assay measures *N*-acetyl-D-glucosamine in matrix of biofilm produced by Staphylococcus epidermidis. *Curr Microbiol* 35, 249–254.

Ullrich, M. S., Schergaut, M., Boch, J. and Ullrich, B., 2000, Temperature-responsive genetic loci in the plant pathogen *Pseudomonas syringae* pv. glycinea. *Microbiology* 146, 2457–2468.

Watnick, P. I. and Kolter, R., 1999, Steps in the development of a Vibrio cholerae El Tor biofilm. *Mol Microbiol* 34, 586–595.

Whitchurch, C. B., Tolker-Nielsen, T., Ragas, P. C. and Mattick, J. S., 2002, Extracellular DNA required for bacterial biofilm formation. *Science* 295, 1487.

Wingender, J., Neu, T. R. and Flemming, H.-C., 1999, *Microbial extracellular polymeric substances*, p. 258. Berlin, Heidelberg, New York: Springer.

Wolfaardt, G. M., Lawrence, J. R. and Korber, D. R., 1999, Function of EPS. In J. Wingender, T. R. Neu and H.-C. Flemming (eds), *Microbial extracellular polymeric substances*, pp. 171–200. Berlin: Springer.

Wozniak, D. J., Wyckoff, T. J., Starkey, M., Keyser, R., Azadi, P., O'Toole, G. A. and Parsek, M. R., 2003, Alginate is not a significant component of the extracellular polysaccharide matrix of PA14 and PAO1 Pseudomonas aeruginosa biofilms. *Proc Natl Acad Sci USA* 100, 7907–7912.

Yu, J., Peñaloza-Vázquez, A., Chakrabarty, A. M. and Bender, C. L., 1999, Involvement of the exopolysaccharide alginate in the virulence and epiphytic fitness of *Pseudomonas syringae* pv. syringae. *Mol Microbiol* 33, 712–720.

Impact of Temperature on the Regulation of Coronatine Biosynthesis in *Pseudomonas syringae*

Y. Braun, A. Smirnova, and M. Ullrich

Abstract The plant pathogenic bacterium *Pseudomonas syringae* pv. glycinea PG4180 synthesizes high levels of the phytotoxin coronatine (COR) at the virulence-promoting temperature of 18°C, but only low amounts at 28°C, the optimal growth temperature. The temperature-dependent COR gene expression is regulated by a modified two-component system, consisting of the histidine protein kinase CorS, a response regulator, CorR, and a third functionally essential protein, CorP. We analyzed at the transcriptional and translational level the expression of *corS* and the *cma* operon involved in COR biosynthesis, after a temperature downshift from 28°C to 18°C. The synthesis of *cma* mRNA was induced within 20 min and then increased steadily and gradually in the 14 h following the shift to 18°C. The synthesis of *corS* mRNA was induced to a lesser extent by the temperature downshift. The induction of *cma* expression was a result of accelerated transcription rather than increased stability of the *cma* transcript at 18°C. Accumulation of the COR biosynthetic protein CmaB correlated with accumulation of *cma* mRNA. However, *cma* transcription was suppressed by inhibition of *de novo* protein biosynthesis.

Keywords Phytotoxin, two-component system, gene expression, temperature

1 Introduction

Bacteria use two-component regulatory systems (TCSs) to adapt cellular functions to changes of diverse environmental parameters. TCSs usually consist of a membrane-bound sensor histidine protein kinase (HPK) that perceives environmental stimuli and a response regulator (RR) that affects gene expression (Calva and Oropeza, 2006). Environmental signals in this context can be osmolarity, pH, light, temperature, CO_2, ammonia, oxygen, metal ions, nutrients, or any host-borne factors (Stock et al., 2000;

School of Engineering and Science, International University Bremen, Jacobs University Bremen as of spring 2007, Bremen, Germany

Author for correspondence: Yvonne Braun; e-mail: y.braun@jacobs-university.de

Beier and Gross, 2006). Although there are many well-characterized bacterial TCSs only very few temperature-sensing TCSs have been studied thus far. Temperature sensing plays a major role in many human pathogenic bacteria where the constant body temperature of the warm-blooded host activates expression of virulence factors and several modes of actions have been described (Hurme and Rhen, 1998). However, the situation is different in plant-pathogenic bacteria, in which low temperature signals often induce expression of virulence genes (Smirnova et al., 2001).

Phytopathogenic bacteria have evolved a number of mechanisms to colonize their host-plants and to evade detection through the plant defense system. Such mechanisms include, e.g. production of exopolysaccharides, effector proteins, cell-wall-degrading enzymes, and phytotoxins. Coronatine (COR) is a non-host specific phytotoxin produced by seven pathovars of *Pseudomonas syringae*: pv. *alisalensis*, *atropurpurea*, *glycinea*, *maculicola*, *morsprunorum*, *porri*, and *tomato* (Budde and Ullrich, 2000). Studies with COR-defective mutants have shown that COR synthesis contributes significantly to lesion expansion, the development of chlorosis, and bacterial multiplication in infected leaves (Bender et al., 1987; Tamura et al., 1998). According to structural analyses and plant response studies COR is believed to mimic plant signaling molecules such as methyl jasmonate, a plant growth regulator of the octadecanoid signaling pathway (Palmer and Bender, 1995). Moreover, COR induces chlorosis, hypertrophy of storage tissue, compression of thyllakoids, thickening of plant cell walls, and the synthesis of ethylene and proteinase inhibitors (Bender et al., 1999). Structurally, COR consists of a polyketide component, coronafacic acid (CFA), which is coupled via amide bond formation to coronamic acid (CMA) an ethylcyclopropyl amino acid derivative.

2   COR Gene Expression and Regulation

P. syringae pv. glycinea PG4180 synthesizes COR in a temperature-dependent manner, with a maximum at 18°C (Ullrich et al., 1995). At 28°C, the optimal growth temperature of *P. syringae*, COR biosynthesis is negligible (Fig. 1). The COR biosynthetic and regulatory genes where shown to be encoded on a 90-kb indigenous plasmid in PG4180 (Bender et al., 1993). Two biosynthetic regions, corresponding to CFA and CMA biosynthesis, are separated by a 3.4-kb regulatory region encoding a modified TCS. This TCS consisting of the RR, CorR, the sensor kinase CorS, and a third component, CorP, was shown to regulate COR biosynthesis (Ullrich et al., 1995). The HPK CorS is believed to respond to a temperature change via autophosphorylation of its conserved histidine residue, and transduces the signal to the cognate RR CorR via phosphorylation of its conserved aspartate residue (Rangaswamy and Bender, 2000). The temperature of 18°C was shown to be an important signal for PG4180 to enhance COR production *in vitro* as well as *in planta* (Palmer and Bender, 1993; Weingart et al., 2004).

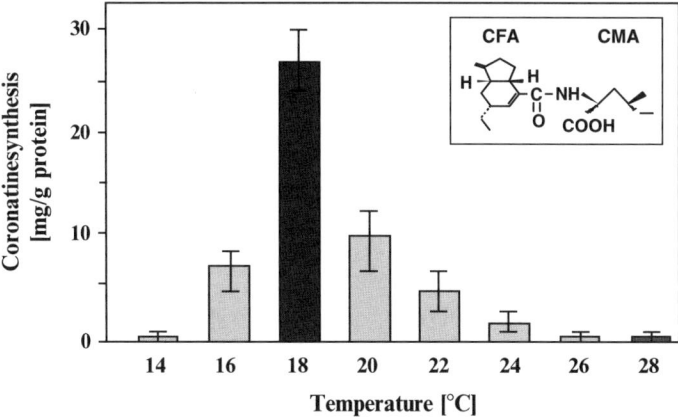

Fig. 1 Effect of temperature on coronatine production in *P. syringae* pv. glycinea PG4180 (Budde et al., 1998)

The production of CFA and CMA in PG4180 is regulated at the transcriptional level. Transcriptional fusions of the CFA and CMA promoter region to a promoterless glucuronidase gene showed maximal transcriptional activities at 18°C and significantly lower activities at 28°C (Bender et al., 1999). In order to investigate how soon the bacteria respond to a temperature change with COR biosynthesis, we examined levels of *cma* and *corS* mRNA after a temperature shift from 28°C to 18°C applying dot-blot analysis. Cultures kept constantly growing at 18°C and 28°C were used as a control to define the minimal and maximal level of expression at the temperatures of interest. The analysis confirmed the temperature-dependent expression of *cma* mRNA obtained earlier using reporter gene fusion. The expression of *corS* was significantly lower than that of *cma* and showed only moderate induction after the downshift in temperature. Further, we demonstrated that there is a significant induction of the *cma* transcription in PG4180 but not of *corS* after the downshift in temperature. The *cma* transcript accumulated steadily over a period of 14h after the shift (Fig. 2). Interestingly, maximum *cma* level only reached 60% of the transcript level of cells constantly incubated at 18°C.

To determine whether a change in mRNA stability contributes to the temperature-induced increase in mRNA levels, we determined half-life of *cma* mRNA at 18°C and 28°C after the addition of the transcriptional inhibitor rifampicin. The half-life of the *cma* transcript decreased from 9.9 min at 18°C to 7.3 min at 28°C. Thus, no significant stabilization at 18°C occurs which would contribute to the accumulation of the transcript at this temperature.

As recently suggested, at low temperature CorS might undergo a conformational change to an active form in order to be efficiently phosphorylated and subsequently initiate transcription of COR biosynthetic genes (Smirnova and Ullrich, 2004). Alternatively, CorS might be synthesized *de novo* and then incorporated into the

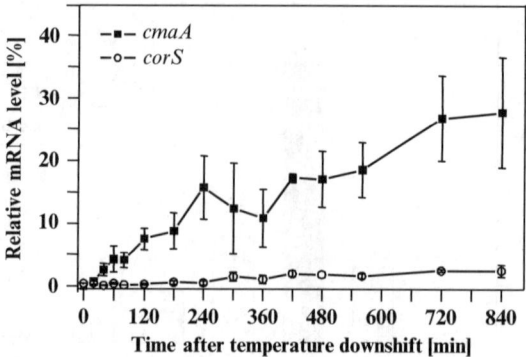

Fig. 2 Synthesis of *cma* and *corS* mRNAs after a temperature downshift from 28°C to 18°C, as examined by RNA dot-blot analysis

membrane in an active conformational state. To examine whether the induction of the *cma* transcription in cultures of PG4180 shifted from 28°C to 18°C required *de novo* protein synthesis, we analyzed levels of *cma* mRNA after addition of the protein biosynthesis inhibitor, chloramphenicol. In parallel, samples were also probed for *ssb* encoding a single-strand binding protein. The levels of *ssb* mRNA were not affected by addition of chloramphenicol (Results not shown). Thus, the inhibition of *de novo* protein synthesis suppressed transcription of *cma* but not of *ssb*. This would indicate that *de novo* synthesis of regulatory proteins is required for the efficient transcription of *cma*.

Nucleotide sequence similarities suggest that the COR biosynthetic protein CmaB is a nonheme dioxygenase that may carry out a hydroxylation or chlorination reaction of the CMA precursor (Couch et al., 2004). To monitor CmaB production after a temperature downshift, protein extracts were subjected to Western blot analyses using CmaB antiserum. After the temperature downshift, levels of CmaB increased gradually and reached 65% of the CmaB levels detected in cultures continuously grown at 18°C (Fig. 3). The accumulation of CmaB was consistent with the increase of *cma* mRNA levels. The level of CmaB in cultures grown at 28°C was significantly lower than in the 18°C culture. Because of the coordinated increase of *cma* mRNA and CmaB levels, we assume that the increased accumulation of CmaB was predominantly caused by accelerated transcription.

The plant pathogenic bacterium *P. syringae* PG4180 often encounters temperature fluctuations during its epiphytic growth on the plant surface. Temperature decreases are associated with rainy or humid weather conditions, which favor bacterial infection of the host plant and affect the subsequent disease development (Dunleavy, 1988). In PG4180, a low-temperature stimulus is known to be sensed and to cause coordinated COR gene expression via the CorRSP-regulatory system (Ullrich et al., 1995). However, important questions that remained to be addressed are how the temperature stimulus is transmitted into COR gene expression and how soon the COR gene expression is induced when temperature changes. That low

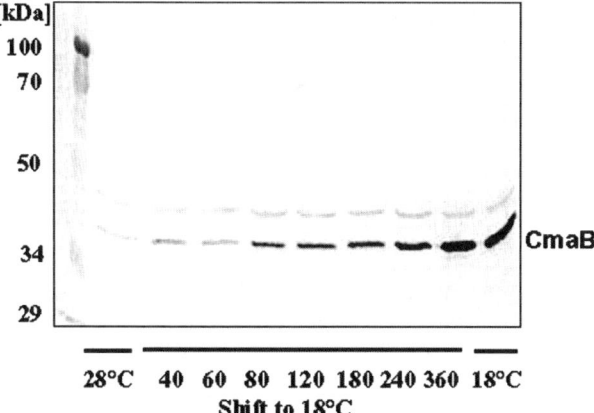

Fig. 3 Effect of temperature on the biosynthesis of CmaB. Western blot analysis of cells constantly incubated at 28°C, cells shifted from 28°C to 18°C (numbers indicate minutes after the temperature downshift), and cells constantly incubated at 18°C

temperatures can induce transcription and protein biosynthesis has been previously reported for cold-shock proteins (Graumann and Marahiel, 1996) and for desaturases in *Bacillus* and *Synechocystis* species (Suzuki et al., 2000; Aguilar et al., 2001). The synthesis of cold-shock proteins is transiently induced and then readjusted to new steady-state levels within 4 h of a temperature downshift (Thieringer et al., 1998). In contrast, in PG4180 the *cma* mRNA and the CmaB protein accumulated steadily and gradually after the shift to 18°C, suggesting that the effect of temperature downshift on *cma* mRNA and CmaB synthesis does not resemble a cold-shock response. It is likely that because temperature fluctuations occur constantly on the plant surface, PG4180 is adapted to alter its gene expression in response to an average daily temperature. A decrease in the average temperature might favor a steady COR gene expression.

3 Conclusions

Our results show that transcription of COR biosynthetic genes – studied for *cma* – is temperature-dependent, whereas the regulatory gene *corS* is expressed constitutively on a low level. No significant temperature-dependent stabilization of *cma* mRNA at 18°C was observed. The *cma* mRNA and the CmaB protein accumulated steadily over several hours after a temperature shift to 18°C, suggesting that the effect of temperature downshift does not resemble a cold-shock response. It rather indicates an adaptation to a long-term temperature decrease, which serves the bacteria as a signal for expression of virulence factors like the phytotoxin coronatine.

References

Aguilar, P.S., Hernandez-Arriaga, A.M., Cybulski, L.E., Erazo, A.C., and de Mendoza, D., 2001, Molecular basis of thermosensing: a two-component signal transduction thermometer in *Bacillus subtilis*. *Embo. J.* 20: 1681–1691.

Beier, D. and Gross, R., 2006, Regulation of bacterial virulence by two-component systems. *Curr. Opin. Microbiol.* 9: 143–152.

Bender, C.L., Stone, H.E., Sims, J.J., and Cooksey, D.A., 1987, Reduced pathogen fitness of Pseudomonas syringae pv. tomato Tn5 mutants defective in coronatine production. *Physiol. Mol. Plant Pathol.* 30: 273–283.

Bender, C.L., Liyanage, H., Palmer, D., Ullrich, M., Young, S., and Mitchell, R., 1993, Characterization of the genes controlling the biosynthesis of the polyketide phytotoxin coronatine including conjugation between coronafacic and coronamic acid. *Gene* 133: 31–38.

Bender, C.L., Alarcon-Chaidez, F., and Gross, D.C., 1999, *Pseudomonas syringae* phytotoxins: mode of action, regulation, and biosynthesis by peptide and polyketide synthetases. *Microbiol. Mol. Biol. Rev.* 6: 266–292.

Budde, I.P. and Ullrich, M.S., 2000, Interactions of *Pseudomonas syringae* pv. glycinea with host and nonhost plants in relation to temperature and phytotoxin synthesis. *Mol. Plant Microbe Interact.* 13: 951–961.

Budde, I.P., Rohde, B.H., Bender, C.L., and Ullrich, M.S., 1998, Growth phase and temperature influence promoter activity, transcript abundance, and protein stability during biosynthesis of the *Pseudomonas syringae* phytotoxin coronatine. *J. Bacteriol.* 180: 1360–1367.

Calva, E. and Oropeza, R., 2006, Two-component signal transduction systems, environmental signals, and virulence. *Microb. Eco.* 51: 166–176.

Couch, R., O'Connor, S.E., Seidle, H., Walsh, C.T., and Parry, R., 2004, Characterization of CmaA, an adenylation-thiolation didomain enzyme involved in the biosynthesis of coronatine. *J. Bacteriol.* 186: 35–42.

Dunleavy, J., 1988, Bacterial, fungal, and viral diseases affecting soybean leaves, p. 40–46. In T.D. Wyllie and D.H. Scott (eds), Soybean Diseases in the North Central Region. American Phytopathological Society, St. Paul, MN.

Graumann, P. and Marahiel, M.A., 1996, Some like it cold: response of microorganisms to cold shock. *Arch. Microbiol.* 166: 293–300.

Hurme, R. and Rhen, M., 1998, Temperature sensing in bacterial gene regulation – what it all boils down to. *Mol. Microbiol.* 30: 1–6.

Palmer, D.A. and Bender, C.L., 1993, Effects of environmental and nutritional factors on production of the polyketide phytotoxin coronatine by *Pseudomonas syringae* pv. glycinea. *Appl. Environ. Microbiol.* 59: 1619–1626.

Palmer, D.A. and Bender, C.L., 1995, Ultrastructure of tomato leaf tissue treated with the pseudomonad phytotoxin coronatine and comparison with methyl jasmonate. *Mol. Plant-Microbe Interact.* 8: 683–692.

Rangaswamy, V. and Bender, C.L., 2000, Phosphorylation of CorS and CorR, regulatory proteins that modulate production of the phytotoxin coronatine in *Pseudomonas syringae*. *FEMS Microbiol. Lett.* 193: 13–18.

Smirnova, A. and Ullrich, M., 2004, Topological and deletion analysis of CorS, a *Pseudomonas syringae* sensor kinase. *Microbiology* 150: 2715–2726.

Smirnova, A., Li, H.Q., Weingart, H., Aufhammer, S., Burse, A., Finis, K., Schenk, A., and Ullrich, M.S., 2001, Thermoregulated expression of virulence factors in plant-associated bacteria. *Arch. Microbiol.* 176: 393–399.

Stock, A.M., Robinson, V.L., and Goudreau, P.N., 2000, Two-component signal transduction. *Annu. Rev. Biochem.* 69: 183–215.

Suzuki, I., Los, D.A., Kanesaki, Y., Mikami, K., and Murata, N., 2000, The pathway for perception and transduction of low-temperature signals in *Synechocystis*. *Embo J.* 19: 1327–1334.

Tamura, K., Zhu, Y., Sato, M., Teraoka, T., Hosokawa, D., and Watanabe, M., 1998, Role of coro-
    natine production by *Pseudomonas syringae* pv. *maculicola* for pathogenicity. *Ann. Phytopathol.
    Soc. Japan* 64: 299–302.
Thieringer, H.A., Jones, P.G., and Inouye, M., 1998, Cold shock and adaptation. *Bioessays* 20:
    49–57.
Ullrich, M., Penaloza-Vazquez, A., Bailey, A.M., and Bender, C.L., 1995, A modified two-com-
    ponent regulatory system is involved in temperature-dependent biosynthesis of the Pseudomonas
    syringae phytotoxin coronatine. *J. Bacteriol.* 177: 6160–6169.
Weingart, H., Stubner, S., Schenk, A., and Ullrich, M.S., 2004, Impact of temperature on *in planta*
    synthesis of the *Pseudomonas syringae* phytotoxin coronatine. *Mol. Plant Microbe Interact.*
    17: 1095–1102.

Role of Flagellin Glycosylation in Bacterial Virulence

Y. Ichinose[1], F. Taguchi[1], K. Takeuchi[2], T. Suzuki[1], K. Toyoda[1], and T. Shiraishi[1]

Abstract *Pseudomonas syringae* pv. *tabaci* 6605 (Pta6605) is a causal agent for wildfire disease in tobacco. Recently, we found that flagellin, a major constituent in the flagella filament of this pathogen, is a potent elicitor of hypersensitive reaction in non-host plant species. We also found that flagellin is required for virulence against the host plant. In this study, we investigated the biochemical features of the glycosyl moiety of flagellin and the phytopathological role of flagellin glycosylation. DNA sequence analysis of the flagellum gene cluster revealed that two genes (*orf1* and *orf2*) encoding a putative glycosyltransferase were located upstream of the *fliC* gene. To investigate the role of flagellin glycosylation, we generated deletion mutants for *orf1* (Δ*orf1*) and *orf2* (Δ*orf2*) in Pta6605 and pv. *glycinea* race 4 (Pgl4). The mutants, Δ*orf1* and Δ*orf2*, of both pathovars produced nonglycosylated or partially glycosylated flagellins, respectively. Inoculation of host plants with these mutant strains confirmed that Δ*orf1* and Δ*orf2* had reduced ability to cause disease. Biochemical and genetic approaches revealed that a total of six serine residues of FliC were glycosylated in Pta6605. The serine residues were replaced individually with Ala by site-directed mutagenesis. All glycosylation-defective mutants including Δ*orf1* and Δ*orf2* and the six Ser/Ala-substituted mutants of pv. *tabaci* retained swimming ability but their swarming ability was reduced. The abilities to adhere to a polystyrene surface and to cause disease in host tobacco plants were also impaired in all Ser/Ala-substituted mutants of Pta6605. When tobacco leaves were inoculated with Pta6605 wild-type, bacteria were embedded and formed a biofilm-like structure in the matrix on the tobacco leaf surface. In contrast, mucoid material was rarely detected in the area surrounding the Δ*orf1* mutant. These results suggest that glycosylation of flagellin in Pta6605 is required for swarming motility, adhesion, biofilm

[1]The Graduate school of Natural Science and Technology, Okayama University, Tsushima 1-1-1, Okayama 700-8530, Japan

[2]National Institute of Agrobiological Sciences, Kannondai 2-1-2, Tsukuba 305-8602, Japan

Author for correspondence: Yuki Ichinose; e-mail: yuki@cc.okayama-u.ac.jp

formation, and bacterial virulence. Furthermore, the ability of nonglycosylated flagellin from Pta6605 to induce a defense response in tobacco cells was greater than that of glycosylated flagellin, suggesting that the glycan moiety of flagellin may mask the elicitor function of the flagellin molecule in its host plant.

Keywords Flagellin, glycosylation, masking

1 Introduction

Protein glycosylation was previously considered to be restricted to eukaryotes. However, in recent years there has been a significant increase in reports of glycosylation in various pathogenic Gram-negative bacteria, including *Aeromonas caviae, Campylobacter coli, C. jejuni, Escherichia coli, Helicobacter pylori, Neisseria meningitides, Pseudomonas aeruginosa*, and *P. syringae* (Logan, 2006; Szymanski and Wren, 2005). Most bacterial glycoproteins appear to be on the surface of the organism, like pili and flagella, indicating that glycans are directly exposed to host cells. Therefore, the glycosylation of pili and flagella may play an important role in interactions with host and non-host organisms, but little is known about the function of glycosylation in bacterial proteins.

Flagellin is a major constituent of the flagella filament. Flagellin is known to possess a microbe-associated molecular pattern (MAMP), which is a molecule or part of a molecule that is recognized by host organisms as a typical "non-self" molecule and is able to activate host immune response in mammals and plants (Felix et al., 1999; Hayashi et al., 2001; Taguchi et al., 2003b). The roles of flagella and flagellin in plant–bacteria interactions were investigated in *Pseudomonas syringae* pv. *tabaci* 6605 (Pta6605), a causal agent of tobacco wildfire disease (Ichinose et al., 2003a). Inoculation experiments with two flagella-defective mutants, ΔfliC and ΔfliD, suggested that the monomer flagellin of Pta6605 is an essential factor in the elicitation of hypersensitive response (HR) in non-host tomato cells, and that flagella are required for complete virulence in the host tobacco (Ichinose et al., 2003b; Shimizu et al., 2003). The MAMP in flagellin of phytopathogenic bacteria is a conserved amino acid region near the N-terminus called flg22 (Felix et al., 1999), and is recognized by plasma membrane receptor FLS2 in *Arabidopsis thaliana* (Gómez-Gómez and Boller, 2000), whereas flagellin in the animal pathogenic bacteria, *Listeria monocytogenes* and *Salmonella dublin* is perceived by its own receptor, TLR5 (Hayashi et al., 2001). The highly conserved N- and C-terminal regions of flagellin in *Salmonella* possess potent pro-inflammatory activities (Eaves-Pyles et al., 2001).

Interestingly, flagellins of *Pseudomonas* species, including the animal pathogen, *P. aeruginosa* (Brimer and Montie, 1998) and the plant pathogen *P. syringae* (Taguchi et al., 2003a), have been reported to be glycosylated.

P. syringae can be classified into more than 50 pathovars by their host plant species. We found that the amino acid sequences of flagellins are identical in Pta6605 and *P. syringae* pv. glycinea race 4 (Pgl4), a pathogen for soybean leaf spot (Taguchi et al., 2003b). However, the elicitor activities of these flagellins are different: flagellin from Pta6605 induces hypersensitive cell death (HR) in soybean but not in tobacco; on the contrary, flagellin from Pgl4 induces cell death in tobacco but not in soybean. Thus, the HR-associated elicitor activity of these flagellins is restricted to non-host plants. Furthermore, the ability to induce an HR is determined by the post-translational glycosylation of flagellin.

In this paper, we introduce the genes required for glycosylation of flagellin and identification of the glycosylated amino acids. Furthermore, we investigated the role of glycosylation in bacterial virulence using various glycosylation-defective mutants.

2   Materials and Methods

2.1   Plants and Bacteria

Tobacco (*Nicotiana tabacum* cv. Xanthi NC) and soybean (*Glycine max* cv. Shirofumi) plants were grown at 25°C. Leaves of 5-week-old tobacco or 3-week-old soybean plants were used for inoculation experiments.

Pseudomonas syringae pathovars (pv.) *tabaci* 6605 (Pta6605) and pv. *glycinea* race 4 (Pgl4) were grown in King's B medium at 27°C. For plant inoculation, bacteria were suspended in 10 mM $MgSO_4$ with 0.02% Silwet L77 at a density of 2×10^8 CFU/ml. Leaves were spray-inoculated with an airbrush.

2.2   Preparation of Flagellin Proteins

Bacteria were cultured overnight in minimal medium (50 mM potassium phosphate buffer, 7.6 mM $(NH_4)_2SO_4$, 1.7 mM $MgCl_2$ and 1.7 mM NaCl, pH 5.7) supplemented with 10 mM of mannitol and fructose (MMMF). After incubation, bacterial cells were harvested and resuspended in 50 mM sodium phosphate buffer (pH 7.0). Flagella were sheared off by vortexing, and flagellin proteins were further purified mainly by centrifugation. Flagellins were used for cell death assay at a final concentration of 0.32 µM (10 µg/ml).

2.3   Generation of Glycosyltransferase-Defective Mutants

We found two putative genes (*orf1* and *orf2*) for glycosyltransferase and a putative gene (*orf3*) for 3-oxoacyl-(acyl carrier protein) synthase III upstream of the flagellin gene, *fliC*, in Pta6605 and pg 14 (Fig. 1). To investigate the possible involvement of these genes in flagellin glycosylation, we generated non-polar

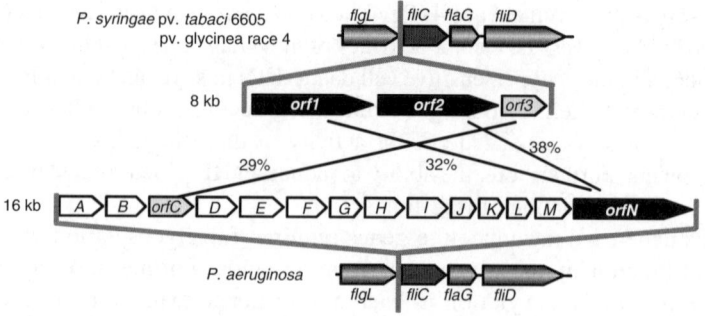

Fig. 1 Physical map of glycosylation island in flagella gene cluster

deletion mutants by PCR-mediated homologous recombination. DNA fragments for both upstream and downstream regions of each *orf* were amplified by PCR with two sets of primers. Each set of amplified DNA fragments was ligated at artificial *Bam*HI sites and inserted into the mobilizable cloning vector, pK18*mobsacB* (Schäfer et al., 1994). The substitution of glycosylated Ser by Ala was carried out using a QuickChange XL site-directed mutagenesis kit in pK18*mobsacB*. The resultant plasmids were transferred into Pta6605 via *E. coli* S17-1 by conjugation, and the mutants were obtained by homologous recombination (Taguchi et al., 2006a; Takeuchi et al., 2003). Thus we obtained Δ*orf1*, Δ*orf2* and Δ*orf3* mutants, six mutants with one Ser/Ala substitution each and a mutant with six Ser/Ala substitutions.

2.4   Identification of Glycosylated Amino Acid Residues

Glycosylated flagellin from wild-type and non-glycosylated flagellin from the Δ*orf1* mutant of Pta6605 were purified, digested with aspartic *N*-peptidase, and the resultant peptides were compared by HPLC. Specific wild-type peaks in HPLC were collected, and amino acid sequences were determined. Various amino acid residues were mutagenized to Ala as candidates for glycosylated residues.

2.5   Motility Tests

For the swimming and swarming assays, bacteria were incubated on 0.3% agar MMMF and 0.5% agar SWM plates, respectively (Taguchi et al., 2006a).

3   Results and Discussions

3.1   Glycosyltransferase Genes Required for Flagellin Glycosylation

In *P. aeruginosa* strain PAK, a cluster of 14 genes (*orfA–orfN*) was found upstream of *fliC* as the determinant of flagellin glycosylation (Fig. 1; Arora et al., 2001). In contrast, we found three genes (*orf1–orf3*) at the same position upstream of *fliC* in Pta6605 and Pgl4 (Taguchi et al., 2006a; Takeuchi et al., 2003). The gene products of *orf1* and *orf2* showed homology to putative glycosyltransferases. *Orf*-specific deletion mutants, Δ*orf1* and Δ*orf2*, lost their ability to glycosylate flagellin completely and partially, respectively (Taguchi et al., 2006a; Takeuchi et al., 2003). However, the molecular mass of flagellin was not affected in the Δ*orf3* mutant. Because the *orf3* product showed homology to 3-oxoacyl-(acyl carrier protein) synthase III, which is required for acyl-chain biosynthesis, a precursor of acylhomoserine lactone, we investigated the ability of the Δ*orf3* mutant to produce acylhomoserine lactones. The wild-type Pta6605 produced large amounts of *N*-hexanoyl-L-homoserine lactone and *N*-(3-oxohexanoyl)-L-homoserine lactone. However, the Δ*orf3* mutant had remarkably reduced ability to produce these compounds, suggesting that the Δ*orf3* mutant is somehow impaired in quorum sensing (Taguchi et al., 2006b).

3.2   MS Analysis of Ser/Ala-Substituted Flagellins

MALDI-TOF MS analysis revealed that the molecular masses of wild-type flagellin in Pta6605 were 32382 Da and 32529 Da. In contrast, the flagellin of the Δ*orf1* mutant was 29145 Da, which is nearly identical to the mass as predicted the amino acid sequence. Thus, one flagellin molecule had approximately 3234 or 3381 Da of glycan. Furthermore, it was clarified that flagellin was heterogeneously glycosylated. To identify peptides that possess glycosylated amino acids these were sequenced. As a result, the six serine residues at positions 143, 164, 176, 183, 193, and 201 were identified as candidates for glycosylated amino acid residues. Generation of site-directed Ser/Ala-substituted mutants revealed that these six serine residues were indeed glycosylated. MALDI-TOF MS analysis revealed that the mass of each glycan is about 540 Da (Taguchi et al., 2006a) (Fig. 2).

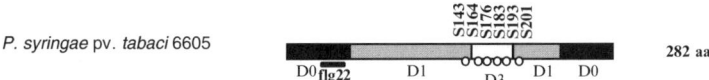

Fig. 2   Flagellin structure and location of glycosylated amino acid residues (*open circles*)

3.3 Virulence of Mutants with Defectively Glycosylated Flagellin

Although all glycosylation-defective mutants in Pta6605 retained swimming ability, the swarming ability was significantly reduced in Δorf1 and Δorf2 mutants and the six Ser/Ala-substituted mutants (Taguchi et al., 2006a). These mutants were also strongly impaired in the ability to adhere to polystyrene surfaces and to cause disease in their host plants. Among the six single Ser/Ala-substituted mutants, the effects of substitution were particularly prominent in the S176A and S183A mutants. These serine residues are expected to be located at the surface of the flagellum. Thus, the glycosyl moiety of flagellin is essential for the virulence of this pathogen. Furthermore, elicitor activities of non-glycosylated flagellins were investigated. Wild-type flagellin from Pta6605 had remarkable ability to induce an HR in cell cultures of soybean but not in those of tobacco (Taguchi et al., 2003b, 2006a). In contrast, the non-glycosylated flagellins from mutant Δorf1 and the six Ser/Ala substituted mutants showed reduced HR induction in soybean cells (Taguchi et al., 2006a) and increased HR induction in tobacco cells (Taguchi et al., unpublished data). These results indicated that the glycosylation of flagellin triggers its HR-inducing ability. Glycosylation might interfere with the elicitor activity of flagellin by masking them from recognition by the host plant surveillance system. Structural analysis of glycosylated and non-glycosylated flagellins in different salt concentrations and pH ranges indicated that glycosylation also increases the stability and strength of the flagellum (Taguchi et al., 2008).

Recently, the importance of flagellin glycosylation in bacterial virulence has been reported in animal pathogens such as P. aeruginosa (Arora et al., 2005) and C. jejuni and C. coli (Guerry et al., 2006). Since the prominently exposed location of glycan results in its direct and immediate exposure to host cells, glycan seems to function in the interactions with host cells, cell adhesion, and/or immune evasion. Additionally, glycan may also have a role in the assembly of flagellins, enhancing stability, and protecting against proteolytic degradation. In the case of Pta6605, flagellin glycosylation seems to contribute to the compatible interactions with a host plant and to physical features of the flagellum structure. Detailed structural analysis of the glycan moiety in different P. syringae pathovars is now on its way and is indispensable to fully understand the function of glycans in bacteria–plant interactions.

Acknowledgements We thank Dr. A. Collmer (Cornell University, USA) and the Leaf Tobacco Research Laboratory of Japan Tobacco Inc. for providing Pgl4 and Pta6605, respectively. This work was supported in part by a Grant-in-Aid for Scientific Research (S) (No. 15108001) and (B) (No. 18380035) from the Ministry of Education, Culture, Sports, Science and Technology of Japan, and the Okayama University COE program "Establishment of Plant Health Science".

References

Arora, S.K., Bangera, M., Lory, S. and Ramphal, R. 2001, A genomic island in *Pseudomonas aeruginosa* carries the determinants of flagellin glycosylation. *Proc. Natl. Acad. Sci. USA* **98**: 9342–9347.

Arora, S.K., Neely, A.N., Blair, B., Lory, S. and Ramphal, R. 2005, Role of motility and flagellin glycosylation in the pathogenesis of *Pseudomonas aeruginosa* burn wound infections. *Infect. Immun.* **73**: 4395–4398.

Brimer, C.D. and Montie, T.C. 1998, Cloning and comparison of fliC genes and identification of glycosylation in the flagellin of *Pseudomonas aeruginosa* a-type strains. *J. Bacteriol.* **180**: 3209–3217.

Eaves-Pyles, T.D., Wong, H.R., Odoms, K. and Pyles, R.B. 2001, *Salmonella* flagellin-dependent proinflammatory responses are localized to the conserved amino and carboxyl regions of the protein. *J. Immunol.* **167**: 7009–7016.

Felix, G., Duran, J.D., Volko, S., and Boller, T. 1999, Plants have a sensitive perception system for the most conserved domain of bacterial flagellin. *Plant J.* **18**: 265–276.

Gómez-Gómez, L. and Boller, T. 2000, FLS2: an LRR receptor-like kinase involved in the perception of the bacterial elicitor flagellin in *Arabidopsis. Mol. Cell* **5**: 1003–1011.

Guerry, P., Ewing, C.P., Schirm, M., Lorenzo, M., Kelly, J., Pattarini, D., Majam, G., Thibault, P. and Logan, S. 2006, Changes in flagellin glycosylation affect *Campylobacter* autoagglutination and virulence. *Mol. Microbiol.* **60**: 299–311.

Hayashi, F., Smith, K.D., Ozinsky, A., Hawn, T.R., Yi, E.C., Goodlett, D.R., Eng, J.K., Akira, S., Underhill, D.M. and Aderem, A. 2001, The innate immune response to bacterial flagellin is mediated by Toll-like receptor 5. *Nature* **410**: 1099–1103.

Ichinose, Y., Shimizu, R., Taguchi, F., Takeuchi, K., Marutani, M., Mukaihara, T., Inagaki, Y., Toyoda, K. and Shiraishi, T. 2003a, Role of flagella and flagellin in plant – *Pseudomonas syringae* interactions. In N. S. Iacobellis (ed.), *Pseudomonas Syringae* and Related Pathogens. Kluwer, Dordrecht, The Netherland, pp. 311–318.

Ichinose, Y., Shimizu, R., Ikeda, Y., Taguchi, F., Marutani, M., Mukaihara, T., Inagaki, Y., Toyoda, K. and Shiraishi, T. 2003b, Need for flagella for complete virulence of *Pseudomonas syringae* pv. *tabaci*: genetic analysis with flagella-defective mutants Δ*fliC* and Δ*fliD* in host tobacco plants. *J. Gen. Plant Pathol.* **69**: 244–249.

Logan, S.M. 2006, Flagellar glycosylation – a new component of the motility repertoire? *Microbiology* **152**: 1249–1262.

Schäfer, A., Tauch, A., Jager, W., Kalinowski, J., Thierbach, G. and Puhler, A., 1994, Small mobilizable multi-purpose cloning vectors derived from the *Escherichia coli* plasmids pK18 and pK19: selection of defined deletions in the chromosome of *Corynebacterium glutamicum*. *Gene* **145**: 69–73.

Shimizu, R., Taguchi, F., Marutani, M., Mukaihara, T., Inagaki, Y., Toyoda, K., Shiraishi, T. and Ichinose, Y. 2003, The Δ*fliD* mutant of *Pseudomonas syringae* pv. *tabaci*, which secretes flagellin monomers, induces a strong hypersensitive reaction (HR) in non-host tomato cells. *Mol. Genet. Genomics* **269**: 21–30.

Szymanski, C.M. and Wren, B.W. 2005, Protein glycosylation in bacterial mucosal pathogens. *Nat. Rev..* **3**: 225–237.

Taguchi, F., Shimizu, R., Inagaki, Y., Toyoda, K., Shiraishi, T. and Ichinose, Y. 2003a, Post-translational modification of flagellin determines the specificity of HR. induction. *Plant Cell Physiol.* **44**: 342–349.

Taguchi, F., Shimizu, R., Nakajima, R., Toyoda, K., Shiraishi, T. and Ichinose, Y. 2003b, Differential effects of flagellins from *Pseudomonas syringae* pv. *tabaci, tomato* and *glycinea* on plant defense response. *Plant Physiol. Biochem.* **41**: 165–174.

Taguchi, F., Takeuchi, K., Katoh, E., Murata, K., Suzuki, T., Marutani, M., Kawasaki, T., Eguchi, M., Katoh, S., Kaku, H., Yasuda, C., Inagaki, Y., Toyoda, K., Shiraishi, T. and Ichinose, Y. 2006a, Identification of glycosylation genes and glycosylated amino acids of flagellin in *Pseudomonas syringae* pv. *tabaci*. *Cell. Microbiol*. **8**: 923–938.

Taguchi, F., Ogawa, Y., Takeuchi, K., Suzuki, T., Toyoda, K., Shiraishi, T. and Ichinose, Y. 2006b, Homologue of 3-oxoacyl-(acyl carrier protein) synthase III gene located in glycosylation island of *Pseudomonas syringae* pv. *tabaci* regulates virulence factors via *N*-acyl homoserine lactone and fatty acid synthesis. *J. Bacteriol*. **188**: 8560–8572.

Takeuchi, K., Taguchi, F., Inagaki, T., Toyoda, K., Shiraishi, T. and Ichinose, Y. 2003, Flagellin glycosylation island in *Pseudomonas syringae* pv. *glycinea* and its role in host specificity. *J. Bacteriol*. **185**: 6658–6665.

Taguchi, F., Shibata, S., Suzuki, T., Ogawa, Y., Aizawa, S., Takeuchi, K. and Ichinose, Y. 2008, Effects of glycosylation on swimming ability and flagella polymorphic transformation of *Pseudomonas syringae* pv. *tabaci* 6605. *J. Bacteriol*. 190: In press.

Genetic Relatedness Among the Different Genetic Lineages of *Pseudomonas syringae* pv. *phaseolicola*

M.E. Führer[1], L. Navarro de la Fuente[1], L. Rivas[1], J.L. Hernandez-Flores[2], R. Garcidueñas-Piña[2], A. Alvarez-Morales[2], and J. Murillo[1]

Abstract *Pseudomonas syringae* pv. *phaseolicola* is an economically important pathogen of bean (*Phaseolus vulgaris* L.) and a research model in phytobacteriology. The analysis of the physiological and genetic variability differentiates the strains isolated from kudzu (*Pueraria lobata*) and the strains that do not produce phaseolo-toxin from most of the common bean strains. We evaluated several techniques to study the variability of *P. syringae* pv. *phaseolicola* and the genetic relatedness among these three groups of strains, in order to understand its epidemiology and the main mechanisms operating the evolution of its virulence gene complement. Insertion sequence typing with IS*801* revealed substantial variability among 32 *P. syringae* pv. *phaseolicola* strains, and allowed the differentiation of three groups of strains. This classification agrees with the grouping resulting from ribotyping and with previous results of other researchers, suggesting that IS*801* fingerprinting is a valuable tool to study genetic variability. Our results support the idea that strains from kudzu form a group differentiable from the other types of strains, and that nontoxigenic strains might represent an ancestor of *P. syringae* pv. *phaseolicola* before it acquired the phaseolotoxin biosynthesis genes

Keywords *P. syringae* pv. *glycinea*, kudzu, *Pueraria lobata*, phaseolotoxin, viru-lence genes

1 Introduction

Pseudomonas syringae pv. *phaseolicola* is an economically important pathogen of bean (*Phaseolus vulgaris* L.) and a research model in phytobacteriology, particu-larly after its genome was sequenced (Smith et al., 1988; Joardar et al., 2005).

[1] Laboratorio de Patología Vegetal, ETS de Ingenieros Agrónomos, Universidad Pública de Navarra. 31006 Pamplona (Spain)

[2] Depto. Ingeniería Genética, CINVESTAV IPN-Unidad Irapuato, México

Author for correspondence: Jesus Murillo; e-mail: jesus.murillo@unavarra.es

M'B. Fatmi et al. (eds.), *Pseudomonas syringae Pathovars and Related Pathogens.*
© Springer Science+Business Media B.V. 2008

Common field isolates of this pathogen from bean show limited variability that, in some instances, can be linked to geographical or plant host of origin (Marques et al., 2000; Güven et al., 2004), which could indicate that commercial seeds are the main source of the pathogen. A particular group of strains are those isolated from kudzu beans (*Pueraria lobata*), which show the same physiological and biochemical properties of bean strains but are differentiated for their ability to use mannitol, to produce ethylene and for specific bands in their ERIC-PCR fingerprint (Völksch and Weingart, 1997; Marques et al., 2000). Additionally, we have described that the majority of the Spanish field isolates of *P. syringae* pv. *phaseolicola* lack the chromosomal pathogenicity island containing the genes for the biosynthesis of the phytotoxin phaseolotoxin (L. Navarro, unpublished results; Rico et al., 2003). This group is also differentiated from the common isolates by three polymorphic bands after arbitrarily primed-PCR, their pattern of insertion of IS*801*, the sequences of the intergenic DNA in the rDNA operons, and because they lack a plasmid-borne pathogenicity island containing effector genes (Oguiza et al., 2004). We are interested in the study of the variability of *P. syringae* pv. *phaseolicola* and the genetic relatedness among these three groups of strains, in order to understand its epidemiology and the main mechanisms operating in the evolution of its virulence gene complement. For this purpose, we selected 32 strains of *P. syringae* pv. *phaseolicola* that were considered representative of the diversity, either because of their origin of isolation or their biochemical or genetic characteristics. We also analyzed three arbitrarily selected strains of *P. syringae* pv. *glycinea*, because this is a pathovar closely related to pv. *phaseolicola*. Finally, the utility of several techniques to evaluate genetic variability and construct a genealogy of these pathogens was examined.

2   Comparison of Nucleotide Sequences

The comparison of nucleotide sequences, especially of essential chromosomal genes, is an easy and efficient way to establish bacterial genealogies. In particular, the comparison of the partial sequences of two to seven housekeeping genes was successfully applied to reveal phylogenies of different species of *Pseudomonas* and also of various *P. syringae* pathovars (Sawada et al., 1999; Yamamoto et al., 2000; Sarkar and Guttman, 2004). In certain cases, however, this strategy was inefficient to reveal intrapathovar diversity due to the high sequence conservation occurring within *P. syringae* pv. *phaseolicola*. For instance, the sequence of partial fragments of the essential genes *gyrB* (534 nt) and *rpoD* (572 nt) was identical among toxigenic and nontoxigenic strains of *P. syringae* pv. *phaseolicola* (Rico et al., 2003). Also, the partial sequence of seven housekeeping genes, 3135 nt in total, was identical among four strains of *P. syringae* pv. *phaseolicola*, including a strain isolated from kudzu (Sarkar and Guttman, 2004). Therefore, little variability through sequence comparison was expected to be revealed.

3 Occurrence of Restriction Fragment Polymorphisms (RFLP)

Genetic diversity among *P. syringae* pv. *phaseolicola* populations by restriction fragment length polymorphism (RFLP) using diverse chromosomal genes as DNA probes was examined. The set of 32 *P. syringae* pv. *phaseolicola* strains examined included the most common toxigenic strains, strains isolated from kudzu plants, and nontoxigenic strains, that together represented nine described races of the pathogen. Additionally, three *P. syringae* pv. *glycinea* strains, which are very closely related to *P. syringae* pv. *phaseolicola* but show a distinctive host rage (Gardan et al., 1992; Marques et al., 2000) were included in this analysis. Cosmid pPgy1563 was identified from a gene library of *P. syringae* pv. *glycinea* 49a/90, and contains in its 28.3-kb insert the *hrp/hrc* cluster in its entirety from the beginning of *avrE* to the middle of *avrPphE*. In Southern blots with genomic DNA digested with *Hin*dIII, the 32 *P. syringae* pv. *phaseolicola* strains displayed an identical pattern of hybridization signal using cosmid pPgy1563 as a probe. However, the three strains of *P. syringae* pv. *glycinea* could be differentiated because they showed a different pattern that resulted from the loss of a single *Hin*dIII site. An example of the patterns of hybridization is shown in Fig. 1. These results indicated a high conservation of the *hrp* cluster in pvs. *phaseolicola* and *glycinea*, and extends our previous finding that the sequence of the exchangeable effector locus situated adjacent to the *hrp* cluster was identical in both, toxigenic and nontoxigenic strains of *P. syringae* pv. *phaseolicola* and *P. syringae* pv. *glycinea* (Oguiza et al., 2004). As expected, the use of other DNA probes, corresponding to the housekeeping genes *dnaG-rpsU*, *gyrB*, *hslU* and *rpoD*, also revealed highly similar hybridization patterns among the different *P. syringae* pv. *phaseolicola* strains (data not shown). In summary, the weak polymorphism observed by RFLP analysis of a variety of genes was not sufficient to delineate significant genealogies.

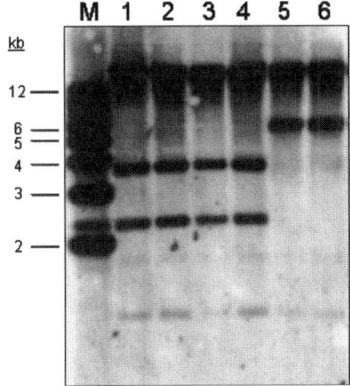

Fig. 1 Conservation of the hrp/hrc cluster among the genetic lineages of *P. syringae* pv. *phaseolicola* and *P. syringae* pv. *glycinea*. Southern hybridization was carried out with genomic DNA digested with *Hin*dIII and using as a probe cosmid pPgy1563, which contains the hrp/hrc cluster from *P.s.* pv. *glycinea* 49a/90. Strains are: *P.s.* pv. *phaseolicola* 1448A (1), CFBP4850 (2), CYL275 (3), CYL314 (4), and P.s. pv. *glycinea* PG4180 (5) and 49a/90 (6). Strain CFBP4850 was isolated from kudzu bean (*Pueraria lobata*), and CYL275 and CYL314 do not posses the pathogenicity island for biosynthesis of phaseolotoxin. M, 1-kb ladder

4   Ribotyping

The RFLP analysis of rRNA genes, known as ribotyping, is often used to differentiate between species or strains. Among its advantages is that it provides results that are comparable to those obtained with other techniques (Bidet et al., 2000); indeed, the analysis of strains from genomospecies III showed that the classifications obtained from an RFLP analysis of ribosomal operons correlated well with the grouping by RAPD and AFLP (Manceau and Horvais, 1997; Clerc et al., 1998). Also, it was previously shown that ribotyping allowed the clear differentiation of *P. syringae* pv. *syringae* from *P. syringae* pv. *phaseolicola*, and that strains of this later pathovar showed diversity (González et al., 2000).

To examine diversity among the set of strains used herein, ribotyping was performed with genomic DNA digested with either *Bgl*I or *Sal*I and using as a DNA probe a 5.2 kb fragment containing a complete rDNA operon from *P. syringae* pv. *phaseolicola* strain 1448A. These hybridization experiments revealed a limited variability, with only 12 polymorphic bands both endonucleases were combined. An example of the ribotyping is shown in Fig. 2. Cluster analysis of these data was done using the unweighted pair-group method with averages (UPGMA) and the Jaccard similarity coefficient, with the program NTSYSpc version 2.11S (Exeter Software, Setauket, NY). This resulted in a dendrogram in which the following four groups could be easily distinguished: (1) containing the toxigenic isolates from bean and other hosts, (2) grouping the nontoxigenic isolates, (3) containing the isolates from kudzu and few other hosts, and (4) containing the three *P. syringae* pv. *glycinea* strains.

5   Insertion Sequence Typing

The construction of genealogies and typing of strains has often been based on the analysis of the number and site of insertion of diverse insertion sequences (Gürtler and Mayall, 2001; Alavi et al., 2007; Gonzalez et al., 2007). Among different

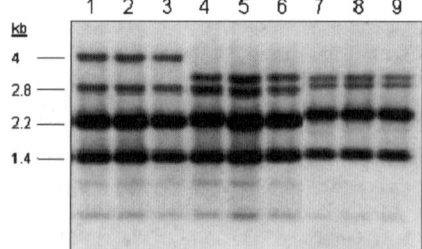

Fig. 2 Ribotypes from strains of the different genetic lineages of *P. syringae* pv. *phaseolicola* and *P. syringae* pv. *glycinea*. Southern hybridization was carried out with genomic DNA digested with BglI and using as a probe a 5.2 kb fragment spanning a whole rDNA operon from *P.s.* pv. *phaseolicola* 1448A. Strains are: *P.s.* pv. *phaseolicola* CFBP4850 (1), CFBP4859 (2), CFBP4860 (3), 1449B (4), CYL275 (5), CYL314 (6), and *P.s.* pv. *glycinea* PG4180 (7), 49a/90 (8) and 16/83 (9). Strains CFBP4850, CFBP4859 and CFBP4860 were isolated from kudzu bean (*Pueraria lobata*), and CYL275 and CYL314 do not posses the pathogenicity island for biosynthesis of phaseolotoxin

pathovars of *P. syringae* there is also significant variation in the type and number of insertion sequences. *P. syringae* pv. *phaseolicola* is characterized by its large number of insertions of these mobile elements (Joardar et al., 2005).

Our group has previously determined that the hybridization fingerprint of genomic DNA to a specific IS*801* probe allowed unequivocal differentiation of toxigenic and nontoxigenic *P. syringae* pv. *phaseolicola* strains (Oguiza et al., 2004). An important advantage of using IS*801* as a marker is that its insertions are likely to be permanent, i.e., that the original copy of the element remains in place after tranposition (Richter et al., 1998). A potential limitation, however, is that there appear to be a number of copies of the element on native plasmids (Oguiza et al., 2004). For instance, the analysis of the complete genome sequence of *P. syringae* pv. *phaseolicola* 1448A (Joardar et al., 2005) identified three full-length copies in the chromosome and one in the largest plasmid, as well at least 15 other copies of variable length distributed on the chromosome and the largest plasmid. From these, there are six copies larger than 1 kb located in the chromosome and four in p1448A-A.

Total genomic DNA from the set of 35 strains analyzed here was digested with either *Hin*dIII or *Pvu*II and analyzed by Southern hybridization to a probe spanning the leftmost 627 nt of IS*801* (Oguiza et al., 2004). A total of 32 (*Hin*dIII) or 26 (*Pvu*II) polymorphic bands could be distinguished in the hybridization fingerprints. In the resulting dendrogram constructed as detailed above the strains could be differentiated in four groups at a level of similarity of approximately 0.48 (Fig. 3). These groups were roughly the same as those differentiated by ribotyping, which suggests the validity of genealogies based on IS*801* typing. The first group

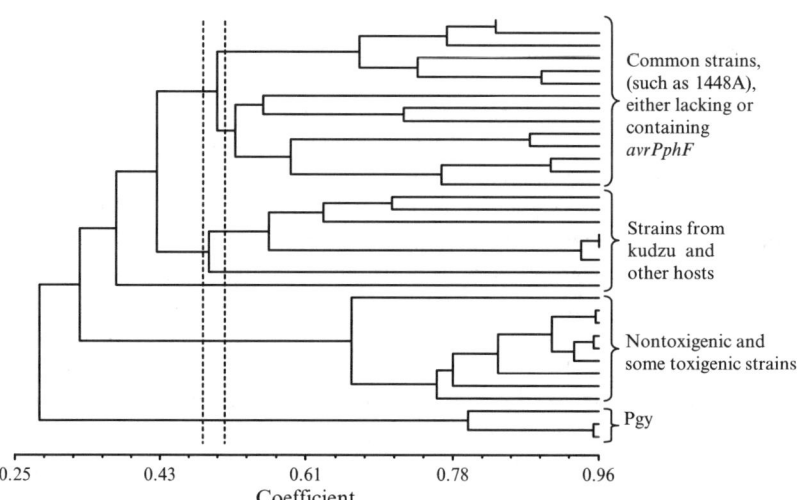

Fig. 3 Dendrogram obtained from comparison of IS *801* fingerprinting patterns from 32 *P. Syringae* pv. *phaseolicola* and 3 *P. Syringae* pv. *glycinea* (Pgy) strains. The dendogram was constructed using program NTSYSpc 2.11 s, with Jaccard's coefficient and the UPGMA method

included the common isolates from bean, which synthesized phaseolotoxin. At a higher level of similarity of approximately 0.5 this group could be further subdivided in two clusters, one of which contained all the strains lacking gene *avrPphF*. Among other phenotypes, its gene product induces a hypersensitive response in cultivars of bean carrying resistance gene *R1*, and the gene is absent from races 2, 3, 4, 6 and 8 of *P. syringae* pv. *phaseolicola* (Tsiamis et al., 2000; Rivas et al., 2005). Our previous results suggested a clonal origin for the races lacking gene *avrPphF* (Rivas et al., 2005), which was confirmed by IS*801* fingerprinting obtained here.

The second group contained strains that were isolated from kudzu bean, as well as other strains isolated from bean, *Desmodium*, or *Lablab purpureus*. Strains from kudzu have been traditionally treated as a separate group of *P. syringae* pv. *phaseolicola* and have even been proposed to be classified into another pathovar (Völksch and Weingart, 1997). Our results also indicated that strains isolated from kudzu conform a homogeneous group, although they are closely related to other strains isolated from different plant hosts. This tight clustering of kudzu isolates is striking despite the fact that even isolates from bean can infect kudzu (Völksch and Weingart, 1997). It might furthermore reflect a better adaptation to the plant host or a specific dissemination mechanism that favours the colonization of the plant by clonal lineages. A distinctive feature of strains from kudzu is their ability to synthesize ethylene, which is due to their possession of a plasmid-borne *efe* gene (Nagahama et al., 1994). However, using specific primers for the coding sequence of the *efe* gene in PCR experiments, we did not observe the presence of this gene in nontoxigenic strains of *P. syringae* pv. *phaseolicola* (Rico et al., 2003) or in any of the other strains analyzed here, except in those isolated from kudzu plants or in *P. syringae* pv. *glycinea*. It is known that the *efe* gene resides on conjugative plasmids (Watanabe et al., 1998), and this asymmetrical distribution might possibly reflect the colonization of similar niches by the strains containing these plasmids.

The nontoxigenic *P. syringae* pv. *phaseolicola* strains examined here were isolated in Australia, Germany and Spain, and all of them clustered together in the group more closely related to *P. syringae* pv. *glycinea*. We showed that these strains do not contain the genes essential for biosynthesis of phaseolotoxin (L. Navarro, unpublished results; Rico et al., 2003; Oguiza et al., 2004), which could have resulted from the loss of these genes. Indeed, the genes coding for the biosynthesis of another toxin, tabtoxin, were reported to be spontaneously lost with a very high frequency (10^{-3} per cfu) from the genome of *P. syringae* BR2 (Kinscherf et al., 1991). Additionally, the ability to synthesize phaseolotoxin is also lost in certain strains of *P. syringae* pv. *phaseolicola* (Völksch et al., 1984; Rico et al., 2003). However and in agreement with our previous data (Oguiza et al., 2004), our results indicated that natural nontoxigenic strains of *P. syringae* pv. *phaseolicola* constitute a distinct group that probably represents the ancestor of this pathovar before it evolutionarily acquired phaseolotoxin biosynthesis genes.

6 Conclusions

Most common typing techniques, such as arbitrarily primed-PCR, RFLP analysis or multi-locus sequencing typing, revealed little intrapathovar diversity in *P. syringae* pv. *phaseolicola*. Fingerprint analysis after hybridization with the insertion sequence IS*801*, however, revealed substantial variation among strains of *P. syringae* pv. *phaseolicola* and *P. syringae* pv. *glycinea*, and allowed their classification in four groups. This classification agrees with previous results (Völksch and Weingart, 1997; Marques et al., 2000; Oguiza et al., 2004) and with the results obtained after ribotyping analysis, suggesting the applicability of insertion sequence typing – at least with IS*801* – to study intrapathovar diversity. The analyses carried out herein supported previous findings that strains of *P. syringae* pv. *phaseolicola* isolated from kudzu are highly related and suggested that nontoxigenic strains might represent an ancestor of toxigenic *P. syringae* pv. *phaseolicola* strains.

Acknowledgements This work was supported with grant AGL-2004-03143, from the Spanish Ministerio de Educación y Ciencia, to J. Murillo, and with grants from CONACYT, to A. Álvarez-Morales. We are grateful to A. de Vicente, M. Lesaux, C. Manceau, M. Scortichini and M. Ullrich for kindly providing us with bacterial strains. We are also grateful to Theresa Osinga for critical reading of the manuscript.

References

Alavi S. M., Poussier S. and Manceau C., 2007, Characterization of IS*Xax1*, a novel insertion sequence restricted to *Xanthomonas axonopodis* pv. phaseoli (variants fuscans and non-fuscans) and *Xanthomonas axonopodis* pv. vesicatoria. *Appl Environ Microbiol* 73: 1678–1682.

Bidet P., Lalande V., Salauze B., Burghoffer B., Avesani V., Delmee M., Rossier A., Barbut F. and Petit J.-C., 2000, Comparison of PCR-ribotyping, arbitrarily primed PCR, and pulsed-field gel electrophoresis for typing *Clostridium difficile. J Clin Microbiol* 38: 2484–2487.

Clerc A., Manceau C. and Nesme X., 1998, Comparison of randomly amplified polymorphic DNA with amplified fragment length polymorphism to assess genetic diversity and genetic relatedness within genomospecies III of *Pseudomonas syringae. Appl Environ Microbiol* 64: 11180–11187.

Gardan L., Bollet C., Abu Ghorrah M., Grimont F. and Grimont P. A. D., 1992, DNA relatedness among the pathovar strains of *Pseudomonas syringae* subsp. *savastanoi* Janse (1982) and proposal of *Pseudomonas savastanoi* sp. nov. *Int J Syst Bacteriol* 42: 606–612.

González A. J., Landeras E. and Mendoza M. C., 2000, Pathovars of *Pseudomonas syringae* causing bacterial brown spot and halo blight in *Phaseolus vulgaris* L. are distinguishable by ribotyping. *Appl Environ Microbiol* 66: 850–854.

González C., Szurek B., Manceau C., Mathieu T., Séré Y. and Verdier V., 2007, Molecular and pathotypic characterization of new *Xanthomonas oryzae* strains from West Africa. *Mol Plant-Microbe Interact* 20: 534–546.

Gürtler V. and Mayall B., 2001, Genomic approaches to typing, taxonomy and evolution of bacterial isolates. *Int J Syst Evol Microbiol* 51: 3–16.

Güven K., Jones J. B., Momol M. T. and Dickstein E. R., 2004, Phenotypic and genetic diversity among *Pseudomonas syringae* pv. phaseolicola. *J Phytopathol* 152: 658–666.

Joardar V., Lindeberg M., Jackson R. W., Selengut J., Dodson R., Brinkac L. M., Daugherty S. C., DeBoy R., Durkin A. S., Giglio M. G., Madupu R., Nelson W. C., Rosovitz M. J., Sullivan S., Crabtree J., Creasy T., Davidsen T., Haft D. H., Zafar N., Zhou L. W., Halpin R., Holley T., Khouri H., Feldblyum T., White O., Fraser C. M., Chatterjee A. K., Cartinhour S., Schneider D. J., Mansfield J., Collmer A. and Buell C. R., 2005, Whole-genome sequence analysis of *Pseudomonas syringae* pv. phaseolicola 1448A reveals divergence among pathovars in genes involved in virulence and transposition. *J Bacteriol* 187: 6488–6498.

Kinscherf T. G., Coleman R. H., Barta T. M. and Willis D. K., 1991, Cloning and expression of the tabtoxin biosynthetic region from *Pseudomonas syringae*. *J Bacteriol* 173: 4124–4132.

Manceau C. and Horvais A., 1997, Assessment of genetic diversity among strains of *Pseudomonas syringae* by PCR-restriction fragment length polymorphism analysis of rRNA operons with special emphasis on *P. syringae* pv. tomato. *Appl Environ Microbiol* 63: 498–505.

Marques A. S. d. A., Corbière R., Gardan L., Tourte C., Manceau C., Taylor J. D. and Samson R., 2000, Multiphasic approach for the identification of the different classification levels of *Pseudomonas savastanoi* pv. *phaseolicola*. *Eur J Plant Pathol* 106: 715–734.

Nagahama K., Yoshino K., Matsuloa M., Sato M., Tanase S., Ogawa T. and Fukuda H., 1994, Ethylene production by strains of the plant-pathogenic bacterium *Pseudomonas syringae* depends upon the presence of indigenous plasmids carrying homologous genes for the ethylene-forming enzyme. *Microbiology* 140: 2309–2313.

Oguiza J. A., Rico A., Rivas L. A., Sutra L., Vivian A. and Murillo J., 2004, *Pseudomonas syringae* pv. phaseolicola can be separated into two genetic lineages distinguished by the possession of the phaseolotoxin biosynthetic cluster. *Microbiology* 150: 473–482.

Richter G. Y., Björklöf K., Romantschuk M. and Mills D., 1998, Insertion specificity and *trans*-activation of IS*801*. *Mol Gen Genet* 260: 381–387.

Rico A., López R., Asensio C., Aizpún M., Asensio-S.-Manzanera C. and Murillo J., 2003, Nontoxigenic strains of *P. syringae* pv. *phaseolicola* are a main cause of halo blight of beans in Spain and escape current detection methods. *Phytopathology* 93: 1553–1559.

Rivas L. A., Mansfield J., Tsiamis G., Jackson R. W. and Murillo J., 2005, Changes in race-specific virulence in *Pseudomonas syringae* pv. phaseolicola are associated with a chimeric transposable element and rare deletion events in a plasmid-borne pathogenicity island. *Appl Environ Microbiol* 71: 3778–3785.

Sarkar S. F. and Guttman D. S., 2004, Evolution of the core genome of *Pseudomonas syringae*, a highly clonal, endemic plant pathogen. *Appl Environ Microbiol* 70: 1999–2012.

Sawada H., Suzuki F., Matsuda I. and Saitou N., 1999, Phylogenetic analysis of *Pseudomonas syringae* pathovars suggests the horizontal gene transfer of *argK* and the evolutionary stability of *hrp* gene cluster. *J Mol Evol* 49: 627–644.

Smith I. M., Dunez J., Lelliott R. A., Phillips D. H. and Archer S. A., 1988, *European Handbook of Plant Diseases*. London: Blackwell.

Tsiamis G., Mansfield J. W., Hockenhull R., Jackson R. W., Sesma A., Athanassopoulos E., Bennett M. A., Stevens C., Vivian A., Taylor J. D. and Murillo J., 2000, Cultivar specific aviru-lence and virulence functions assigned to *avrPphF* in *Pseudomonas syringae* pv. *phaseolicola*, the cause of bean halo-blight disease. *EMBO J* 19: 3204–3214.

Völksch B., Laplace F. and Fritsche W., 1984, Studies on the variability of the phaseolotoxin pro-duction by *Pseudomonas syringae* pv. phaseolicola. *Zentralbl Mikrobiol* 139: 109–118.

Völksch B. and Weingart H., 1997, Comparison of ethylene-producing *Pseudomonas syringae* strains isolated from kudzu (*Pueraria lobata*) with *Pseudomonas syringae* pv. *phaseolicola* and *Pseudomonas syringae* pv. *glycinea*. *Eur J Plant Pathol* 103: 795–802.

Watanabe K., Nagahama K. and Sato M., 1998, A conjugative plasmid carrying the *efe* gene for the ethylene-forming enzyme isolated from *Pseudomonas syringae* pv. *glycinea*. *Phytopathology* 88: 1205–1209.

Yamamoto S., Kasai H., Arnold D. L., Jackson R. W., Vivian A. and Harayama S., 2000, Phylogeny of the genus *Pseudomonas*: intrageneric structure reconstructed from the nucleotide sequences of *gyrB* and *rpoD* genes. *Microbiology* 146: 2385–2394.

WLIP and Analogues of Tolaasin I, Lipodepsipeptides from *Pseudomonas reactans* and *Pseudomonas tolaasii*: A Comparison of Their Activity on Natural and Model Membranes

R. Paletti[1], M. Coraiola[1], A. Cimmino[2], P. Lo Cantore[3],
A. Evidente[2], N.S. Iacobellis[3], and M. Dalla Serra[1]

Abstract Tolaasin I and the White Line Inducing Principle (WLIP) are lipodepsipeptides (LDPs) produced by virulent strains of *Pseudomonas tolaasii* and *P. reactans* which are the causal agents of "brown blotch" disease of *Agaricus bisporus* and of the "yellowing" of *Pleurotus ostreatus*. Both tolaasin I and WLIP have haemolytic activity, permeabilize lipid vesicles and appear to be involved in mushroom tissue alterations by forming transmembrane pores. Herein, we describe the activity of tolaasin II and five natural tolaasin analogues on natural and model membranes. Tolaasin I and tolaasin II shared similar activity; the other tolaasin analogues were instead remarkably less active (up to 15 times when compared to tolaasin I) both on natural and model membranes. Additionally, the activity of mixtures formed by tolaasin I and WLIP on vesicles is reported. Preliminary evidences of an interaction of the two LDPs leading to an antagonistic action became apparent.

Keywords Lipodepsipeptides, *Pseudomonas tolaasii*, *Pseudomonas reactans*, liposomes, haemolysis, transmembrane pores

1 Introduction

Brown blotch of *Agaricus bisporus* and the yellowing of *Pleurotus ostreatus* have been recently shown to be complex diseases. In fact, besides *Pseudomonas tolaasii*, known for long as the causal agent of the disease (Tolaas, 1915), also other bacteria,

[1] CNR-FBK Istituto di Biofisica, Unità di Trento, Povo (Trento), Italy

[2] Dipartimento di Scienze del Suolo, Pianta, dell'Ambiente e delle Produzioni Animali, Università di Napoli "Federico II", Italy

[3] Dipartimento di Biologia, Difesa e Biotecnologie Agro Forestali, Università della Basilicata, Potenza, Italy

Author for correspondence: Mauro Dalla Serra; e-mail: mdalla@itc.it

M'B. Fatmi et al. (eds.), *Pseudomonas syringae Pathovars and Related Pathogens.*   183
© Springer Science+Business Media B.V. 2008

including *P. reactans*, cause the disease symptoms even though in a different way (Godfrey et al., 2001; Munsch et al., 2002; Iacobellis and Lo Cantore, 2003). Furthermore, *P. reactans* has recently been demonstrated as the causal agent of yellowing of *Pleurotus eryngii* (Iacobellis et al., 2003). *P. reactans* is a bacterial entity not yet classified and is known from the literature for its use in the white line assay for the identification of *P. tolaasii* (Wong and Preece, 1979). The fact that avirulent morphological variants of *P. reactans* were shown to lack WLIP production (Iacobellis and Lo Cantore, 2003) suggested that WLIP may play an important role in virulence and pathogenicity of the bacterium. Tolaasin I appeared to cause mushroom tissue alterations by disrupting the cell membrane through the formation of transmembrane pores (Brodey et al., 1991; Rainey et al., 1992; Hutchison and Johnstone, 1993; Cho and Kim, 2003; Coraiola et al., 2006; Lo Cantore et al., 2006). The potential involvement of both bacteria in disease development on mushrooms and the specific interaction occurring between the above secondary metabolites, made it interesting to study these molecules in a comparative way. Tolaasin I is a lipodepsipeptide with a molecular weight of 1985 Da composed of 18 amino acids, 11 of which are in D form, with the lactone ring formed between the hydroxyl group of D-Thr[14] and the C-terminal L-Lys[18]. Besides tolaasin I and tolaasin II (Nutkins et al., 1991) five other analogues, named tolaasin A–E, are produced by *P. tolaasii*. The primary structures and antimicrobial activity of the analogues have been recently reported (Bassarello et al., 2004). WLIP is a lipodepsipeptide (LDP) with a molecular weight of 1125 Da composed of an N-terminal β-hydroxydecanoic acid and a peptide moiety of nine amino acids, six of which are in D form. The molecule contains a lactone ring, including seven amino acids, between D-allo-threonine (D-Thr[3]) and the C-terminal L-isoleucine (Ile[9]) (Mortishire-Smith et al., 1991b). To understand the basic LDPs structural features needed for functionality, the effect of tolaasin analogues on natural and model membranes have been evaluated herein. In the attempt to get insights into the nature of peptide–peptide interactions, experiments in the concurrent presence of tolaasin I and WLIP have been performed.

2   Materials and Methods

2.1   Chemicals

Lipids used were egg phosphatidylcholine (PC) and sphingomyelin (SM) from Avanti Polar Lipids (Alabaster, Alabama, USA). Calcein, EDTA, and Sephadex G-50 were from Sigma (Milan, Italy), Triton X-100 from Merck (USA). Rabbit Red Blood Cells (RRBC) were purchased from Zootecnica Il Gabbiano (Italy). Buffer A was 20 mM Tris-HCl, 100 mM NaCl, 1 mM EDTA, pH 7.0. The isolation and purification of LDPs were performed as described earlier (Mortishire-Smith et al., 1991a; Bassarello et al., 2004; Lo Cantore et al., 2006).

2.2   Calcein Release Assay

Large unilamellar vesicles (LUVs) loaded with 80 mM calcein were used to measure membrane permeabilization as reported previously (Coraiola et al., 2006). The efflux of calcein into the external medium was detected by an increase in fluorescence, which was measured with a fluorimetric microplate reader (Fluostar, BMG, Offenburg, Germany) as described in (Dalla Serra et al., 1999). The percentage of calcein released (R%), was then calculated as follows:

$$R\% = 100 \ (F_{fin} - F_{in})/(F_{max} - F_{in}) \tag{1}$$

where F_{in} and F_{fin} represented the initial and final (after 45 min) value of fluorescence before and after toxin addition, respectively. F_{max} was determined by adding 1 mM Triton X-100 and used as 100% of release. Before each assay, WLIP was dissolved in DMSO in order to favour its solubility; the same amount of DMSO tested alone did not induce any permeabilization of LUVs (data not shown).

2.3   Haemolytic Assay

Haemolytic activity of LDPs was determined by measuring the turbidity change at 650 nm after toxin addition to RRBC suspensions. A 96-well microplate reader (UVmax, Molecular Devices, Sunnyvale, California) was used as described in (Dalla Serra et al., 1999). LDP solutions were twofold serially diluted from well to well with buffer A. The experiment started after supplementing each well with 100 µl of RRBC suspended in the same buffer (final volume 200 µl/well). The apparent initial absorbance of intact RRBC was about 0.1 OD, which corresponded to 2.4×10^6 cells/well. The percentage of haemolysis (H%) was calculated as follows:

$$H\% = 100 \ (A_i - A_f)/(A_i - A_w) \tag{2}$$

where A_i and A_f are the absorbance at the beginning and at the end of the experiment (4 h long) and A_w the minimum value obtained after hypotonical lysis with pure water. All experiments were performed at room temperature.

2.4   Effects of the Co-presence of Tolaasin I and WLIP on LUVs

Tolaasin I and WLIP were mixed together in different molar ratios, incubated for 24 or 72 h at 4°C, and the activity was tested on LUVs composed of PC/SM (1:1 mol). The

Limpel's equation (Limpel et al., 1962) was used to determine the presence of either synergistic or antagonistic interactions between the two toxins present in the mixtures:

$$E_e = [R\%(t) + R\%(W)] - [R\%(t)\ R\%(W)]/100 \tag{3}$$

where E_e is the expected effect from a simple additive response of the two compounds, $R\%(t)$ and $R\%(W)$ are the observed effects (in percentage) of the two compounds (i.e. tolaasin I and WLIP) when applied singularly and at the same amount used in the mixture. If the observed E_o effect is greater than the expected one, then synergism is said to occur. On the same basis if the observed effect is lower than the expected one, then the mixture exhibits an antagonistic effect. The LDPs concentrations in the linear region of the dose–response relationship were used (e.g. the values reported in Table 2). To express the intensity of the interaction of compounds in the mixture we introduced a new parameter, the RC (Reduction Coefficient) calculated by the following equation:

$$RC = 100\ (E_e - E_o)/E_e \tag{4}$$

3   Results and Discussion

3.1   Calcein Release and Haemolytic Assay

It is known that some structural features of LDPs are important for their pore-forming ability: for instance, the activity of syringomycin E (SRE), an LDP produced by *P. syringae* pv. *syringae* is abolished when the lactone ring is hydrolysed (Dalla Serra et al., 2003). *P. tolaasii* was shown to produce *in vitro* tolaasin I, tolaasin II (Nutkins et al., 1991) and five minor metabolites, tolaasins A, B, C, D, and E whose structures were recently reported (Bassarello et al., 2004) and are shown in Fig. 1.

The permeabilizing activity of the above substances on LUVs composed of PC/SM (1:1 mol) and their ability to lyse RRBC, expressed as the reciprocal of the concentration of toxin causing the release of 50% calcein or the 50% of haemolysis, respectively, are reported in Table 1. All LDPs tested were able to permeabilize LUVs and to induce haemolysis of RRBC.

Tolaasin II, differing from tolaasin I only in the replacement of Hse_{16} with Gly_{16}, showed an activity which was very similar to that of tolaasin I.

Tolaasin D, containing a Leu residue in place of the Ile in position 15, showed only a slight decrease in activity. Tolaasin E simultaneously containing the tolaasin II and tolaasin D modifications caused a significant decrease in permeabilizing activity and a complete loss of haemolytic activity. Tolaasin C, the acyclic version of tolaasin I derived by the hydrolysis of its lactone ring, showed a major loss of activity but was not completely inactive. This was in contrast to what usually happens to other LDPs with similar modifications (Dalla Serra et al., 2003). Tolaasin

Tolaasin I

aa1   aa2 aa3 aa4   aa5 aa6 aa7   aa8 aa9 aa10 aa11 aa12 aa13   aa14   aa15 aa16 aa17 aa18
L1–ΔBut–Pro–Scr–Leu–Val–Scr–Leu–Val–Val–Gln–Leu–Val–ΔBut–aThr–Ile–Hse–Dab–Lys

LDP	ACYL CHAIN	aa14	aa15	aa16	aa17	aa18
Tolaasin I	L1		Ile	Hse		
Tolaasin II	L1			Gly		
Tolaasin A	L1					
Tolaasin B	L1					
Tolaasin C	L1		Val			
Tolaasin D	L1		Leu			
Tolaasin E	L1		Leu	Gly		
Tolaasin C	L1	Acyclic version				

L1: $CH_3 (CH_2)_4 CH (OH) CH_2 CO$
L2: $HOOC (CH_2)_3 CO$

Fig. 1 Structure of the analogues of tolaasin I, produced by *P. tolaasii* (Bassarello et al., 2004). ΔBut, dehydrobutyrine; Dab, diaminobutyric acid; Hse, homoserine; aThr, allo-threonine

Table 1 Permeabilizing and haemolytic effect of tolaasin I and its alogues. C_{50} indicates the LDP concentration causing a 50% release of calcein from PC/SM (1:1 mol) vesicles or 50% haemolysis of rabbit red blood cells

LDP	Permeabilizing activity	Haemolytic activity
Tolaasin I	2.19 ± 0.83	0.017 ± 0.002
Tolaasin II	2.64 ± 0.86	0.015 ± 0.002
Tolaasin A	0.18 ± 0.02	n.a (225)[a]
Tolaasin B	0.18 ± 0.03	n.a. (225)[a]
Tolaasin C	0.19 ± 0.03	0.006
Tolaasin D	0.94 ± 0.11	0.006
Tolaasin E	0.13 ± 0.03	n.a (225)[a]

[a]n.a., non-active (highest concentration tested)

A and B were less active on both vesicles and RRBC. The replacement of a single amino acid or other minor changes in tolaasin I structure sometimes extensively modified the pore-forming ability. This clearly indicated that the structural features of LDPs are important for their function on target membrane(s). The fact that *Pseudomonas tolaasii* (like other *Pseudomonas* spp.) produces different types of LDPs in vitro suggested a possible synergism among them and this may be important in the pathogenicity of the bacterium and/or in its ecological fitness.

Table 2 Activity of mixtures of tolaasin I and WLIP tested on PC/SM (1:1 mol) vescicles after 24 or 72 h of incubation

Incubation time (@ 4°C)	Mixture (tol I: WLIP)	[tol I + WLIP] (μM)	tol I E_o (%)[a]	WLIP E_o (%)[a]	E_e (%)[b]	Mixture E_o (%)[a]	RC[c]
24 h	1:1	0.2 + 0.2	89	20	91	81	12
	1:1	0.2 + 0.2	89	20	91	73	19
72 h	1:2	0.2 + 0.4	89	47	94	79	16
	1:5	0.2 + 1.0	89	84	98	86	13

[a] The observed effect calculated from calcein release experiments
[b] The expected effect due to the additive responses of the two compounds
[c] The reduction coefficient used to express the degree of interaction between compounds in the mixture (Eq. 4)

3.2   Effects of the Co-presence of Tolaasin I and WLIP on LUVs

The facts that both *P. tolaasii* and *P. reactans* appeared to participate in the development of disease symptoms on mushrooms and that their secondary metabolites interacted in the "white line" assay, prompted us to investigate the activity of mixtures of tolaasin I and WLIP in different molar ratio on LUVs composed of PC/SM (1:1 mol). We calculated the E_e (expected effect) from additive responses of the two compounds and compared it with the E_o (observed effect) (Table 2). In all mixtures, E_o was always lower than E_e which excluded the possibility of a synergistic action of both LDPs and rather supported an antagonistic effect. The RC is maximal when tolaasin I and WLIP were mixed in an equi-molar ratio. Next, the question on the nature of the measured antagonistic interaction was addressed with respect to whether it happens in solution, with an aggregating mechanism similar to that responsible for the "white line" precipitate, or directly at the membrane site. Some preliminary data obtained in our laboratory with inactive WLIP analogues (data not shown) supported the second hypothesis and allow speculations on the formation of a mixed aggregate, where both LDPs interact and assemble onto the membrane to form a non active unit.

Acknowledgements This work was financially supported by Fondazione Cassa di Risparmio di Trento e Rovereto and by Provincia Autonoma di Trento (PAT, project SyrTox).

References

Bassarello, C., Lazzaroni, S., Bifulco, G., Lo Cantore, P., Iacobellis, N. S., Riccio, R., Gomez-Paloma, L., and Evidente, A., 2004, Tolaasins A-E, five new lipodepsipeptides produced by *Pseudomonas tolaasii*. *J. Nat. Prod.*, **67**, 811–816.
Brodey, C. L., Rainey, P. B., Tester, M., and Johnstone, K., 1991, Bacterial blotch disease of the cultivated mushroom is caused by an ion channel forming lipodepsipeptide toxin. *Mol. Plant Microbe Interact.*, **4**, 407–411.

Cho, K. H. and Kim, Y. K., 2003, Two types of ion channel formation of tolaasin, a *Pseudomonas* peptide toxin. *FEMS Microbiol. Lett.*, **221**, 221–226.

Coraiola, M., Lo Cantore, P., Lazzaroni, S., Evidente, A., Iacobellis, N.S., and Dalla Serra, M., 2006, Tolaasin I and WLIP, Lipodepsipeptides from *Pseudomonas tolaasii* and *P. reactans*, permeabilize model membranes. *Biochim. Biophys. Acta*, **1758**, 1713–1722.

Dalla Serra, M., Fagiuoli, G., Nordera, P., Bernhart, I., Della Volpe, C., Di Giorgio, D., Ballio, A., and Menestrina, G., 1999, The interaction of lipodepsipeptide toxins from *Pseudomonas syringae* pv. *syringae* with biological and model membranes: a comparison of syringotoxin, syringomycin and syringopeptins. *Mol. Plant Microbe Interact.*, **12**, 391–400.

Dalla Serra, M., Menestrina, G., Coraiola, M., and Grgurina, I., 2003, Interaction of syringomycin E structural analogues with biological and model membranes. In Iacobellis, S. N. (Ed.), *Pseudomonas syringae and related pathogens. Biology and Genetic*, Kluwer, Dordrecht (NL), pp. 207–215.

Godfrey, S. A., Marshall, J. W., and Klena, J. D., 2001, Genetic characterization of *Pseudomonas* 'NZI7' – a novel pathogen that results in a brown blotch disease of *Agaricus bisporus*. *J. Appl. Microbiol.*, **91**, 412–420.

Hutchison, M. L. and Johnstone, K., 1993, Evidence for the involvement of the surface active properties of the extracellular toxin tolaasin in the manifestation of the brown blotch disease symptoms by *Pseudomonas tolaasii* on *Agaricus bisporus*. *Physiol. Mol. Plant Pathol.*, **42**, 373–384.

Iacobellis, N. S. and Lo Cantore, P., 2003, *Pseudomonas reactans* a new pathogen of cultivated mushrooms. In Iacobellis, N. S. (Ed.), *Pseudomonas syringae and Related Pathogens. Biology and Genetic*, Kluwer, Dordrecht (NL), pp. 595–605.

Limpel, L. E., Schuldt, P. H., and Lamont, D., 1962, Weed control by dimethyl tetrachloroterephthalate alone and in certain combinations. *Proc. North. Weed Control Conf.*, **16**, 48–53.

Lo Cantore, P., Lazzaroni, S., Coraiola, M., Dalla Serra, M., Cafarchia, C., Evidente, A., and Iacobellis, N. S., 2006, Biological Characterisation of WLIP produced by *Pseudomonas reactans* strain NCPPB1311. *Mol. Plant Microbe Interact.*, **19**, 1113–1120.

Mortishire-Smith, R. J., Drake, A. F., Nutkins, J. C., and Williams, D. H., 1991a, Left handed alpha-helix formation by a bacterial peptide. *FEBS Lett.*, **278**, 244–246.

Mortishire-Smith, R. J., Nutkins, J. C., Packman, L. C., Brodey, C. L., Rainey, P. B., Johnstone, K., and Williams, D. H., 1991b, Determination of the structure of an extracellular peptide produced by the mushroom saprophytic *Pseudomonas reactans*. *Tetrahedron*, **47**, 3645–3654.

Munsch, P., Alatossava, T., Marttinen, N., Meyer, J. M., Christen, R., and Gardan, L., 2002, *Pseudomonas costantinii* sp. *nov.*, another causal agent of brown blotch disease, isolated from cultivated mushroom sporophores in Finland. *Int. J. Syst. Evol. Microbiol.*, **52**, 1973–1983.

Nutkins, J. C., Mortishire-Smith, R. J., Packman, L. C., Brodey, C. L., Rainey, P. B., Johnstone, K., and Williams, D. H., 1991, Structure determination of tolaasin, an extracellular lipodepsipeptide produced by the mushroom pathogen *Pseudomonas tolaasii* Paine. *J. Am. Chem. Soc.*, **113**, 2621–2627.

Rainey, P. B., Brodey, C. L., and Johnstone, K., 1992, Biology of *Pseudomonas tolaasii*, cause of brown blotch disease of the cultivated mushroom. *Adv. Plant Pathol.*, **8**, 95–117.

Tolaas, A. G., 1915, A bacterial disease of cultivated mushrooms. *Phytopathology*, **5**, 51–53.

Wong, W. C. and Preece, T. F., 1979, Identification of *Pseudomonas tolaasii*: the white line in agar and mushroom tissue block rapid pitting tests. *J. Appl. Bacteriol.*, **47**, 401–407.

Competitive Index in Mixed Infection: A Sensitive and Accurate Method to Quantify Growth of *Pseudomonas syringae* in Different Plants

A.P. Macho, A. Zumaquero, I. Ortiz-Martín, and C.R. Beuzón

Abstract The ability of *Pseudomonas syringae* to replicate in susceptible host plants depends on a large number of virulence factors, and is directly related to its pathogenicity. Absence of one or more of these factors frequently causes a reduction in bacterial growth in the host and an attenuation of the induced symptoms. Determining the role that these factors play in the infection process requires sensitive methods to quantify virulence. In animal models, mixed infections have allowed the development of this type of analysis; from measuring otherwise undetectable virulence defects of type III effector mutants to allowing genetic analysis of multiple mutants to determine functional relationships between different genes. Virulence factors of pathogens such as *Salmonella* or *Streptococcus* have often been assigned to regulons or to secretion or transport systems. Furthermore, the establishment of mixed infections in different animal models has been key to the development of high-throughput methods of screening for virulence factors (e.g. signature-tagged mutagenesis). In this work, we have established that determining competitive indexes within mixed infections is an accurate method to quantify virulence and carry out genetic analysis of *Pseudomonas syringae* interaction with the plant. If the right inoculation dose is used, the strains within the mixed infection neither compete for the resources nor interfere with each other, thus multiplying within the plant in a manner that reflects their respective pathogenic capabilities. We have set up the method using *P. syringae* pv *phaseolicola* in its interaction with bean plants. In addition, we have tested the application of the method to other established models using *P. syringae* pv *tomato* in both its interactions with tomato plants and *Arabidopsis*, and *P. syringae* pv *syringae* in its interaction with bean plants. We found that this method is highly sensitive and reproducible thereby greatly reducing experimental variation and consequently the number of plants required to obtain significant results. The method allows a variety of applications some of which are discussed.

Departamento de Biología Celular, Genética y Fisiología, Área de Genética, Universidad de Málaga, Campus de Teatinos, Málaga E-29071, Spain

Author for correspondence: Carmen Beuzón; e-mail: beuzon@uma.es

M'B. Fatmi et al. (eds.), *Pseudomonas syringae Pathovars and Related Pathogens.*
© Springer Science+Business Media B.V. 2008

Keywords Virulence determinants, plant defences, hypersensitive response, type III secretion, effector

1 Introduction

Populations of pathogenic and non-pathogenic bacteria develop in characteristically different ways within the plant tissue, depending on the type of interaction established with the plant (Ercolani and Crosse, 1966; Omer and Wood, 1969). Typically, in compatible interactions bacterial pathogens multiply rapidly, reaching large populations that are associated with symptom induction and disease development. The ability of *Pseudomonas syringae* to replicate within compatible hosts is directly related to its pathogenicity and depends on a large number of virulence factors. Amongst these factors, the Hrp type III secretion system (TTSS) and its secreted effector proteins are required both for development of infection in compatible hosts, and for induction of hypersensitive response in incompatible hosts (Alfano and Collmer, 1997). Lack of a functional TTSS causes a reduction in bacterial growth within compatible hosts accompanied by an attenuation of the induced disease symptoms. Some type III-secreted effectors have been recently shown to be involved in suppressing plant defence responses (Jakobek et al., 1993; Tsiamis et al., 2000; Abramovitch et al., 2003; Espinosa et al., 2003; Hauck et al., 2003; Jamir et al., 2004; Kim et al., 2005; de Torres et. al., 2006) Accordingly, mutants lacking one of these effectors could trigger the induction of plant defences, which are either non-host or gene-for-gene resistance responses. Both types of defence responses determine a severe reduction on bacterial growth within the plant, and can be associated to either a macroscopic hypersensitive response, or to absence of symptoms (Ritter and Dangl, 1995; Jackson et al., 1999; Abramovitch et al., 2003; Espinosa et al., 2003). In a similar manner, loss of plant factors may affect bacterial growth, by altering the type of interaction established (Yu et al., 1993; Lu et al., 2001). Thus, measuring growth of pathogens *in planta* is key to carry out genetic analysis on the effect that both bacterial and plant factors play during the establishment of their interaction. This requires accurate and sensitive methods to quantify bacterial growth.

Bacterial growth is commonly measured in *P. syringae*-plant interactions by determining bacterial colony forming units (cfu) from a fixed-sized leaf sample several days post-inoculation (dpi) by either pressure infiltration, dipping, or spraying (Ercolani and Crosse, 1966; Whalen et al., 1991; Tornero and Dangl, 2001). However, experimental plant-to-plant variations are difficult to eliminate and frequently hinder accurate quantification of small differences. As genetic analyses of virulence factors involved in plant-pathogen interactions get increasingly complex, fine quantification of growth differences becomes desirable. In animal models, the use of competitive index in mixed infections allows accurate quantification of virulence (Freter et al., 1981; Taylor et al., 1987), and has proved very helpful in the identification and characterisation of type III-secreted effectors (Beuzón et al., 2000; Ruiz-Albert et al., 2002). It has also allowed to determine functional relationships between effectors as well as other virulence factors by simultaneous genetic analysis of multiple mutants (Baümler et al., 1997; Shea et al., 1999; Beuzón et al., 2000, 2001; Beuzón and Holden, 2001; Ruiz-Albert et al., 2002).

Although antecedents on the use of mixed infections in the analysis of plant-pathogen interactions date from the 1960s, their use as virulence assays has been hindered by two facts. Firstly, populations of non-pathogenic bacteria reach higher levels when co-inoculated with pathogenic bacteria, and secondly, triggering of HR by an incompatible pathogen hampers growth of a co-inoculated compatible one, unless the latter is capable of inhibiting HR development (Omer and Wood, 1969; Young, 1974; Robinette and Matthysse, 1990; López-Solanilla et al., 2001). Furthermore, as reported by Llama-Palacios and co-workers (Llama-Palacios et al., 2002) using mixed infections of different *Erwinia chrysantemi* mutant derivatives in potato disks, fully virulent mutants could be less competitive than the wild type when co-inoculated, due to differences in fitness. However, all these studies have been carried out using a considerably high inoculation dose, in which interference between the strains is predictable. Our hypothesis was that, in a manner similar to what happens in animal hosts (Beuzón and Holden, 2001), it should be possible to prevent interferences between strains by adjusting the inoculation dose. We hypothesised that use of lower inoculation doses would allow bacteria to disperse within the leaf tissue in such a way that each bacterium would replicate according to its own ability and thus establish a favourable interaction with the plant. In such an experimental setting virulence of each strain within the mixed infection is expected to remain the same as in an individual infection. Therefore, variation from experiment to experiment should be drastically reduced due to the control strain being present within every sample as seen in animal models (Beuzón and Holden, 2001).

2 Materials and Methods

2.1 Growth Conditions

Bacteria were grown at 37°C (*Escherichia coli*) or 28°C (*Pseudomonas syringae*) with aeration in LB medium supplemented with ampicillin (100 µg ml⁻¹ for *E. coli* strains; 300 µg ml⁻¹ for liquid cultures and 500 µg ml⁻¹ for plates, for *P. syringae* strains), kanamycin (15 µg ml⁻¹), chloramphenicol (20 µg ml⁻¹ for *E. coli* strains; 6 µg ml⁻¹ for *P. syringae* pv. *phaseolicola* strains), nitrofurantoin (20 µg ml⁻¹), rifampicin (15 µg ml⁻¹), cycloheximide (50 µg ml⁻¹) as appropriate.

2.2 Generation of Mutants

Mutants, AZJ09 and AZJ10, were generated by conjugation and the transposon insertion was confirmed by PCR and hybridisation analysis (data not shown), as previously described (Koch et al., 2001). Wild type growth and induction of

Table 1 CI and RIR analysis of *P.syringae* strains

Infections	CI	RIR
1448a – Bean		
WT vs *hrcV*	0.0023 ± 0.0007[a]	0.0030 ± 0.0038[a]
WT vs Tn7	1.1057 ± 0.1786[b]	0.8719 ± 0.7365[b]
DC3000 – *Arabidopsis*		
WT vs *hrcC*	0.0044 ± 0.0043[a]	0.0028 ± 0.0027[a]
WT vs Tn7	1.1218 ± 0.3185[b]	1.0795 ± 1.0767[b]
WT vs WT/ pAvrRpt2	0.0080 ± 0.0061	0.0201 ± 0.0191
DC3000 – Tomato		
Wt vs *hrcC*	0.0099 ± 0.0005[a]	0.0125 ± 0.0116[a]
Wt vs Tn7	0.8639 ± 0.0796[b]	1.6832 ± 0.6229[b]
B728a – Bean		
Wt vs *hrcC*	$4 \times 10^5 \pm 2 \times 10^{5}$[a]	$1 \times 10^5 \pm 9 \times 10^{6}$[a]
Wt vs Tn5	0.5844 ± 0.3986[b]	0.4053 ± 0.2733[b]

[a]Significantly different from 1.0 ($p < 0.05$)
[b]Not significantly different from 1.0 ($p < 0.05$)

symptoms by the resulting mutants was also confirmed (data not shown, Table 1). AZJ11 was generated as previously described (Andersen et al., 1998). A single transposon insertion was confirmed by Southern hybridisation analysis (data not shown). Full virulence and wild type growth of mutant AZJ11 was also confirmed in bean plants (data not shown, Table 1). Mutant IOM1 was generated by allelic exchange as previously described for other TTSS mutations (Ortiz-Martín et al., 2006). Briefly, a fragment of 2.1 Kb, that included the *hrcV* ORF (2088 bp) was amplified by PCR from genomic DNA of *P. syringae* pv. phaseolicola 1448a using primers *hrcV1* (5'-CCTCTAGAACTATAGCTAGCAACTTTCTGAACATG GT CGC-3') and *hrcV2* (5'-GCAATGATCATCATGACAGATCATCTTCAAGGTC GAAGAG-3') and cloned into the *EcoR*V restriction site of pKAS32 (Skorupski and Taylor, 1996). A 1-kb internal fragment of the *hrcV* ORF (nucleotides 507–1546) was deleted by PCR amplification using outward-facing primers 21 (5'-AAT CCTCGAGAGATACAGCGAGTGCTCACGCTGACG-3') and 22 (5'-AATCCTCG AGTCAGGTGGCG CGCAGATCACTGTCG-3'). These two primers introduce a new *Xho*I restriction site in the *hrcV* ORF into which the *aphT* gene, which confers resistance to kanamycin, was cloned. The new plasmid was used for allelic exchange as described (Ortiz-Martín et al., 2006), and the resulting mutant was confirmed by Southern blot hybridisation analysis of genomic DNA, using the appropriate restriction enzymes and the *aphT* ORF as the probe (data not shown).

2.3   Plant Experiments

Seeds of *Phaseolus vulgaris* cultivar Canadian Wonder (Chase Organics Ltd, Hersham, Surrey, UK) were germinated and grown with 16 h light-8 h dark cycles. For inoculum preparation, bacterial lawns were grown on LB plates for 48 h at 28°C and resuspended in 2 ml of 10 mM Cl$_2$Mg. The OD$_{600}$ was adjusted to 0,1 (5×10^7 cfu/ml) and serial dilutions were made to reach the final inoculum dose (5×10^4 cfu/ml). Fully expanded leaves of eight weeks-old bean plants (cv. Canadian Wonder) were inoculated with 200 µl of the appropriate bacterial suspension using a 2-ml syringe without needle, rendering an initial inoculation dose of 10^4 cfu per infiltrated leaf. Samples were taken from infiltrated leaves at the appropriate dpi using a 10-mm diameter cork borer. Five disks were taken per plant, placed into 1 ml of 10 mM Cl_2Mg, and homogenized by mechanical disruption. Serial dilutions of the resulting bacterial suspensions were plated onto LB plates supplemented with 50 µg/ml of cycloheximide.

Seeds of *Arabidopsis thaliana* accession Col-0 were germinated and grown in growth chambers with 8 h-light 16 h-dark cycles at 22°C. Four to five-week-old plants were inoculated with a 10^4 cfu ml^{-1} bacterial suspension using a 2-ml syringae without needle. Leaf disks were taken with a 10-mm diameter cork borer from infected tissue at 4 dpi. Three leaf disks were taken per plant, placed into 1 ml of sterile distilled water, and homogenized by mechanical disruption. Serial dilutions of the resulting bacterial suspension were plated onto selective plates supplemented with cycloheximide.

Seeds of *Licopersicum sculentum* cultivar Money maker were germinated and grown with 16 h light-8 h dark cycles. Four to five-week-old plants were inoculated with a 10^4 cfu ml^{-1} bacterial suspension using a 2-ml syringae without needle. Leaf disks were taken with a 10-mm diameter cork borer from infected tissue at 4 dpi. Three leaf disks were taken per plant, placed into 1 ml of sterile distilled water, and homogenized by mechanical disruption. Serial dilutions of the resulting bacterial suspension were plated onto selective plates supplemented with cycloheximide.

2.4 Mixed Infections, Competitive Index Assays and Relative Increase Ratio Assays

For competitive index (CI) assays, a mixed inoculum containing equal cfu of wild type and mutants was inoculated by infiltration into leaf tissue. Serial dilutions of the inoculum were plated onto LB agar and LB agar with the corresponding antibiotic to confirm dose and relative proportion between the strains, and thus obtain the input mutant/wt ratio that should be approximately 1. After a number of dpi that varied with the infection model used, bacteria were recovered from the infected leaves as described above and enumerated using respective LB agar plates.

For relative increase ratio (RIR) assays, individual inocula of the strains were inoculated into different plants and samples were taken after a fixed number of dpi to determine output cfu. Five hundred µl of a mixed inoculum at 5×10^4 cfu/ml, containing equal amounts of wild type and mutant bacteria was inoculated into 4.5 ml of LB medium and grown for 24 h at 28°C with aeration to determine *in vitro* CIs. Serial dilutions were then plated onto LB-agar and LB-Km-agar to calculate the relative proportion between the strains.

CI and ^{iv}CI are the mean of three independent experiments. Six plants (3 per strain) were used to obtain the mean RIR value. All experiments have been repeated at least twice obtaining results similar to those shown here. In all three indexes error bars represent standard deviation. Each CI, [iv]CI and RIR was analysed using the Student's test and the null hypothesis: mean index is not significantly different from 1.0. P values of 0.05 or less were considered significant.

3   Results and Discussion

3.1   Establishing CI Assays as a Virulence Assay for **P. syringae** pv. **phaseolicola**

In this work, we set out to establish the basis of the use of competitive index (CI) in mixed infections as an accurate virulence assay to carry out genetic analysis of *P. syringae* interactions with different hosts. We started the work by using the interaction between *P. syringae* pv. *phaseolicola* (hereafter referred to as Pph) and *Phaseolus vulgaris* as our initial model. We selected the fully sequenced strain 1448a (Teverson, 1991), a wild type representative of Race 6 that is fully virulent in all bean cultivars tested (Teverson, 1991). For these experiments, we also generated a non-pathogenic 1448a mutant, IOM1, carrying a non-polar knockout of the *hrcV* gene. HrcV is an essential and highly conserved component of the Hrp TTSS (Cornelis and Van Gijsegem, 2000).

Virulence in the Pph-bean model is most commonly assayed by pod assays that give a clear, albeit qualitative measure of bacterial virulence (Harper et al., 1987). Examples of bacterial growth measurements in this model can also be found in the literature (Rahme et al., 1991; Jackson et al., 1999). The most common inoculation dose previously used was 10^7–10^8 cfu. However, a number of reports using such high inoculation doses showed interference between co-inoculated strains: Either complementation of an otherwise attenuated strain or restriction of growth of a virulent strain due to triggering host defences by an incompatible one (Omer and Wood, 1969; Young, 1974; Robinette and Matthysse, 1990). We tested this inoculation dose for mixed infections, and found both, complementation of mutant growth by the presence of a wild type strain and restriction of wild type growth due induction of hypersensitive response by an incompatible strain (data not shown). Thus, since an accurate virulence assay requires each strain to grow as it would during an individual infection, we decided to lower the inoculation dose by a 1,000-fold to reduce the likelihood of interference between the strains. To confirm the reproducibility of the bacterial counts obtained using this dose and to establish the best time-point to measure the growth difference between the wild type and the *hrp* mutant, we monitored the growth of 1448a and IOM1 ($\Delta hrcV$) after independent inoculations of bean plants. Wild type populations reached the highest value at 10 dpi. The maximal differences between the wild type and the mutants were found between 10 and 14 dpi and the minimal variation between samples was at 14 dpi (data not shown). The CI is defined as the output mutant-to-wild type ratio, divided by the equivalent ratio in the input (Freter et al., 1981; Taylor et al., 1987) (Fig. 1).

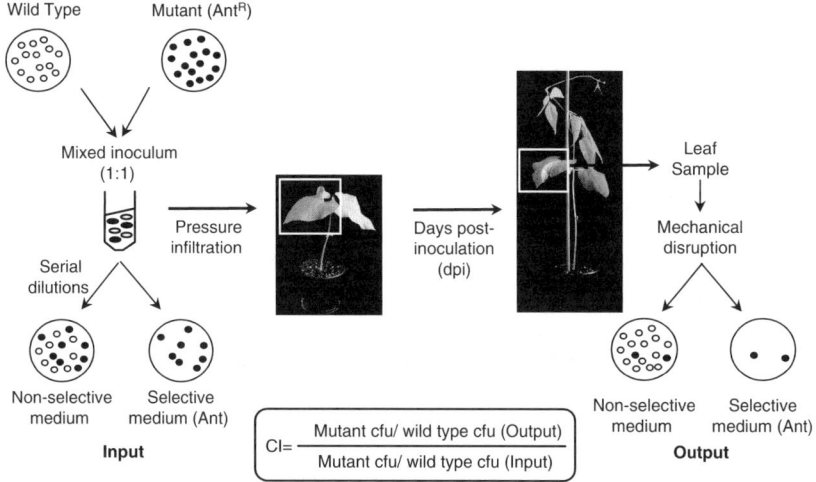

Fig. 1 Schematic representation of Competitive Index (CI) assays

Thus, two equally virulent strains should give a CI of nearly 1, whereas an attenuated strain should give a CI significantly smaller than 1, getting closer to 0 as the attenuation of the strains increases. To allow an easier comparison between the variations observed in individual *versus* mixed infections, we calculated a CI-like index that we named *r*elative *i*ncrease *r*atio (RIR), using cfu counts from the growth of each strain within individual infections. RIR was calculated as the cfu ratio between mutant and wild type strain after growth during individual infections, divided by the cfu ratio between the strains in their respective initial inocula.

Table 1 shows the CI and RIR resulting from analysis of bacterial growth in mixed infections of 1448a and IOM1 (Δ*hrcV*). The CI of IOM1 (Δ*hrcV*) showed a 100- to 1,000-fold attenuation in growth compared to that of 1448a. This level of attenuation is consistent with that detected during individual growth, confirming our working hypothesis that within a mixed infection with lower inoculation dose, each bacterium replicates according to its own ability to establish a favourable interaction with the plant. These results also indicate that 5×10^4 cfu/ml is an appropriate inoculation dose, which is low enough to prevent complementation of mutant growth by the presence of the wild type within the mixed infection. Thus, the virulence of each strain remains the same as in an individual infection, but plant-to-plant variation decreased due to the wild type being present within each inoculum thus acting as an internal control. The RIR resulting from comparison of IOM1 with 1448a is very similar to the CI. However, standard deviations from CI results were almost 1,000-fold smaller than that obtained from RIR assaya (Table 1). The very large variation obtained from individual experiments could certainly be reduced by increasing the number of plants assayed, since only three plants per strain were used. However, only three plants were necessary to obtain a meaningful CI with minimal variation. It is worth pointing out that total population sizes were equally variable in mixed infection experiments and individual infections. The key difference between CI and RIR is that the comparison between growth of the mutant and the wild type is carried out within the same plant in the case of the CI but in different plants in the case of the RIR.

A parallel competitive *in vitro* growth experiment was carried out in LB medium to determine if the growth difference detected between mutant and wild type *in planta* could also be detected *in vitro*. A CI-like index (hereafter referred to as *in vitro* competitive index - ^{iv}CI) was obtained as the cfu ratio of mutant/wild type plated from the culture after incubation divided by the mutant/wild type ratio in the initial inoculum. The ^{iv}CI obtained was not significantly different from 1 (data not shown) indicating that the difference in growth between mutant and wild type was specific to *in planta* growth.

Another pre-requisite for the use of CI in mixed infections as a measure for virulence in a particular host-pathogen interaction is that two equally virulent strains must give a CI not significantly different from 1. In other words, the inoculation dose must be such as to avoid bacterial fitness to cause differences in growth within a mixed infection (Llama-Palacios et al., 2002). To test this under our experimental conditions, we generated a fully virulent 1448a derivative carrying a mini-Tn7-*gfp1* (conferring resistance to kanamycin) designated AZJ09. Table 1 shows that the CI of AZJ09 is not significantly different from 1, indicating that mixed infections in this model also fulfil the requirement. Similarly, RIR was not significantly different from 1, although the standard deviation was 10-fold higher than that of the CI (Table 1). As expected ^{iv}CI showed no difference in growth between the two strains (data not shown).

3.2   Establishing CI Assays as a Virulence Assay in Other Models

Our successful proof-of-principle reported above prompted us to test whether the CI assay was also applicable to other plant model systems. For this, we selected the fully sequenced *P. syringae* pv. *tomato* strain DC3000 (Cuppels, 1986; Buell et al., 2003) in its interaction with both, *Arabidopsis thaliana* (The Arabidopsis Genome Initiative, 2000) and its natural host *Lycopersicon esculentum* (tomato). Table 1 shows that using an inoculation dose of 5×10^4 cfu/ml the DC3000-*Arabidopsis* interaction model does fulfil the two requirements. The CI obtained for the fully virulent miniTn7-labelled DC3000 derivative, AZJ10, is not significantly different from 1. In contrast, a DC3000 knock-out mutation within *hrcC* (Mudgett and Staskawicz, 1999), which codes for a structural component of the Hrp TTSS (Deng et al., 1998), gives a CI 100- to 1000-fold smaller than 1 (Table 1), consistent with previously published results (Fouts et al., 2003). Concurrently, RIR for AZJ10 was not significantly different from 1, and RIR of the DC3000 *VhrcC* mutant showed a similar attenuation as the CI. Comparably, Table 1 shows that using the established

inoculation dose, the DC3000-tomato interaction model also fulfils the two requirements for the use of CI as virulence assay. The CI obtained for the miniTn7-labelled
DC3000 derivative was not significantly different from 1 and the CI for DC3000
∇hrcC was 100-fold smaller than 1 (Table 1), also consistent with previous results
(Fouts et al., 2003).

To finalize the testing of the suitability of this type of assay, we assayed B728a,
the fully sequenced *P. syringae* pv. *syringae* (hereafter referred to as Psy) strain
(Feil et al., 2005; Loper and Lindow, 1987) in its interaction with bean plants. We
failed to obtain a mini-Tn7 derivative of B728a (data not shown). Instead, we generated a mini-Tn5 derivative designated AZJ11. As an attenuated strain we assayed
a B728a Δ*hrcC* mutant (Hirano et al., 1999). Initially, we used the above-mentioned
inoculation dose for this pathosystem. However, a general collapse of the infiltrated
leaf tissue was noticed as soon as 2 dpi, and the growth of B728a Δ*hrcC* was partially complemented by the presence of the wild type strain in the mixed infection.
Consequently, the inoculation doses were lowered to 10^3 and to 5×10^3 cfu/ml in
two differents experiments. Results obtained using the lower dose (10^3 cfu/ml) were
too variable to give a reliable index in this model (data not shown). Consequently,
an inoculation dose of 5×10^3 cfu/ml was selected for the assay using Psy. This
way, we obtained a CI not significantly different from 1 for the fully virulent B728a
derivative, AZJ11, and a CI 10^4–10^5 times smaller for the B728a Δ*hrcC* mutant
(Table 1) clearly confirming previously published results (Deng et al., 1998). RIRs
were in both cases similar to the CIs (Table 1) suggesting that the herein described
assay can be applied to the third pathosystem. The fact that a lower inoculation dose
was necessary to apply CI assays for the Psy-Bean model illustrated that although
an inoculation dose of 5×10^4 cfu/ml is a good starting point for CI analysis in a
new model, it may need to be adjusted according to the specific requirements of
each novel pathosystem.

3.3   Establishing CI Assays for Avirulent Strains

Type III-secreted effector proteins can trigger a hypersensitive response in hosts
carrying the appropriate R-gene (Alfano and Collmer, 1997). AvrRpt2 is such an
effector originally isolated from *P. syringae* pv *tomato* strain JL1065 (Whalen et al.,
1991) and required for virulence on tomato (Lim and Kunkel, 2005). *P. syringae*
strains expressing *avrRpt2* rapidly induce host defence responses on *Arabidopsis*
plants carrying a functional *RPS2* gene that limit their growth and disease development, rendering the strains avirulent on these plants (Kunkel et al., 1993; Yu et al.,
1993). Since previous reports showed that an HR-triggering high dosage of an
incompatible strain hampers growth of a compatible one in mixed infection (Omer
and Wood, 1969; Robinette and Matthysse, 1990), CI analysis could be misleading
by showing no difference in growth between wild type and the avirulent strain.
Thus, our established dose needed to be tested for interference between an avirulent
strain and a virulent one in mixed infection. For this purpose, we determined the CI

and RIR for DC3000 *versus* DC3000 expressing *avrRpt2*. CI and RIR were similar and showed a 100-fold reduction on the growth of DC3000 expressing *avrRpt2* compared to that of DC3000 (Table 1), proving the validity of the assay to carry out analyses of avirulent strains.

3.4   Concluding Remarks

The results presented herein showed that it is possible to find a inoculation dose for which each bacterium within a mixed infection of a plant host will have the same probability to multiply as during single infections. This finding can be exploited to use competitive index in mixed infections as a simple and quick virulence assay. CI analysis of just three samples, provide an accurate and reproducible quantification of the virulence of all tested strains. All experiments were repeated at least twice giving similar results (data not shown). Variation within CIs was considerably smaller than that of RIR in the majority of the assays, except in those cases where RIR variation was particularly small. The smaller variation of this type of assay suggested that – similarly to what happens in animal pathogens – CI assays may be useful to detect small differences in attenuation. This is particularly true when experimental and plant-to-plant variations may render quantification by single infections insensitive. These results offer the option to use CI assays for the genetic analysis of multiple mutants as previously been established for animal pathogens. Moreover, our results established the basis for the application of high-throughput techniques such as signature-tagged mutagenesis (Hensel et al., 1995), that depend on growth selection within complex mixed infections to identify novel virulence factors.

Acknowledgements We are very grateful to A. Charkowsky, S. Hirano, D. Cuppels, J. Murillo, and P. Tornero for kindly providing strains and plasmids. We are also grateful to J. Ruiz-Albert, E.R. Bejarano and C. Ramos, for critical reading of the manuscript and P. Tornero for helpful discussion. We thank F. Gordillo for statistical advice and T. Duarte and C. Soto for technical assistance. C.R. Beuzón was supported by the "Ramón y Cajal" Program from the Ministerio de Ciencia y Tecnología. I. Ortiz-Martín was supported by a Fellowship from Junta de Andalucía. This work was supported by a project grant from the Ministerio de Educación y Ciencia (Spain) to C.R. Beuzón.

References

Abramovitch, R.B., Kim, Y.J., Chen, S., Dickman, M.B. and Martin, G.B., 2003, *Pseudomonas* type III effector AvrPtoB induces plant disease susceptibility by inhibition of host programmed cell death. *EMBO J.* 22, 60–69.

Alfano, J.R. and Collmer, A., 1997, The type III (Hrp) secretion pathway of plant pathogenic bacteria: trafficking harpins, Avr proteins, and death. *J. Bacteriol.* 179, 5655–5662.

Andersen, J.B., Sternberg, C., Poulsen, L.K., Bjorn, S.P., Givskov, M. and Molin, S., 1998, New unstable variants of green fluorescent protein for studies of transient gene expression in bacteria. *Appl. Environ. Microbiol.* 64, 2240–2246.

Baümler, A.J., Tsolis, R.M., Valentine, P.J., Ficht, T.A. and Heffron, F., 1997, Synergistic effect of mutations in *invA* and *lpfC* on the ability of *Salmonella typhimurium* to cause murine typhoid. *Infect. Immun.* 65, 2254–2259.

Beuzón, C.R. and Holden, D.W., 2001, Use of mixed infections with *Salmonella* strains to study virulence genes and their interactions *in vivo. Microbes Infect.* 3, 1345–1352.

Beuzón, C.R., Meresse, S., Unsworth, K.E., Ruiz-Albert, J., Garvis, S., Waterman, S.R., Ryder, T. A., Boucrot, E. and Holden, D.W., 2000, *Salmonella* maintains the integrity of its intracellular vacuole through the action of SifA. *EMBO J.* 19, 3235–3249.

Beuzón, C.R., Unsworth, K.E. and Holden, D.W., 2001, *In vivo* genetic analysis indicates that PhoP-PhoQ and the *Salmonella* pathogenicity island 2 type III secretion system contribute independently to *Salmonella enterica* serovar Typhimurium virulence. *Infect. Immun.* 69, 7254–7261.

Buell, C.R., Joardar, V., Lindeberg, M., Selengut, J., Paulsen, I.T., Gwinn, M.L., Dodson, R.J., Deboy, R.T., Durkin, A.S., Kolonay, J.F., Madupu, R., Daugherty, S., Brinkac, L., Beanan, M.J., Haft, D.H., Nelson, W.C., Davidsen, T., Zafar, N., Zhou, L., Liu, J., Yuan, Q., Khouri, H., Fedorova, N., Tran, B., Russell, D., Berry, K., Utterback, T., Van Aken, S.E., Feldblyum, T.V., D'Ascenzo, M., Deng, W.L., Ramos, A.R., Alfano, J.R., Cartinhour, S., Chatterjee, A.K., Delaney, T.P., Lazarowitz, S.G., Martin, G.B., Schneider, D.J., Tang, X., Bender, C.L., White, O., Fraser, C.M. and Collmer, A., 2003, The complete genome sequence of the *Arabidopsis* and tomato pathogen *Pseudomonas syringae* pv. *tomato* DC3000. *Proc. Natl. Acad. Sci.* 100, 10181–10186.

Cornelis, G.R. and Van Gijsegem, F., 2000, Assembly and function of type III secretory systems. *Annu. Rev. Microbiol.* 54, 735–774.

Cuppels, D.A., 1986, Generation and characterization of Tn*5* insertion mutations in *Pseudomonas syringae* pv. *tomato. Appl. Environ. Microbiol.* 51, 323–327.

de Torres, M., Mansfield, J.W., Grabov, N., Brown, I.R., Ammouneh, H., Tsiamis, G., Forsyth, A., Robatzek, S., Grant, M. and Boch, J., 2006, *Pseudomonas syringae* effector AvrPtoB suppresses basal defence in *Arabidopsis. Plant J.* 47, 368–382.

Deng, W.L., Preston, G., Collmer, A., Chang, C.J. and Huang, H.C., 1998, Characterization of the *hrpC* and *hrpRS* operons of *Pseudomonas syringae* pathovars *syringae, tomato,* and *glycinea* and analysis of the ability of *hrpF, hrpG, hrcC, hrpT,* and *hrpV* mutants to elicit the hypersensitive response and disease in plants. *J. Bacteriol.* 180, 4523–4531.

Ercolani, G.L. and Crosse, J.E., 1966, The Growth of *Pseudomonas phaseolicola* and Related Plant Pathogens *in vivo. J. Gen. Microbiol.* 45, 429–439.

Espinosa, A., Guo, M., Tam, V.C., Fu, Z.Q. and Alfano, J.R., 2003, The *Pseudomonas syringae* type III-secreted protein HopPtoD2 possesses protein tyrosine phosphatase activity and suppresses programmed cell death in plants. *Mol. Microbiol.* 49, 377–387.

Feil, H., Feil, W.S., Chain, P., Larimer, F., DiBartolo, G., Copeland, A., Lykidis, A., Trong, S., Nolan, M., Goltsman, E., Thiel, J., Malfatti, S., Loper, J.E., Lapidus, A., Detter, J.C., Land, M., Richardson, P.M., Kyrpides, N.C., Ivanova, N. and Lindow, S.E., 2005, Comparison of the complete genome sequences of *Pseudomonas syringae* pv. *syringae* B728a and pv. *tomato* DC3000. *Proc. Natl. Acad. Sci.* 102, 11064–11069.

Fouts, D.E., Badel, J.L., Ramos, A.R., Rapp, R.A. and Collmer, A., 2003, A *Pseudomonas syringae* pv. tomato DC3000 Hrp (Type III secretion) deletion mutant expressing the Hrp system of bean pathogen *P. syringae* pv. *syringae* 61 retains normal host specificity for tomato. *Mol. Plant Microbe Interact.* 16, 43–52.

Freter, R., Allweiss, B., O'Brien, P.C., Halstead, S.A. and Macsai, M.S., 1981, Role of chemotaxis in the association of motile bacteria with intestinal mucosa: *in vitro* studies. *Infect. Immun.* 34, 241–249.

Harper, S., Zewdie, N., Brown, I.R. and Mansfield, J.W., 1987, Histological, physiological and genetical studies of the responses of leaves and pods of *Phaseolus vulgaris* to three races of

Pseudomonas syringae pv. *phaseolicola* and to *Pseudomonas syringae* pv. *coronafaciens*. *Physiol. Mol. Plant Pathol*. 31, 153–172.

Hauck, P., Thilmony, R. and He, S.Y., 2003, A *Pseudomonas syringae* type III effector suppresses cell wall-based extracellular defense in susceptible *Arabidopsis* plants. *Proc. Natl. Acad. Sci*. 100, 8577–8582.

Hensel, M., Shea, J.E, Gleeson, C., Jones, M.D., Dalton, E. and Holden D.W., 1995, Simultaneous identification of bacterial virulence genes by negative selection. *Science* 269, 400–403.

Hirano, S.S., Charkowski, A.O., Collmer, A., Willis, D.K. and Upper, C.D., 1999, Role of the Hrp type III protein secretion system in growth of *Pseudomonas syringae* pv. *syringae* B728a on host plants in the field. *Proc. Natl. Acad. Sci*. 96, 9851–9856.

Jackson, R.W., Athanassopoulos, E., Tsiamis, G., Mansfield, J.W., Sesma, A., Arnold, D.L., Gibbon, M.J., Murillo, J., Taylor, J.D. and Vivian, A., 1999, Identification of a pathogenicity island, which contains genes for virulence and avirulence, on a large native plasmid in the bean pathogen *Pseudomonas syringae* pathovar *phaseolicola*. *Proc. Natl. Acad. Sci*. 96, 10875–10880.

Jakobek, J.L., Smith, J.A. and Lindgren, P.B., 1993, Suppression of bean defense responses by *Pseudomonas syringae*. *Plant Cell* 5, 57–63.

Jamir, Y., Guo, M., Oh, H.S., Petnicki-Ocwieja, T., Chen, S., Tang, X., Dickman, M.B., Collmer, A. and Alfano, J.R., 2004, Identification of *Pseudomonas syringae* type III effectors that can suppress programmed cell death in plants and yeast. *Plant J*. 37, 554–565.

Kim, M.G., da Cunha, L., McFall, A.J., Belkhadir, Y., DebRoy, S., Dangl, J.L. and Mackey, D., 2005, Two *Pseudomonas syringae* type III effectors inhibit RIN4-regulated basal defense in *Arabidopsis*. *Cell* 121, 749–759.

Koch, B., Jensen, L.E. and Nybroe, O., 2001, A panel of Tn7-based vectors for insertion of the *gfp* marker gene or for delivery of cloned DNA into Gram-negative bacteria at a neutral chromosomal site. *J. Microbiol. Methods* 45, 187–195.

Kunkel, B.N., Bent, A.F., Dahlbeck, D., Innes, R.W. and Staskawicz B.J., 1993, *RPS2*, an *Arabidopsis* disease resistance locus specifying recognition of *Pseudomonas syringae* strains expressing the avirulence gene *avrRpt2*. *Plant Cell* 5, 865–875.

Llama-Palacios, A., Lopez-Solanilla, E. and Rodriguez-Palenzuela, P., 2002, The *ybiT* gene of *Erwinia chrysanthemi* codes for a putative ABC transporter and is involved in competitiveness against endophytic bacteria during infection. *Appl. Environ. Microbiol*. 68, 1624–1630.

Lim, M.T.S. and Kunkel, B.N., 2005, The *Pseudomonas syringae avrRpt2* gene contributes to virulence on Tomato. *Mol Plant Microbe Interact*. 18, 626–633.

Loper, J.E. and Lindow, S.E., 1987, Lack of evidence for *in situ* fluorescent pigment production by *Pseudomonas syringae* pv. *syringae* on bean leaf surfaces. *Phytopathology* 77, 1449–1454.

López-Solanilla, E., Llama-Palacios, A., Collmer, A., García-Olmedo, F. and Rodriguez-Palenzuela, P., 2001, Relative effects on virulence of mutations in the *sap*, *pel*, and *hrp* loci of *Erwinia chrysanthemi*. *Mol. Plant Microbe Interact*. 14, 386–393.

Lu, M., Tang, X. and Zhou, J.M., 2001, *Arabidopsis* NHO1 is required for general resistance against *Pseudomonas* bacteria. *Plant Cell* 13, 437–447.

Mudgett, M.B. and Staskawicz, B.J., 1999, Characterization of the *Pseudomonas syringae* pv. *tomato* AvrRpt2 protein: demonstration of secretion and processing during bacterial pathogenesis. *Mol. Microbiol*. 32, 927–941.

Omer, M.E.H. and Wood, R.K.S., 1969, Growth of *Pseudomonas phaseolicola* in susceptible and resistant bean plants. *Ann. Appl. Biol*. 63, 103–116.

Ortiz-Martín, I., Macho, A.P., Lambersten, L., Ramos, C. and Beuzon, C.R., 2006, Suicide vectors for antibiotic marker exchange and rapid generation of multiple knockout mutants by allelic exchange in Gram-negative bacteria. *J. Microbiol. Methods* 67, 395–407.

Rahme, L.G., Mindrinos, M.N. and Panopoulos, N.J., 1991, Genetic and transcriptional organization of the hrp cluster of *Pseudomonas syringae* pv. *phaseolicola*. *J. Bacteriol*. 173, 575–586.

Ritter, C. and Dangl, J.L., 1995, The *avrRpm1* gene of *Pseudomonas syringae* pv. *maculicola* is required for virulence on *Arabidopsis*. *Mol. Plant Microbe Interact*. 8, 444–453.

Robinette, D. and Matthysse, A.G., 1990, Inhibition by *Agrobacterium tumefaciens* and *Pseudomonas savastanoi* of development of the hypersensitive response elicited by *Pseudomonas syringae* pv. *phaseolicola. J. Bacteriol.* 172, 5742–5749.

Ruiz-Albert, J., Yu, X.J., Beuzón, C.R., Blakey, A.N., Galyov, E.E. and Holden, D.W., 2002, Complementary activities of SseJ and SifA regulate dynamics of the *Salmonella typhimurium* vacuolar membrane. *Mol. Microbiol.* 44, 645–661.

Shea, J.E., Beuzón, C.R., Gleeson, C., Mundy, R. and Holden, D.W., 1999, Influence of the *Salmonella typhimurium* pathogenicity island 2 type III secretion system on bacterial growth in the mouse. *Infect. Immun.* 67, 213–219.

Skorupski, K. and Taylor, R.K., 1996, Positive selection vectors for allelic exchange. *Gene* 169, 47–52.

Taylor, R.K., Miller, V.L., Furlong, D.B. and Mekalanos, J.J., 1987, Use of *phoA* gene fusions to identify a pilus colonization factor coordinately regulated with cholera toxin. *Proc. Natl. Acad. Sci.* 84, 2833–2837.

Teverson, D.M., 1991, Genetics of pathogenicity and resistance in the halo-blight disease of beans in Africa. Ph.D. thesis. Birmingham, UK, University of Birmingham.

The Arabidopsis Genome Initiative, 2000, Analysis of the genome sequence of the flowering plant *Arabidopsis thaliana. Nature* 408, 796–815.

Tornero, P. and Dangl, J.L., 2001, A high-throughput method for quantifying growth of phytopathogenic bacteria in *Arabidopsis thaliana. Plant J.* 28, 475–481.

Tsiamis, G., Mansfield, J.W., Hockenhull, R., Jackson, R.W., Sesma, A., Athanassopoulos, E., Bennett, M.A., Stevens, C., Vivian, A., Taylor, J.D. and Murillo, J., 2000, Cultivar-specific avirulence and virulence functions assigned to *avrPphF* in *Pseudomonas syringae* pv. *phaseolicola*, the cause of bean halo-blight disease. *EMBO J.* 19, 3204–3214.

Whalen, M.C., Innes, R.W., Bent, A.F. and Staskawicz, B.J., 1991, Identification of *Pseudomonas syringae* pathogens of *Arabidopsis* and a bacterial locus determining avirulence on both *Arabidopsis* and soybean. *Plant Cell* 3, 49–59.

Young, J.M., 1974, Development of bacterial populations *in vivo* in relation to plant pathogenicity. *N. Z. J. Agric. Res.* 17, 105–113.

Yu, G.L., Katagiri, F. and Ausubel, F.M., 1993, *Arabidopsis* mutations at the RPS2 locus result in loss of resistance to *Pseudomonas syringae* strains expressing the avirulence gene *avrRpt2. Mol. Plant Microbe Interact.* 6, 434–443.

Genomic Analysis of *Pseudomonas syringae* Pathovars: Identification of Virulence Genes and Associated Regulatory Elements Using Pattern-Based Searches and Genome Comparison

M. Lindeberg[1], D.J. Schneider[2], S. Cartinhour[2], and A. Collmer[1]

Abstract The availability of complete genome sequences for three *P. syringae* pathovars (*P. s.* pv. *tomato* DC3000, *P. s.* pv. *phaseolicola* 1448A, and *P. s.* pv. *syringae* B728a) provides researchers with unprecedented opportunities to explore factors contributing to pathogenicity and virulence in these pathogens. One method employed for genome sequence analysis involves identification of genes and regulatory elements through association with conserved sequence motifs. Using this approach, ~50 and ~30 genes for T3SS-dependent Hop/Avr effector proteins have been identified in Pst DC3000 and Psp 1448A, respectively, on the basis of their proximity to conserved "hrp box" regulatory motifs and the presence of Hop-associated patterns at the N-terminus of the encoded proteins. These analyses have also led to a comprehensive catalog of the *hrpL* regulons of all three pathovars. Similar pattern-based searches have led to comprehensive identification of other regulatory elements in the Pst DC3000 genome, including binding sites for RpoD, RpoE, RpoN, RpoS, and Fur. Genome comparison represents a second powerful tool for identification of the genes and regulatory elements involved in diverse stages of the plant–pathogen association. For example, genome comparison of the three sequenced *P. syringae* pathovars sheds light on genes contributing to their different pathogenic strategies. More broadly, genes distinguishing the plant-associated pseudomonads from pseudomonads occupying other niches can be identified as described in the study comparing the Pst DC3000 genome with that of the saprophyte *P. putida* and the animal pathogen *P. aeruginosa*. Using this approach, 44 "LSRs" (lineage-specific regions) were identified in the Pst DC3000 genome. Enriched for genes encoding known virulence factors, the LSRs also contain a large number of genes encoding uncharacterized "hypothetical" proteins with potential roles in the bacteria–plant interaction. Comparisons among genomes of plant-associated pseudomonads and enterobacteriaceae may also yield insight into the underlying, broadly conserved functions required for bacteria to associate

---

[1] Cornell University, Ithaca, NY, 14850, USA

[2] US Department of Agriculture, Agricultural Research Service, Ithaca, NY 14853, USA

Author for correspondence: Magdalen Lindeberg; e-mail: ML16@cornell.edu

M'B. Fatmi et al. (eds.), *Pseudomonas syringae Pathovars and Related Pathogens.*
© Springer Science + Business Media B.V. 2008

with their plant hosts. However, the effective use of genomic data for any analysis requires that genome annotation be accurate, up-to-date, and enhanced as needed by improvements such as consistent nomenclature (as done for the effectors) or addition of gene ontology (GO) terms. Furthermore, the genome annotation, results of large-scale genomic analyses, and relevant tools must be readily accessible to the research community. To this end, the Pseudomonas–Plant Interaction (PPI) web site (http://pseudomonas-syringae.org) was developed and continues to be maintained for the benefit of the *P. syingae* research community.

Keywords *Pseudomonas syringae*, Hop, effector, iron, genomic island

1   Introduction

Pseudomonas syringae represents a group of economically important bacterial pathogens exhibiting diverse interactions with their plant hosts. Pathovars of *P. syringae* vary in their capacity for epiphytic survival, the nature of the symptoms they elicit, and their host range, with the outcome of the host–pathogen interaction dependent upon the interplay of pathogen and host factors involved in virulence and resistance. Comprehensive characterization of factors contributing to virulence and survival in the plant-associated environment has been greatly limited by redundancy of function, low expression levels, and absence of easily measurable phenotypes. However, the availability of complete genome sequences for three *P. syringae* pathovars has opened new investigative directions, including the use of pattern-based searches for identification of novel factors associated with known regulatory and targeting motifs, and genome comparison for identification of regions correlated with specific properties of the strain or species in which it is located.

The three sequenced strains of *Pseudomonas syringae*: *P. syringae* pv. *tomato* strain DC3000 (Pst DC3000) (Buell et al., 2003), *P. syringae* pv. *syringae* strain B728a (Psy B728a) (Feil et al., 2005), and *P. syringae* pv. *phaseolicola* strain 1448A (Psp 1448A) (Joardar et al., 2005) comprise an attractive set for genomic analysis. These strains represent each of the three major monophyletic groups within the *P. syringae* species (Sawada et al., 2002; Sarkar and Guttman, 2004), and each exhibits a different host range and pathogenic strategy. Pst DC3000 is the causal agent of bacterial speck in tomato and Arabidopsis and is an important model system for exploration of virulence factors (Preston, 2000). Psp 1448A causes halo blight in bean, an economically significant disease in developing countries, and is a model for exploration of race-cultivar interactions (Jackson et al., 1999). Psy B728a, the cause of brown spot on bean, is a model for the study of epiphytic fitness (Hirano and Upper, 2000; Monier and Lindow, 2003; Marco et al., 2005).

Use of genome sequence data for exploration of virulence determinants was initiated upon release of the first draft sequences, with the most intensive efforts focused on identification of novel Type III effectors or Hops (*h*rp *o*uter *p*rotein).

Hops represent an ideal class of proteins for functional genomic characterization given the significance of their role in the host–pathogen interaction, their large number, and the difficulty of identifying them by traditional approaches. Prior to genome sequencing, Hops were identified on the basis of phenotype, exemplified by the Avr proteins, expression of which induces a hypersensitive or "avirulent" response on plants carrying the corresponding *R* gene, or by their genetic proximity to genes of associated function, such as those found near the cluster of *hrp* genes encoding the T3SS structural genes (Alfano et al., 2000). However, the availability of complete genome sequences made possible the identification of novel proteins by an iterative strategy of alternating bioinformatic and experimental analyses. In its general form, this approach involves (i) identification of conserved patterns or sequence motifs associated with the genes of interest, (ii) scanning the genome for these motifs, (iii) experimental evaluation of the candidates identified by the search, and (iv) refinement of the search model by incorporation of the experimentally confirmed sequences.

2 Identification of Hop Effector Proteins in *P. syringae* Pathovars Using Functional Genomics

Hop proteins are well-suited to identification by a pattern-based approach given their regulation by HrpL, an alternative sigma factor that is part of a larger well-characterized, regulatory cascade (Xiao et al., 1994; Grimm et al., 1995; Hutcheson et al., 2001), and their selective trafficking through the T3SS. Using experimentally confirmed examples of the HrpL-binding site, or "hrp box", a hidden Markov model (HMM) was developed and employed for identification of uncharacterized hrp boxes in the Pst DC3000 genome. By examining the ORFs following the predicted hrp boxes for N-terminal sequence patterns associated with T3SS-dependent translocation, candidate effector genes were identified and evaluated for secretion or translocation (Fouts et al., 2002; Petnicki-Ocwieja et al., 2002). The conserved nature of the *hrp* box and/or translocation motif were similarly employed in effector-mining investigations by Guttman et al. (2002) and Zweisler-Vollick et al. (2002). Collectively, these various investigations expanded the number of Pst DC3000 proteins exhibiting *hrpL*-dependence and T3SS-dependent translocation to nearly 40 (Collmer et al., 2002).

The second phase of Pst DC3000 T3SS-related investigations was characterized by a number of significant developments including: (i) expansion of the number of Hop candidates tested, (ii) refinement of the HMM model used for hrp box identification, (iii) identification of specific Hop chaperones, (iv) improvements in the assays used for evaluating translocation, and (v) development of improved guidelines for *hop* classification and nomenclature. In order to achieve a more comprehensive list of *hop* genes, candidates were selected on the basis of their proximity to HrpL-regulated promoters, the presence of T3SS signals, proximity to known *hop* genes and/or predicted function in eukaryotic cells (Schechter et al., 2004, 2006; Vinatzer et al., 2005). Systematic evaluation of these candidates expanded the inventory of *hop/avr* genes and pseudogenes to 48 and helper genes, including harpins and pilus-associated proteins, to eight. Several of the *hop* candidates and *hop* orthologs evaluated were disrupted by transposon insertions in either the coding DNA or region between the hrp box and coding start site, consistent with a high degree of fluidity in the repertoire of active Hop proteins.

The durability of the HMM model for identification of hrpL binding sites was confirmed by Ferreira et al. (2006) using an ORF-specific microarray to identify genes differentially regulated by HrpL (Fereirra et al., 2006). Gibbs sampling of the regions upstream of up-regulated genes revealed that the hrp promoter represents the only identifiable regulatory motif. This study also generated a comprehensive list of genes and putative operons preceded by HrpL-reponsive promoters, including not only the type III effectors, but also twin-argenine transport (TAT) substrates and proteins involved in the synthesis or metabolism of phytohormones, phytotoxins, and myo-inositol. Differential expression of several sigma factors and transcription factors was observed in the absence of HrpL expression including RpoD, RpoS, SigW, and members of the SlyA/MarR, TetR, sigma-70 and sigma-24 families (Fereirra et al., 2006).

Characterization of hrpL-regulated operons has also provided clues as to location of genes encoding Hop chaperones. These genes are typically found proximal to the coding region of the Hop in whose translocation they assist, and the encoded proteins are characterized by low molecular mass and acidic isoelectric points. Nine Shc (specific hop chaperone) proteins have thus far been identified in Pst DC3000, with varying degrees of characterization (van Dijk et al., 2002; Badel et al., 2003; Shan et al., 2004; Wehling et al., 2004; Guo et al., 2005).

The intensive focus of numerous labs on Hop identification has necessitated technical improvement and standardization of the methods used for their characterization. One of the more significant technical developments has been adaptation of the adenylate cyclase (Cya) reporter system for evaluating translocation in *P. syringae* (Schechter et al., 2004). The Cya reporter has two main advantages over the more widely used AvrRptII-based reporter system in that its translocation can be quantified, and as a protein that is not itself an effector, its presence is unlikely to confound analyses of the effectors to which it is fused. While AvrRptII continues to be used, translocation assays with both reporters are now routinely conducted using fusions with full-length *hop* genes to minimize false positives. Standardization of the guidelines for *hop* designation and nomenclature also represent a significant development within the field (Lindeberg et al., 2005) and are described in a subsequent section.

Application of the bioinformatic methods developed for Pst DC3000 to the genomes of Psp 1448a and Psy B728a has led to rapid and comprehensive analysis of the *hrp*-regulon and *hop* genes in these two strains (Lindeberg et al., 2006). As reported by Vencato et al. (2006), use of HMM and weight-matrix models for identification of hrpL promoters, coupled with analysis of downstream ORFs for type III targeting-associated motifs, resulted in the identification of 44 high probability Hrp promoters. HrpL-dependence for 13 of these was confirmed using RT-PCR, and 6 were shown to be translocated by the T3SS using the Cya translocation reporter system. Among those regulated by hrpL in Psp 1448A were genes encoding a member of the ApbE family of proteins and an alcohol dehydrogenase. HrpL-regulated orthologs of these two genes were subsequently identified in Pst DC3000, and mutations shown to reduce bacterial growth in host Arabidopsis leaves (Vencato et al., 2006). The Psp 1448A *hop* inventory identified using bioinformatic methods has proven very similar to that arrived at using near-saturation DFI screening (Chang et al., 2005), confirming the utility of bioinformatics as a rapid and efficient means of identifying coding regions associated with specific regulatory elements. However, in contrast with phenotype-dependent experimental characterization, bioinformatic screens for hrp boxes do not identify genes that are indirectly activated by HrpL.

Application of HMM and weight-matrix models to the genome of Psy B728a resulted in a comprehensive list of 74 high probability hrp boxes, 16 of which precede *hop* genes or *hop*-containing operons (Lindeberg et al., 2006). None of the remaining hrp boxes precede proteins with recognizable type III translocation motifs, suggesting that the current list of functional *hops* and *hop* orthologs in Psy B728a is more or less complete. Translocation of the majority of confirmed Psy B728a Hops has been demonstrated using AvrRptII fusions (Vinatzer et al., 2005, 2006).

3 Identification of Additional Iron-Responsive Regulatory Elements in Pst DC3000

The hrpL regulon is just one part of a complex regulatory network controlling virulence and fitness in *P. syringae*. Of particular interest are the regulatory pathways responding to iron bioavailability, implicated in expression of several pathogenicity associated factors including the T3SS, biosynthetic pathways for the phytotoxins coronatine and various siderophores, iron transport systems, and multiple sigma factors (Expert et al., 1996; Visca et al., 2002). Iron has also been shown to play a role in quorum-sensing and biofilm formation/stability (Vasil and Ochsner, 1999; Andrews et al., 2003). Investigation of regulatory networks involved in iron homeostasis in *P. syringae* is currently the focus of the USDA-ARS *P. syringae* Systems Biology Group in Ithaca, New York, with particular emphasis on identification of binding sites for the ferric uptake regulator, Fur, the sigma factor PvdS, and sigma factors RpoD and RpoN.

Studies in *P. aeruginosa* have revealed a key role for the ferric uptake regulator, Fur, in iron regulation. In the presence of Zn^{+2} and Fe^{+2}, Fur forms a homodimer that binds to DNA motifs known as "Fur boxes", repressing expression of downstream transcriptional units in conditions of high iron availability (Litwin and Calderwood, 1993; Escolar et al., 1999). Rudimentary models of iron-responsive gene expression in Pst DC3000 and other pseudomonads were constructed from microarray data on iron-responsive genes in *P. aeruginosa* (Ochsner et al., 2002) by Gibbs sampling of the regions upstream of iron-responsive genes. The resulting

HMM was calibrated and used to scan available pseudomonad genome sequences. Many of the sites identified in Pst DC3000 are upstream of genes orthologous to iron-regulated genes in *P. aeruginosa*. Subsequent mutagenesis of Pst DC3000 with a mini-Tn5-*lux* reporter transposon and growth in the presence or absence of ferric citrate confirmed iron-responsiveness of many genes downstream of predicted Fur boxes. Preliminary studies suggest that approximately 5% of the genes in Pst DC3000 are regulated by iron (D. Schneider, personal communication).

P. syringae encodes 15 sigma factors, 10 of which are members of the ECF group of sigma-70 sigma factors. Of these, five (including hrpL) are stress response factors and the remaining five, members of the iron responsive group, downstream of confirmed Fur boxes. Named for its role in synthesis and uptake of the siderophore pyoverdine, the iron-responsive sigma factor PvdS binds to a bipartite "iron starvation" or IS-box, previously characterized in *P. aeruginosa*. Unlike the Fur box, the PvdS binding site characterized in *P. aeruginosa* is insufficient for identification of IS-boxes in *P. syringae*. Characterization of PvdS-regulated genes in Pst DC3000 was therefore initiated using promoter trapping experiments to identify PvdS-responsive genes, followed by Gibbs sampling of the intergenic regions upstream of the PvdS-responsive genes. Over 30 candidate PvdS binding sites were identified, with PvdS-dependent expression of 15 confirmed using a *lux*-reporter system (B. Swingle, personal communication). One of the PvdS-dependent genes, *gidA*, is of particular interest owing to its involvement in production of the antibiotics syringomycin and syringopeptin as well as swarming behavior in Psy B728a (Kinscherf and Willis, 2002).

Similar approaches have been employed for identification of binding sites for the RpoN sigma factor, known to influence expression of several virulence factors in various *P. syringae* pathovars. A model of RpoN-dependent promoters was assembled from RpoN-dependent promoters experimentally identified various pseudomonads, and improved in specificity by incorporation of AT-rich structure outside the -24 and -12 boxes typically used to define RpoN promoters. Predicted RpoN promoters were found upstream of the *hrpL* gene as well as genes related to flagellar biosynthesis and nitrogen metabolism (D. Schneider, personal communication). In combination with experimental confirmation of predicted motifs, ongoing characterization of the complete transcriptome of Pst DC3000 under iron limiting conditions is expected to reveal novel insights into the complex network of iron-responsive regulatory motifs in *P. syringae*.

4.   Genome Comparison Reveals Extensive Variation in the Distribution and Relative Locations of *hop* Genes

Identification of novel genes and gene families with roles in virulence and adaptation to the plant niche remains an important goal for the *P. syringae* research community; however, the vast majority of genes are not associated with virulence-associated regulatory elements and a significant percentage are of unknown function, providing no clue as to their biochemical function. Fortunately, genome comparison can be used to shed light on genes and regions correlated with species-specific and strain-specific phenotypes.

The genomes of the three sequenced *P. syringae* strains are highly conserved with approximately 80% of genes orthologous and over 70% syntenically conserved. Among the syntenically conserved orthologs are numerous virulence genes including those encoding the Type II and Type III secretion pathways, alginate biosynthesis, and pyoverdin biosynthesis. To evaluate whether the locations of *hop* genes were similarly conserved among the three genomes, their arrangement was initially assessed in Pst DC3000. Of the 51 characterized *hop* genes and pseudogenes, 35 were clustered with one or more other *hop* genes, where clusters are defined as *hop* genes separated from the next nearest *hop* by no more than 10 intervening ORFs. Ten *hop* clusters were identified in Pst DC3000 and locations of the conserved, component *hops* examined in Psp 1448a and Psy B728a. Of the 10 Pst DC3000 *hop* clusters, only cluster VI, corresponding to the conserved effector locus, was found in all three strains, consistent with previous reports that acquisition of the main *hrp* island and conserved effector locus predates division of *P. syringae* into separate pathovars (Rohmer et al., 2004). Cluster V is conserved in Psy B728a and pairs of *hop* genes in clusters III and IV are found similarly linked in Psp 1448A, but overall, there is little relative conservation in the repertoire and arrangement of individual *hops* among the three genomes.

5 Analysis of Genome Structure in the Three *P. syringae* Genomes

Although individual *hop* genes show little conservation, their collective distribution may reflect underlying patterns of genome organization. Prokaryotic genomes are typically composed of a mosaic of vertically and horizontally transferred genetic material or genomic islands (Hacker and Kaper, 2000; Gal-Mor and Finlay, 2006). These regions are of particular interest, owing to their association with virulence-related genes (aka pathogenicity islands), and high incidence of uncharacterized hypothetical and conserved hypothetical genes (Hsiao et al., 2005). Consequently, identification of genomic islands in the *P. syringae* genomes represents a potentially powerful approach for revealing locations of candidate virulence determinants.

Several approaches are commonly used for prediction of genomic islands, including: (i) use of atypical sequence composition such as abnormal %G + C, dinucleotide bias or codon usage, (ii) association with annotation features indicative of genetic transfer such as integrases, transposases, phage, or genes highly similar to ones found in phylogenetically distant species, and (iii) comparison of related genomes for regions of divergence. While features in the annotation and regions of divergence among related genomes are readily obtainable, systematic assessment of variations in sequence composition required additional computational analysis. To this end, Alien Hunter, an island-locating program developed at the Sanger Center (Vernikos and Parkhill, 2006) was used to analyze the Pst DC3000 genome, resulting in the identification of 135 islands, accounting for approximately 30% of the genome. Reliability of the predictions was subsequently evaluated by examining predicted islands for the presence of mobile genetic elements, REP sequences, and *hop* genes.

Mobile genetic elements (MGEs) are the features most consistently associated with regions of horizontal transfer, and include genes encoding prophage, plasmids, and transposable elements. Seven percent of the ORFs in Pst DC3000 genome are annotated as MGEs, and 75% of these genes are located within the islands predicted by Alien Hunter. In contrast, the short, repetitive, intragenic sequences known are REP sequences are typically associated with the genomic backbone and can be used as markers for non-horizontally transferred regions (Paulsen et al., 2005). Of the 368 REP sequences identified in Pst DC3000 using a Hidden Markov Model (D, Schneider, unpublished), fewer than 5% were found to be associated with the predicted islands.

Previous studies have shown that many *hop* genes were likely acquired through horizontal transfer (Rohmer et al., 2004). Analysis of the locations of helpers and *hop* genes and pseudogenes reveals that 43/48 hops and 4/7 helpers are associated with the islands predicted by Alien Hunter. Of the five *hops* and three helpers not present in predicted islands, five are known to be ancient in their association with the *P. syringae* species (Rohmer et al., 2004), revealing one of the shortcomings of this approach; namely, that horizontally transferred genetic material ameliorates over time, reducing the efficiency of detection for more ancient transfer events. Analysis of the dinucleotide content and %GC of the remaining three using IslandPath (http://www.pathogenomics.sfu.ca/islandpath/current/IPindex.pl) (Hsiao et al., 2003) reveals that two of the three show no variation relative to sequence properties of the genomic backbone.

Given that the Alien Hunter predictions reliably correlate with features associated with horizontal transfer, island predictions were generated for Psp 1448A and Psy B728a so that relative conservation of island contents and location could be compared with Pst DC3000. Analyses yielded 112 and 135 islands for Pph 1448A and Psy B728a, accounting for 21% and 25% of their respective genomes. As observed for Pst DC3000, annotated mobile genetic elements were found to be preferentially associated with the predicted islands. Island location was deemed conserved if islands were flanked by greater than 10 kb of conserved, aligned DNA, and content was scored as conserved if there was >70% alignment between predicted islands, partially conserved of there was 30–70% alignment of contents and not conserved, if there was <30% alignment.

Patterns of island conservation observed between Pst DC3000 and Pph 1448A were similar to those observed upon comparison of the Pst DC3000 and Psy B728a predictions and the remaining, albeit preliminary, analyses will be limited to the Pst DC3000 islands with similar profiles for both comparisons. The Pst DC3000 islands with no conservation relative to corresponding regions in Psp 1448A and Psy B728a had the highest island scores, reflecting higher levels of difference in sequence properties relative to the genome backbone. These same islands were also home to the majority of mobile genetic elements and *hop* genes.

   Of particular interest are the islands that were conserved in their relative locations but not conserved in their content. In addition to having a high percentage of MGEs and *hop* genes, this subset also showed a high level of correspondence with Pst DC3000 lineage specific regions identified through comparison of the Pst DC3000 genome with the genomes of *P. aeruginosa* PAO1 and *P. putida* KT2440. This finding suggests that insertion sites for horizontally transferred genetic material may be broadly conserved throughout the genus. Features in the genome accounting for the sites of preferential insertion are as yet uncharacterized, but preliminary analyses indicate that tRNAs, previously shown to be hot spots for integration of foreign DNA, are preferentially associated with this same group of islands (Blum et al., 1994; Ritter et al., 1995; Jackson et al., 2000). A summary of these findings is shown in Fig. 1.

   The ultimate goal of genomic comparison among the three sequenced strains is the identification of genes with a high likelihood of involvement in species-specific and/or strain-specific host–pathogen interaction. Characterization and refinement of candidate pathogenicity islands represents a significant step in this direction, but future analyses will also focus on identification of genes shared by two or more *P. syringae* strains as well as those unique to the individual strains, with those in horizontally transferred regions receiving particular attention. The USDA-ARS *P. syringae* Systems Biology Group is presently generating a global transcript map in addition to a near-saturating, fully sequenced, transposon reporter insertion library for Pst DC3000. The transcript map will permit researchers to better distinguish

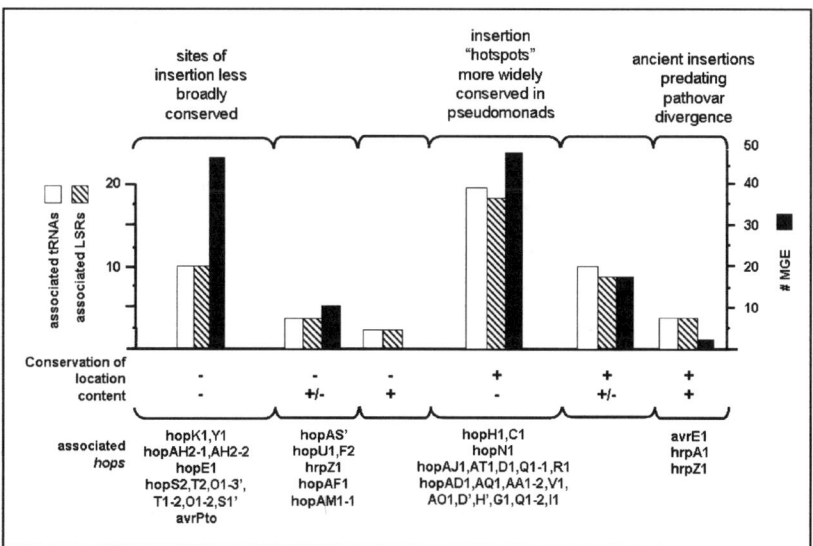

Fig. 1 Pst DC3000 genomic islands predicted by Alien Hunter (Vernikos and Parkhill, 2006) categorized by conservation of location and content with genomic islands predicted in Psp 1448A and Psy B728a. Also shown are the number of mobile genetic elements (MGEs), tRNA genes, lineage specific regions (LSRs), and hop genes associated with each category of genomic island. Interpretations of the different island types are shown at the top

"real" genes from those that have been incorrectly annotated. And the insertion library will be an invaluable tool for evaluating roles of individual genes at multiple stages in the plant–bacterium interaction.

6 Community Resources for Accessing Sequence and Genomics Data

The advent of genome sequencing coupled with the volume of data being generated through functional genomics demand that data be actively managed to enhance its accessibility to the research community. To meet this need, the Pseudomonas–Plant Interaction (PPI) web site (http://www.pseudomonas-syringae.org/) maintains general resources, including tutorials and explanatory material, for navigating and viewing the genome data.

In order for the body of annotation to retain its usefulness, it must be actively maintained and updated to reflect the findings of ongoing research and analyses. To this end, updates are routinely gathered by literature search and personal communication for submission to the databases at Genbank and TIGR-CMR, with log files of recent changes to the three *P. syringae* genomes maintained at the PPI web site. Typical updates include changes to gene names, addition of hrp promoter sequences, changes in coordinates, and addition and subtraction of ORFs to reflect proteomic characterization of Pst DC3000.

As previously mentioned, the proliferation of new Hops following release of the Pst DC3000 genome sequence necessitated development of consistent criteria for defining and naming Hop proteins. Guidelines for Hop classification, phylogenetic analysis, and nomenclature represent a major part of the PPI web site, in addition to a database listing all *hop* genes identified in *P. syringae* together with previous names, accession numbers, sequences, and experimental evidence supporting their identity as *hops*. Significantly, the nomenclature system developed for the *P. syringae* Hop proteins has since been adopted by researchers characterizing type III effectors in *E. coli* 0157 (Tobe et al., 2006).

Acknowledgements This work was supported by the NSF Plant Genome Research Program Cooperative Agreement DBI-0077622.

References

Alfano, J.R., Charkowski, A.O., Deng, W.-L., Badel, J.L., Petnicki-Ocwieja, T., van Dijk, K., and Collmer, A. (2000) The *Pseudomonas syringae* Hrp pathogenicity island has a tripartite mosaic structure composed of a cluster of type III secretion genes bounded by exchangeable effector and conserved effector loci that contribute to parasitic fitness and pathogenicity in plants. *Proc. Natl. Acad. Sci. USA* 97: 4856–4861.

Andrews, S.C., Robinson, A.K., and Rodriguez-Quinones, F. (2003) Bacterial iron homeostasis. *FEMS Microbiol. Rev.* 27: 215–237.

Badel, J.L., Nomura, K., Bandyopadhyay, S., Shimizu, R., Collmer, A., and He, S.Y. (2003) *Pseudomonas syringae* pv. *tomato* DC3000 HopPtoM (CEL ORF3) is important for lesion formation but not growth in tomato and is secreted and translocated by the Hrp type III secretion system in a chaperone-dependent manner. *Mol. Microbiol.* 49: 1239–1251.

Blum, G., Ott, M., Lischewski, A., Ritter, A., Imrich, H., Tschape, H., and Hacker, J. (1994) Excision of large DNA regions termed pathogenicity islands from tRNA-specific loci in the chromosome of an *Escherichia coli* wild-type pathogen. *Infect. Immun.* 62: 606–614.

Buell, C.R., Joardar, V., Lindeberg, M., Selengut, J., Paulsen, I.T., Gwinn, M.L., Dodson, R.J., Deboy, R.T., Durkin, A.S., Kolonay, J.F., Madupu, R., Daugherty, S., Brinkac, L., Beanan, M.J., Haft, D.H., Nelson, W.C., Davidsen, T., Zafar, N., Zhou, L., Liu, J., Yuan, Q., Khouri, H., Fedorova, N., Tran, B., Russell, D., Berry, K., Utterback, T., Van Aken, S.E., Feldblyum, T.V., D'Ascenzo, M., Deng, W.L., Ramos, A.R., Alfano, J.R., Cartinhour, S., Chatterjee, A.K., Delaney, T.P., Lazarowitz, S.G., Martin, G.B., Schneider, D.J., Tang, X., Bender, C.L., White, O., Fraser, C.M., and Collmer, A. (2003) The complete genome sequence of the Arabidopsis and tomato pathogen *Pseudomonas syringae* pv. *tomato* DC3000. *Proc. Natl. Acad. Sci. USA* 100: 10181–10186.

Chang, J.H., Urbach, J.M., Law, T.F., Arnold, L.W., Hu, A., Gombar, S., Grant, S.R., Ausubel, F.M., and Dangl, J.L. (2005) A high-throughput, near-saturating screen for type III effector genes from *Pseudomonas syringae*. *Proc. Natl. Acad. Sci. USA* 102: 2549–2554.

Collmer, A., Lindeberg, M., Petnicki-Ocwieja, T., Schneider, D.J., and Alfano, J.R. (2002) Genomic mining type III secretion system effectors in *Pseudomonas syringae* yields new picks for all TTSS prospectors. *Trends Microbiol.* 10: 462–469.

Escolar, L., Perez-Martin, J., and de Lorenzo, V. (1999) Opening the iron box: transcriptional metalloregulation by the Fur protein. *J. Bacteriol.* 181: 6223–6229.

Expert, D., Enard, C., and Masclaux, C. (1996) The role of iron in plant host–pathogen interactions. *Trends Microbiol.* 4: 232–237.

Feil, H., Feil, W.S., Chain, P., Larimer, F., DiBartolo, G., Copeland, A., Lykidis, A., Trong, S., Nolan, M., Goltsman, E., Thiel, J., Malfatti, S., Loper, J.E., Lapidus, A., Detter, J.C., Land, M., Richardson, P.M., Kyrpides, N.C., Ivanova, N., and Lindow, S.E. (2005) Comparison of the complete genome sequences of *Pseudomonas syringae* pv. *syringae* B728a and pv. *tomato* DC3000. *Proc. Natl. Acad. Sci. USA* 102: 11064–11069.

Fereirra, A., Gordon, J., Martin, G.B., Vencato, M., Collmer, A., Wehling, M.D., Alfano, J.R., Moreno-Hagelsieb, G., Lamboy, W.F., DeClerck, G.A., Schneider, D.J., Myers, C.R., and Cartinhour, S. (2006) Whole-genome expression profiling defines the HrpL regulon of *Pseudomonas syringae* pv *tomato* DC3000 allows *de novo* reconstruction of the Hrp cis element, and identifies novel co-regulated genes. *Mol. Plant Microbe. Interact.* 19: 1167–1179.

Fouts, D.E., Abramovitch, R.B., Alfano, J.R., Baldo, A.M., Buell, C.R., Cartinhour, S., Chatterjee, A.K., D'Ascenzo, M., Gwinn, M.L., Lazarowitz, S.G., Lin, N.C., Martin, G.B., Rehm, A.H., Schneider, D.J., van Dijk, K., Tang, X., and Collmer, A. (2002) Genomewide identification of *Pseudomonas syringae* pv. *tomato* DC3000 promoters controlled by the HrpL alternative sigma factor. *Proc. Natl. Acad. Sci. USA* 99: 2275–2280.

Gal-Mor, O. and Finlay, B.B. (2006) Pathogenicity islands: a molecular toolbox for bacterial virulence. *Cell. Microbiol.* 8: 1707–1719.

Grimm, C., Aufsatz, W., and Panopoulos, N.J. (1995) The *hrpRS* locus of *Pseudomonas syringae* pv. *phaseolicola* constitutes a complex regulatory unit. *Mol. Microbiol.* 15: 155–165.

Guo, M., Chancey, S.T., Tian, F., Ge, Z., Jamir, Y., and Alfano, J.R. (2005) *Pseudomonas syringae* type III chaperones ShcO1, ShcS1, and ShcS2 facilitate translocation of their cognate effectors and can substitute for each other in the secretion of HopO1–1. *J. Bacteriol.* 187: 4257–4269.

Guttman, D.S., Vinatzer, B.A., Sarkar, S.F., Ranall, M.V., Kettler, G., and Greenberg, J.T. (2002) A functional screen for the type III (Hrp) secretome of the plant pathogen *Pseudomonas syringae*. *Science* 295: 1722–1726.

Hacker, J. and Kaper, J.B. (2000) Pathogenicity islands and the evolution of microbes. *Annu. Rev. Microbiol.* 54: 641–679.

Hirano, S.S. and Upper, C.D. (2000) Bacteria in the leaf ecosystem with emphasis on *Pseudomonas syringae*-a pathogen, ice nucleus, and epiphyte. *Microbiol. Mol. Biol. Rev.* 64: 624–653.

Hsiao, W., Wan, I., Jones, S.J., and Brinkman, F.S. (2003) IslandPath: aiding detection of genomic islands in prokaryotes. *Bioinformatics* 19: 418–420.

Hsiao, W.W., Ung, K., Aeschliman, D., Bryan, J., Finlay, B.B., and Brinkman, F.S. (2005) Evidence of a large novel gene pool associated with prokaryotic genomic islands. *PLoS Genet.* 1: e62.

Hutcheson, S.W., Bretz, J., Sussan, T., Jin, S., and Pak, K. (2001) Enhancer-binding proteins HrpR and HrpS interact to regulate *hrp*-encoded type III protein secretion in *Pseudomonas syringae* strains. *J. Bacteriol.* 183: 5589–5598.

Jackson, R.W., Athanassopoulos, E., Tsiamis, G., Mansfield, J.W., Sesma, A., Arnold, D.L., Gibbon, M.J., Murillo, J., Taylor, J.D., and Vivian, A. (1999) Identification of a pathogenicity island, which contains genes for virulence and avirulence, on a large native plasmid in the bean pathogen *Pseudomonas syringae* pathovar phaseolicola. *Proc. Natl. Acad. Sci. USA* 96: 10875–10880.

Jackson, R.W., Mansfield, J.W., Arnold, D.L., Sesma, A., Paynter, C.D., Murillo, J., Taylor, J.D., and Vivian, A. (2000) Excision from tRNA genes of a large chromosomal region, carrying *avrPphB*, associated with race change in the bean pathogen, *Pseudomonas syringae* pv. *phaseolicola*. *Mol. Microbiol.* 38: 186–197.

Joardar, V., Lindeberg, M., Jackson, R.W., Selengut, J., Dodson, R., Brinkac, L.M., Daugherty, S C., Deboy, R., Durkin, A.S., Giglio, M.G., Madupu, R., Nelson, W.C., Rosovitz, M.J., Sullivan, S., Crabtree, J., Creasy, T., Davidsen, T., Haft, D.H., Zafar, N., Zhou, L., Halpin, R., Holley, T., Khouri, H., Feldblyum, T., White, O., Fraser, C.M., Chatterjee, A.K., Cartinhour, S., Schneider, D.J., Mansfield, J., Collmer, A., and Buell, C.R. (2005) Whole-genome sequence analysis of *Pseudomonas syringae* pv. *phaseolicola* 1448A reveals divergence among pathovars in genes involved in virulence and transposition. *J. Bacteriol.* 187: 6488–6498.

Kinscherf, T.G. and Willis, D.K. (2002) Global Regulation by gidA in *Pseudomonas syringae*. *J. Bacteriol.* 184: 2281–2286.

Lindeberg, M., Stavrinides, J., Chang, J.H., Alfano, J.R., Collmer, A., Dangl, J.L., Greenberg, J.T., Mansfield, J.W., and Guttman, D.S. (2005) Proposed guidelines for a unified nomenclature and phylogenetic analysis of type III hop effector proteins in the plant pathogen *Pseudomonas syringae*. *Mol. Plant Microbe Interact.* 18: 275–282.

Lindeberg, M., Cartinhour, S., Myers, C.R., Schechter, L.M., Schneider, D.J., and Collmer, A. (2006) Closing the circle on the discovery of genes encoding Hrp regulon members and type III secretion system effectors in the genomes of three model *Pseudomonas syringae* strains. *Mol. Plant Microbe Interact.* 19: 1151–1158.

Litwin, C.M. and Calderwood, S.B. (1993) Role of iron in regulation of virulence genes. *Clin. Microbiol. Rev.* 6: 137–149.

Marco, M.L., Legac, J., and Lindow, S.E. (2005) *Pseudomonas syringae* genes induced during colonization of leaf surfaces. *Environ. Microbiol.* 7: 1379–1391.

Monier, J.M. and Lindow, S.E. (2003) Differential survival of solitary and aggregated bacterial cells promotes aggregate formation on leaf surfaces. *Proc. Natl. Acad. Sci. USA* 100: 15977–15982.

Ochsner, U.A., Wilderman, P.J., Vasil, A.I., and Vasil, M.L. (2002) GeneChip expression analysis of the iron starvation response in *Pseudomonas aeruginosa*: identification of novel pyoverdine biosynthesis genes. *Mol. Microbiol.* 45: 1277–1287.

Paulsen, I.T., Press, C.M., Ravel, J., Kobayashi, D.Y., Myers, G.S., Mavrodi, D.V., DeBoy, R.T., Seshadri, R., Ren, Q., Madupu, R., Dodson, R.J., Durkin, A.S., Brinkac, L.M., Daugherty, S. C., Sullivan, S.A., Rosovitz, M.J., Gwinn, M.L., Zhou, L., Schneider, D.J., Cartinhour, S.W., Nelson, W.C., Weidman, J., Watkins, K., Tran, K., Khouri, H., Pierson, E.A., Pierson, L.S., 3rd, Thomashow, L.S., and Loper, J.E. (2005) Complete genome sequence of the plant commensal *Pseudomonas fluorescens* Pf-5. *Nat. Biotechnol.* 23: 873–878.

Petnicki-Ocwieja, T., Schneider, D.J., Tam, V.C., Chancey, S.T., Shan, L., Jamir, Y., Schechter, L.M., Janes, M.D., Buell, C.R., Tang, X., Collmer, A., and Alfano, J.R. (2002) Genomewide identification of proteins secreted by the Hrp type III protein secretion system of *Pseudomonas syringae* pv. *tomato* DC3000. *Proc. Natl. Acad. Sci. USA* 99: 7652–7657.

Preston, G.M. (2000) *Pseudomonas syringae* pv. *tomato*: the right pathogen, of the right plant, at the right time. *Mol. Plant Pathol.* 1: 263–275.

Ritter, A., Blum, G., Emody, L., Kerenyi, M., Bock, A., Neuhierl, B., Rabsch, W., Scheutz, F., and Hacker, J. (1995) tRNA genes and pathogenicity islands: influence on virulence and metabolic properties of uropathogenic *Escherichia coli*. *Mol. Microbiol.* 17: 109–121.

Rohmer, L., Guttman, D.S., and Dangl, J.L. (2004) Diverse evolutionary mechanisms shape the type III effector virulence factor repertoire in the plant pathogen *Pseudomonas syringae*. *Genetics* 167: 1341–1360.

Sarkar, S.F. and Guttman, D.S. (2004) Evolution of the core genome of *Pseudomonas syringae*, a highly clonal, endemic plant pathogen. *Appl. Environ. Microbiol.* 70: 1999–2012.

Sawada, H., Kanaya, S., Tsuda, M., Suzuki, F., Azegami, K., and Saitou, N. (2002) A phylogenomic study of the OCTase genes in *Pseudomonas syringae* pathovars: the horizontal transfer of the argK-tox cluster and the evolutionary history of OCTase genes on their genomes. *J. Mol. Evol.* 54: 437–457.

Schechter, L.M., Roberts, K.A., Jamir, Y., Alfano, J.R., and Collmer, A. (2004) *Pseudomonas syringae* type III secretion system targeting signals and novel effectors studied with a Cya translocation reporter. *J. Bacteriol.* 186: 543–555.

Schechter, L.M., Vencato, M., Jordan, K.L., Schneider, S.E., Schneider, D.J., and Collmer, A. (2006) Multiple approaches to a complete inventory of *Pseudomonas syringae* pv *tomato* DC3000 type III secretion system effector proteins. *Mol. Plant Microbe Interact.* 19: 1180–1192.

Shan, L., Oh, H.S., Chen, J., Guo, M., Zhou, J., Alfano, J.R., Collmer, A., Jia, X., and Tang, X. (2004) The HopPtoF locus of *Pseudomonas syringae* pv. *tomato* DC3000 encodes a type III chaperone and a cognate effector. *Mol. Plant Microbe Interact.* 17: 447–455.

Tobe, T., Beatson, S.A., Taniguchi, H., Abe, H., Bailey, C.M., Fivian, A., Younis, R., Matthews, S., Marches, O., Frankel, G., Hayashi, T., and Pallen, M.J. (2006) An extensive repertoire of type III secretion effectors in Escherichia coli O157 and the role of lambdoid phages in their dissemination. *Proc. Natl. Acad. Sci. USA* 103: 14941–14946.

van Dijk, K., Tam, V.C., Records, A.R., Petnicki-Ocwieja, T., and Alfano, J.R. (2002) The ShcA protein is a molecular chaperone that assists in the secretion of the HopPsyA effector from the type III (Hrp) protein secretion system of *Pseudomonas syringae*. *Mol. Microbiol.* 44: 1469–1481.

Vasil, M.L. and Ochsner, U.A. (1999) The response of Pseudomonas aeruginosa to iron: genetics, biochemistry and virulence. *Mol. Microbiol.* 34: 399–413.

Vencato, M., Tian, F., Alfano, J.R., Buell, C.R., Cartinhour, S., DeClerck, G.A., Guttman, D.S., Stavrinides, J., Joardar, V., Lindeberg, M., Bronstein, P.A., Mansfield, J.W., Myers, C.R., Collmer, A., and Schneider, D.J. (2006) Bioinformatics-enabled inventory of the Hrp regulon and type III secretion system effector proteins of *Pseudomonas syringae* pv. *phaseolicola* 1448A. *Mol. Plant Microbe Interact.* 19: 1193–1206.

Vernikos, G.S. and Parkhill, J. (2006) Interpolated variable order motifs for identification of horizontally acquired DNA: revisiting the *Salmonella* pathogenicity islands. *Bioinformatics* 22: 2196–2203.

Vinatzer, B.A., Jelenska, J., and Greenberg, J.T. (2005) Bioinformatics correctly identifies many type III secretion substrates in the plant pathogen *Pseudomonas syringae* and the biocontrol isolate *P. fluorescens* SBW25. *Mol. Plant Microbe Interact.* 18: 877–888.

Vinatzer, B.A., Teitzel, G.M., Lee, M.W., Jelenska, J., Hotton, S., Fairfax, K., Jenrette, J., and Greenberg, J.T. (2006) The type III effector repertoire of *Pseudomonas syringae* pv. *syringae* B728a and its role in survival and disease on host and non-host plants. *Mol. Microbiol.* 62: 26–44.

Visca, P., Leoni, L., Wilson, M.J., and Lamont, I.L. (2002) Iron transport and regulation, cell signalling and genomics: lessons from *Escherichia coli* and *Pseudomonas*. *Mol. Microbiol.* 45: 1177–1190.

Wehling, M.D., Guo, M., Fu, Z.Q., and Alfano, J.R. (2004) The *Pseudomonas syringae* HopPtoV protein is secreted in culture and translocated into plant cells via the type III protein secretion system in a manner dependent on the ShcV type III chaperone. *J. Bacteriol.* 186: 3621–3630.

Xiao, Y., Heu, S., Yi, J., Lu, Y., and Hutcheson, S.W. (1994) Identification of a putative alternate sigma factor and characterization of a multicomponent regulatory cascade controlling the expression of *Pseudomonas syringae* pv. *syringae* Pss61 *hrp* and *hrmA* genes. *J. Bacteriol.* 176: 1025–1036.

Zwiesler-Vollick, J., Plovanich-Jones, A.E., Nomura, K., Bandyopadhyay, S., Joardar, V., Kunkel, B.N., and He, S.Y. (2002) Identification of novel hrp-regulated genes through functional genomic analysis of the *Pseudomonas syringae* pv. *tomato* DC3000 genome. *Mol. Microbiol.* 45: 1207–1218.

Gene Ontology (GO) for Microbe–Host Interactions and Its Use in Ongoing Annotation of Three *Pseudomonas syringae* Genomes via the *Pseudomonas*–Plant Interaction (PPI) Web Site

C.W. Collmer[1,2], M. Lindeberg[1], and A. Collmer[1]

Abstract Genome-scale sequencing of plant pathogens is an increasingly important tool for exploring host–pathogen interactions. While comparative *structural* analyses of three *Pseudomonas syringae* pathovars that differ in pathogenicity (*P. s.* pv. *tomato* DC3000, *P. s.* pv. *phaseolicola* 1448A, and *P. s.* pv. *syringae* B728a) offer valuable insights, a second type of analysis based on the ability to find and compare gene products with similar *function* (but possibly different structure) across even diverse pathogens is now possible through genome annotation using the controlled vocabularies of the Gene Ontology Consortium (GO; http://www.geneontology.org). As members of the Plant-Associated Microbe Gene Ontology (PAMGO) Interest Group (http://pamgo.vbi.vt.edu), we have been developing precisely defined GO terms that describe the biological processes involved in a microbe's interactions with its host (e.g. GO:0044409, entry into host; GO:0044002, acquisition of nutrients from host). Using these terms for annotating the genes of prokaryotic (or eukaryotic) pathogens (or mutualists) that attack plant (or animal) hosts promises to offer a new mechanism for viewing the diversity of microbe strategies for overcoming common host challenges (e.g. finding all gene products across diverse microbes that are annotated to GO:0044414, suppression of host defenses). However, because genome annotation and analysis must continue over time to match experimentation and discovery, genome projects must provide portals for continually updating genome sequence and annotation data. The Pseudomonas–Plant Interaction (PPI) web site (http://pseudomonas-syringae.org) exemplifies the "hub model" of web-based information management for small-scale genome projects. This model maximizes access to the most current analytical tools (e.g. the Artemis Genome Viewer) through links to primary off-site resources, while tailoring instructional tutorials, specific information, and organization to the needs of the *P. syringae* community. The web site also serves as a portal for submission of updated genome sequence or GO annotation data as it becomes available.

[1] Cornell University, Ithaca, NY 14850, USA

[2] Wells College, Aurora, NY 13026, USA

Author for correspondence: Candace W. Collmer; e-mail: ccollmer@wells.edu

M'B. Fatmi et al. (eds.), *Pseudomonas syringae Pathovars and Related Pathogens.*   221
© Springer Science + Business Media B.V. 2008

Keywords Plant pathogenesis, bacterial pathogen

Full-genome sequencing has the potential to unveil formerly hidden microbial secrets, if the right questions can be formulated and the tools to answer them are available. Thus, important and obvious questions in relationship to plant pathogenesis could be posed once full sequences were available for three *Pseudomonas* species – *P. syringae* pv. *tomato* DC3000, a plant pathogen; *P. aeruginosa* PA01, an animal pathogen; and *P. putida* KT2440, a non-pathogen. Equally interesting questions arose with the availability of genomic sequences for three *P. syringae* pathovars that differ in pathogenicity – *P. s.* pv. *tomato* DC3000, which causes bacterial speck on tomato and Arabidopsis; *P. s.* pv. *phaseolicola* 1448A, the causal agent of halo blight on bean; and *P. s.* pv. *syringae* B728a, which causes brown spot on bean. To address these questions, genome-scale analyses are essential. However, the tools for addressing genome-wide questions that focus specifically on plant pathogenesis have until recently been rather limited. That limitation has in part derived from the lack of universally understandable terms that can be used to tag and fully annotate genes in microbes that function in pathogenesis. In this paper, we report progress to date in working with the Gene Ontology (GO) Consortium (http://www.geneontology.org) to develop precisely defined GO terms that describe the biological processes involved in a microbe's interactions with its host, including those of pathogenesis. In addition, we describe features of the Pseudomonas–Plant Interaction (PPI) web site (http://pseudomonas-syringae.org) that allow for the viewing as well as the continuous updating of genome sequence and annotation data as experimentation and discovery continue on three *P. syringae* pathovars.

In the process of genome-wide analysis, it is relatively straightforward to do informative *structural* comparisons by aligning genes and even whole genomes; such a comparison could show, for example, that a particular gene active in one genome is inactivated by insertion of a transposable element in others. However, the ability to find and compare gene products with similar *function* (regardless of structural similarity) across genomes has been much harder. Comparisons based on the functions of gene products require not only that functions be assigned to genes in the process of annotation, but also that the vocabulary for describing such functions be universally understandable. Indeed, the annotation of the full genome sequence of *P. syringae* pv. *tomato* DC3000 in collaboration with scientists at The Institute for Genomic Research (TIGR) included the assignment of genes to broad, prokaryotic role categories derived from those of Riley (1993); a comparison of the relative number of genes assigned to these different cellular roles revealed that DC3000 differs significantly from both the animal pathogen *P. aeruginosa* PA01 and the non-pathogen *P. putida* KT2440 in having greater transport capacity for sugars, but less for amino acids (Buell et al., 2003). Recognition of the utility of such comparisons coupled with the need for a much larger number of functional categories had already led to TIGR membership in the Gene Ontology Consortium, and it was TIGR scientists who spear-headed the development of GO terms appropriate for prokaryotic gene functions.

Since 1998, the GO Consortium has been developing three ontologies – Molecular Function, Biological Process, and Cellular Component – each comprised of a hierarchical-like arrangement of terms whose definitions and interrelationships are precisely defined (Gene Ontology Consortium, 2001). As the original developers of GO were associated with the mouse, yeast (*Saccharomyces*), and fly (*Drosophila*) genome projects, the early versions of the ontologies were focused on eukaryotic cells; as mentioned above, terms related to prokaryotes were integrated into the continually developing ontologies as their need for the annotation of prokaryotic genomes became apparent. Terms assigned during annotation to a particular gene from each of the three ontologies tell what the gene's product does (its molecular function), where it does it (its cellular component), and why it does it (as a part of what larger biological process). Each GO term (or phrase) within these ontologies has a unique GO identifier, which once annotated to a gene product in one organism then allows searches for genes with the same identifier (e.g. function or process) in other organisms, regardless of whether the products of these different genes share structural similarity. While the three GO ontologies have been continually expanding since 1998, at the time the *P. syringae* pv. *tomato* DC3000 genome was annotated, there were few GO terms appropriate for annotating gene products implicated in the virulence of plant pathogens.

However, the need for such terms, and for GO annotation in general, as a tool for answering critical questions about plant pathogenesis through cross-genome comparisons is easily apparent. Without GO terms, searching across genomes for gene products with similar functions is difficult unless they also share structural features. For example, while gene names can be searched, they are often based on idiosyncratic phenotypes, which obscure gene function (or address only one of several). Secondly, while one can search databases for specific function terms in Keywords or Comments fields, such searches are doomed to be incomplete because of inconsistent terminology (e.g. "attachment," "adhesion," "prepenetration activity" could be used by different authors to describe similar events). In addition, there is presently no way to systematically add new information on the function of gene products in a manner that supports comparison across genomes. A good example of the problem is found in the genes encoding the type II protein secretion pathway in bacterial plant pathogens. The outer membrane pore protein is known as XcpD in *Pseudomonas*, OutD in *Erwinia*, and XpsD in *Xanthomonas*. Worse, there can be common components involved in type II protein secretion and type IV pilin biogenesis, and one cannot tell from the name designation whether a given protein serves one or both systems. Because both systems are important in pathogenesis, they have been extensively studied in several different organisms. But new information on the functions of secretion system components is not accessible in any systematic way through current sequence-oriented searches of databases. Information available on a gene product in one organism could prove useful for understanding a comparable one in another if their functional relationships were easily recognizable.

For instance, learning experimentally that a gene product is involved in type IV pilin biogenesis could prompt a search for gene products in other organisms known to be involved in the same process; finding that one of these was also annotated for a role in type II protein secretion could suggest a meaningful experimental question relating to the gene product of the original organism.

It was the recognition of the potential utility of such searches for answering critical questions about plant pathogenesis, and of the need for more GO terms related to pathogenesis, that led in 2004 to the formation of the Plant-Associated Microbe Gene Ontology (PAMGO) interest group. As members of that group, we have been working with the GO Consortium to develop precisely-defined GO terms that describe the biological processes involved in a microbe's interactions with its host (Gene Ontology Consortium, 2006). The PAMGO interest group (http://pamgo.vbi. vt.edu) comprises scientists from six institutions working on genome projects of both prokaryotic and eukaryotic microbial plant pathogens – bacteria, oomycetes, fungi, and nematodes. Our original goal was to develop higher level biological process terms appropriate for annotating genes in these organisms that had been implicated in their pathogenic interactions with plants. However, we soon realized the broader utility of developing terms that were as general as possible, relevant not only for pathogens of all kingdoms, but also for the whole range of host–microbe interactions (from mutualism through parasitism) and for all hosts, animals as well as plants. Given that microbes of all types (whether prokaryotic or eukaryotic) that approach a plant to initiate a relationship (whether ultimately mutualistic or pathogenic) must all begin by recognizing their host, the utility of a GO term that could be used to find all annotated microbe genes involved in host recognition seemed obvious. In addition, the fact that at least some biological processes involved in pathogenesis are shared by both plant and animal pathogens (e.g. the injection of pathogen-encoded effectors into host cells via a type III secretion system) underscored the desirability of broad GO terms that could serve both communities.

Using such broad GO terms for annotating the genes of prokaryotic (or eukaryotic) pathogens (or mutualists) that interact with plant (or animal) hosts promises to offer a new mechanism for viewing the diversity of microbe strategies for overcoming common host challenges (e.g. finding all gene products across diverse microbes that are annotated to the [PAMGO-developed] term GO:0044414, suppression of host defenses). However, not all GO terms are broad ones; GO terms are imbedded in the ontologies within "tree" structures (directed acyclic graphs, or DAGs) that show their relationships as "children" of broader terms or "parents" of more specific ones. Thus, while microbial gene products that effect recognition of either a plant or an animal host could all be collected by a search using the broad term "recognition of host," that broad term could have more specific "children" terms that describe either plant-specific or animal-specific recognition processes appropriate for the annotation of some gene products but not others. In addition, such a hierarchical-like structure allows for the annotation of genes with terms denoting broad functional categories, or to terms with more specificity as knowledge accrues.

PAMGO members began in 2004 with a proposal to the GO Consortium for about thirty higher level (broad) biological process terms describing the range of interactions between microbes and their hosts. This proposal generated intensive discussion before, during, and after its presentation at a GO Content meeting in August of 2004. Following the consideration of three alternative "tree" options at a GO Consortium meeting in September, a final set of 35 new terms was submitted to GO in December, 2004, and accepted into the GO Biological Process ontology in January 2005 (Gene Ontology Consortium, 2006). The process of developing these terms included broad-ranging discussion across the wider GO community about the definitions of terms such as pathogenesis and symbiosis. In the end, a broad definition of the parent term symbiosis was adopted, indicating an interaction between organisms including parasitism, commensalism, and mutualism, and acknowledging that the three are not always discreet categories of interactions but rather occur on a continuum of interaction ranging from parasitism to mutualism. Most of these terms are shown in Fig. 1 as children of the parent term GO:0044404, symbiotic interaction between host and other organism.

In addition to providing precise definitions for all terms, as well as the option of annotating a gene product at whatever level of specificity is warranted by the available information (e.g. "binding" versus "protein kinase binding"), annotation using the Gene Ontology also requires the concomitant assignment of an appropriate Evidence Code. The thirteen Evidence Codes indicate the type of data on which each annotation is based, which in turn provides some measure of the relative degree of certainty in the annotation. For instance, an annotation that is "Inferred from Direct Assay" is more reliable than one "Inferred from Electronic Annotation."

As genome annotation using the new GO terms began in 2005–2006 for genes implicated in the virulence of *Pseudomonas syringae* pathovars, *Erwinia chrysanthemi* 3937, and *Phytophthora* spp., it was immediately obvious that more specific GO terms were needed to capture information available in the published literature. Accordingly, a three-day jamboree for GO term creation in July 2006, which involved five PAMGO members and two members of the GO editorial board, produced a draft list of several hundred GO terms that are more specific children to the original 35 broader terms. While these newly proposed terms describe many biological processes involved in the interaction between a microbe and its host, a large number are children of GO:0044003, modification of host morphology or physiology. Examples of new, more specific terms include, for example, "induction by other organism of host defensive cell wall thickening," "modification of host morphology or physiology via protein secreted by type III secretion system," and "induction by other organism of host phytoalexin production." These new terms are still in the process of review but should soon be integrated into the GO and be available for the continuing annotation of the genomes of plant pathogens. And as research continues to provide new insights and understanding, the GO ontologies will continue to grow.

physiological process
    cellular physiological process
    interaction between organisms ♦
        cell killing
        competition with other, non-host, organism ♦
        multi-species biofilm formation ♦
            multi-species biofilm formation in or on host organism ♦
            multi-species biofilm formation on inanimate substrate ♦
        <u>symbiosis, mutualism through parasitism</u> ♦
            <u>symbiotic interaction between host and other organism</u> ♦
                acquisition of nutrients from host ♦
                adhesion to host ♦
                    cytoadherence to microvasculature
                    multi-species biofilm formation in or on host organism ♦
                    single-species biofilm formation in or on host organism ♦
                avoidance of host defenses ♦
                    evasion of host defenses ♦
                        evasion of host defense response
                            evasion of host immune response
                            viral host defense evasion
                    suppression of host defenses ♦
                dissemination or transmission of an organism from a host ♦
                    dissemination or transmission of an organism from a host by a vector ♦
                    viral transmission
                entry into host ♦
                    entry into host cell ♦
                    entry into host through host barriers ♦
                    entry into host through natural portals ♦
                growth on or near host surface ♦
                growth within host ♦
                induction of host defense response ♦
                    viral induction of host immune response
                modification of host morphology or physiology ♦
                    disruption of host cells ♦
                    induction in host of a tumor, nodule, or growth ♦
                    viral host cell process manipulation
                    viral transformation
                movement within host ♦
                    migration within host ♦
                    viral spread within host
                <u>pathogenesis</u>
                recognition of host ♦
                translocation of molecules into host ♦
                    translocation of DNA into host ♦
                    translocation of proteins or peptides into host ♦
                virus-host interaction
            symbiotic interaction with other, non-host organism ♦

Fig. 1 Newly developed GO terms for the Biological Process ontology, showing the parent-child relationships among terms. All terms marked by a star were developed by PAMGO and added to GO in January 2005. The unmarked terms were already in GO but were integrated as children to newly developed PAMGO terms when appropriate

Because genome annotation and analysis must continue over time to match experimentation and discovery, whole genome sequencing projects must provide portals for not only presenting but also continually updating genome sequence and annotation data. To meet this need for the *Pseudomonas syringae* pathovars, a web site was developed to provide general resources for navigating and viewing the genome data and for communicating updates to the annotation. Few smaller-scale genome projects have access to the extensive foundational resources that support web sites for the major model organisms, characterized as they are by extensive on-site databases and a large support staff with specialized skills. The Pseudomonas–Plant Interaction (PPI) web site (http://pseudomonas-syringae.org), in contrast, exemplifies the "hub" model of web-based information management, offering a model by which researchers can create and maintain web sites to leverage the data from smaller scale projects. The two chief hallmarks of the "hub" model are (i) emphasis on the use of pre-existing, primary databases and analytical tools whenever possible, and (ii) maintenance by personnel familiar with the organism and embedded within the research community. The direct use of primary databases and tools has many advantages, including the fact that this approach reduces the expense of developing and maintaining on-site databases in a personnel-intensive manner, and that by directly accessing primary databases and tool development sites, users are ensured of access to the most up-to-date resources. Furthermore, this approach allows the "hub" model web site to be maintained primarily by biological researchers themselves, who can design and adapt the site to meet the evolving needs of the research community while focusing limited time and personnel power on high-priority organism-specific resources.

In its current form, the PPI site is a central hub for genome research on the three *P. syringae* pathovars with a special focus on navigating the existing annotation, viewing the genome, and communicating annotation updates. The GenBank genome annotation records for each strain are used as the primary database, with updates to those annotation records being forwarded there directly. In addition, the recent, intense interest in Hop effector proteins has made development of an on-site Hop database a high priority for the *P. syringae* research community. The Hop database currently maintains a regularly updated list of characterized proteins together with information related to their nomenclature, phylogeny, and phenotypic characteristics. Users interested in viewing the genome are directed to the Artemis genome viewer and Artemis Comparison Tool developed at Sanger rather than to an on-site genome viewer, with the PPI site focusing on Artemis-related resources tailored to the needs of the *P. syringae* research community. Resources include tutorials for viewing the *P. syringae* genomes in Artemis, and Artemis-readable files that allow researchers to view *hop* genes and other virulence factors of special interest to researchers by location in the genome. The PPI web site also serves as a portal for submission of updated genome annotations, such as translational start sites, which come from either personal communication or review of the literature. These are then forwarded to GenBank and the Comprehensive Microbial Resource (CMR) at TIGR. GO annotation data also can be communicated through the PPI web site and from there to TIGR, where it becomes part of annotated genome updates sent from TIGR to the GO Consortium databases, ensuring broad accessibility.

Acknowledgements The work of the Plant-Associated Microbe Gene Ontology (PAMGO) interest group has been supported by NSF/USDA award #EF-0523736. Summer work by C.W.C. was supported by NSF award #DBI-0077622.

References

Buell, C.R., Joardar, V., Lindeberg, M., Selengut, J., Paulsen, I.T., Gwinn, M.L., Dodson, R.J., Deboy, R.T., Durkin, A.S., Kolonay, J.F., Madupu, R., Daugherty, S., Brinkac, L., Beanan, M.J., Haft, D.H., Nelson, W.C., Davidsen, T., Liu, J., Yuan, Q., Khouri, H., Fedorova, N., Tran, B., Russell, D., Berry, K., Utterback, T., Vanaken, S.E., Feldblyum, T.V., D'Ascenzo, M., Deng, W.-L., Ramos, A.R., Alfano, J.R., Cartinhour, S., Chatterjee, A.K., Delaney, T.P., Lazarowitz, S.G., Martin, G.B., Schneider, D.J., Tang, X., Bender, C.L., White, O., Fraser, C.M., and Collmer, A., 2003, The complete sequence of the Arabidopsis and tomato pathogen *Pseudomonas syringae* pv. *tomato* DC3000. *Proc. Natl. Acad. Sci. USA* 100(18): 10181–10186.
Gene Ontology Consortium, 2001, Creating the gene ontology resource: design and implementation. *Genome Res.* 11(8): 1425–1433.
Gene Ontology Consortium, 2006, The gene ontology (GO) project in 2006. *Nucleic Acids Res.* 34(Database Issue): D322–D326.
Riley, M., 1993, Functions of the gene products of *Escherichia coli*. *Microbiol. Rev.* 57(4): 862–952.

Exploring the Functions of Proteins Secreted by the Hrp Type III Secretion System of *Pseudomonas syringae*

A. Collmer[1], B.H. Kvitko[1], J.E. Morello[1], K.R. Munkvold[1], H.-S. Oh[1], and C.-F. Wei[2]

Abstract *P. syringae* pathogenesis is dependent on the Hrp type III secretion system (T3SS), which injects effector (Avr/Hop) proteins into plant cells. To better understand how the T3SS functions in pathogenesis, we are exploring the delivery and activity of these proteins. The model strain *P. syringae* pv. *tomato* DC3000 appears to strongly express and inject approximately 30 effectors into plant cells. We are taking three approaches to identifying informative biological phenotypes. First, we are expressing these effectors in a galactose-inducible manner in yeast (*Saccharomyces cerevisiae*) to identify those that inhibit or kill this model eukaryote. Among the several effectors whose expression is toxic to yeast, we are focusing on HopAA1-1, which appears to localize to yeast mitochondria. C-terminal truncations of HopAA1-1 that diminish killing in yeast also diminish killing of plant cells following *Agrobacterium*-mediated transient expression. Our observations with HopAA1-1 and other effectors support the utility of yeast for exploring effector activities. Second, we are using *P. fluorescens* expressing cloned clusters of *P. syringae hrp/hrc* genes to explore the ability of individual effectors to suppress the basal resistance that is otherwise induced by this nonpathogen. Our focus in this work is on proteins that appear to function in the apoplast, such as HopP1, a harpin with a lytic transglycosylase domain. Third, we are constructing DC3000 polymutants in which multiple effector genes are deleted, and we are testing these mutants for their ability to grow and cause disease in Arabidopsis, tomato, and *Nicotiana benthamiana*. Our focus in this work is on *hopQ1-1*, whose deletion enables DC3000 (which normally produces virtually no symptoms on *N. benthamiana*) to cause extensive necrosis. Deletion of *hopQ1-1* has no apparent effect on DC3000 interactions with Arabidopsis and tomato, but deletion of other combinations of effectors results in marked reductions in virulence in these hosts. The overall goal of this work is to understand how the T3SS enables *P. syringae* to suppress defenses and cause disease in a host-specific manner in plants.

[1] Department of Plant Pathology, Cornell University, Ithaca NY 14853, USA

[2] Graduate Institute of Biotechnology, National Chung Hsing University, Taichung 40224, Taiwan

Author for correspondence: Alan Collmer; e-mail: arc2@cornell.edu

M'B. Fatmi et al. (eds.), *Pseudomonas syringae Pathovars and Related Pathogens.*
© Springer Science + Business Media B.V. 2008

Keywords Genomics, *hrp* genes, hypersensitive response and pathogenesis

1   Introduction

Pseudomonas syringae pv. *tomato* (*Pto*) DC3000 is a model pathogen of tomato and Arabidopsis whose genome has been sequenced to completion and is now the subject of extensive, ongoing, functional annotation (Buell et al., 2003) (http://pseudomonas-syringae.org). The type III secretion system (T3SS), which injects effector proteins into plant cells, is key to the ability of DC3000 to defeat basal defenses and to grow and cause disease lesions in host plants, but effectors can also act as avirulence determinants with the potential to limit the host range of *P. syringae* strains (Alfano and Collmer, 2004; Nomura et al., 2005; Abramovitch et al., 2006). Indeed, pathovars of *P. syringae* are considered able to cause disease in only a few plant species. For example, *Pto* DC3000 does not cause disease in the model plant *Nicotiana benthamiana*, which is considered to be a nonhost for this pathovar.

The effector repertoire of DC3000 has been extensively characterized (Lindeberg et al., 2006). The pathogen appears to deploy two classes of proteins, which are generically referred to as Hops (*Hrp* *o*uter *p*roteins) (Lindeberg et al., 2005). Approximately 30 Hops are considered to be true effectors that function primarily within plant cells, whereas four harpin-like Hops are thought to function primarily in the apoplast. Apparent redundancies in both classes of Hops complicate investigation of the functions of the individual proteins. Importantly, mutations in individual *hop* genes typically have little or no virulence phenotype. We have used three strategies to overcome this problem in our search for Hop functions. First, individual Hops were expressed in various biological systems to identify gain-of-function phenotypes. Second, polymutants lacking multiple Hops were constructed to reduce redundancy and thereby reveal loss-of-function phenotypes. Third, additional assays were developed to identify subtle phenotypes related to pathogenesis. To illustrate these strategies, we will highlight work on three Hop proteins, the true effectors HopAA1-1 and HopQ1-1 and the harpin HopP1.

2   Results and Discussion

2.1   *The Pto DC3000 Effector Repertoire*

More than half of the effector genes in DC3000 occur in ten clusters of two or more genes. These clusters are evident when the repertoire is ranked by PSPTO locus numbers (Schechter et al., 2006) or visualized on the genome (http://pseudomonas-syringae.org). Table 1 presents the clusters and also lists the remaining *hop/avr*

Table 1 Effector genes in *P. syringae* pv. *tomato* DC3000[a]

Cluster	Effector genes
I	*hopU1, hopF2*
II	*hopH1, hopC1*
III	*hopAJ1,* (*hopAT'*)
IV	*hopD1, hopQ1-1, hopR1*
V	(*hopAG::ISPssy*), (*hopAH1*), (*hopAI1*)
VI	*hopN1, hopAA1-1, hopM1, avrE1*
VII	(*hopAH2-1*), (*hopAH2-2*)
VIII	*hopS2,* (*hopT2*), (*hopO1-3'*),
	hopT1-2, hopO1-2, (*hopS1::ISPssy*)
IX	*hopAA1-2, hopV1, hopAO1,* (*hopD::IS52*),
	(*hopH::ISPsyr4*), *hopG1,* (*hopQ1-2*)
pDC3000A	*hopAM1-2, hopX1, hopO1-1, hopT1-1*
Orphans	*hopK1, hopY1, hopAM1-1, hopB1, hopAF1,*
	hopAB2, avrPto1, hopE1, hopI1, hopA1

[a]Clusters are numbered in the order of their location on the DC3000 chromosome. Putative pseudogenes are indicated with parentheses. Cluster VI is also known as the conserved effector locus. Orphans are listed in the order in which they appear in the genome

genes that are thought to be injected into plant cells by the T3SS (Lindeberg et al., 2006). Putative pseudogenes, weakly expressed genes, and genes encoding apoplast Hops are excluded from the list of orphans.

2.2 *HopAA1-1*

Cell killing is associated with the activity of many of the type III effectors produced by both plant and animal pathogens. We hypothesized that some effectors will promote cell death and that they will do so through universal eukaryotic targets. This hypothesis suggests the utility of screening the DC3000 effector repertoire for the ability of individual effectors to kill or inhibit the growth of yeast (*Saccharomyces cerevisiae*). Yeast was transformed with vector pYES-DEST52 (Invitrogen) expressing effectors in a galactose-inducible manner. When transformed yeast cells were shifted from glucose to galactose medium, we found that those expressing several effectors were reduced or abolished in their ability to grow.

We focused on two of these effectors, HopN1 and HopAA1-1. We had previously shown that HopN1 was a cysteine protease and that two amino acids are particularly important to the biological activity of HopN1: C172 and H282 (López-Solanilla et al., 2004). We accordingly constructed mutations affecting these two amino acids in the *hopN1* gene expressed in pYES-DEST52 and found both mutations to strongly reduce the ability of the protein to inhibit the growth of yeast. These results provide further evidence that the biochemical activities of *P. syringae* Hops can be productively studied in yeast, as was demonstrated with AvrRpt2 (Coaker et al., 2005).

The biochemical activity of HopAA1-1 is not known. HopAA1-1 is a potentially universal *P. syringae* effector that is encoded in the conserved effector locus (Alfano et al., 2000). DC3000 has a homolog, HopAA1-2, which is encoded elsewhere in the genome. A DC3000 mutant lacking both *hopAA1-1* and *hopAA1-2* was found to be quantitatively reduced in its ability to initiate colony development in Arabidopsis leaves, as observed with GFP-labeled bacteria in confocal microscopy (Badel et al., 2002). We have found that HopAA1-1 is localized to yeast mitochondria and that the failure of yeast expressing this effector to grow can be attributed to cell killing. Mitochondrial localization was observed by immunofluorescence microscopy employing antibody to a V5-epitope tag fused to the C-terminus of HopAA1-1, with mitochondrial porin serving as a marker for colocalization analysis. The ability of HopAA1-1 to kill yeast was determined by loss of active uptake of the dye FUN-1 into vacuoles (Millard et al., 1997). Compared with a LacZ control, HopAA1-1 caused more than a 50% reduction in dye uptake.

We also constructed derivatives of the 486-amino-acid HopAA1-1 protein with progressive C-terminal and N-terminal deletions. The loss of either 100 amino acids from the C-terminus or 51 amino acids from the N-terminus abolished cell killing in yeast. We then used *Agrobacterium*-mediated transient expression to produce native and mutant derivatives of HopAA1-1 within *N. benthamiana* leaf cells. The *hopAA1-1* derivatives were expressed from the 35S promoter of pLN462 and carried C-terminal hemagglutinin (HA) epitope tags, which enabled us to confirm their production by immunoblotting. Native HopAA1-1 elicited cell death in this test, whereas truncated versions lacking either the first 51 amino acids of the N-terminus or the last 100 amino acids of the C-terminus did not. These results suggest that, as with HopN1, the same biochemical activity that is required for HopAA1-1 to kill yeast is required for the protein to kill plant cells. Yeast is a model eukaryote with an extensive molecular toolbox, and it has been successfully used to explore the functions of type III effectors of animal pathogens (Lesser and Miller, 2001). We are now investigating the mechanism by which HopAA1-1 kills yeast cells.

2.3   HopP1

DC3000 appears to deploy four harpin-like proteins. Harpins have the following properties: they are expressed in a HrpL-dependent manner, secreted by the T3SS, lack cysteine residues, and are able to elicit the hypersensitive response when infiltrated into the apoplast of tobacco and several other test plants (Alfano and Collmer, 2004). The HrpZ1 and HrpW1 harpins had previously been reported (Charkowski et al., 1998; He et al., 1993). HrpW1 has two distinct domains. The N-terminal half of the protein is glycine-rich, like HrpZ, but shows no homology with proteins of known function. The C-terminal half is highly similar to pectate lyases. HrpW1 binds specifically to pectate, but no pectolytic activity has been reported (Charkowski et al., 1998). DC3000 produces two more proteins with the same overall properties and two-domain structure of HrpW1. These are HopAK1 and

HopP1. The C-terminal half of HopAK1 is similar to pectin lyases, and the C-terminal half of HopP1 is similar to lytic transglycosylases. Furthermore, the N-terminal half of HopP1 is homologous to the N-terminal half of HrpW1. The ability of purified harpins to elicit plant defenses has been extensively studied (Alfano and Collmer, 2004), but much less is known about the function of harpins when deployed by *P. syringae* in natural infections. The ability of purified harpins to generate pores in membranes suggests that they may have a role in translocation of effectors (Lee et al., 2001). However, we have not observed individual harpin-deficient mutants to be abolished in their ability to translocate AvrPto or other Hops based on tests using C-terminal fusions of these proteins with the Cya reporter (*Bordetella pertussis* adenylate cyclase).

To search for other potential activities of harpins, we have used *P. fluorescens* carrying pLN18, which expresses the *P. syringae* pv. *syringae* 61 T3SS genes without any effector (Jamir et al., 2004). This system enables the effects of a test Hop to be observed without the confounding influence of other Hop proteins. In particular, we have used this system to explore the ability of several effectors to suppress basal resistance (Oh and Collmer, 2005). *P. fluorescens* (pLN18) elicits basal resistance as indicated by several assays. First, *N. benthamiana* leaf areas that have been infiltrated with bacteria at 10^8 cfu/ml show reduced uptake into their xylem of petiole-fed Neutral Red. Second, when these regions are challenged 6 h later with DC3000, which would normally elicit the hypersensitive response in *N. benthamiana*, no tissue collapse develops. We have found that in both of these tests, if the *P. fluorescens* (pLN18) cells are also expressing HopP1, then basal resistance appears to be suppressed. That is, there is stronger uptake of Neutral Red and the DC3000 challenge inoculation is able to elicit the hypersensitive response. We are now exploring the mechanism by which HopP1 appears to suppress basal resistance and the potential involvement of other harpins in defense suppression.

2.4 HopQ1-1

To reduce redundancy in the effector repertoire and to explore the role of effectors in nonhost resistance – particularly with *N. benthamiana* – we have deleted several effector gene clusters from DC3000. The DC3000 effector clusters are presented in Table 1; we chose to delete four of them containing a total of 13 active effector genes (Wei et al., 2007). Cluster II contains *hopH1* and *hopC1*. Both effectors appear to be strongly produced in DC3000 (Chang et al., 2005; Ferreira et al., 2006). Cluster IV contains *hopD1*, *hopQ1-1*, and *hopR1*. HopQ1-1 was a candidate avirulence determinant for nonhost *N. benthamiana* because it elicited the hypersensitive response in that species when delivered by T3SS-proficient *P. fluorescens* (pLN18) (Schechter et al., 2004). Cluster IX carries *hopAA1-2*, *hopV1*, *hopAO1*, and *hopG1*. Weak phenotypes have been associated with individual *hopAA1-2* and *hopAO1* mutations (Badel et al., 2002; Espinosa et al., 2003). We also exploited CUCPB5138, from which the two native plasmids in DC3000 – pDC3000A and pDC3000B – have been deleted (Buell et al., 2003). The former plasmid carries *hopAM1-2*, *hopX1*, *hopO1-1* and *hopT1-1*. We deleted the chromosomal clusters, singly and in combination, using a system whose components are a Gateway-ready derivative of pRK415 and Gateway PCR clones that contain a FRT-Gm^R cassette ligated into sequences flanking the deleted target gene. The cassette-marked region can be rapidly recombined into pRK415, transformed into *P. syringae*, and then introduced into the chromosome by homologous recombination. The cassette is then removed from the genome using the yeast Flp recombinase, which leaves the deletion marked only by FRT scars (Hoang et al., 1998).

To determine the effects of these mutations on the interactions of DC3000 with *N. benthamiana* we inoculated leaves at 10^4 cfu/ml using a blunt syringe. At this low level of inoculum, *P. syringae* pathovars are not expected to produce symptoms on nonhosts. Indeed, wild-type DC3000 did not produce necrotic lesions in these inoculations. Nor did CUCPB5138 or the mutants lacking clusters II and IX. However, the mutant lacking effector cluster IV caused extensive necrosis 8 days after inoculation. Furthermore, the cluster IV mutant grew as well as the tobacco wildfire pathogen *P. syringae* pv. *tabaci* 11528 and produced comparable symptoms in *N. benthamiana*. To further explore the disease that is caused in *N. benthamiana* by the cluster IV mutant, we inoculated *N. benthamiana* by dipping in a suspension of bacteria at 10^5 cfu/ml in 0.01% Silwet L-77 in water and observed symptoms 8 days later. Wild-type DC3000 caused no disease, but the cluster IV mutant caused spreading necrosis as well as numerous small necrotic lesions that were remarkably similar to the bacterial speck symptoms caused in tomato by DC3000. To further test the role of HopQ1-1 as an avirulence determinant for DC3000 in *N. benthamiana*, we deleted just the *hopQ1-1* gene from DC3000. The resulting mutant was now virulent in *N. benthamiana* and produced symptoms slightly faster than the cluster IV deletion mutant. Thus, HopQ1-1 appears to be the avirulence determinant responsible for the nonhost resistance interaction of *Pto* DC3000 with *N. benthamiana* (Wei et al., 2007).

Polymutants lacking effector cluster IV plus other clusters were also analyzed. The loss of clusters II and IX reduced virulence, as indicated by less necrosis following blunt-syringe inoculation and by a shift from spreading necrosis to speck-only symptoms following dip inoculation. However, deleting all three chromosomal clusters and pDC3000A did not abolish the ability of the mutants to grow and produce disease lesions in *N. benthamiana*.

The effector polymutants were also tested for their ability to grow and produce symptoms in Arabidopsis Col-0 and tomato cv. Moneymaker. Plants were inoculated by vacuum infiltration with inoculum at 10^5 cfu/ml in water with 0.01% Silwet L-77 and then monitored for bacterial growth and symptom production. The mutant lacking cluster IV was not discernibly different from wild-type DC3000. However, in both plants, mutants lacking either clusters II or IX were reduced in lesion formation. Interestingly in both plants, the mutant lacking cluster II was reduced in growth *in planta*, whereas the mutant lacking cluster IX was not.

Polymutants lacking clusters II, IV, IX, and pDC3000A were strongly reduced, but not abolished, in growth and lesion formation.

Two aspects of these findings warrant further discussion. First, by simply deleting *hopQ1-1*, we have developed a useful new disease model that enables viral-induced gene silencing, a powerful tool that works particularly well in *N. benthamiana* (Baulcombe, 1999), to be applied directly to DC3000 pathogenesis. Second, the observation that the respective phenotypes of the cluster II and IX mutations are the same in both Arabidopsis and tomato suggests that the virulence targets of effectors may be the same in all plants. This observation, in turn, provides more support for a prevailing view that host range is determined primarily by interactions of effector repertoires with the anti-effector surveillance systems of plants. As noted in earlier studies with *P. syringae*, a given pathovar may possess multiple avirulence determinants that are recognized by a nonhost species (Kobayashi et al., 1989). Also, a *P. syringae* strain may become avirulent if it loses an effector that normally masks the avirulence activity of another effector in the repertoire (Jackson et al., 1999). By combining the use of polymutants, individually expressed effectors in T3SS-proficient *P. fluorescens*, and test plants with diverse defense genotypes, we may begin to sort out the interplay of avirulence determinants, suppressors, and plant factors in controlling host specificity in *Pto* DC3000.

3 Conclusions

The examples given above are representative of progress being made by many labs in exploring the functions of type III effectors, which are now seen to be central to *P. syringae* virulence (Grant et al., 2006). The enormous diversity of *P. syringae* interactions with plants in the field has been noted (Hirano and Upper, 2000), and we now see that the molecular underpinnings of these interactions are equally complex. These interactions are also dynamic. For example, spontaneous loss of an effector with avirulence activity during exposure to a resistant plant has been observed (Pitman et al., 2005). Given our observation that loss of a single effector can extend the host range of *Pto* to a new species it is puzzling that *P. syringae* host specificity at the species–pathovar level appears relatively stable in the field. One explanation is that although a DC3000 *hopQ1-1* mutant may cause disease in *N. benthamiana* plants in the laboratory, the mutant lacks multiple adaptations for virulence on *N. benthamiana* that might be important in the field. These adaptations could involve nutrition, tolerance of antimicrobials, or factors associated with entry and dissemination from leaves. To fully understand *P. syringae*–plant interactions, we will ultimately need to expand our knowledge beyond effector repertoires and gain a better understanding of the interactions of *P. syringae* and plants in agricultural and natural ecosystems.

Acknowledgements This work is supported by NSF grants MCB-0544066 and DBI-0605059.

References

Abramovitch, R.B., Anderson, J.C., and Martin, G.B. (2006) Bacterial elicitation and evasion of plant innate immunity. *Nat. Rev. Mol. Cell Biol.* **7**: 601–611.

Alfano, J.R., Charkowski, A.O., Deng, W.-L., Badel, J.L., Petnicki-Ocwieja, T., van Dijk, K., and Collmer, A. (2000) The *Pseudomonas syringae* Hrp pathogenicity island has a tripartite mosaic structure composed of a cluster of type III secretion genes bounded by exchangeable effector and conserved effector loci that contribute to parasitic fitness and pathogenicity in plants. *Proc. Natl. Acad. Sci. USA* **97**: 4856–4861.

Alfano, J.R. and Collmer, A. (2004) Type III secretion system effector proteins: double agents in bacterial disease and plant defense. *Annu. Rev. Phytopathol.* **42**: 385–414.

Badel, J.L., Charkowski, A.O., Deng, W.-L., and Collmer, A. (2002) A gene in the *Pseudomonas syringae* pv. *tomato* Hrp pathogenicity island conserved effector locus, *hopPtoA1*, contributes to efficient formation of bacterial colonies in planta and is duplicated elsewhere in the genome. *Mol. Plant Microbe Interact.* **15**: 1014–1024.

Baulcombe, D.C. (1999) Fast forward genetics based on virus-induced gene silencing. *Curr. Opin. Plant Biol.* **2**: 109–113.

Buell, C.R., Joardar, V., Lindeberg, M., Selengut, J., Paulsen, I.T. et al. (2003) The complete sequence of the Arabidopsis and tomato pathogen *Pseudomonas syringae* pv. *tomato* DC3000. *Proc. Natl. Acad. Sci. USA* **100**: 10181–10186.

Chang, J.H., Urbach, J.M., Law, T.F., Arnold, L.W., Hu, A., Gombar, S., Grant, S.R., Ausubel, F.M., and Dangl, J.L. (2005) A high-throughput, near-saturating screen for type III effector genes from *Pseudomonas syringae*. *Proc. Natl. Acad. Sci. USA* **102**: 2549–2554.

Charkowski, A.O., Alfano, J.R., Preston, G., Yuan, J., He, S.Y., and Collmer, A. (1998) The *Pseudomonas syringae* pv. tomato HrpW protein has domains similar to harpins and pectate lyases and can elicit the plant hypersensitive response and bind to pectate. *J. Bacteriol.* **180**: 5211–5217.

Coaker, G., Falick, A., and Staskawicz, B. (2005) Activation of a phytopathogenic bacterial effector protein by a eukaryotic cyclophilin. *Science* **308**: 548–550.

Espinosa, A., Guo, M., Tam, V.C., Fu, Z.Q., and Alfano, J.R. (2003) The *Pseudomonas syringae* type III-secreted protein HopPtoD2 possesses protein tyrosine phosphatase activity and suppresses programmed cell death in plants. *Mol. Microbiol.* **49**: 377–387.

Ferreira, A.O., Myers, C.R., Gordon, J.S., Martin, G.B., Vencato, M., Collmer, A., Wehling, M.D., Alfano, J.R., Moreno-Hagelsieb, G., Lamboy, W.F., DeClerck, G., Schneider, D.J., and Cartinhour, S.W. (2006) Whole-genome expression profiling defines the HrpL regulon of *Pseudomonas syringae* pv. *tomato* DC3000, allows *de novo* reconstruction of the Hrp cis element, and identifies novel co-regulated gene. *Mol. Plant Microbe Interact.* **19**: 1167–1179.

Grant, S.R., Fisher, E.J., Chang, J.H., Mole, B.M., and Dangl, J.L. (2006) Subterfuge and manipulation: type III effector proteins of phytopathogenic bacteria. *Annu. Rev. Microbiol.* **60**: 425–449.

He, S.Y., Huang, H.-C., and Collmer, A. (1993) *Pseudomonas syringae* pv. *syringae* harpin$_{Pss}$: a protein that is secreted via the Hrp pathway and elicits the hypersensitive response in plants. *Cell* **73**: 1255–1266.

Hirano, S.S. and Upper, C.D. (2000) Bacteria in the leaf ecosystem with emphasis on *Pseudomonas syringae* – a pathogen, ice nucleus, and epiphyte. *Microbiol. Mol. Biol. Rev.* **64**: 624–653.

Hoang, T.T., Karkhoff-Schweizer, R.R., Kutchma, A.J., and Schweizer, H.P. (1998) A broad-host-range Flp-FRT recombination system for site-specific excision of chromosomally-located DNA sequences: application for isolation of unmarked *Pseudomonas aeruginosa* mutants. *Gene* **212**: 77–86.

Jackson, R.W., Athanassopoulos, E., Tsiamis, G., Mansfield, J.W., Sesma, A., Arnold, D.L., Gibbon, M.J., Murillo, J., Taylor, J.D., and Vivian, A. (1999) Identification of a pathogenicity island, which contains genes for virulence and avirulence, on a large native plasmid in the bean pathogen *Pseudomonas syringae* pathovar phaseolicola. *Proc. Natl. Acad. Sci. USA* **96**: 10875–10880.

Jamir, Y., Guo, M., Oh, H.-S., Petnicki-Ocwieja, T., Chen, S., Tang, X., Dickman, M.B., Collmer, A., and Alfano, J.R. (2004) Identification of *Pseudomonas syringae* type III secreted effectors that suppress programmed cell death in plants and yeast. *Plant J.* **37**: 554–565.

Kobayashi, D.Y., Tamaki, S.J., and Keen, N.T. (1989) Cloned avirulence genes from the tomato pathogen *Pseudomonas syringae* pv. *tomato* confer cultivar specificity on soybean. *Proc. Natl. Acad. Sci. USA* **86**: 157–161.

Lee, J., Klusener, B., Tsiamis, G., Stevens, C., Neyt, C., Tampakaki, A.P., Panopoulos, N.J., Noller, J., Weiler, E.W., Cornelis, G.R., Mansfield, J.W., and Nurnberger, T. (2001) HrpZ$_{Psph}$ from the plant pathogen *Pseudomonas syringae* pv. *phaseolicola* binds to lipid bilayers and forms an ion-conducting pore in vitro. *Proc. Natl. Acad. Sci. USA* **98**: 289–294.

Lesser, C.F., and Miller, S.I. (2001) Expression of microbial virulence proteins in *Saccharomyces cerevisiae* models mammalian infection. *EMBO J.* **20**: 1840–1849.

Lindeberg, M., Stavrinides, J., Chang, J.H., Alfano, J.R., Collmer, A., Dangl, J.L., Greenberg, J.T., Mansfield, J.W., and Guttman, D.S. (2005) Proposed guidelines for a unified nomenclature and phylogenetic analysis of type III Hop effector proteins in the plant pathogen *Pseudomonas syringae*. *Mol. Plant Microbe Interact.* **18**: 275–282.

Lindeberg, M., Cartinhour, S., Myers, C.R., Schechter, L.M., Schneider, D.J., and Collmer, A. (2006) Closing the circle on the discovery of genes encoding Hrp regulon members and type III secretion system effectors in the genomes of three model *Pseudomonas syringae* strains. *Mol. Plant Microbe Interact.* **19**: 1151–1158.

López-Solanilla, E., Bronstein, P.A., Schneider, A.R., and Collmer, A. (2004) HopPtoN is a *Pseudomonas syringae* Hrp (type III secretion system) cysteine protease effector that suppresses pathogen-induced necrosis associated with both compatible and incompatible plant interactions. *Mol. Microbiol.* **54**: 353–365.

Millard, P.J., Roth, B.L., Thi, H.P., Yue, S.T., and Haugland, R.P. (1997) Development of the FUN-1 family of fluorescent probes for vacuole labeling and viability testing of yeasts. *Appl. Environ. Microbiol.* **63**: 2897–2905.

Nomura, K., Melotto, M., and He, S.Y. (2005) Suppression of host defense in compatible plant-*Pseudomonas syringae* interactions. *Curr. Opin. Plant Biol.* **8**: 361–368.

Oh, H.-S., and Collmer, A. (2005) Basal resistance against bacteria in *Nicotiana benthamiana* leaves is accompanied by reduced vascular staining and suppressed by multiple *Pseudomonas syringae* type III secretion system effector proteins. *Plant J.* **44**: 348–359.

Pitman, A.R., Jackson, R.W., Mansfield, J.W., Kaitell, V., Thwaites, R., and Arnold, D.L. (2005) Exposure to host resistance mechanisms drives evolution of bacterial virulence in plants. *Curr. Biol.* **15**: 2230–2235.

Schechter, L.M., Roberts, K.A., Jamir, Y., Alfano, J.R., and Collmer, A. (2004) *Pseudomonas syringae* type III secretion system targeting signals and novel effectors studied with a Cya translocation reporter. *J. Bacteriol.* **186**: 543–555.

Schechter, L.M., Vencato, M., Jordan, K.L., Schneider, S.E., Schneider, D.J., and Collmer, A. (2006) Multiple approaches to a complete inventory of *Pseudomonas syringae* pv. *tomato* DC3000 type III secretion system effector proteins. *Mol. Plant Microbe Interact.* **19**: 1180–1192.

Wei, C.-F., Kvitko, B.H., Shimizu, R., Crabill, E., Alfano, J.R., Lin, N.-C., Martin, G.B., Huang, H.-C., and Collmer, A. (2007) A *Pseudomonas syringae* pv. *tomato* DC3000 mutant lacking the type III effector HopQ1–1 is able to cause disease in the model plant *Nicotiana benthamiana*. *Plant J.* **51**: 32–46

Conservation of the Pathogenicity Island for Biosynthesis of the Phytotoxin Phaseolotoxin in *Pseudomonas syringae* Pathovars

L. Navarro De La Fuente[1], M.E. Führer[1], S. Aguilera[2],
A. Álvarez-Morales[2], and J. Murillo[1]

Abstract Several pathovars of *Pseudomonas syringae* produce extracellular toxins that often increase their virulence towards plants. Phaseolotoxin is a modified tripeptide that inhibits the key plant enzyme, ornithine carbamoyl transferase (OCTase), and is produced by the *P. syringae* pvs. phaseolicola and actinidiae as well as by a single strain of *P. syringae* pv. syringae, CFBP3388. Genes required for the biosynthesis of phaseolotoxin map to a ca. 28-kb genomic region designated as the Pht cluster, which is included in a larger region characteristic for a pathogenicity island. The Pht cluster included *argK*, a gene coding for an OCTase which confers resistance to phaseolotoxin. Since the sequence of *argK* is identical among strains of *P. syringae* pv. phaseolicola and *P. syringae* pv. actinidiae, others suggested that the *argK-tox* cluster might have been horizontally acquired. The comparison of the published sequence of four Pht clusters shows a very high level of identity (around 99.8%), although the few occurring nt changes often result in important changes in the corresponding deduced gene products. The sequence and the overall organization of the Pht cluster and flanking regions is conserved in all examined strains of pathovars phaseolicola and actinidiae that produced phaseolotoxin. However, PCR experiments indicated that the sequence and/or organization of the Pht cluster are only poorly conserved in *P. syringae* pv. syringae CFBP3388. The sequence of a 2.4-kb fragment from the Pht cluster from this strain, spanning from *phtO* to *amtA*, showed an 83% overall identity with the corresponding sequence of *P. syringae* pv. phaseolicola.1448A. Collectively, our results indicated that the pathogenicity island containing the Pht cluster has evolutionary invaded *P. syringae* several times.

Keywords Actinidiae, *amtA*, antimetabolite toxins, *argK*, integrases, phaseolicola

[1] Laboratorio de Patología Vegetal, ETS de Ingenieros Agrónomos, Universidad Pública de Navarra 31006 Pamplona, Spain

[2] Depto. Ingeniería Genética, CINVESTAV IPN-Unidad Irapuato, México

Author for correspondence: Jesus Murillo; e-mail: jesus.murillo@unavarra.es

M'B. Fatmi et al. (eds.), *Pseudomonas syringae Pathovars and Related Pathogens.*
© Springer Science + Business Media B.V. 2008

1   Introduction

Pathovars of *Pseudomonas syringae* can produce one or several non-host specific toxins that are very diverse in structure and mechanisms of action (Bender et al., 1999). One of these is phaseolotoxin, a modified tripeptide composed of ornithine, alanine and homoarginine, linked to the inorganic group (N′-sulfodiaminophosphinyl) (Mitchell, 1976b). The toxin affects a wide variety of organisms, from plants to bacteria, because it inhibits the key enzymes ornithine carbamoyl transferase (OCTase), which is essential for the biosynthesis of arginine, and ornithine decar-boxylase, which participates in the biosynthesis of polyamines (Bachmann et al., 1998; Langley et al., 2000).

Production of phaseolotoxin is less common and has been described only for *P. syringae* pv. phaseolicola, which infects bean (*Phaseolus vulgaris.* L), in *P. syringae* pv. actinidiae, infecting kiwi (*Actinidia deliciosa*), and in a single strain of *P. syringae* pv. syringae, CFBP3388, isolated from vetch (*Vicia sativa*) (Mitchell, 1976a; Tourte and Manceau, 1995; Tamura et al., 2002). As with other toxins, there are natural isolates of these three pathovars that do not produce the toxin and that do not contain the DNA responsible for its synthesis (Tourte and Manceau, 1995; Han et al., 2003; Rico et al., 2003) indicating that the ability to produce the toxin has been acquired after pathovar delineation. Genes required for the biosynthesis of phaseolotoxin are organizad as the so-called Pht cluster and map to a larger genomic region that has the characteristics of a pathogenicity island (Genka et al., 2006; Aguilera et al., 2007) designated herein as Pht-PAI (Fig. 1). The Pht cluster comprises 22 unidirectional genes organized in three transcriptional units and a gene in opposite orientation, *argK*, that codes for an OCTase which confers resist-ance to phaseolotoxin (Mosqueda et al., 1990; Aguilera et al., 2007). Since the

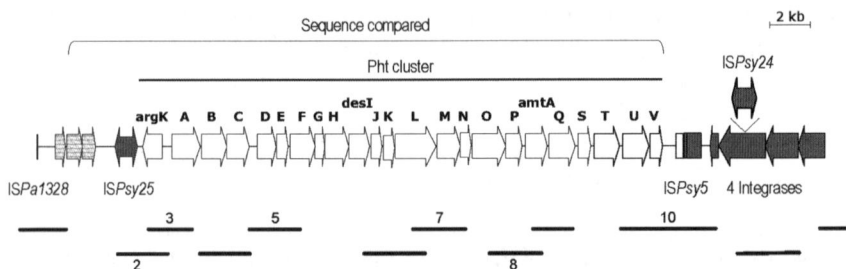

Fig. 1 Structure of the pathogenicity island for biosynthesis of phaseolotoxin (Pht-PAI) from *P. syringae* pv. phaseolicola NPS3121. Genes involved in biosynthesis of and resistance to phase-olotoxin, that collectively conform the Pht cluster, are denoted as white arrows and designated following the proposal of Aguilera et al. (2007). Mobile elements are represented as grey single or double arrows (complete elements) or as boxes (incomplete elements). The sequenced com-pared in Table 1 is indicated at the top. In strain MAFF302282 only, there is a copy of IS *Psy24* inserted in one of the integrases at the right border (Genka et al., 2006), as indicated. Lines at the bottom indicate the approximate size and location of the PCR products assayed to evaluate the conservation of the Pht-PAI

sequence of *argK* is identical among strains of *P. syringae* pv. phaseolicola and *P. syringae* pv. actinidiae, it was suggested that the *argK*-tox cluster might have been horizontally acquired (Sawada et al., 1999). This assumption received further support from the recent sequencing and characterization of the Pht-PAI (Genka et al., 2006; Aguilera et al., 2007). However, it remained unclear how the Pht-PAI has moved among *P. syringae* pathovars and if it is equally conserved among the bacterial population that produces phaseolotoxin.

2 Conservation of Pht-PAI Among Strains of Pathovars Phaseolicola and Actinidiae

Phaseolotoxin production has been traditionally associated to strains of pathovar phaseolicola, from which it took its name (Mitchell, 1976a). The description of phaseolotoxin by strains of pathovar actinidiae (Tamura et al., 1989, 2002) generated much interest and facilitated the study of the genetics and population dynamics of the genes involved in phaseolotoxin biosynthesis. Pathovars phaseolicola and actinidiae were shown by different analyses to be phylogenetically separated and were assigned to different genomospecies (Gardan et al., 1999; Sawada et al., 1999; Sarkar and Guttman, 2004). The existence of nearly identical Pht-PAI in both pathovars clearly showed that this DNA has been recently acquired by horizontal transfer (Sawada et al., 1999, 2002; Genka et al., 2006). The analysis of genes *argK* and *desI* from the Pht cluster indicated that they probably were foreign to *Pseudomonas* and were acquired from a Gram-positive bacterium (Sawada et al., 1999, 2002).

2.1 Comparison of the Four Sequenced Pht-PAI

Recently, Pht-PAI has been sequenced partially or in its entirety from *P. syringae* pv. phaseolicola strains NPS3121 and MAFF302282 and from *P. syringae* pv. actinidiae KW-11 (Genka et al., 2006; Aguilera et al., 2007). Additionally, its sequence is also available from the genome sequence of *P. syringae* pv. phaseolicola 1448A (Joardar et al., 2005). Sequence conservation along a 30234–30245 nt fragment spanning the Pht cluster (Fig. 1) was evaluated with multiple alignments constructed with MultiPipMaker (Schwartz et al., 2003) and pair alignments of the sequence from NPS3121 as a standard with each of the other three sequences using Stretcher (Myers and Miller, 1988). From these comparisons (Table 1), it is apparent that the Pht cluster is highly conserved. The sequence from strain NPS3121 showed 35 nt differences to the corresponding sequence from 1448A (Table 1; Aguilera et al., 2007), 33 nt differences with that of MAFF302282, and only 31 nt differences with that of *P. syringae* pv. actinidiae KW-11 corresponding to more than 99.8% of similarity.

The corresponding differences observed in protein products involved in the biosynthesis of phaseolotoxin are summarized in Fig. 2. For clarity, we are using the

Table 1 Nucleotide changes occurring in the sequenced Pht-PAI from strains of *P. syringae* pv. *phaseolicola* and *P. syringae* pv. *actinidiae* with respect to the sequence of *P. syringae* pv. *phaseolicola* NPS3121[a]

Strain[b]	Mismatch	Deletions (nt)	Insertions (nt)	Total
All	16	5	4	25
Pph 1448A	–	10[c]	–	10
Pph MAFF302282	5	2	1	8
Pac KW-11	5	1	–	6

[a]The compared regions are 30245 nt in NPS3121, and 30234 in the other three strains; the extent of the compared sequence is shown in Fig. 1

[b]"All" means that the same changes occur in strains 1448A, MAFF302282 and KW-11; Pph, *P. syringae* pv. phaseolicola; Pac, *P. syringae* pv. actinidiae

[c]The deletion of 10 nt in 1448A correspond to a 1 nt deletion and to a separate continuous 9 nt deletion; the later eliminating the stop codon of *phtM* and creating a chimera of *phtM* and *phtN*

gene designation proposed by Aguilera et al. (2007). Interestingly, some of the nt changes among the different sequences of the Pht cluster result in substantial changes in the deduced products. Accordingly, PhtD is shortened by 43 aa in its C-terminal end in strain MAFF302282; PhtJ is shortened by 36 aa in its N-terminal end in NPS3121, and PhtM and PhtN form a predicted chimeric protein in 1448A, due to the occurrence of a 9 nt deletion that eliminates the last two codons, including the stop codon, of *phtM*. However, although they obviously do not prevent the biosynthesis of the toxin, it is not known if these modifications affect the production of the toxin in any way. Finally, and due to a single nt deletion, the deduced product of *amtA* shows a different N-terminus in strains NPS3121 and MAFF302282. This appears to be the exception, because partial sequencing of this region from strains *P. syringae* pv. phaseolicola ISPaVe596, CFBP4850 and CFBP4852, these last two isolated from kudzu (*Pueraria lobata*), and *P. syringae* pv. actinidiae NCPPB3871, showed that they had same sequence as that of 1448A and KW-11 (Fig. 2).

The remaining of the Pht-PAI is also highly conserved among strains 1448A, MAFF302282 and KW-11, the major difference being an insertion of ISPsy24 in one of the integrases flanking the right border of the Pht-PAI in strain MAF302282 (Fig. 1; Genka et al., 2006). By DNA hybridization and PCR we confirmed that this insertion is neither present in the other three strains of *P. syringae* pv. phaseolicola (CFBP4852, HRI1375A and PK2) nor in the three strains of *P. syringae* pv. actinidiae (NCPPB3738, NCPPB3739 and NCPPB3871). Therefore, this particular insertion might occur with low frequency in the population of toxigenic strains.

2.2 Conservation of the Pht-PAI in Other Strains

The conservation of the organization of the Pht-PAI in other strains of *P. syringae* pv. phaseolicola, including isolates from kudzu, and *P. syringae* pv. actinidiae was evaluated by PCR using 12 primer pairs directed to genes adjacent to the Pht-PAI or overlapping with its borders (Fig. 1). All the toxigenic strains examined of *P. syringae* pathovars actinidiae and phaseolicola produced amplicons of equal size

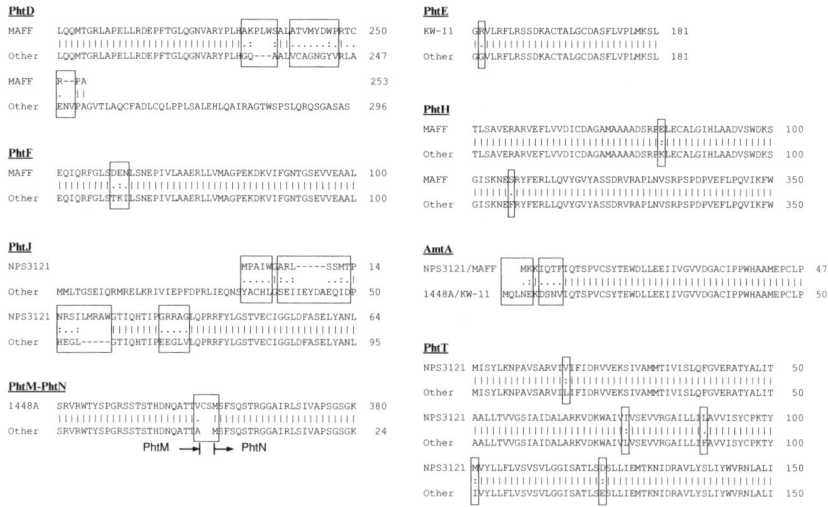

Fig. 2 Deduced products of genes involved in phaseolotoxin biosynthesis that show changes among the four sequenced clusters. Sequences compared were from *P. syringae* pv. phaseolicola NPS3121 (acc. no. DQ141263), 1448A (acc. no. CP000058), MAFF302282 (abbreviated as MAFF; acc. no. AB237164), and *P. syringae* pv. actinidiae KW-11 (acc. no. AB237163). Only relevant fragments, showing all the observed changes, are displayed. The differences, all with respect to strain NPS3121, are due to a single nt deletion, for PhtD; a single nt change, for PhtE; a nt deletion and a nt insertion, for PhtF; 2 nt difference, for PhtH; s single nt change, a nt deletion and two separate single nt insertion, for PhtJ; deletion of a continuous 9 nt fragment, resulting in a chimeric product for PhtM and PhtN; a single nt deletion, for AmtA; and six separate single nt changes, for PhtT

with all primer pairs assayed. Additionally, the majority of the amplicons produced identical patterns of digestion with *Hae*III for all the strains assayed. However, three of the amplicons (amplicons 2, 5, and 10 in Fig. 1) showed variability in at least one of the strains (Fig. 3), indicating the existence of further sequence variation in the population of toxigenic bacteria.

Incidentally, all assayed strains produced identical amplicons for the borders of Pht-PAI, indicating that it is inserted in the same genomic location in all of them. Sequencing of some of these amplicons confirmed this and showed that the Pht-PAI was in all cases bordered by the tetranucleotide TACG as it was previously described (Genka et al., 2006).

3 Conservation of the Pht-PAI in *P. syringae* pv. *syringae*

More than 12 years ago, Tourte and Manceau (1995) isolated an atypical phaseolotoxin-producing strain from a vetch plant. Their careful analyses clearly demonstrated that this strain, designated CFBP3388, belongs to *P. syringae* pv. syringae and indeed produced phaseolotoxin. *P. syringae* pv. syringae is phylogenetically separated from pathovars actinidiae and phaseolicola and was assigned to a different genomospecies

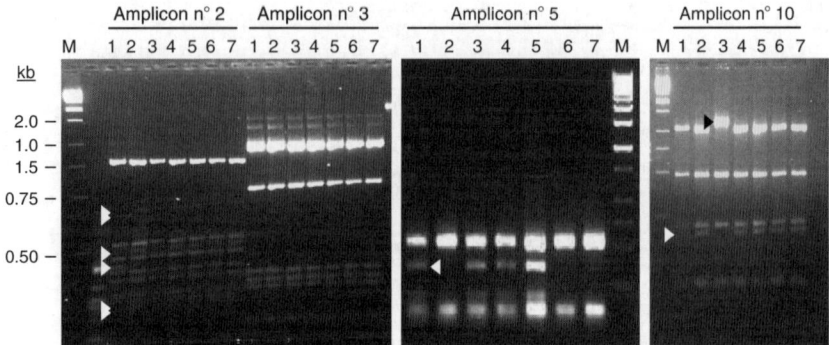

Fig. 3 Sequence variability in the Pht cluster. The amplicons shown in Fig. 1 as number 2 (1943 nt), 3 (2082 nt), 5 (2378 nt) and 10 (2040 nt) were digested with HaeIII and separated in agarose gels. Strains are *P. syringae* pv. actinidiae NCPPB3871 (lanes 1); *P. syringae* pv. phaseolicola ISPaVE596 (lanes 2), HRI1375A (lanes 3), CFBP4850 (lanes 4), CFBP4852 (lanes 5), NPS3121 (lanes 6), and 1448A (lanes 7). M, 1 kb molecular weight marker. Polymorphic bands in the digestions of amplicons 2, 5 and 10 are indicated by white or black arrowheads; the restriction pattern of amplicon 3 was identical for all the strains assayed

(Gardan et al., 1999; Sawada et al., 1999; Sarkar and Guttman, 2004). This situation is not inusual because it is fairly common to find strains of several unrelated pathovars of *P. syringae* that produce the same toxin; typical examples are coronatine, that is produced by pathovars actinidiae, atropurpurea, glycinea, maculicola, morsprunorum and tomato, or tabtoxin, produced by pathovars coronafaciens, garcae and tabaci (Bender et al., 1999; Han et al., 2003).

Unexpectedly, the genes involved in biosynthesis of phaseolotoxin are divergent in strain CFBP3388 as compared to *P. syringae* pvs. phaseolicola and actinidiae. Unlike these two later pathovars PCR amplification of genes adjacent to the phaseolotoxin cluster of CFBP3388 only produced amplicons 7 and 8 (Fig. 1). Currently, we can not discern if this is due to a different organization of the Pht cluster in CFBP3388 or to sequence variations in the primer annealing sites. Also, Tourte and Manceau (1995) demonstrated the amplification of part of gene *desI* using previously designed primers (Prosen et al., 1993).

A 2439 nt sequence spanning nearly all of amplicon number 8 from CFBP3388 covers the complete sequence of *phtP* (756 nt), partial sequences of *phtO* (860 nt; whose end overlaps 10 nt with phtP) and *amtA* (666 nt), and the intergenic region between *phtP* and *amtA* (167 nt) (Figs. 1 and 4). This sequence was very divergent and showed an overall identity of 83% when compared to the corresponding sequence of strain 1448A (Fig. 4). The identity between the deduced products was 83.9% for *phtO* (last 283 aa); 84.6% for *phtP*, and 91% for *amtA* (first 220 aa). The region of lowest similarity was the intergenic region between *phtP* and *amtA* (Fig. 4). This region contained four different insertions of 3, 1, 19, and 14 nt, respectively, in CFBP3388. We are currently completing sequencing of a larger fragment of the Pht-PAI from CFBP3388, which is syntenic with that of strain 1448A and shows a comparable level of similarity. Besides this sequence divergence,

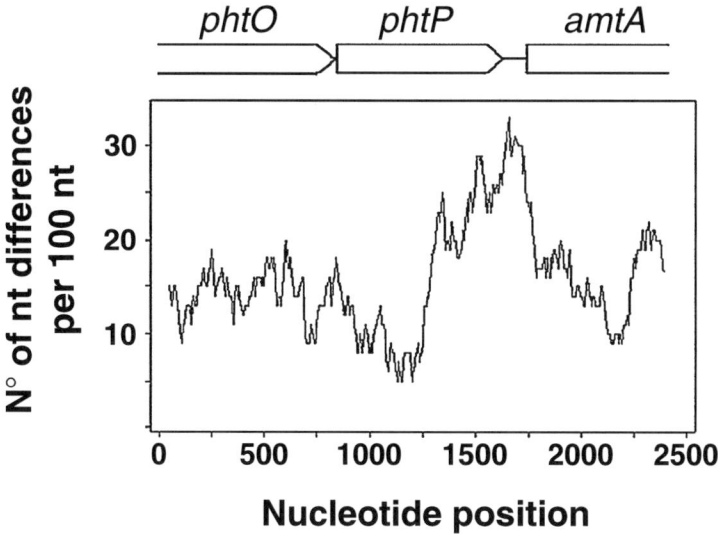

Fig. 4 Sliding window analysis of nucleotide variability [Pi (π)]. The graph shows the comparison, using the computer program DnaSP (Rozas et al., 2003), of the ca. 2.4 kb sequence of amplicon number 8 (see Fig. 1) from strains *P. syringae* pv. phaseolicola 1448A and pv. syringae CFBP3388. Comparison was made with a window of 100 nt measured every 5 nt

our previous results also indicated that the Pht-PAI is inserted into a different position in the genome of strain CFBP3388 (Data not shown).

4 Conclusions

Production of phaseolotoxin is exhibited by strains of at least three pathovars of *P. syringae*: actinidiae, phaseolicola, and syringae. The DNA involved in biosynthesis of the toxin is included in a ca. 38-kb putative pathogenicity island, which has clearly been acquired by horizontal gene transfer by strains of pathovars actinidiae and phaseolicola. The sequence of the Pht-PAI is nearly identical in all examined strains of pvs. actinidiae and phaseolicola. However, a small number of nucleotide changes often resulted in drastic changes in the deduced protein products. This might indicate a high speed of evolution of these gene regions. The sequence divergence that was uncovered in strain *P. syringae* pv. syringae CFBP3388 is striking but currently hard to explain, although it clearly indicated that the Pht-PAI might have colonized *P. syringae* repeatedly along its evolution.

Acknowledgements This work was supported with grant AGL-2004-03143, from the Spanish Ministerio de Educación y Ciencia, to J. Murillo, and with grants from CONACYT, to A. Álvarez-Morales. We are grateful to Antonio de Vicente, Marion Lesaux, Charles Manceau and Marco Scortichini for kindly providing us with bacterial strains. We are also grateful to Theresa Osinga for critical reading of the manuscript.

References

Aguilera S., López-López K., Nieto Y., Garcidueñas-Piña R., Hernández-Guzmán G., Hernández-Flores J. L., Murillo J. and Alvarez-Morales A. (2007). Functional characterization of the gene cluster from *Pseudomonas syringae* pv. phaseolicola NPS3121 involved in the synthesis of phaseolotoxin. J Bacteriol 189: 2834–2843.

Bachmann A. S., Matile P. and Slusarenko A. J. (1998). Inhibition of ornithine decarboxylase activity by phaseolotoxin: implications for symptom production in halo blight of French bean. Physiol Mol Plant Pathol 53: 287–299.

Bender C. L., Alarcón-Chaidez F. and Gross D. C. (1999). *Pseudomonas syringae* phytotoxins: mode of action, regulation, and biosynthesis by peptide and polyketide synthetases. Microbiol Mol Biol Rev 63: 266–292.

Gardan L., Shafik H., Belouin S., Broch R., Grimont F. and Grimont P. A. D. (1999). DNA relatedness among the pathovars of *Pseudomonas syringae* and description of *Pseudomonas tremae* sp. nov. and *Pseudomonas cannabina* sp. nov. (ex Sutic and Dowson 1959). Int J Syst Bacteriol 49: 469–478.

Genka H., Baba T., Tsuda M., Kanaya S., Mori H., Yoshida T., Noguchi M. T., Tsuchiya K. and Sawada H. (2006). Comparative analysis of *argK*-tox clusters and their flanking regions in phaseolotoxin-producing *Pseudomonas syringae* pathovars. J Mol Evol 63: 401–414.

Han H. S., Oak E. J., Koh Y. J., Hur J.-S. and Jung J. S. (2003). Characterization of *Pseudomonas syringae* pv. actinidiae isolated in Korea and genetic relationship among coronatine-producing pathovars based on cmaU sequences. Acta Hort 610: 403–408.

Joardar V., Lindeberg M., Jackson R. W., Selengut J., Dodson R., Brinkac L. M., Daugherty S. C., DeBoy R., Durkin A. S., Giglio M. G., Madupu R., Nelson W. C., Rosovitz M. J., Sullivan S., Crabtree J., Creasy T., Davidsen T., Haft D. H., Zafar N., Zhou L. W., Halpin R., Holley T., Khouri H., Feldblyum T., White O., Fraser C. M., Chatterjee A. K., Cartinhour S., Schneider D. J., Mansfield J., Collmer A. and Buell C. R. (2005). Whole-genome sequence analysis of *Pseudomonas syringae* pv. phaseolicola 1448A reveals divergence among pathovars in genes involved in virulence and transposition. J Bacteriol 187: 6488–6498.

Langley D. B., Templeton M. D., Fields B. A., Mitchell R. E. and Collyer C. A. (2000). Mechanism of inactivation of ornithine transcarbamoylase by Nδ-(N'-sulfodiaminophosphinyl)-L-ornithine, a true transition state analogue? Crystal structure and implications for catalytic mechanism. J Biol Chem 275: 20012–20019.

Mitchell R. E. (1976a). Bean halo-blight toxin. Nature 260: 75–76.

Mitchell R. E. (1976b). Isolation and structure of a chlorosis-inducing toxin of *Pseudomonas phaseolicola*. Phytochemistry 15: 1941–1947.

Mosqueda G., Van den Broeck G., Saucedo O., Bailey A., Alvarez-Morales A. and L. H.-E. (1990). Isolation and characterization of the gene from *Pseudomonas syringae* pv. phaseolicola encoding the phaseolotoxin-insensitive ornithine carbamoyltransferase. Mol Gen Genet 222: 461–466.

Myers E. W. and Miller W. (1988). Optimal alignments in linear space. Comput Appl Biosci 4: 11–17.

Prosen D., Hatziloukas E., Schaad N. W. and Panopoulos N. J. (1993). Specific detection of *Pseudomonas syringae* pv. phaseolicola DNA in bean seed by polymerase chain reaction-based amplification of a phaseolotoxin gene region. Phytopathology 83: 965–970.

Rico A., López R., Asensio C., Aizpún M., Asensio-S.-Manzanera C. and Murillo J. (2003). Nontoxigenic strains of *P. syringae* pv. phaseolicola are a main cause of halo blight of beans in Spain and escape current detection methods. Phytopathology 93: 1553–1559.

Rozas J., Sanchez-DelBarrio J. C., Messeguer X. and Rozas R. (2003). DnaSP, DNA polymorphism analyses by the coalescent and other methods. Bioinformatics 19: 2496–2497.

Sarkar S. F. and Guttman D. S. (2004). Evolution of the core genome of *Pseudomonas syringae*, a highly clonal, endemic plant pathogen. Appl Environ Microbiol 70: 1999–2012.

Sawada H., Kanaya S., Tsuda M., Suzuki F., Azegami K. and Saitou N. (2002). A phylogenomic study of the OCTase genes in *Pseudomonas syringae* pathovars: the horizontal transfer of the *argK*-tox cluster and the evolutionary history of OCTase genes on their genomes. J Mol Evol 54: 437–457.

Sawada H., Suzuki F., Matsuda I. and Saitou N. (1999). Phylogenetic analysis of *Pseudomonas syringae* pathovars suggests the horizontal gene transfer of *argK* and the evolutionary stability of hrp gene cluster. J Mol Evol 49: 627–644.

Schwartz S., Elnitski L., Li M., Weirauch M., Riemer C., Smit A., Program N. C. S., Green E. D., Hardison R. C. and Miller W. (2003). MultiPipMaker and supporting tools: alignments and analysis of multiple genomic DNA sequences. Nucleic Acids Res 31: 3518–3524.

Tamura K., Imamura M., Yoneyama K., Kohno Y., Takikawa Y., Yamaguchi I. and Takahashi H. (2002). Role of phaseolotoxin production by *Pseudomonas syringae* pv. actinidiae in the formation of halo lesions of kiwifruit canker disease. Physiol Mol Plant Pathol 60: 207–214.

Tamura K., Takikawa Y., Tsuyumu S. and Goto M. (1989). Characterization of the toxin produced by *Pseudomonas syringae* pv. actinidiae, the causal bacterium of kiwifruit canker. Ann Phytopathol Soc Japan 55: 512.

Tourte C. and Manceau C. (1995). A strain of *Pseudomonas syringae* which does not belong to pathovar phaseolicola produces phaseolotoxin. Eur J Plant Pathol 101: 483–490.

Syringolin A: Action on Plants, Regulation of Biosynthesis, and Phylogenetic Occurrence of Structurally Related Compounds

B. Schellenberg, C. Ramel, and R. Dudler

Abstract Syringolin A, the product of a mixed non-ribosomal peptide/polyketide synthetase, is secreted by *Pseudomonas syringae* pv. *syringae* under *in planta* conditions and is one of the molecular determinants recognized by nonhost plant species. Spray application of syringolin A onto powdery mildew-infected wheat and *Arabidopsis* has the remarkable effect of reprogramming epidermal cells that are colonized by the powdery mildew fungi *Blumeria graminis* f. sp. *tritici* and *Erysiphe cichoracearum*, respectively, in a compatible interaction to undergo hypersensitive cell death. No hypersensitive cell death is observed if the compound is applied onto uninfected plants. Transcriptome analyses in wheat and *Arabidopsis* with regard to powdery mildew inoculation and/or syringolin A spraying lead to a hypothesis about how syringolin A may accomplish to induce the hypersensitive reaction (HR) in colonized cells. The model is supported by transcriptome analysis of an *Arabidopsis* mutant in which HR is not induced upon syringolin A spraying of powdery mildew-infected plants. Cloning of the syringolin A synthetase genes has allowed us to build a detailed model of syringolin A synthesis based on the gene structure. This model in turn enabled us to clone the genes responsible for the synthesis of glidobactins (syn. cepafungins), antibiotics with a structure related to syringolin A that were reported to have antitumor activity, from an unknown species belonging to the order Burkholderiales. Comparisons to the approximately 700 complete eubacterial genomic sequences known resulted in the identification of a small but very intriguing group of pathogenic bacteria postulated to produce glidobactin-like molecules.

Keywords Powdery mildew resistance, effector, glidobactin, non-ribosomal peptide synthetase, polyketide synthetase

Institute of Plant Biology, University of Zurich, Zurich, Switzerland

Author for correspondence: Robert Dudler; e-mail: rdudler@botinst.uzh.ch

M'B. Fatmi et al. (eds.), *Pseudomonas syringae Pathovars and Related Pathogens.*   249
© Springer Science+Business Media B.V. 2008

1 Introduction

Syringolin A (Fig. 1A) is the major variant of a family of structurally related small cyclic peptides that are secreted by many strains of the phytopathogenic bacterium *Pseudomonas syringae* pv. *syringae in planta* or when grown under conditions mimicking an *in planta* environment (Wäspi et al., 1998, 1999). Syringolin A has no toxic activity towards fungi and bacteria as far as tested (Wäspi et al., 1998). Syringolin A can act as an elicitor of acquired resistance in wheat against the powdery mildew pathogen *Blumeria graminis* f. sp. *tritici* and in rice against the blast fungus *Pyricularia oryzae* when sprayed onto leaves before inoculation (Wäspi et al., 1998). However, it also exhibits the remarkable property to trigger hypersensitive cell death at infection sites of powdery mildew-infected wheat leaves if sprayed two days after inoculation (Wäspi et al., 2001), i.e. syringolin A transforms a compatible interaction into an incompatible one. Effective concentrations in spray applications range from 5 to 100 μM in an aqueous solutions containing 0.5% (v/v) of the non-ionic detergent Tween-20 (Wäspi et al., 2001). However, the response to syringolin A as measured by gene induction in cultured rice cells, which do not have a cuticle, were shown to be elicited already in the nanomolar concentration range (Hassa et al., 2000). Killing the fungus in a compatible interaction with a fungicide (cyprodinil) does not induce hypersensitive cell death to a significant extent. Syringolin A treatment of uninfected plants does not induce hypersensitive cell death or any other visible phenotype at the concentrations given above, and pathogenesis-related genes, which are commonly activated by elicitors, are not activated by syringolin A on such plants. It was speculated that the mode of action of syringolin A may involve breaking of a suppression of the host's defense responses that the fungus exerts in a compatible interaction. This was suggested by the observation that some pathogenesis-related (PR) genes which are activated initially following inoculation of wheat with a compatible powdery mildew isolate but then are shut down in the course of the next few days in spite of the fact that the fungus continues to overgrow its host, were reactivated after syringolin A spraying (Wäspi et al., 2001).

The action of syringolin A on *Arabidopsis* plants infected with a compatible isolate of the powdery mildew fungus *Golovinomyces* (formerly *Erysiphe*) *cichoracearum* is very similar to the one observed in wheat (Michel et al., 2006). Spray application of a 20 μM syringolin A solution two days after infection led to the complete arrest of fungal growth at most infection sites. As in wheat, arrest of fungal growth was accompanied by autofluorescence at infection sites. Only in rare cases an infection site showed no autofluorescence and the fungus continued growth. Killing the powdery mildew fungus with the fungicide cyprodinil did not lead to autofluorescence at most infection sites (88%). Syringolin A application on uninfected *Arabidopsis* never caused a visible phenotype if concentrations below 40 μM were employed. Slight yellowing of leaves became visible at concentrations above 100 μM. Thus, similar to what was observed in wheat, syringolin A application can convert a compatible interaction with powdery mildew to an incompatible one also in *Arabidopsis*.

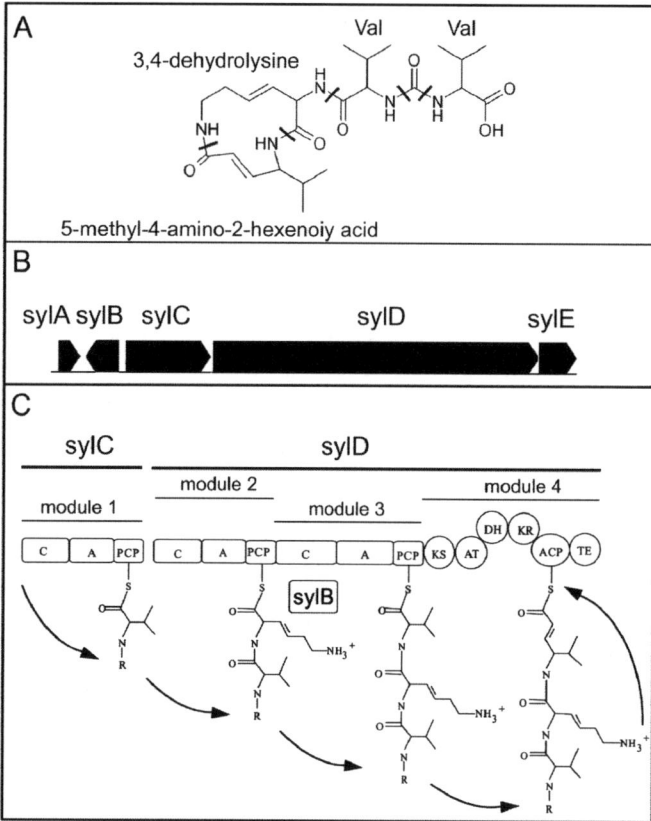

Fig. 1 (A) Structure of syringolin A. (B) Schematic representation of the syringolin A synthetase gene cluster. *SylA* encodes a LuxR-type transcription factor, SylB a modifying enzyme thought to desaturate the lysine residue in syringolin A, SylC and SylD encode non-ribosomal peptide synthetase (NRPS) and polyketide synthetase (PKS) modules, and SylE encodes the putative syringolin exporter. (C) Biosynthesis model of the peptide part of syringolin A. C, condensation domain; A, activation domain; PCP, peptide carrier protein; KS, β-ketoacyl synthase domain; AT, acyl transferase domain; DH, dehydratase domain; KR, β-ketoreductase domain; ACP, acyl carrier protein; TE, thioesterase domain

2   Results and Discussion

2.1   *Effects of Syringolin A on the Transcriptome of Infected and Uninfected Wheat and Arabidopsis Plants*

To monitor transcriptional changes in wheat, 307 cDNA clones representing 158 unigenes from powdery mildew infected, syringolin A sprayed wheat leaves were cloned by a suppression subtractive hybridization cloning procedure. These cDNAs were microarrayed onto glass slides together with 1,088 cDNA-AFLP clones obtained from powdery mildew-infected wheat (Bruggmann et al., 2005). Microarray hybridization experiments were performed with probes derived from leaves, epidermal tissue, and mesophyll preparations of mildewed or uninfected wheat plants 12 and 24 h after syringolin A or control treatment. Normalized hybridization signals averaged over three repetitions were changed at least two-fold (error probability $p < 0.05$) for cDNAs representing 100 different transcripts (Michel et al., 2006). The majority of these (95) accumulated after syringolin A treatment of uninfected plants on whole leaves. None of these were epidermis-specific. Surprisingly, all transcripts accumulating in infected leaves after syringolin A treatment formed a subset of the latter set, i.e. none accumulated specifically only in infected plants. Two thirds of the corresponding genes fell into four gene ontology (GO) biological and molecular function groups (Ashburner et al., 2000): 22 genes involved in carbohydrate metabolism, 11 genes encoding proteins involved in ubiquitin-dependent protein catabolism, 11 genes involved in the response to abiotic stimuli, and 19 genes with unknown function (Michel et al., 2006).

Similar hybridization experiments were also performed in triplicate in *Arabidopsis* using the Affymetrix ATH1 whole genome GeneChip. The results indicated a conserved mode of action of syringolin A as similar gene groups were transiently induced in *Arabidopsis* by syringolin A spraying. 2394 genes were at least twofold induced ($p < 0.05$) after syringolin A spraying of uninfected plants, where transcript levels were lowered at least twofold for 2,873 genes (Michel et al., 2006). With a few exceptions, all 455 transcripts that accumulated in powdery mildew-infected plants after syringolin A spraying formed a subset of the ones accumulating in uninfected plants. The 12 exceptions were only weakly induced and seem not to be significant (Michel et al., 2006). Surprisingly, no transcript was identified whose level was significantly lowered in infected plants after syringolin A spraying. Prominent GO term groups include genes associated with the proteasomal degradation pathway, mitochondrial and other heat shock genes, genes involved in mitochondrial alternative electron pathways, and genes encoding glycolytic and fermentative enzymes. In summary, in both species no good candidate genes whose corresponding transcripts accumulated exclusively in infected plants after syringolin A spraying were identified. On the contrary, in both species the observed transcriptional response to syringolin A was considerably weaker in infected plants as compared to uninfected plants. These results led to the working hypothesis that cell death observed at infection sites may result from a parasite-induced suppression of the transcriptional response and thus to insufficient production of protective proteins necessary for the recovery of these cells from whatever insult is imposed by syringolin A (Michel et al., 2006). Alternatively, syringolin A may eliminate fungal suppression of hypersensitive cell death.

2.2 Biosynthesis, Regulation, and Mode of Action of Syringolin A

As a first step to elucidate the biological function of syringolin A, we sought to clone the genes responsible for its synthesis. Syringolin A is a tripeptide that contains a

12-membered ring structure consisting of the unique non-proteinogenic amino acids 5-methyl-4-amino-2-hexenoic acid and 3,4-dehydrolysine (Fig. 1A). The latter is connected by a peptide bond to a valine which in turn is linked to a second valine residue via an unusual ureido group. The minor structural variants syringolin B–F differ from syringolin A by the substitution of the 3,4-dehydrolysine by lysine, or of one or the other valine by an isoleucine residue, and by combinations of these substitutions (Wäspi et al., 1999). The structure and abundance relations of the syringolin variants suggested that these peptides are synthesized by the same non-ribosomal peptide synthetase (NRPS) (Wäspi et al., 1998, 1999). Using a PCR approach with degenerate primers we succeeded to clone genes (sylA–sylE) encoding (at least part of) the syringolin synthetase (sylC, sylD), which consists of both NRPS and polyketide synthetase (PKS) modules (Fig. 1B) (Amrein et al., 2004). SylA encodes a LuxR-type transcription factor necessary for syringolin A synthesis, while sylB is hypothesized to play a role in a modification step and sylE encodes an exporter (Amrein et al., 2004). Based on the gene structure we derived a syringolin A biosynthesis model (Fig. 1C) that would account for the biosynthesis of the tripeptide without the ureido group and the valine attached to this group. The attachment of the latter two moieties is still enigmatic at present. Similar ureido groups joining two amino acids have been first found in a few cyclic peptides: in keramide A and konbamide (Kobayashi et al., 1991), cyclic peptides extracted from marine sponge species of the genus Theonella, but which likely are synthesized by bacteria associated with the sponges (Piel, 2004; Dudler and Eberl, 2006), in mozamamides A and B, also isolated from theonellide sponges (Schmidt et al., 1997), in ferintoic acids from the cyanobacterium Microcystis aeruginosa (Williams et al., 1996), and in anabaenopeptins from the cyanobacteria Anabaena flos-aquae and Oscillatoria agahdhii (Harada et al., 1995). To our knowledge, the pathway of joining two amino acids via an ureido group is not known for any of these molecules.

A sequence comparison of the relevant homologous regions in the completely sequenced genome of P. syringae pv. syringae strain B728a, which also contains a syringolin synthetase gene cluster of nearly identical nucleotide sequence, with the genome sequence of P. syringae pv. tomato DC3000, which does not produce syringolins, revealed that the syringolin synthetase genes form a cassette that is missing in the latter genome and that contains precisely the sylA–sylE gene cluster.

The biosynthesis of syringolin A is dependent on the gacA/gacS two-component system (Reimmann et al., 1995; Wäspi et al., 1998) that also regulates syringomycin biosynthesis and pathogenicity (Hrabak and Willis, 1992). Furthermore, sylD has been shown to belong to the salA regulon (Lu et al., 2005). P. syringae pv. syringae B301D-R produces syringolins in liquid standing cultures in a defined medium (SRM_{AF}) that was optimized for the production of syringomycin (Gross, 1985) and mimics in planta conditions. In rich media and in shaken SRM_{AF} cultures no syringolin A is produced. We have identified promoter fragments that are necessary and sufficient for the transcription of the sylB and sylC genes using lacZ reporter gene constructs (Dudler, unpublished data). These experiments also suggest that the promoters of sylB and in particular of sylC decrease their activities (measured as specific activities of the reporter gene product β-galactosidase) in bacterial cultures at optical densities above 0.1 (550 nm). Strains carrying a *sylA::lacZ* reporter gene showed no β-galactosidase activity above background, suggesting that this gene is transcribed at low levels. The density dependence of syringolin A accumulation suggests that syringolin A biosynthesis may be negatively regulated by the quorum sensing system.

Syringolins are molecular determinants of *P. syringae* pv. *syringae* recognized, or reacted to, by non-host plants like wheat and *Arabidopsis*. However, we believe it is likely that syringolins provide a selective advantage for the bacteria on host plants, i.e. that they are virulence factors. We have constructed mutants in the B301D-R strain each carrying an insertion in one of the *syl* genes. The *sylA*, *sylC*, and *sylD* insertion mutants all did not produced detectable amounts of syringolin A. In addition, a *sylC* insertional knock-out mutant was constructed in strain B728a. B301D-R was isolated from a diseased pear fruit (Xu and Gross, 1988) and B728a from bean (Willis et al., 1990). Preliminary results of assays of syringolin-negative mutants and the wild-type strain indicate that the mutant causes less disease symptoms than the wild type, suggesting that syringolin A indeed is a virulence factor.

2.3 Syringolin A-like Molecules in Other Bacterial Species

The structure of syringolin A is closely related to a group of anti-tumor antibiotics termed glidobactins or cepafungins (Fig. 2A; hereafter referred to as glidobactins) that have been isolated from the Gram-negative bacteria *Polyangium brachysporum* and *Pseudomonas* sp., respectively (Oka et al., 1988a, 1988b; Shoji et al., 1990). These compounds also contain a 12-membered ring consisting of 4-amino-2-pentenoic acid and 4-hydroxylysine that is linked to an L-threonine residue which in turn is acylated by different unsaturated fatty acids. The acyl group is important for the antifungal activity, as its removal was reported to eliminate toxicity to fungi (Oka et al., 1988c). This is consistent with the observation that syringolin A, which is not acylated, does not exhibit antifungal activity *in vitro*. The model for the biosynthesis of the tripeptide body of syringolin A (Fig. 1C) can nicely account also for the peptide body of glidobactins if it is postulated that the first amino acid activated is a threonine (instead of valine as in syringolin A), the second a lysine, and the third an alanine (instead of a valine as in syringolin A). The latter would be condensed to a malonate by a PKS module, which would result in the unique 4-amino-2-pentenoic acid after the decarboxylation and reduction steps. Thus, we hypothesized that the latter amino acid may be synthesized by the product of a *sylD* homologue in the glidobactin-producing bacterium *P. brachysporum*. *P. brachysporum* has been deposited at the German Collection of Microorganisms and Cell Cultures (DSMZ; http://www.dsmz.de/) as accession DSM 7029. However, its taxonomic identification as *Polyangium brachysporum* is unjustified according to the DSMZ. We have determined the 16S RNA gene sequence of DSM 7029. The sequence places DSM 7029 into the family *Incertae sedis* of the order Burkholderiales (β-proteobacteria).

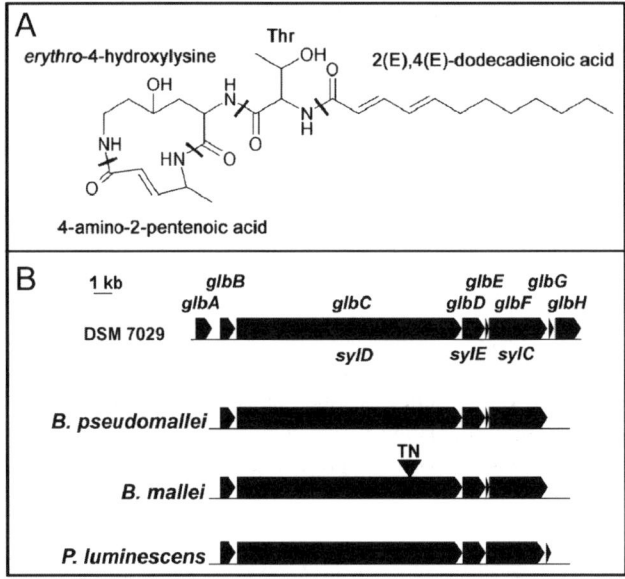

Fig. 2 (A) Structure of glidobactin A. (B) Schematic representation of the glidobactin A gene cluster of strain DSM 7029 and homologous gene clusters in the bacteria indicated. Gene names are indicated above the DSM 7029 gene cluster representation, the names of homologous syringolin A synthetase genes of *P. syringae pv. syringae* are indicated below. TN denotes the site of transposon induced gene rearrangements in *B. mallei*

Sequence comparison of the *P. syringae* pv. *syringae sylD* gene with the nearly 700 available bacterial genome sequences revealed homologous genes in three taxa in addition to *P. syringae* pv. *syringae*: in all ten currently sequenced strains of *Burkholderia pseudomallei*, the causal agent of melioidosis, in *Burkholderia mallei*, the causal agent of glanders, where it matches a disrupted non-functional gene in all nine currently sequenced strains, and in *Photorhabdus luminescens*, an insect pathogen. The intact *sylD* homologues in *B. pseudomallei* and *P. luminescens* are flanked on the downstream side by *sylE* homologues (efflux transporter gene), which in turn are flanked by *sylC* homologues (Fig. 2B). Thus, the gene order is not the same as in the *syl* gene cluster, where *sylC* is located upstream of *sylD*. We designed nested degenerate oligonucleotides from conserved signature sequences flanking the border between the second NRPS module and the PKS module of the sylD protein (Fig. 1C) and its homologues, in the hope that this particular arrangement would be unique for the class of cyclic peptides containing a 12-membered ring. We were successful to amplify an approximately 700 bp fragment by PCR using these primers with genomic DNA isolated from DSM 7029 as a template. This fragment was used to isolate the complete gene cluster from DSM 7029 (Fig. 2B). Insertion mutants confirmed that this cluster encoded the glidobactin A synthetase (Schellenberg et al., 2007). This analysis suggests that *B. pseudomallei* and *P. luminescens* are able to produce glidobactin-like compounds. The taxonomic occurrence of gene clusters encoding syringolin/glidobactin-like molecules suggest that these genes were acquired horizontally.

3   Conclusions

Syringolin A belongs to the unique class of compounds containing a 12-membered ring structure which have an intriguing distribution in nature as far as it is currently known. Elucidation of the biology and mode of action of this class not only increases knowledge about how a plant pathogen interacts with plants, but may also have implications for other pathogens such as *B. pseudomallei*, the causing agent of melioidosis. Finally, knowledge of the gene architecture encoding synthetases for products of this class allows to predict their occurrence from genome and metagenome sequence data.

References

Amrein, H., Makart, S., Granado, J., Shakya, R., Schneider-Pokorny, J., and Dudler, R. (2004). Functional analysis of genes involved in the synthesis of syringolin A by *Pseudomonas syringae* pv. *syringae* B301D-R. Mol. Plant Microbe Interact. *17*, 90–97.

Ashburner, M., Ball, C. A., Blake, J. A., Botstein, D., Butler, H., Cherry, J. M., Davis, A. P., Dolinski, K., Dwight, S. S., Eppig, J. T. et al. (2000). Gene Ontology: tool for the unification of biology. Nat. Genet. *25*, 25–29.

Bruggmann, R., Abderhalden, O., Reymond, P., and Dudler, R. (2005). Analysis of epidermis- and mesophyll-specific transcript accumulation in powdery mildew-inoculated wheat leaves. Plant Mol. Biol. *58*, 247–267.

Dudler, R. and Eberl, L. (2006). Interaction between bacteria and eukaryotes via small molecules. Curr. Opin. Biotechnol. *17*, 268–273.

Gross, D. C. (1985). Regulation of syringomycin synthesis in *Pseudomonas syringae* pv. *syringae* and defined conditions for its production. J. Appl. Bacteriol. *58*, 167–174.

Harada, K., Fujii, K., Shimada, T., Suzuki, M., Sano, H., Adachi, K., and Carmichael, W. W. (1995). Two cyclic peptides, anabaenopeptins, a third group of bioactive compounds from the cyanobacterium *Anabaena flos-aquae* NRC-525-17. Tetrahedron Lett. *36*, 1511–1514.

Hassa, P., Granado, J., Freydl, E., Waspi, U., and Dudler, R. (2000). Syringolin-mediated activation of the Pir7b esterase gene in rice cells is suppressed by phosphatase inhibitors. Mol. Plant Microbe Interact. *13*, 342–346.

Hrabak, E. M., and Willis, D. K. (1992). The *lemA* gene required for pathogenicity of *Pseudomonas syringae* pv. *syringae* on bean is a member of a family of two-component regulators. J. Bacteriol. *174*, 3011–3020.

Kobayashi, J., Sato, M., Murayama, T., Ishibashi, M., Walchi, M. R., Kanai, M., Shoji, J., and Ohizumi, Y. (1991). Konbamide, a novel peptide with calmodulin antagonistic activity from the okinawan marine sponge *Theonella* sp. J. Chem. Soc., Chem. Commun. *15*, 1050–1052.

Lu, S. E., Wang, N., Wang, J. L., Chen, Z. J., and Gross, D. C. (2005). Oligonucleotide microarray analysis of the *salA* regulon controlling phytotoxin production by *Pseudomonas syringae* pv. *syringae*. Mol. Plant Microbe Interact. *18*, 324–333.

Michel, K., Abderhalden, O., Bruggmann, R., and Dudler, R. (2006). Transcriptional changes in powdery mildew infected wheat and *Arabidopsis* leaves undergoing syringolin-triggered hypersensitive cell death at infection sites. Plant Mol. Biol. *62*, 561–578.

Oka, M., Nishiyama, Y., Ohta, S., Kamei, H., Konishi, M., Miyaki, T., Oki, T., and Kawaguchi, H. (1988a). Glidobactins A, B and C, new antitumor antibiotics. I. Production, isolation, chemical properties and biological activity. J. Antibiot. *41*, 1331–1337.

Oka, M., Yaginuma, K., Numata, K., Konishi, M., Oki, T., and Kawaguchi, H. (1988b). Glidobactins A, B and C, new antitumor antibiotics. II. Structure elucidation. J. Antibiot. *41*, 1338–1350.

Oka, M., Numata, K., Nishiyama, Y., Kamei, H., Konishi, M., Oki, T., and Kawaguchi, H. (1988c). Chemical modification of the antitumor antibiotic glidobactin. J. Antibiot. *41*, 1812–1822.

Piel, J. (2004). Metabolites from symbiotic bacteria. Nat. Prod. Rep. *21*, 519–538.

Reimmann, C., Hofmann, C., Mauch, F., and Dudler, R. (1995). Characterization of a rice gene induced by *Pseudomonas syringae* pv. *syringae*: Requirement for the bacterial *lemA* gene function. Physiol. Mol. Plant Pathol. *46*, 71–81.

Schellenberg, B., Bigler, L., and Dudler, R. (2007). Identification of genes involved in the biosynthesis of the cytotoxic compound glidobactin from a soil bacterium. Environ. Microbiol. *9*(7), 1640–1650

Schmidt, E. W., Harper, M. K., and Faulkner, D. J. (1997). Mozamides A and B, cyclic peptides from a theonellid sponge from mozambique. J. Nat. Prod. *60*, 779–782.

Shoji, J., Hinoo, H., Kato, T., Hattori, T., Hirooka, K., Tawara, K., Shiratori, O., and Terui, Y. (1990). Isolation of cepafungins I, II and III from *Pseudomonas* species. J. Antibiot. *43*, 783–787.

Wäspi, U., Blanc, D., Winkler, T., Ruedi, P., and Dudler, R. (1998). Syringolin, a novel peptide elicitor from *Pseudomonas syringae* pv. *syringae* that induces resistance to *Pyricularia oryzae* in rice. Mol. Plant Microbe Interact. *11*, 727–733.

Wäspi, U., Hassa, P., Staempfli, A., Molleyres, L.-P., Winkler, T., and Dudler, R. (1999). Identification and structure of a family of syringolin variants: unusual cyclic peptides from *Pseudomonas syringae* pv. *syringae* that elicit defense responses in rice. Microbiol. Res. *154*, 1–5.

Wäspi, U., Schweizer, P., and Dudler, R. (2001). Syringolin reprograms wheat to undergo hypersensitive cell death in a compatible interaction with powdery mildew. Plant Cell *13*, 153–161.

Williams, D. E., Craig, M., Holmes, C. F. B., and Andersen, R. J. (1996). Ferintoic acids A and B, new cyclic hexapeptides from the freshwater cyanobacterium *Microcystis aeruginosa*. J. Nat. Prod. *59*, 570–575.

Willis, D. K., Hrabak, E. M., Rich, J. J., Barta, T. M., Lindow, S. E., and Panopoulos, N. J. (1990). Isolation and characterization of a *Pseudomonas syringae* pathovar *syringae* mutant deficient in lesion formation on bean. Mol. Plant Microbe Interact. *3*, 149–156.

Xu, G. W. and Gross, D. C. (1988). Physical and functional analyses of the *syrA* and *syrB* genes involved in syringomycin production by *Pseudomonas syringae* pv. *syringae*. J. Bacteriol. *170*, 5680–5688.

An RND-Type Multidrug Efflux Pump from *Pseudomonas syringae*

S. Stoitsova, M.S. Ullrich, and H. Weingart

Abstract Multidrug efflux (MDE) transporters are major contributors to bacterial resistance towards antibiotics and naturally occurring toxic substances. The goal of our research is to identify and characterize MDE pumps in the plant pathogen *Pseudomonas syringae* and to gain in-depth knowledge about their regulation, structure, mechanism of transport and natural functions. MDE pumps may play an important role in the adaptation of *P. syringae* to its respective host plants by protecting them against plant antimicrobials.

In Gram-negative bacteria, members of the RND family are the most relevant in respect of resistance to antimicrobial compounds. In *Pseudomonas aeruginosa*, a clinically important pathogen, the RND-type pump MexAB has been recognized as one of the major efflux system which confers resistance to a broad range of antibiotics. We identified homologues of MexAB in *P. syringae* and generated *mexAB*-deficient mutants. Determination of minimum inhibitory concentrations revealed that mutation of this RND transporter dramatically reduced tolerance to various antibiotics and toxins. Moreover, the *mexAB*-deficient mutants were significantly reduced in their ability to grow *in planta*. These results demonstrate that this transporter system plays an important role in multidrug resistance and virulence of *P. syringae*.

Keywords Multidrug efflux, plant pathogen, RND, secondary transporter, MIC, resistance mechanism

1 Introduction

Bacteria have developed various ways to resist the toxic effects of antibiotics. Antimicrobial compounds can either be modified or degraded by specific enzymes. Other mechanisms of resistance involve alteration of the drug target or changes in

Jacobs University Bremen, School of Engineering and Science, Bremen, Germany

Author for correspondence: Helge Weingart; e-mail: h.weingart@iu-bremen.de

M'B. Fatmi et al. (eds.), *Pseudomonas syringae Pathovars and Related Pathogens.*
© Springer Science+Business Media B.V. 2008

the bacterial surface that reduce entry of drugs into the cell. Alternatively, toxic compounds can be extruded by membrane-associated transport systems. Some transporters are specific and mediate the extrusion of a given drug or class of drugs. In contrast, so-called multidrug efflux (MDE) transporters can transport a wide range of structurally unrelated compounds.

Bacterial multidrug transporters are members of a limited number of families. On the basis of energy source and structural relationships, these transporters can be divided into two major classes. The first class is represented by ATP-binding cassette (ABC) transporters that utilize the energy of ATP hydrolysis to pump drugs out of the cell (Lage, 2003). The second class consists of secondary transporters that mediate drug efflux in a coupled exchange with protons or sodium ions along a concentration gradient as symport or antiport (Paulsen et al., 1996). Members of this class are the major facilitator superfamily (MFS), the small multidrug resistance (SMR) family, the multidrug and toxic compound extrusion (MATE) family, and the resistance-nodulation-cell division (RND) family.

In Gram-negative bacteria, members of the RND family are the most relevant in terms of resistance to clinically important agents (Poole, 2004). The RND transporter forms together with a periplasmic, so-called membrane fusion protein (MFP) and a channel-forming outer membrane protein a functional pump. This complex allows drug transport across both the inner and outer membrane of Gram-negative bacteria directly into the external medium (Zgurskaya et al., 2003). The RND pumps often have very wide substrate specificities (Poole, 2004).

In *Pseudomonas aeruginosa*, a clinically important pathogen, the RND-type pump MexAB has been recognized as one of the major efflux system which confers resistance to a broad range of antimicrobial compounds (Poole and Srikumar, 2001). We have identified homologues of MexAB in the plant pathogen *Pseudomonas syringae*.

2   Results

2.1   Identification of RND-Type Transporter in P. Syringae pv. tomato DC3000

At least 11 possible operons homologous to *P. aeruginosa mex* operons were found in the *P. syringae* pv. *tomato* DC3000 genome by searching first with the *P. aeruginosa* MexB amino acid sequence and then with *P. syringae* Mex sequences. A phylogenetic tree demonstrated the relationship of the RND-type pumps from the plant pathogen within a subset of well-characterized RND transporters from the human pathogen *P. aeruginosa* (Fig. 1). One transporter from *P. syringae* pv. *tomato* DC3000 (PSPTO4304) showed high similarity (79% identity) to the MexB transporter of *P. aeruginosa*. In *P. aeruginosa*, MexAB-OprM has been recognized as one of the major efflux systems. Therefore, a knockout mutant was generated to study the role of this transporter in *P. syringae* pv. *tomato* DC3000.

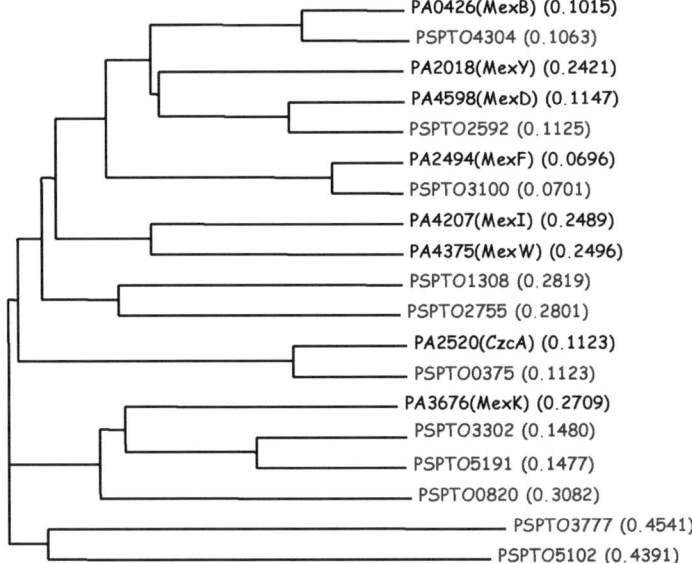

Fig. 1 Phylogenetic tree showing the relationship of the RND-type transporter from *Pseudomonas syringae* within a subset of well-characterized RND transporters from the human pathogen *Pseudomonas aeruginosa*. The tree was prepared using the Vector NTI software (InforMax)

2.2 Characterization of the mexAB-Negative Mutant of P. syringae pv. tomato DC3000

In order to demonstrate that MexAB is an active efflux system, the accumulation of the fluorescent dye ethidium bromide in cells of the wild-type and the *mexAB*-negative mutant was analyzed. Ethidium bromide which changes its spectroscopic properties upon entering the cell, is particularly suitable for uptake experiments because it fluoresce weakly in aqueous environments, but become strongly fluorescent when it intercalates between DNA. Ethidium bromide was added to the bactevethidium bromide in cells of the wild-type than in *mexAB*-negative cells, indicating that ethidium bromide is a substrate for the MexAB transporter.

Addition of CCCP to the assay mixture increased cellular ethidium bromide levels in both types of cells. CCCP (carbonylcyanide m-chlorophenylhydrazone) is an uncoupler of oxidative phosphorylation and leads to breakdown of the proton gradient on the membrane. The final levels of intracellular ethidium bromide after the addition of CCCP were almost identical in the two strains, indicating that the accumulation levels of ethidium bromide in both strains are the same under de-energized conditions. An important point is that the ethidium bromide accumulation level in wt cells was much lower than that in cells of *mexAB* mutant before the addition of CCCP. This indicates that cells of wt possess energy-dependent

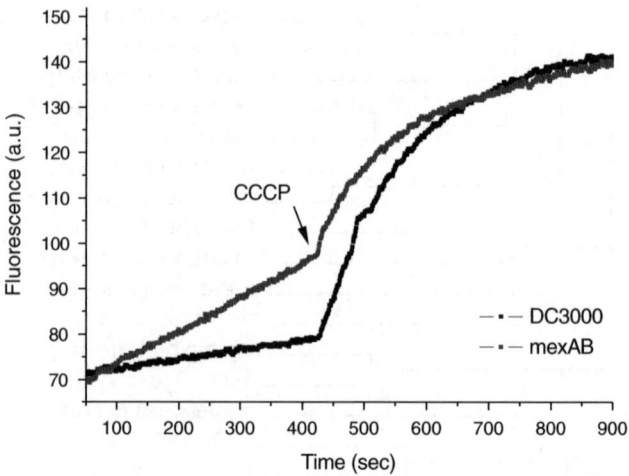

Fig. 2 Changes in ethidium fluorescence were measured with an excitation wavelength at 500 nm and an emission wavelength at 580 nm. Ethidium bromide (4 μM) was added to the assay mixture to initiate the assay, and CCCP (final 100 μM) was added at the time point indicated by an arrow

ethidium bromide efflux activity. Thus, we conclude that MexAB from *P. syringae* pv. *tomato* DC3000 is an energy-dependent drug efflux pump.

Many RND efflux systems are characterized by broad substrate specificity. The resistance of a *mexAB*-negative mutant was compared with that of the wild-type by determination of minimal inhibitory concentrations (MIC). MIC is defined as the lowest concentration of an antibiotic that completely stops visible cell growth. Table 1 summarizes the MICs of the wild-type and the *mexAB*-negative mutant. Deletion of MexAB resulted in increased susceptibilities to a variety of antibiotics and agents, including β-lactams, fluoroquinolones, macrolides, surface-acting agent, several dyes and to same plant-derived antimicrobials. These results suggest that MexAB plays an important role in the resistance of *P. syringae* pv. *tomato* DC3000 towards toxic compounds.

2.3   Contribution of MexAB to Virulence of *P. Syringae* pv. *tomato* DC3000

The importance of the MexAB efflux pump for virulence of *P. syringae* pv. *tomato* strain DC3000 was assayed by comparing bacterial multiplication in tomato and *Arabidopsis thaliana* plants inoculated with *mexAB*-negative mutant and wild-type strain. The *mexAB*-negative mutant was significantly reduced in its ability to grow *in planta*. However, the degree of reduction was plant-dependent. Population sizes of the *mexAB* mutant were 10- and 15-fold lower on tomato plants and more than 20-fold lower on *Arabidopsis thaliana* (Fig. 3).

Table 1 The MIC (μg/ml) was determined by the microdilution assay in Mueller-Hinton broth

	Berberine	Naringenin	Phloretin	Ampicillin	Carbenicillin	Piperacillin	Cefoperazone	Nalidixic acid	Ciprofloxacin	Norfloxacin	Chloramphenicol	Tetracycline	Erythromycin	Clindamycin	Puromycin
DC3000	>1000	>1000	>1000	62.5	500	15.6	31	25	0.06	0.5	50	0.5	12.5	>1000	125
ΔmexAB	125	250	250	0.78	3.1	0.3	0.62	3.1	0.01	0.03	1.5	0.03	3.1	250	15.6

	Novobiocin	Daunorubicin	Nitrofurantoin	Tetraphenylphosphonium chloride	Benzalkonium chloride	Trimethoprim	Fusidic acid	Mitomycin C	Fusaric acid	Ethidium bromide	Acridine Orange	Rhodamine 6G	Crystal violet	Acriflavin
DC3000	>1000	>100	500	>1000	25	125	500	1.25	1000	125	250	500	25	31
ΔmexAB	31	12.5	125	62.5	6.25	31	62.5	0.15	62.5	6.25	31	31	1.5	2.5

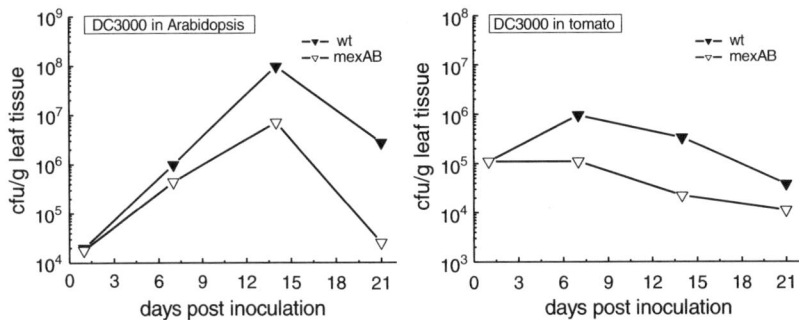

Fig. 3 Population dynamics of *Pseudomonas syringae* pv. *tomato* strain DC3000 wild-type and of the *mexAB*-negative mutant in *Arabidopsis thaliana* and tomato

The mutant caused disease symptoms typical of the wild-type strain, however, consistent with the lower population size of the *mexAB*-negative mutant, only very weak symptoms were observed on leaves inoculated with the mutant.

3   Conclusions

A homologue of MexAB was identified in the plant pathogen *P. syringae* pv. *tomato* strain DC3000. Mutation of *mexAB* dramatically reduced tolerance to hydrophilic and amphiphilic antibiotics, dyes, surface-acting agents, and plant-derived antimicrobials comparable with the reported broad-substrate spectrum of MexAB from *Pseudomonas aeruginosa*. Moreover, disruption of the *mexAB* gene resulted in a dramatically reduced virulence. The *mexAB*-negative mutant was significantly impaired in multiplication after inoculation into its host plants tomato and *Arabidopsis thaliana*. These results demonstrate that this transporter system plays an important role in multidrug resistance and virulence in *P. syringae*.

References

Lage, H., 2003, ABC-transporters: implications on drug resistance from microorganisms to human cancers. Int J Antimicrob Agents 22:188–199.

Paulsen, I. T., Brown, M. H., and Skurray, R. A., 1996, Proton-dependent multidrug efflux systems. Microbiol Rev 60:575–608.

Poole, K. and Srikumar, R., 2001, Multidrug efflux in *Pseudomonas aeruginosa*: components, mechanisms and clinical significance. Curr Top Med Chem 1:59–71.

Poole, K., 2004, Efflux-mediated multiresistance in Gram-negative bacteria. Clin Microbiol Infect 10:12–26.

Zgurskaya, H. I., Krishnamoorthy, G., Tikhonova, E. B., Lau, S. Y., and Stratton, K. L., 2003, Mechanism of antibiotic efflux in Gram-negative bacteria. Front Biosci 8:862–873.

Regulation of the Levansucrase Genes from *Pseudomonas syringae* pv. *glycinea* at the Level of Transcription

D. Zhurina[1,2], A. Srivastava[1], H. Weingart[1], P. Buttigieg[1], and M. Ullrich[1,2]

Abstract In the plant pathogenic bacterium Pseudomonas syringae pv. glycinea PG4180 the exopolysaccharide levan is synthesized by two levansucrases, LscB and LscC. Their respective genes, lscB and lscC, are expressed temperature-dependently, with maximum transcription at 18°C. Temperature-responsive expression relies on a particular promoter structure of the levansucrase genes as well as on regulators, which mediate this process. Structure of the lscB upstream sequence was analyzed by nested deletion analysis in order to map the minimal promoter sequence and presumable thermo- and substrate sensory regions. Current results suggest that at least roughly 1 kb upstream of lscB is necessary to express levansucrase. It was also shown that not all P. syringae strains exhibit the same thermo-responsive synthesis of the levansucrases as PG4180. Comparison of the levansucrases promoter sequences from different Pseudomonas strains to identify the regulator-binding sites is currently in progress. In order to reveal trans-acting regulators, which are responsible for levansucrases expression, lscB and lscC genes together with their presumable native promoters were brought separately into the heterologous host Pseudomonas putida KT2440 strain. Neither lscC nor lscB induced levan formation in P. putida. However, three cosmids from a PG4180 genomic library were found to induce levan formation in P. putida (lscC+) and P. putida (lscB+). Restriction pattern of these cosmids revealed several common fragments, which together comprise approximately 12 kb. Transposon in vitro mutagenesis was performed for the identification of the particular levansucrase gene(s), responsible for levansucrase gene expression. Nucleotide sequencing of the region containing the potential regulator is underway.

Keywords Exopolysaccharide, nested deletion analysis, heterologous expression

---

[1] School of Engineering and Sciences, Jacobs University Bremen, Bremen, Germany

[2] Max-Planck-Institute for Marine Microbiology, Bremen, Germany

Author for correspondence: M. Ullrich; e-mail: m.ullrich@jacobs-university.de

M'B. Fatmi et al. (eds.), *Pseudomonas syringae Pathovars and Related Pathogens.*   265
© Springer Science+Business Media B.V. 2008

1   Introduction

In the phytopathogenic bacterium *Pseudomonas syringae* pv. glycinea PG4180 synthesis of the exopolysaccharide levan is mediated by the enzyme levansucrase. In this organism the levansucrase gene is present in three copies: *lscA*, *lscB* and *lscC*, from which *lscA* is transcriptionally inactive while *lscB* and *lscC* are functional and reside on a 60-kb plasmid and the chromosome, respectively (Li and Ullrich, 2001). Levansucrases B and C were shown to be expressed and secreted temperature-dependently (Li et al., 2006), both *in vitro* and *in planta*. In order to investigate, how these processes are regulated, detailed analysis of levansucrases promoter regions as well as revealing trans-activating factors were performed.

2   Results

2.1   Promoter Analysis of the lscB Gene

In order to determine the minimal promoter region upstream of the *lscB* gene, several deletion sub-clones were generated by nested deletion technique (Erase-a-base kit, Promega, Germany). Here, a 7.2-kb *lscB* containing clone was chosen as a starting length of DNA, with approximately 3 kb region upstream of the levansucrase gene. Two restriction sites were selected, namely *Xba*I (as a susceptible site to S1 nuclease digestion) and *Sac*I (as a resistant site to S1 nuclease activity) in the multiple cloning site. DNA fragments with variable truncated upstream sequence were cloned into pBBR1MCS-3 shuttle vector in the opposite direction to the vector-borne *lacZ* promoter. Then, the generated deletion subclones were trans-conjugated into the *lsc*-deficient PG4180 mutant M6 and the phenotypes were monitored. Four sub-clones, namely −1517, −996, −920, −666 (the number represents the number of nucleotides upstream to the translational start site of *lscB*) were found to still mediate production of levan on MG agar plates supplemented with 5% sucrose (Fig. 1). This indicated that the promoter is still intact and sufficient enough to synthesize the transcript. Smaller sub-clones did not mediate levan synthesis to mutant M6. In our future work, we will further zoom in into the promoter region of *lscB*.

2.2   Identification of Regulator(S) for the Levansucrases from P. syringae by Using Its Heterologous Expression in P. putida

It was shown that the well-known two-component regulatory system GacS/GacA has no impact on expression of either levansucrase gene (Li and Ullrich, 2001). That is why specific regulator(s), which mediate this process, had to be determined.

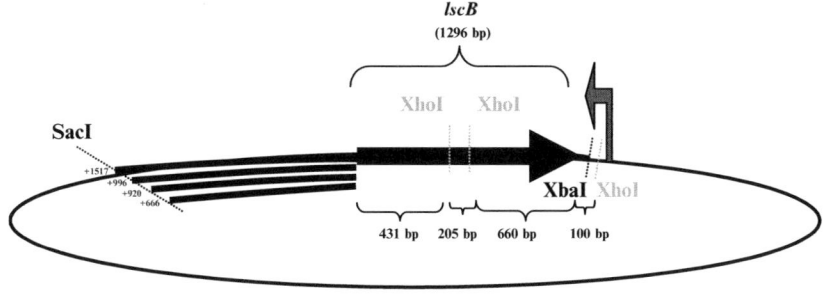

lscB
(1296 bp)

XhoI   XhoI

SacI
+1517
+996
+920
+666

XbaI   XhoI

431 bp   205 bp   660 bp   100 bp

pBBR1MCS-3 (Tc); 5228 bp

Fig. 1 Pictorial representation of various deletion subclones (drawn not to the scale)

The idea of heterologous expression is based on the assumption that in *P. putida* KT2440, which naturally does not contain levansucrase genes in the genome, correspondingly no regulators for *lscC* and *lscB* expression or secretion are present. That is why screening of the genomic library from *P. syringae* PG4180 in *P. putida* should reveal the regulators necessary for these processes.

To perform the heterologous expression, DNA fragments containing either *lscB* on a 7.2-kb *Eco*RV fragment or *lscC* on a 5.5-kb *Sal*I fragment were cloned into the vector pBBR1MCS-2 and subsequently brought into the strain KT2440 by tri-parental mating. Constructs were such that *lscB* and *lscC*, respectively, were oriented in opposite to the vector-borne P$_{lac}$ promoter to guarantee expression of both genes from their native promoters. No levan formation was detected in *P. putida* (*lscB*+) and *P. putida* (*lscC*+). Subsequently, a genomic cosmid library of *P. syringae* PG4180 was shot-gun conjugated into *P. putida* (*lscB*+). Transconjugants were screened on MG agar plates supplemented with 5% sucrose. Out of approximately 2000 transconjugants screened only three were found to form dome-shaped colonies due to levan formation.

Further testing of the cosmids is represented on Fig. 2. First, each of the three cosmids was isolated from the levan-producing clones and transformed into *E. coli* DH5α. Second, isolated cosmids were conjugated into *P. putida* (*lscB*+) to check if they do restore the levan-producing phenotype and into *P. putida* (*lscC*+) to study influence of these cosmids on expression of the *lscC* gene. Last, each cosmid was conjugated in *P. putida* WT to check for levan production. Screening results indicate that the isolated cosmids indeed induce levansucrase genes expression when brought into *P. putida* (*lscC*+) or P. putida (*lscB*+). Moreover, levan formation was possible only when either of the *lsc* genes was present *in trans*, indicating that cosmid inserts do not contain any of the levansucrase genes.

Endonuclease treatment of all three cosmids revealed several common fragments. In total they comprise approximately 12 kb (Fig. 3); on this DNA fragment a single regulator gene or a small operon, responsible for levansucrases expression in *P. putida* and presumably in the natural host *P. syringae*, might be located.

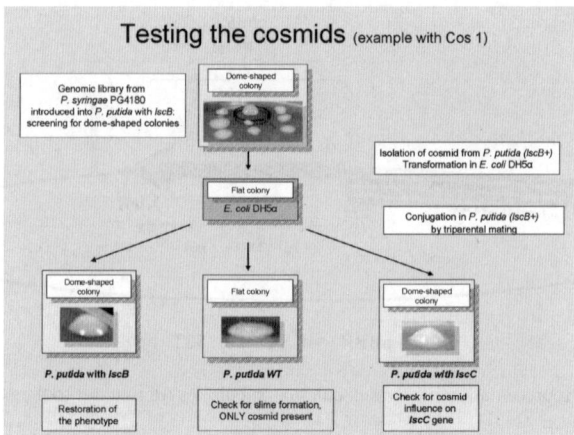

Fig. 2 Scheme for testing the isolated cosmids

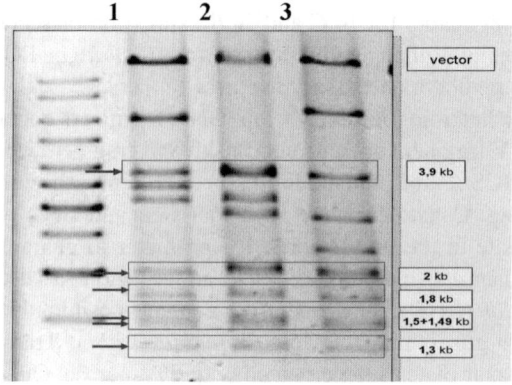

Fig. 3 Comparison of the *PstI* restriction patterns for cosmids 1, 2 and 3; common bands are marked by arrows and put into red boxes, sizes of the common fragment are indicated

In order to reveal, which gene(s) induce expression of the levansucrase genes in *P. putida*, *in vitro* transposon mutagenesis was performed for each cosmid. A pool of mutagenized cosmids, each of them having a single random transposon insertion, was obtained. Disruption of the levansucrase regulator on a cosmid should result in absence of levan formation in *P. putida* ($lscB^+$) or *P. putida* ($lscC^+$). This approach yielded the Tn-mutagenized cosmid 3 (cos3-Tn), which no longer induced levan production in *P. putida* ($lscC^+$) (Fig. 4a). Cos3-Tn was isolated and its restriction pattern was compared to the intact cosmid 3 (Fig. 4b). It was shown that transposon insertion is in the 3.9 kb *PstI* fragment, which is common for all three cosmids (Figs. 3, 4b).

Sequencing of the transposon-hit region, which is assumed to be crucial for levansucrases genes expression in *P. putida* ($lscC^+$) and presumably in PG4180, is currently underway.

a)   b)

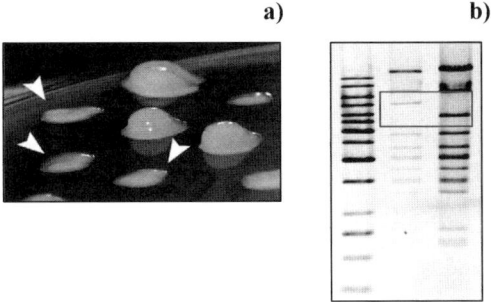

Fig. 4 (**a**) Identification of levan-negative *P. putida* (lscC⁺, cos3-Tn) on MG agar supplemented with 5% sucrose. Levan-negative clones are marked with white arrows. (**b**) *PstI* restriction pattern of cosmid 3 (lane2) and cos3-Tn (lane 1). Increase in size of the 3.9 kb band (framed in red) is due to the transposon insertion

3   Conclusions

Our results suggest that at least the 666 bps upstream sequence of *lscB* gene is necessary for its transcription. Differential expression of levansucrase genes, found in other *Pseudomonas* strains, might be due to the different organization of the respective promoter regions and other transcriptional regulators, which mediate this process. By means of heterologous expression a potential regulatory gene for PG4180 *lscC* and *lscB* genes was found. Further studies on regulation of levansucrases genes are underway.

References

Li, H. and M. Ullrich, 2001, Characterization and mutational analysis of three allelic lsc genes encoding for the extracellular enzyme levansucrase in *Pseudomonas syringae*. J. Bacteriol. 183: 3282–3292.

Li, H., A. Schenk, A. Srivastava, D. Zhurina, and M. Ullrich, 2006, Thermo-responsive expression and differential secretion of the extracellular enzyme levansucrase in the plant pathogenic bacterium *Pseudomonas syringae* pv. *glycinea*. FEMS Microbiol. Lett. 265: 178–185.

Evaluation of Phenotypic and Genetic Techniques to Analyze Diversity of *Pseudomonas syringae* pv. *syringae* Strains Isolates from Mango Trees

J.A. Gutiérrez-Barranquero[1], E. Arrebola[1], A. Pérez-García[1], J.C. Codina[1], J. Murillo[2], A. De Vicente[1], and F.M. Cazorla[1]

Abstract Bacterial apical necrosis of mango, produced by *Pseudomonas syringae* pv. *syringae* (*Pss*), is the main disease affecting mango production in the Mediterranean area. Surveys carried out in the main areas of cultivation ascertained the presence of Pss strains and resulted in a collection of Pss strains from different seasons and locations (including mainland Spain and Canary Islands, Portugal, Italy and Israel). To study the diversity relationships among these *Pss* strains, different phenotypic and genetic techniques were evaluated by using a selection of representative *Pss* strains isolated from mango tissues. The use of physiological tests were based on conventional identification techniques (API tests), toxins production based on biological tests, and analysis of copper resistance. The genetic diversity studies were mainly based on repetitive PCR fingerprinting using ERIC, BOX and REP primers set with UPGMA analysis, and 16S rDNA gene sequencing and ARDRA analysis. Additionally, the native plasmid profiles of these representative strains were determined, and the presence of some genes of interest were detected by hybridization analysis in the most abundant plasmid (62 kb). Preliminary results indicate a considerable phenotypic diversity. Analysis of genetic techniques resulted in repetitive PCR fingerprintings using some primers sets, showing higher diversity than the other techniques used.

Keywords PCR-RFLP of 16S rDNA, rep-PCR, native plasmids, copper resistance

[1]Grupo de Microbiología y Patología Vegetal, Departamento de Microbiología, Facultad de Ciencias, Universidad de Málaga s/n. 29071-Málaga, Spain

[2]Laboratorio de Patología Vegetal, ETS de Ingenieros Agrónomos, Universidad Pública de Navarra. 31006-Pamplona, Spain

Author for correspondence: F.M. Cazorla; e-mail: cazorla@uma.es

M'B. Fatmi et al. (eds.), *Pseudomonas syringae Pathovars and Related Pathogens.*
© Springer Science+Business Media B.V. 2008

1   Introduction

Bacterial apical necrosis of mango (*Mangifera indica*), caused by *Pseudomonas syringae* pv. *syringae* (*Pss*), is a serious disease of this crop in the Mediterranean area. The disease affects buds, leaves, and stems; necrosis of flower and vegetative buds on commercial trees during winter dormancy is the most destructive symptom of the disease (Cazorla et al., 1998). Growers in Spain, Portugal and Israel make extensive use of registered copper compounds, mostly Bordeaux mixture, to control bacterial apical necrosis (Cazorla et al., 2002a). In the last few years, however, growers have noticed a reduction in the efficacy of the treatments which, in turn, has led to an increase in the dosage and frequency of copper applications and appearance of alternative treatments (Cazorla et al., 2006). Simultaneously, new locations of bacterial apical necrosis disease have been reported along the Mediterranean area, at Israel (Pinkas et al., 1996) and Italy (Torta et al., 2003).

From previous studies, a collection of *Pss* strains isolated during different seasons (since 1992) from the main Mediterranean producing areas (mainly Spain and Portugal, but also Italy and Israel) have been generated in our laboratory. Then comparison of characteristics among these strains would be of great interest in order to unravel the diversity among isolated strains, and to obtain important epidemiological information.

In this study, conventional identification and secondary metabolite-based tests were combined with some genetic techniques, including repetitive PCR fingerprinting, sequence of the 16S rDNA, detection of native plasmids and specific genes harboured in them. These results will contribute to evaluate methods of measuring the diversity among *Pss* strains isolated from mango tissues, and to select the most appropriate methods for measuring diversity in further studies.

2   Materials and Methods

2.1   *Bacterial Strains and Culture Conditions*

Strains of *P. syringae* pv. *syringae* (*Pss*) used in this work are listed in Table 1. *P. syringae* strains were grown in King's medium B (KMB) at 27°C. All the strains used in this work were obtained and identified in previous studies; they were isolated from mango trees in different years (from 1992 to 2002) and from different geographical locations in the South of the Iberian Peninsula (Portugal and Spain), Canary Islands, Italy and Israel. *Pseudomonas syringae* pv. *syringae* B728a, *P. syringae* pv. *tomato* DC3000 and *P. syringae* pv. *phaseolicola* 1448A strains were used as references (Table 1).

Table 1 Characteristics of the selected *Pseudomonas syringae* pv. *syringae* strains used in this study. *P. syringae* pv. *syringae* B728a, *P. syringae* pv. *phaseolicola* 1448A and *P. syringae* pv. *tomato* DC3000 were used as reference strains

Strains	Location	Year of isolation	Source/reference
P. syringae pv. *syringae* isolated from mango			
UMAF0049	Algarrobo (Málaga, Spain)	1992	Cazorla et al., 2002a
UMAF0081	Algarrobo (Málaga, Spain)	1992	Cazorla et al., 2002a
UMAF0158	Algarrobo (Málaga, Spain)	1993	Cazorla et al., 2002a
UMAF0167	Torrox (Málaga, Spain)	1994	Cazorla et al., 2002a
UMAF0176	Benajarafe (Málaga, Spain)	1994	Arrebola et al., 2003
UMAF0170	Estepona (Málaga, Spain)	1993	Cazorla et al., 2002a
UMAF1003	Lepe (Huelva, Spain)	1997	Arrebola et al., 2003
UMAF1060	Redondela (Huelva, Spain)	1998	Arrebola et al., 2003
UMAF2801	La Palma (Canary I., Spain)	2000	E. Rodríguez
ICMP14923	Sicily (Italy)	2002	N. Iacobellis
Ps10	Israel	1999	A. Litcher
Reference strains			
P. syringae pv. syringae B728a			Hirano et al., 1999
P. syringae pv. phaseolicola 1448A			Teverson, 1991
P. syringae pv. tomato DC3000			Cuppels, 1986

2.2 Assessing Phenotypic Diversity

2.2.1 Carbohydrates Metabolism and Physiological Tests

A number of phenotypic and physiological tests were performed to characterize the selected *Pss* strains by using the API20NE and API50CH test systems (BioMerieux, Mercy L'Etoyle, France) following the manufacturer's recommendations. After incubation at 25°C at different times, response variations in the tests were revealed by a change in colour or an increase of turbidity on the sample.

2.2.2 Copper Resistance

Copper resistance was assayed in the selected bacterial strains by plating them onto MGY media (Bender and Cooksey, 1986) amended with 0.8 mM of copper sulphate. Those strains growing in presence of 0.8 mM of copper sulphate were considered as copper-resistant strains (Cazorla et al., 2002a).

2.2.3 Detection of *Pss* Toxin

The syringomycin complex production by *Pss* strains was determined by growth inhibition tests on potato dextrose agar (PDA) against *Geotrichum candidum*

(Gross and De Vay, 1977) and *Rhodotorula pilimanae* (Iacobellis et al., 1992). The antimetabolite toxin production was assayed by the indicator technique previously described (Gasson, 1980) with minor modifications (Arrebola et al., 2003), involving growth inhibition of *Escherichia coli* on Pseudomonas minimal medium (PMS). Briefly, a double layer of indicator microorganism was made using *Escherichia coli* CECT831. After solidification, strains of *P. syringae* pv. *syringae* to be tested were stabbed and plates were incubated at 22°C for 24 h and at 37°C for an additional 24-h period. To confirm the biochemical step that is the putative target of the mangotoxin, the same plate bioassay was carried out, but adding to the double layer 100 µl of a 6-mM solution of ornithine or *N*-acetyl-ornithine (Arrebola et al., 2003).

2.3 Assessing Genetic Diversity

2.3.1 Characterization of Native Plasmids of *Pss* Strains

Plasmid DNA was isolated according to a modified alkaline lysis method (Zhou et al., 1990) and separated by electrophoresis on 0.6% agarose gels (Sambrook and Russell, 2001). Plasmid size was estimated by comparison with plasmids from *P. syringae* pv. *tomato* PT23. Purified plasmid DNA digested with *Eco*RI was separated by electrophoresis on 1% agarose gels and examined for restriction fragment length polymorphisms (Cazorla et al., 2002a).

To reveal the presence of specific genes on the more abundant 62-kb plasmid, approximately 5 µg of plasmid DNA was digested with *Eco*RI, separated by electrophoresis in 1% agarose gels, transferred to nylon membranes, and hybridized with radioactive probes following standard procedures (Sambrook and Russell, 2001). Southern blot analyses were performed using labelled probes from the *oriV* replication origin obtained from *P. syringae* pv. *tomato* PT23 (Murillo and Keen, 1994), the *rulAB* genes which can confers tolerance to UV radiation obtained from pAKC, containing part of the *rulAB* genes from *P. syringae* pv. *tomato* PT23 (Sundin and Bender, 1996), and the *copABCD* operon which can confer copper resistance obtained from pCOP2, containing the *copABCD* genes from *P. syringae* pv. *tomato* PT23 (Bender and Cooksey, 1987). Obtention of these genetic markers have been described previously (Cazorla et al., 2002b).

2.3.2 Repetitive PCR Genomic Fingerprinting

Total genomic DNA to be used in these experiments was prepared by using a modification of a lysis cell process (Versalovic et al., 1994). Briefly, a bacterial colony was picked up from a KMB plate and resuspended in 30 µl of sterile water (Reagent grade) in a 1.5 ml plastic tube sealed. After incubation at 100°C for 10 min, cell debris were discarded after centrifugation during 1 min at 13,000 × g, and the supernatants were used to obtain the repetitive PCR fingerprinting. Seven primer sets were evaluated (BOXA1, ERIC, $(CAG)_5$, $(GTG)_5$, M13/pUC forward and reverse, and IS50; Louws et al., 1994; Rademaker et al., 2000; Scortichini et al., 2006; Weingart and Volksch, 1997). PCR tests were carried out using the reagents and the programs described previously (Bhattacharya et al., 2003; Louws et al., 1994; Oguiza et al., 2004; Rasschaert et al., 2005). After separation and visualization of the PCR products, band sizes were assigned by comparison to DNA standard marker (1 kb, Qiagen, Spain). Presence and absence of a determined band was reported in each strain, generating a distance matrix using the Dice's coefficient, and resulting in a dendrogram constructed by the UPGMA method using NTSYSpc Software (Exeter Software, New York).

2.3.3 16S rDNA Analysis

To obtain the partial 16S rDNA sequence, colony PCR was performed on the selected strains, using the primers 41F (5'-GCT CAG ATT GAA CGC TGG CG-3') and 1486r-N (5'-GCT ACC TTG TTA CGA CTT CAC CCC-3'), described by Stackebrandt and Goodfellow (1991). Briefly, a single colony from a 24- to 48-h culture on a KMB plate was suspended in 20 µl of sterile water. Subsequently, 1.0 ml of a 0.05 M NaOH and 0.25% SDS (sodium dodecyl sulphate) solution was added and the mixture was incubated at 95°C for 15 min. Ten µl of a tenfold dilution of this suspension in sterile water was used as a template in a PCR reaction with a hot start for 120 s at 95°C, followed by ten cycles consisting of 40 s at 94°C, 60 s at a decreasing temperature gradient of 1°C per cycle starting from 60°C, 120 s at 72°C, and subsequently, 25 cycles consisting of 40 s at 94°C, 60 s at 50°C, 120 s at 72°C, and a final extension of 72°C for 120 s. The resulting PCR fragment was purified (GFX PCR DNA and gel band purification kit, GE Healthcare Europe GmbH, Germany) and was used directly for sequencing (Macrogen Inc., Seoul, Korea). The sequence was analyzed using DNAman software (Lynnon Biosoft, Quebec, Canada). Homology studies were carried out using National Center for Biotechnology Information GenBank Blast software (Bethesda, Maryland, USA). In addition, the 16S rDNA sequences obtained, were aligned by the ClustalW Software present on line in the web of European Bioinformatics Institute. Tree were constructed with TreeView Software (Hayashimoto et al., 2005; Smit et al., 2001).

For ARDRA analysis, four restriction enzymes were used (*Alu*I, *Hae*III, *Hinf*I and *Rsa*I) with the partial 16S rDNA amplified. Digestion was performed as previously described (Porteous et al., 2002). The restriction products were separated by 2.5% agarose gel electrophoresis in TAE buffer, and visualized after staining with ethidium bromide.

3   Results and Discussions

The test systems API20NE and API50CH showed that in general these *Pss* strains have identical physiological responses (nitrate reduction, glucose metabolism, arginine dihydrolase, oxidase, indol production, etc.) and very similar pattern of carbohydrates metabolism with only few differences summarized in Table 2, when

Table 2 Phenotypic characteristics of *P. syringae* pv. *syringae* strains isolated from mango: carbohydrates used as sole C source, toxins produced (mangotoxin and syringomycin) and copper resistance

Strains	Carbohydrates metabolism[a]											Toxins production[b]		Copper resistance[c]
	GLY	ERY	RIB	DXYL	INO	MAN	SOR	MEL	SAC	TRE	DARL	Mtox	Sm	
P. syringae pv. *syringae*														
UMAF0049	+	+	+	+	+	+	+	+	+	−	+	+	+	R
UMAF0081	+	+	+	−	+	+	+	−	+	−	+	+	−	R
UMAF0158	+	−	+	+	+	+	+	−	+	−	−	+	+	S
UMAF0167	+	+	+	+	+	+	+	+	+	−	−	nd[d]	+	R
UMAF0176	+	−	+	+	+	+	+	−	+	−	+	+	+	S
UMAF0170	+	−	+	+	+	+	+	−	+	−	−	+	+	R
UMAF1003	−	−	−	+	−	+	−	−	+	−	−	+	+	S
UMAF1060	−	−	−	+	+	+	−	−	+	−	−	+	+	S
UMAF2801	+	−	−	+	+	+	+	−	+	−	+	+	+	S
ICMP14923	−	−	−	+	−	−	−	−	−	−	−	nd	−	S
Ps10	+	−	−	+	−	+	−	−	+	−	−	+	+	R
Reference strains														
Pss B728a	+	−	+	+	+	+	+	−	−	+	−	−	+	R
Psp 1448A	−	−	+	−	−	−	−	−	−	−	−	−	−	R
Pst DC3000	−	−	+	−	−	−	−	−	−	−	−	−	−	S

[a] Carbohydrates used as sole C source analyzed by API 50CH. GLY: Glycerol, ERY: Erythritol, RIB: D-Ribose, DXLY: D-Xylose, INO: Inositol, MAN: Mannitol, SOR: Sorbitol, MEL: D-Melibiose, SAC: D-Saccharose (sucrose), TRE: D-Threalose, and DARL: D-Arabitol.

[b] Mtox: mangotoxin; Sm: syringomycin

[c] R: resistant, S: sensitive. Resistance to copper was determined by using mannitol-glutamate-yeast extract medium (MGY), amended with 0.8 mM of CuSO4

[d] No data

they were compared with the *Pss* strain B728a or among them. It is remarkable that nearly all the selected strains have a specific carbohydrates utilization pattern, at least in the use of one assayed C source, displaying diversity at this level (Table 2). Related with the production of different toxins and copper-resistance, also differences were found among the selected strains. So, analysis of copper-resistance in these 11 selected *Pss* strains showed that approximately half of them were copper-resistant and half copper-sensitive (Table 2). However, in relation with the toxin production, the diversity was low, and most of the strains produce mangotoxin and syringomycin (Table 2). The detection of native plasmids (Table 3) revealed that only one strain had two plasmids, one of 88 kb and other of 62 kb. The strain *Pss* UMAF167 showed the presence of one plasmid of 120 kb; whereas the other strains had only one plasmid of 62 kb (n = 6) or do not have plasmids (n = 3). The 62-kb plasmid are very abundant in *Pss* strains (64% of assayed strains), and all of them belonged to the family of pPT23A-like plasmids, having a homologous

Table 3 Plasmids profile of *P. syringae* pv. *syringae* strains isolated from mango. The *Eco*RI restriction pattern of 62-kb plasmid DNA and the hybridization with gene probes of *rul*AB genes, *ori*V and *cop*ABCD operon are showed

Strains	Plasmid Profiles (kb)	62-kb *Eco*RI restriction pattern[a]	62-kb plasmid hybridization with		
			*ori*V	*rul*AB	*cop*ABCD
P. syringae pv. *syringae*					
UMAF0049	88/62	1	+	+	+
UMAF0081	62	1	+	+	+
UMAF0158	62	4	+	+	−
UMAF0167	120[b]		−	−	+
UMAF0176	62	4	+	+	−
UMAF0170	62	2	+	+	+
UMAF1003	Plasmidless				
UMAF1060	62	3	+	+	−
UMAF2801	Plasmidless				
ICMP14923	62	nd[c]	nd	nd	nd
Ps10	Plasmidless				
Reference strains					
Pss B728a	Plasmidless				
Psp 1448A	131				
Pst DC3000	67				

[a] Restriction patterns of 62-kb plasmid as described in Cazorla et al., 2002a

[b] Hybridization experiments were also performed with 120-kb plasmid from *Pss* UMAF0167

[c] No data

replication origin to *ori*V, and also all of the 62-kb plasmids harbour homologous genes to *rul*AB (Table 3). These abundance of the 62-kb plamids suggest that it could be conserved in these *Pss* strains. However, differences among the 62-kb plasmids in *Pss* strains isolated from mango tissues have been determined as previously reported (Cazorla et al., 2002a). They presented a high diversity of restriction patterns when they were digested with *Eco*RI, generating at least four different patterns (Table 3). Additionally, the copper-resistant strains showed homologous genes to the *cop*ABCD operon in the 62-kb plasmids (Table 3), but not in 62-kb plasmid from copper-sensitive strains, suggesting a direct relationship between copper resistance and plasmids and a significant source of diversity (Cazorla et al., 2002a).

The repetitive PCR fingerprinting showed a high ability to differentiate genetically *Pss* strains (Table 4). For this approach, only bands of size between 2500 pb and 500 pb were considered. Every primer set resulted in a different dendrogram. The most resolutive of the genetic techniques to assess diversity was obtained using the ERIC primers set (Fig. 1). By this technique, nearly all the strains have a specific fingerprinting (only one pattern was shared by different strains). With this technique, it was possible to separate the reference strains

Table 4 Genetic diversity of *P. syringae* pv. *syringae* strains based on Rep-PCR and ARDRA analysis with different primers and restriction enzymes, respectively

Molecular techniques	Total number of bands[a]	Number of different patterns	Patterns number observed from 11 *Pss* strains in[b]		
			Only 1 strain	2 strains	≥3 strains
Rep-PCR					
ERIC	3–6	10	9	1	
BOX	3–6	7	3	4	
(CAG)5	7–8	3		1	2
(GTG)5	7–10	5	2	3	1
M13/pUC -forward	–[c]	–			
M13/pUC- reverse	–	–			
IS50	–	–			
ARDRA					
*Hae*III	5	1			1
*Alu*I	2	3	1	1	1
*Hinf*I	4	1			1
*Rsa*I	4	1			1

[a] Range of bands number observed in the studied *Pss* strains

[b] Number of different patterns from 11 *Pss* strains

[c] The obtained results with this technique were not repetitive

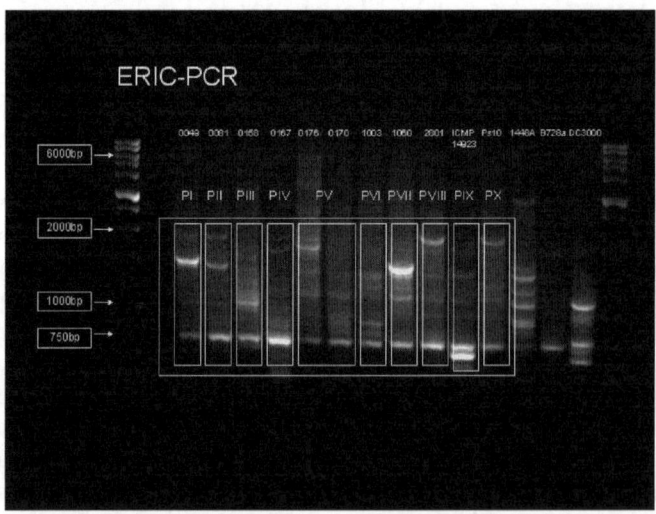

Fig. 1 Representative agarose gel electrophoresis applying ERIC-PCR of DNA obtained from 11 *Pseudomonas syringae* pv. syrinagae strains isolated from mango and the reference strains. Different patterns are marked by squares numbered from PI to PX

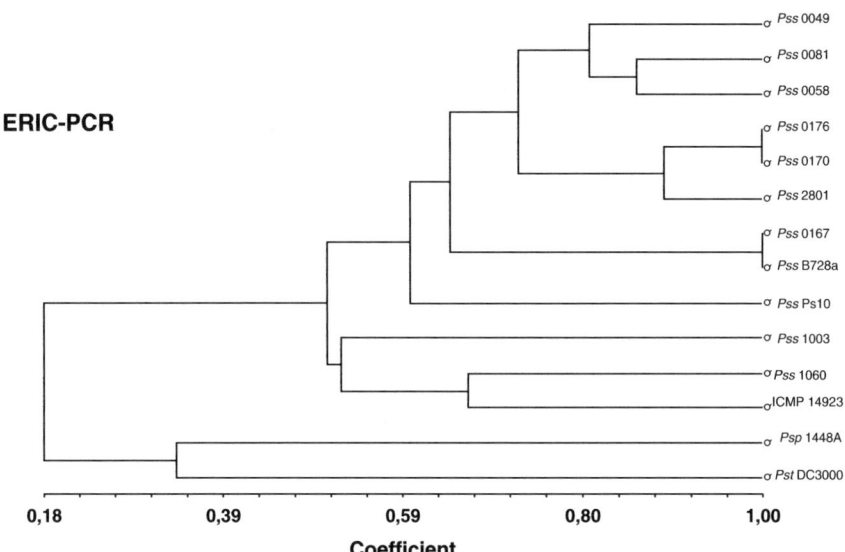

Fig. 2 Dendrogram corresponding to the ERIC-PCR analysis showed in Fig. 1. Cluster analysis was performed by UPGMA on matrix calculated with the Dice's coefficient

P. syringae pv. *tomato* DC3000 and *P. syringae* pv. *phaseolicola* 1448A from the *Pss* group (Fig. 2). The other techniques that also provide a good resolution were obtained using BOXA1, $(CAG)_5$, and $(GTC)_5$ primers sets and ARDRA analysis with *Alu*I (Table 4). The ARDRA technique using the other three restriction enzymes showed a common pattern shared by all the *Pss* strains tested, and the 16S rDNA sequence do not showed differences between the *Pss* tested and neither with strains of other pathovars.

4   Conclusions

From all the different techniques evaluated in this work, some of the rep-PCR techniques (ERIC or BOX primer sets), analysis of native plasmids and the phenotypic analysis could be useful to evaluate diversity in *Pss* strains isolated from mango. Other techniques, like ARDRA or 16S rDNA sequence resulted in poor diversity among the selected *Pss* strains.

Acknowledgements This work was partially supported by Junta de Andalucia (Grupo AGR-0169) and from Plan Nacional de I + D del Ministerio de Educación y Ciencia (AGL-2005-06347-CO3-01). We want to thank Dr. N. S. Iacobellis, Dr. A. Litcher and E. García Rodríguez for kindly providing us with some of the bacterial strains.

References

Arrebola E., Arrebola, E., Cazorla, F.M., Durán, V.E., Rivera, E., Olea, F., Codina, J.C., Pérez-García, A., and de Vicente, A. (2003) Mangotoxin: a novel antimetabolite toxin produced by *Pseudomonas syringae* inhibiting ornithine/arginine biosynthesis. *Physiological and Molecular Plant Pathology* 63:117–127.

Bender C.L. and Cooksey D.A. (1986) Indigenous plasmids in *Pseudomonas syringae* pv. *tomato*: conjugative transfer and role in copper resistance. *Journal of Bacteriology* 165:534–541.

Bender C.L. and Cooksey D.A. (1987) Molecular cloning of copper resistance genes from *Pseudomonas syringae* pv. *tomato*. *Journal of Bacteriology* 169: 470–474.

Bhattacharya D., Sarma P.M., Krishnan S., Mishra S. and Lal B. (2003) Evaluation of genetic diversity among *Pseudomonas citronellolis* strains isolated from oily sludge-contaminated sites. *Applied and Environmental Microbiology* 69: 1435–1441.

Cazorla F.M., Torés J.A., Olalla L., Pérez-García A., Farré J.M. and de Vicente A. (1998) Bacterial apical necrosis of mango in southern Spain: a disease caused by *Pseudomonas syringae* pv. *syringae*. *Phytopathology* 88: 614–620.

Cazorla F.M., Arrebola E., Sesma A., Pérez-García A., Codina J.C., Murillo J. and de Vicente A. (2002a) Copper resistance in *Pseudomonas syringae* strains isolated from mango is encoded mainly by plasmids. *Phytopathology* 92(8): 909–916.

Cazorla F.M., Arrebola E., Abad C., Codina J.C., Pérez-García A., and de Vicente A. (2002b) Epiphytic fitness of *Pseudomonas syringae* pv. *syringae* on mango trees is increased by 62-kb plasmids. In: N.S. Iacobellis, A. Collmer, S.W. Hutcheson, J.W. Mansfield, C.E. Morris, J. Murillo, N.W. Schaad, D.E. Stead, G. Surico and M.S. Ullrich (eds), *Pseudomonas syringae and Related Pathogens. Biology and Genetic*. Kluwer, Dordrecht, The Netherlands, pp. 79–88

Cazorla F.M., Arrebola E., Olea F., Velasco L., Hermoso J.M., Pérez-García A., Torés J.A., Farré J.M. and de Vicente A. (2006) Field evaluation of treatments for the control of the bacterial apical necrosis of mango (*Mangifera indica*) caused by *Pseudomonas syringae* pv. *syringae*. *European Journal of Plant Pathology* 116: 279–288.

Cuppels D.A. (1986). Generation and characterization of *Tn5* insertion mutations in *Pseudomonas syringae* pv. *tomato*. *Applied and Environmental Microbiology* 52: 323–327.

Gasson M.J. (1980) Indicator technique for antimetabolic toxin production by phytopatogenic species of *Pseudomonas*. *Applied and Environmental Microbiology* 39: 25–29.

Gross D.C. and De Vay S.E. (1977) Production and purification of syringomycin, a phytotoxin produced by a *Pseudomonas syringae*. *Physiological Plant Pathology* 11:13–28.

Hayashimoto, N., Takakura, A. and Itoh, T. (2005) Genetic diversity on 16S rDNA sequence and phylogenic tree analysis in *Pasteurella pneumotropica* strains isolated from laboratory animals. *Current Microbiology* 51(4): 239–243.

Hirano S.S., Charkowski A.O., Collmer A., Willis D.K. and Upper C.D. (1999) Role of the Hrp type III protein secretion system in growth of *Pseudomonas syringae* pv. *syringae* B728a on host plants in the field. *Proceedings of the Natural Academy of Sciences USA* 96: 9851–9856.

Iacobellis N.S., Lavermicocca P., Grgurina I., Simmaco M. and Ballio A. (1992) Phytotoxic properties of *Pseudomonas syringae* pv. *syringae* toxins. *Physiological and Molecular Plant Pathology* 40: 107–16.

Louws F.J., Full bright D.W., Stephens C.T. and De Brujin F.J. (1994) Specific genomic fingerprints of phytopathogenic *Xanthomonas* and *Pseudomonas* pathovars and strains generated with repetitive sequences and PCR. *Applied and Environmental Microbiology* 60: 2286–2295.

Murillo J. and Keen N.T. (1994) Two native plasmids of *Pseudomonas syringae* pathovar *tomato* strain PT23 share a large amount of repeated DNA, including replication sequences. *Molecular Microbiology* 12: 941–950.

Oguiza A., Rico A., Rivas L.A., Sutra L., Vivian A. and Murillo J. (2004) *Pseudomonas syringae* pv. *phaseolicola* can be separated into two genetic lineages distinguished by the possession of the phaseolotoxin biosynthetic cluster. *Microbiology* 150: 473–482.

Pinkas Y., Maymon M. and Smolewich Y. (1996) Bacterial black blight of mango. *Alon Hanotea* 50: 475.

Porteous L.A., Widmer F. and Seidler R.J. (2002) Multiple enzyme restriction fragment length polymorphism analysis for high resolution distinction of *Pseudomonas* (*sensu stricto*) 16S rRNA genes. *Journal of Microbiological Methods* 51: 337–348.

Rademaker J.L., Hoste B., Louws F.J., Kersters K., Swings J., Vauterin L., Vauterin P. and De Bruijn F.J. (2000) Comparison of AFLP and rep-PCR genomic fingerprinting with DNA-DNA homology studies: *Xanthomonas* as a model system. *International Journal of Systematic and Evolutionary Microbiology* 50: 665–677.

Rasschaert G., Houf K., Imberechts H., Grijspeerdt K., De Zutter L., and Heyndrickx M. (2005) Comparison of five repetitive-sequence-based PCR typing methods for molecular discrimination of *Salmonella enterica* isolates. *Journal of Clinical Microbiology* 43(8): 3615–3623.

Sambrook J. and Russell D.W. (2001) Molecular Cloning: A Laboratory Manual. Cold Spring Harbor Laboratory, Cold Spring Harbor, NY.

Scortichini M., Natalini E. and Marchesi U. (2006) Evidence for separate origins of the two *Pseudomonas avellanae* lineages. *Plant Pathology* 55: 451–457.

Smit E., Leeflang P., Gommans S., Van Den Broek J. Van Mil S. and Wernars K. (2001) Diversity and seasonal fluctuations of the dominant members of the bacterial soil community in a wheat field as determined by cultivation and molecular methods. *Applied and Environmental Microbiology* 67(5): 2284–2291.

Stackebrandt E. and Goodfellow M. (1991) Nucleic Acid Techniques in Bacterial Systematics. Wiley, Chichester, NH.

Sundin G.W. and Bender C.L. (1996) Molecular analysis of closely related copper- and streptomycin-resistance plasmids in *Pseudomonas syringae* pv. *syringae*. *Plasmid* 35: 98–107.

Teverson D.M. (1991) Genetics of pathogenicity and resistance in the halo blight disease of beans in Africa. Ph.D. thesis. University of Birmingham, Birmingham, UK.

Torta L., Lo Piccolo S., Burruano S., Lo Cantore P. and Iacobellis N.S. (2003) Necrosi apicale del mango (Mangifera indica L.) causata da *Pseudomonas syringae* pv. *syringae* van Hall in Sicilia. *Informatore Fitopatologico* 11: 44–46.

Versalovic J., Schneider M., De Brujin F.J. and Lupski J.R. (1994) Genomic fingerprinting of bacteria using repetitive sequence-based polymerase chain reaction. *Methods in Molecular and Cellular Biology* 5: 25–40.

Weingart H. and Volksch B. (1997) Ethylene production by *Pseudomonas syringae* pathovars in vitro and in planta. *Applied and Environmental Microbiology* 63: 156–161.

Zhou C., Yang Y. and Jong A. (1990) Miniprep in ten minutes. *BioTechniques* 8: 172–173.

Characterization of *Pseudomonas syringae* Strains Isolated from Diseased Horse-chestnut Trees in Belgium

A. Bultreys[1], I. Gheysen[1], and V. Planchon[2]

Abstract *Pseudomonas syringae* was isolated from lesions found in cortical tissues of Belgian horse-chestnut (*Aesculus hippocastanum*) trunks. A collection of about 50 strains was established from 6 sites in Brussels and 11 sites in Wallonia, but the pathogen was not found in the South-East of Belgium. The strains were identified by pyoverdin tests, induction of potato rot, the *cfl* PCR test, and REP-PCR. The investigated strains were highly virulent on horse-chestnut detached twigs collected in winter and summer, although the propagation outside the inoculated zones was reduced in summer. The accentuated propagation during winter was confirmed by inoculation of 20 young horse-chestnut trunks: in the end of winter and early spring, a mean progression in the cortical tissues of 3.45 mm in 48 days was observed (0.05 mm for the controls), whereas the progression during spring and summer was reduced by a factor seven. In this latter period, canker formations were observed; also, *P. syringae* was still isolable from naturally infected older trunks. Comparison of REP-PCR profiles were carried out with 60 pathovars of *P. syringae* and with *Pseudomonas viridiflava*, *Pseudomonas meliae*, *Pseudomonas ficuserectae*, *Pseudomonas cannabina* and *Pseudomonas tremae*. REP- and ERIC-PCR analyses indicated the relatedness of the horse-chestnut strains, although site-related genetic groups were observed, as well as genetic similarities and differences with *P. syringae* pv. *morsprunorum* race 1, pv. *aesculi*, pv. *cunninghamiae* and pv. *daphniphylli*; both the pathovars *morsprunorum* race 1 and *aesculi* possessed the *cfl* gene and all the investigated horse-chestnut strains produced coronatine. Only strains from horse-chestnut and pathovar *aesculi* were similarly virulent on *A. hippocastanum*. The data indicate that *P. syringae* strains are living on *A. hippocastanum* for a long time and would agree with their grouping with *P. syringae* pv. *aesculi* inducing leaf spots on *Aesculus indica*, but this pathovar would then be genetically heterogeneous.

[1] Département Biotechnologie Centre Wallon de Recherches Agronomiques, Chaussée de Charleroi 234, B-5030 Gembloux, Belgium

[2] Section Biométrie, Gestion des données et Agrométéorologie, Centre Wallon de Recherches Agronomiques, B-5030 Gembloux, Belgium

Author for correspondence: Alain Bultreys; e-mail: bultreys@cra.wallonie.be

M'B. Fatmi et al. (eds.), *Pseudomonas syringae Pathovars and Related Pathogens.*
© Springer Science + Business Media B.V. 2008

Keywords Horse-chestnut, canker, *Pseudomonas syringae*, morsprunorum, aesculi, emerging disease

1   Introduction

Since December 2004, horse-chestnut trees (*Aesculus hippocastanum*, originating from the Balkans) showing cankers and/or lesions in the cortical tissues with sometimes tarry exudations have been studied by the Walloon Agricultural Research Centre in Brussels and in Wallonia (Southern part of Belgium). The same syndrome has been reported in the Netherlands and United Kingdom (Janse et al., 2006; Webber et al., 2006). The disease can induce dead of branches and trees. It is of great concern for cities like Brussels and for regional administrations in charge of plantations. Indeed, horse-chestnut is an attractive tree that is frequently planted alongside roadways and avenues, as well as in parks. Bacterial analyses were initiated in January 2006 because of the non-confirmation of the involvement of an initially suspected *Phytophthora* (Bultreys et al., 2006). *Pseudomonas syringae* was readily isolated from lesions in the cortical tissues of horse-chestnut.

2   Materials and Methods

2.1   Isolation and Identification

Strain isolations were performed onto King B medium (King et al., 1954) from the margins of fresh or dry lesions found on horse-chestnut trees located in 6 sites in Brussels and 11 sites in Wallonia. These sites had been detected during a survey in Brussels and Wallonia (Bultreys et al., 2006). Identification was through pyoverdin-based tests, including visual, spectrophotometric and high performance liquid chromatography (Bultreys et al., 2001, 2003), the induction of potato rot, and repetitive extragenic palindromic (REP)-PCR. REP-PCR performed on lysed cells (5×10^5 lysed cells per reaction) prepared as previously described (Bultreys and Gheysen, 1999). The PCR conditions were modified from Pooler et al. (1996) and Kingsley et al. (2002); in 25-µl total volume: 1 U *Taq* DNA polymerase (Qiagen) used with its buffer, 200 µM of each dNTP, 2.5 mM $MgCl_2$, and 2 µM of the REP primers (Versalovic et al., 1991). The program was: an initial denaturation at 94°C for 5 min; 30 cycles at 94°C for 1 min, 44°C for 1 min, and 72°C for 2 min; and a final elongation at 72°C for 10 min. The electrophoresis conditions were those used by Cubero and Graham (2002). The REP-PCR profiles were compared with the profiles of the first strains isolated, but each strain was conserved independently of its profile. The modified PCR test detecting the *cfl* gene involved in coronatine synthesis was also used in identification (Bereswill et al., 1994; Bultreys and Gheysen, 1999)

2.2   Strain Characterization

Coronatine production was detected using a biological test on Spunta potato slides (Bultreys, 2001). Analysis of the *cfl* gene by PCR and comparison of REP-PCR profiles were carried out for 60 pathovars of *P. syringae*, including the pathovar *aesculi* inducing leaf spots on *Aesculus indica* in India (Durgapal and Singh, 1980), and the pathotype strains of the related *Pseudomonas* species *P. viridiflava*, *P. meliae*, *P. ficuserectae*, *P. cannabina* and *P. tremae*. Enterobacterial repetitive intergenic consensus (ERIC)-PCR analyses were also carried out on the pathovars that appeared related by REP-PCR. The PCR conditions were those used for REP-PCR but the Tm temperature was 52°C and the ERIC primers (Hulton et al., 1991) were used.

2.3   Pathogenicity Tests

Pathogenicity tests were conducted on horse-chestnut and sweet cherry 1-year-old detached twigs and on trunks of young horse-chestnut trees. Bacteria grown in King B agar medium for 24 h were used to produce bacterial suspensions in water of about 2.6×10^8 CFU per ml. The tests on 1-year-old horse chestnut twigs were performed in winter and summer. The twigs of the winter test were collected on young *A. hippocastanum* trees grown in pots; those of the summer tests on young trees (trunk diameter: 10–12 cm) of *A. hippocastanum* Baumanii transplanted in Gembloux in March 2006. The test on 1-year-old sweet cherry twigs was performed in winter; the twigs of the Napoleon variety were taken from an orchard in Gembloux. The twigs were superficially washed and disinfected, cut to an appropriate length and sealed with paraffin jelly. Cortical tissues were exposed in an about 1.5×0.5 cm area by removing superficial bark and 20 µl of either sterile water or bacterial suspension were delicately spread on the exposed tissues. The twigs were put independently in sterile glass tubes containing absorbent cotton-wool saturated with water. The tubes were sealed with cellophane and incubated at 15°C in the dark for 20–30 days. The reading was performed by visual estimation of the area presenting a browning resembling that observed during isolations from naturally infected trees; the twigs were distributed in classes of damage: 0, no damage, up to 4 all the area damaged. The strains investigated on either horse-chestnut or cherry were MY2-1, MY2-2, MY2-3, MY3-1, MY3-2, MY3-3, CH3, N2 and SB5-1, which are *P. syringae* strains from horse-chestnut, MY3-4, an unidentified white-colored bacterium isolated from horse-chestnut together with *P. syringae*, *P. syringae* pv. *morsprunorum* race 1 PmC36, pv. *aesculi* CFBP 2894, pv. *cunninghamiae*

CFBP 4218 and pv. *daphniphylli* CFBP 4219. Five or ten twigs were used per test. Anova analysis (SAS software version 9.1.3) was used for strain comparisons based on classes of virulence. Levene's test was applied for variance comparisons, and the test of Tukey for mean comparisons between strains.

Twenty trunks of young horse-chestnut trees were inoculated in March 2006: in two places on the same trunk, the superficial bark was lifted up in a 0.5 × 0.5 mm area to expose cortical tissues; a 20-μl drop of either sterile water or bacterial suspension of *P. syringae* strain MY3-1 were delicately spread on the exposed tissues; then, the superficial bark was replaced and fixed with grafting rubber. Disease symptoms were measured after 48 and 186 days. In both types of test, isolations were performed from progression outside the inoculated zones and the purified bacteria were tested for pyoverdin produced to confirm that they belonged to the *P. syringae* group.

3   Results and Discussion

3.1   *Isolation and Identification*

Sites and cities in the Walloon and Brussels-Capital Regions were contacted or visited to detect diseased horse-chestnut trees (Fig. 1A–1C), Fluorescent *Pseudomonas* were easily isolated, often in pure culture, from the margins of freshly attacked tissues found in 6 isolation sites in Brussels and 11 isolation sites in Wallonia (Fig. 1D). A collection of about 50 strains was established, but despite contacts with cities and active research, the disease was not found in the South-East of Belgium. Whether the disease is present in this region will be further investigated in the future.

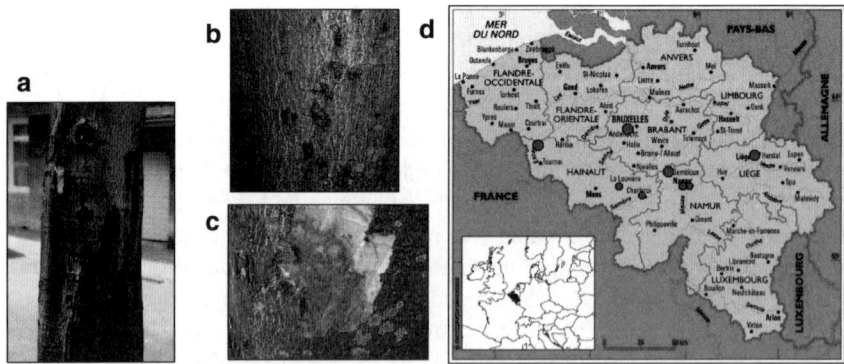

Fig. 1 The horse-chestnut disease in Belgium. **a** Bleeding cankers on trunk. **b** Exudations on trunks. **c** Lesions in cortical tissues. **d** Belgian sites (large dots) where *P. syringae* was isolated from diseased trees. The larger dots indicate more than one isolation site at the location

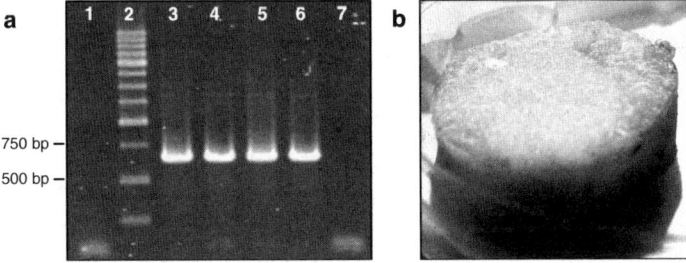

Fig. 2 Coronatine tests. **a** Detection of a 655-bp fragment of the *cfl* gene involved in coronatine synthesis by PCR: lane 1, water; lane 2 DNA-marker XVI; lanes 3–6, *P. syringae* strains from horse-chestnut MY3-2, MY3-1, MY2-2, MY2-1; lane 7, coronatine-non-producing *P. tremae* CFBP 3229. **b** Visualization of coronatine production on a potato slice: induction of callus production by a coronatine producing *P. syringae* strain

Fluorescent *Pseudomonas* were easily isolated, often in pure culture, from freshly attacked tissues. This was apparently true independent of the season since isolations were successful in winter, spring and summer. Also, similar strains were isolated from leaves. The isolates were initially identified as *P. syringae* by the HPLC detection of pyoverdin and by their inability to induce potato rot. The REP-PCR profiles of the first isolated strains were then used as references in REP-PCR identification. No isolated *P. syringae* strain produced a REP-PCR fingerprint radically different from the references. It was also noticed that the isolated strains gave positive responses in the *cfl* PCR test (Fig. 2A) and this test was then also used in identification. It was then confirmed that all the isolated *P. syringae* strains from horse-chestnut possessed the *cfl* gene. Only a simplified visual pyoverdin test, REP-PCR and the *cfl* PCR test were finally used in identification.

3.2 Strain Characterization

All horse-chestnut strains induced callus formation on potato slices (Fig. 2B), a phenotype which is related to the production of coronatine (Sakai et al., 1979); in a previous study (Bultreys, 2001), this biological test showed a good correlation with the possession of the *cfl* gene by *P. syringae* strains.

The ERIC-PCR analyses indicated the relatedness of the *P. syringae* strains isolated from horse-chestnut, although site-related small variations were observed (Fig. 3). At least three and maybe five genetic groups were apparent based on the variation found in one particular place of the profiles. Interestingly, in one isolation site one genetic group was most generally dominant.

The REP-PCR analyses also indicated similarities and differences between the horse-chestnut strains and strains of *P. syringae* pv. *morsprunorum* race 1, pv. *aesculi*, pv. *cunninghamiae* and pv. *daphniphylli* (Fig. 4A). ERIC-PCR analyses did not enabled to distinguish between the different genetic groups detected by REP-PCR

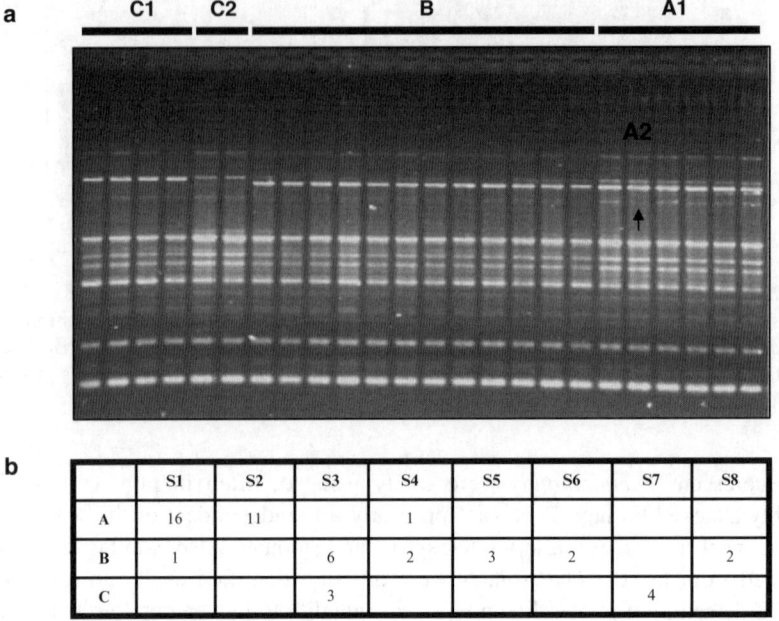

Fig. 3 REP-PCR analyses among horse-chestnut strains. **a** Genetic similarities and differences among *P. syringae* strains isolated from horse-chestnut trees in Belgium. **b** Distribution of Belgian strains isolated from eight sites (S1–S8) among the three principal genetic groups (A–C) showing relations between the genetic groups and the isolation sites

among the Belgian horse-chestnut strains, and they confirmed the similarities and differences with the pathovars *morsprunorum* race 1, *aesculi*, *cunninghamiae* and *daphniphylli* (Fig. 4B). The presence of the *cfl* gene and the probably linked production of coronatine, was observed in different pathovars not previously known to possess this gene: the pathovars *aesculi*, *alisalensis*, *berberidis*, *ulmi*, *porri*, *spinaceae*, and *zizaniae*; as well as in the species *P. cannabina*. However, among the pathovars showing similarities with the horse-chestnut strains, only the pathovars *morsprunorum* race 1 and *aesculi* possessed the *cfl* gene, the production of coronatine by *P. syringae* pv. *morsprunorum* race 1 being well known (Bereswill et al., 1994).

3.3 Pathogenicity Tests

The first six isolated *P. syringae* strains MY2-1, MY2-2, MY2-3, MY3-1, MY3-2 and MY3-3 were investigated in January and February 2006 for their virulence on horse-chestnut detached twigs, in comparison with the unidentified white-colored bacterium MY3-4 isolated together with *P. syringae* from diseased cortical tissues. All the *P. syringae* strains proved highly virulent and statistically similarly virulent,

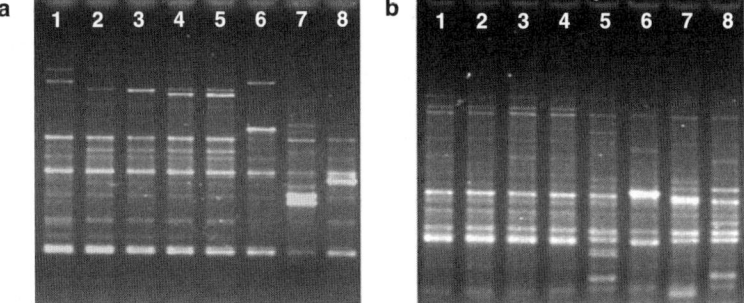

Fig. 4 REP- and ERIC-PCR analyses among pathovars and species. **a** REP-PCR analyses for the *P. syringae* strains from horse-chestnut SB5-1, N2, CH3, Bxl4 from different genetic groups (lanes 2–5) and for *P. syringae* pv. *morsprunorum* race 1 LMG 2222 (lane 1), pv. *cunninghamiae* CFBP 4218 (lane 6), pv. *daphniphylli* CFBP 4219 (lane 7), and pv. *aesculi* CFBP 2894 (lane 8). **b** ERIC-PCR analyses for *P. syringae* strains from horse-chestnut SB5-1, N2, CH3, Bxl4 from different genetic groups (lanes 1–4) and for *P. syringae* pv. *morsprunorum* race 1 LMG 2222 (lane 5), pv. *aesculi* CFBP 2894 (lane 6), pv. *cunninghamiae* CFBP 4218 (lane 7), and pv. *daphniphylli* CFBP 4219 (lane 8)

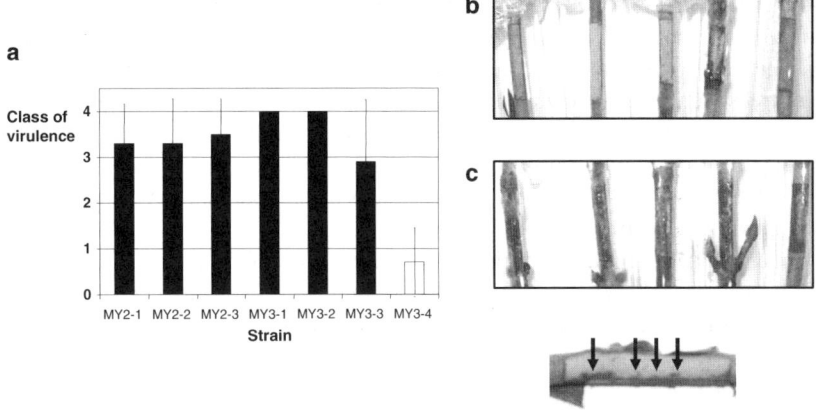

Fig. 5 Pathogenicity test on horse-chestnut detached twigs. **a** Mean and standard deviation for six *P. syringae* strains (black bars) and one unidentified white-colored bacterium (white bar). **b** Few symptoms were induced by the non pathogenic strain MY3-4: the inoculated zone most generally remained totally green. **c** Lesions induced by a *P. syringae* strain: the inoculated zones turned brown and expansion was noted outside the inoculated zone

whereas MY3-4 induced little damage (Fig. 5). Only in the case of the *P. syringae* strains was a progression of the damages outside the inoculated zone systematically observed. Also, *P. syringae* could be systematically re-isolated from these progression zones, validating the Koch's postulate for the strain abilities to induce damages in horse-chestnut cortical tissues.

    The ability to cause disease during summer was confirmed by the virulence of
the strain MY3-1 on horse-chestnut twigs collected in Augustus 2006 (mean virulence class of 3.75 ± 0.75 compared to 0.5 ± 0.35 for the water control), but no
progression outside the inoculated zone was noticed in that test. Increased susceptibility during winter was confirmed by inoculation of 20 horse-chestnut trunks in
March 2006: in the end of winter and early spring, a mean progression of
3.45 ± 1.21 mm in the cortical tissues of ten inoculated trunks was observed after
48 days, compared to 0.05 ± 0.15 mm for the water controls on these trunks,
whereas the progression noted after 186 days on the ten other inoculated trunks was
only 4.8 ± 1.21 mm. The observed progression during spring and summer was
therefore reduced by a factor seven when compared to winter (Fig. 6A). In summer,
higher frequency of canker formation was observed at inoculation sites on trunks
than in the water controls (Fig. 6B and 6C) and, interestingly, typical longitudinal
canker formation was sometime noticed (Fig. 6D). This result confirmed that the
presence of virulent *P. syringae* strains in the cortical tissues of horse-chestnut in
trunks can result in canker formation similar to that observed under natural conditions. However, *P. syringae* was only re-isolated once from cankers in the end of
summer, whereas it was systematically re-isolated from the damaged cortical tissues in early spring. This would mean that canker formation is a reaction of the tree
and that the survival of *P. syringae* in summer in young trees is limited. The high
difficulty of isolating *P. syringae* from cankers was also noticed on naturally
infected older trees. Also, the successful isolations of *P. syringae* in summer from
older trees were not obtained from cankers, but from damaged cortical tissues that
externally presented no other symptom than small exudations on trunk resembling
those presented in Fig. 1B. This would mean that, at least in older trees, *P. syringae*
can be present in horse-chestnut cortical tissues in summer without canker presence
and, apparently, without canker formation.
    Given the high similarity of the REP-PCR profiles of the horse-chestnut strains
and *P. syringae* pv. *morsprunorum* race 1 (Fig. 4A), the virulence of the strains
MY3-1, MY3-2 and MY2-3 on sweet cherry detached twigs was investigated.

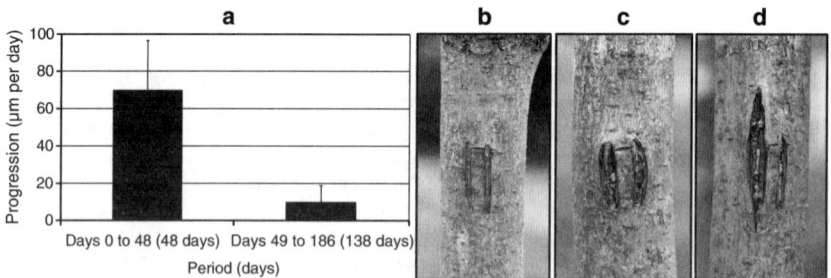

Fig. 6 Pathogenicity tests on horse-chestnut trunks. **a** Difference in progression in cortical tissues in the periods 'end of winter/early spring' (days 0–48) and 'spring/summer' (days 49–186).
b–d Cankers observed on trunks in the end of summer, 186 days after inoculation, for a water
inoculated site **b** and for *P. syringae* inoculated sites **c** and **d**

However, these strains proved to be non-virulent when compared to water and the virulent strain *P. syringae* pv. *morsprunorum* race 1 PmC36 (not shown). Horse-chestnut strains representative of different genetic groups and the pathotypes strains of *P. syringae* pv. *aesculi*, pv. *cunninghamiae* and pv. *daphniphylli* were investigated for their virulence on horse-chestnut twigs in September 2006. Only representative of the different horse-chestnut groups and of pathovar *aesculi* were similarly virulent on detached *A. hippocastanum* twigs (Fig. 7).

The genetic and pathogenicity data, as well as the production of coronatine and the fact that both type of strains induce disease in the *Aesculus* genus would agree with the grouping of the Belgian isolates of *P. syringae* from *A. hippocastanum* with *P. syringae* pv. *aesculi*, known to induce leaf spots on *A. indica* (Durgapal and Singh 1980). However, the genetic analyses presented here indicate clearly that this pathovar would then be genetically heterogeneous since all the Belgian strains can be grouped in an ERIC group 2 whereas the pathotype strain of *P. syringae* pv. *aesculi* showed a profile than can be named ERIC group 1 (Fig. 4b); also, it should also be confirmed that the Belgian *A. hippocastanum* isolates can induce leaf spot on *A. indica*. The conclusions are rather in agreement with published results obtained in United Kingdom where it was noticed that *P. syringae* strains isolated from *A. hippocastanum* had the same gyrase B gene (*gyrB*) sequence as *P. syringae* pv. *aesculi* (Webber et al., 2006). Whether *gyrB* sequence data would be less informative than the REP- and ERIC-PCR analyses presented here (Fig. 4) to detect differences between the *P. syringae* isolates from *A. hippocastanum* and *A. indica* is a possibility, but a comparison of strains from different countries should confirm that the *A. hippocastanum* pathogens are identical in all European countries; indeed closest similarities of isolates with *P. tremae* and *P. syringae* pv. *ulmi* have also been reported (Janse et al., 2006).

The REP-PCR (Fig. 3) data indicate that a genetic diversity already exists within the Belgian strains from *A. hippocastanum*, which indicates that these strains have

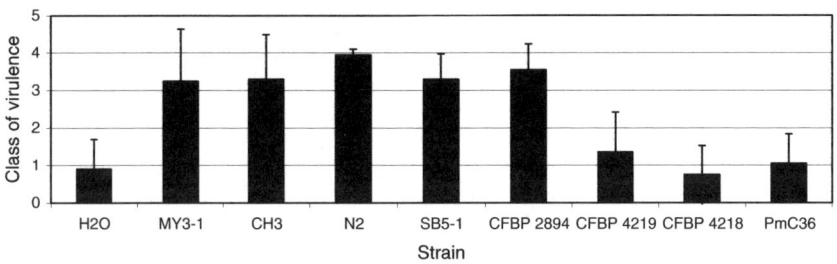

Fig. 7 Pathogenicity on horse-chestnut detached twigs of representatives of genetic groups of *P. syringae* from *A. hippocastanum* (MY3-1, CH3, N2 and SB5-1) and the genetically related strains *P. syringae* pv. *aesculi* CFBP 2894, pv. *daphniphylli* CFBP 4219, pv. *cunninghamiae* CFBP 4218 and pv. *morsprunorum* race 1 PmC36. Statistical analysis indicated no difference of variance for the different strains. On the other hand, significant differences between means were observed (p < 0.0001) and the test of Tukey showed clearly two homogeneous groups of strains differing by their virulence: N2, CFBP 2894, CH3, SB5-1 and MY3-1on one side, and CFBP 4219, PmC36, H20 and CFBP 4218 on the other side

already evolved on this host. The REP- and ERIC-PCR genetic data (Fig. 4) also indicate that all the Belgian strains differ more markedly from the Indian strain of *P. syringae* pv. *aesculi* isolated from *A. indica* in India than from each other, but also that there is a clear similarity of both types of strain with the pathovar *morsprunorum* race 1, which is also similar at the phytotoxin production level. Globally, these observations are rather in agreement with an evolution of the strains from a common ancestor by specialization on their respective hosts during evolution. This is in disaccord with the common feeling encountered among Belgian administrators that the horse-chestnut canker disease could be due to the recent apparition of a new pathogen; this opinion being justified by the apparent sudden outbreak of the disease reported these last years. By contrast, it seems probable that *P. syringae* strains live for a long time on *A. hippocastanum*. This highlights the need to determine the factors that could have resulted, these last years, in either a stronger propagation of the disease or an aggravation of the symptoms.

Acknowledgements We thank the persons, too numerous to be cited here, from regional and communal administrations or institutions for help in the localization of diseased trees in Brussels and Wallonia, as well as J.-M. Hautenauven and J. Zini for help in the collect of diseased horse-chestnut samples.

This work was supported by the Walloon Agricultural Research Centre, subsided by the Ministry of Agriculture of the Walloon Region.

References

Bereswill S., Bugert P., Völksch B., Ullrich M., Bender C.L. and Geider K. (1994) Identification and relatedness of coronatine-producing *Pseudomonas syringae* pathovars by PCR analysis and sequence determination of the amplification products. *Appl. Environ. Microbiol.* 60: 2924–2930.

Bultreys A. (2001) Ph.D. thesis. Utilisation de la production de métabolites secondaires de types phytotoxine et sidérophore en vue de la caractérisation et de l'identification de *Pseudomonas syringae*, pp 61–64. Université Catholique de Louvain (Ed), Louvain-la-Neuve.

Bultreys A. and Gheysen I. (1999) Biological and molecular detection of toxic lipodepsipeptide-producing *Pseudomonas syringae* strains and PCR identification in plants. *Appl. Environ. Microbiol.* 65: 1904–1909.

Bultreys A., Gheysen I., Maraite H. and de Hoffmann E. (2001) Characterization of fluorescent and nonfluorescent peptide siderophores produced by *Pseudomonas syringae* strains and their potential use in strain identification. *Appl. Environ. Microbiol.* 67: 1718–1727.

Bultreys A., Gheysen I., Wathelet B., Maraite H. and de Hoffmann E. (2003) High-performance liquid chromatography analyses of pyoverdin siderophores differentiate among phytopathogenic fluorescent *Pseudomonas* species. *Appl. Environ. Microbiol.* 69: 1143–1153.

Bultreys A., Gheysen I., Chandelier A., Zini J., Planchon V., Papart A.-T. and Decoux V. (2006) *Pseudomonas syringae*, the possible responsible of horse-chestnut tree disease in Belgium. Proceedings of the 11th International Conference on Plant Pathogenic Bacteria, pp 139. Elphinstone J., Weller S., Thwaites R., Parkinson N., Stead D. and Saddler G. (Eds), Edinburgh.

Cubero J. and Graham J.H. (2002) Genetic relationship among worldwide strains of *Xanthomonas* causing canker in citrus species and design of new primers for their identification by PCR. *Appl. Environ. Microbiol.* 68: 1257–1264.

Durgapal, J.C. and Singh B. (1980) Taxonomy of pseudomonads pathogenic to horse-chestnut, wild fig and cherry in India. *Indian Phytopath*. 33: 533–535.

Hulton C.S.J., Higgins C.F. and Sharp P.M. (1991) ERIC sequences: a novel family of repetitive elements in the genomes of *Escherichia coli*, *Salmonella typhimurium* and other enteric bacteria. *Mol. Microbiol*. 5: 825–834.

Janse J.D., van Beuningen A.R, van Vaerenbergh J., Speksnijder A.G.C.L., Heyrman J., de Vos P. and Maes M. (2006) An overview of some emerging diseases, including a taxonomic study of a *Pseudomonas* sp. causing bleeding canker of horse-chestnut (*Aesculus hippocastanum*). Proceedings of the 11th International Conference on Plant Pathogenic Bacteria, pp 11–12. Elphinstone J., Weller S., Thwaites R., Parkinson N., Stead D. and Saddler G. (Eds), Edinburgh.

King E.O., Ward M.K. and Raney D.E. (1954) Two simple media for the demonstration of pyocyanin and fluorescein. *J. Lab. Clin. Med*. 44: 301–307.

Kingsley M., Straub T.M., Call D.R., Daly D.S., Wunschel S.C. and Chandler D.P. (2002) Fingerprinting closely related *Xanthomonas* pathovars with random nonamer oligonucleotide microarrays. *Appl. Environ. Microbiol*. 68: 6361–6370.

Pooler M.R., Ritchie D.F. and Hartung J.S. (1996) Genetic relationships among strains of *Xanthomonas fragariae* based on random amplified polymorphic DNA PCR, repetitive extragenic palindromic PCR, and enterobacterial repetitive intergenic consensus PCR data and generation of multiplexed PCR primers useful for the identification of this phytopathogen. *Appl. Environ. Microbiol*. 62: 3121–3127.

Sakai R., Nishiyama K., Ichihara A., Shiraishi K. and Sakamura S. (1979) The relation between bacterial toxic action and plant growth regulation. Recognition and Specificity in Plant Host-Parasite Interactions, pp 165–179. Daly J.M. and Uritani I. (Eds), University Park Press, Baltimore, MD.

Versalovic J., Koeuth T. and Lupski J.R. (1991) Distribution of repetitive DNA sequences in eubacteria and application to fingerprinting of bacterial genomes. *Nucleic Acids Res*. 19: 6823–6831.

Webber J., Rose J., Parkinson N.M., Stanford H. and Elphinstone J.G. (2006) Characterisation of a possible causal agent of horse chestnut bleeding canker in the UK. Proceedings of the 11th International Conference on Plant Pathogenic Bacteria, pp 140. Elphinstone J., Weller S., Thwaites R., Parkinson N., Stead D. and Saddler G. (Eds), Edinburgh.

Interactions of Pseudomonads with Mushrooms and Other Eukaryotic Hosts

P. Burlinson[1], J. Knaggs, J. Hodgkin[2], C. Pears[2], and G.M. Preston[1]

Abstract Pseudomonads are ubiquitous Gram-negative bacteria with the potential to interact with diverse organisms in the environment. We are investigating the molecular basis of pathogenesis in mushroom-pathogenic Pseudomonas (NZ strains) and also evaluating NZ interactions with other organisms found in the mushroom environment, notably the nematode Caenorhabditis elegans and the amoeba Dictyostelium discoidium. Strain Pseudomonas sp. NZI7 is a tolaasin-producing mushroom pathogen that causes blotch disease of Agaricus bisporus, inhibits nematode and Dictyostelium predation and inhibits the growth of gram-positive and gram-negative bacteria. We have generated a collection of miniTn5::gfp::lux transposon mutants of strain NZI7, which have been used to identify genes and gene clusters whose loss of function alters one or more of these interactions. We are currently characterizing these in terms of their function, expression and regulation, and the effect that their disruption has upon bacterial fitness and colonization. Our screens have implicated tolaasin as a major factor in mushroom pathogenesis and antibacterial activity, but not in NZI7–nematode interactions. The two-component sensor kinase GacS plays a central role in regulating expression of tolaasin and other anti-microbial factors, including nematode-inhibition genes. We have also shown that NZI7 and other mushroom-pathogenic Pseudomonas contain type III secretion (T3SS) genes similar to those found in plant pathogenic P. syringae.

Keywords Mushroom-pathogenic *Pseudomonas*, *Agaricus bisporus*, *Caenorhabditis elegans*, *Dictyostelium discoidium*, tolaasin, GacS

[1] Department of Plant Sciences, University of Oxford, South Parks Road, Oxford, UK

[2] Department of Biochemistry, University of Oxford, South Parks Road, Oxford, UK

Author for correspondence: Gail Pruton; e-mail: gail.preston@plants.ox.ac.uk

M'B. Fatmi et al. (eds.), *Pseudomonas syringae Pathovars and Related Pathogens.*
© Springer Science+Business Media B.V. 2008

1   Introduction

The molecular basis of parasitism by *Pseudomonas* species towards plants, mush-rooms and humans has formed the focus of extensive research aimed at identifying the mechanisms involved in allowing them to occupy a given niche. The advent of genome sequencing is currently accelerating our journey towards a greater under-standing of the genetic differences between strains that underlie their ability to exploit different habitats. There are currently ten *Pseudomonas* genomes available on the web, representing human (Stover et al., 2000; Lee et al., 2006), insect (Vodovar et al., 2006) and plant pathogens (Feil et al., 2005) as well as saprophytes and strains involved in biocontrol, with others pending. The advent of rapid and cheap genome sequencing technology (Margulies et al., 2005) promises to expand this list further, bringing with it increased challenges of analyzing such data. Comparatively little attention has been focused towards understanding what pathogenic *Pseudomonas* do when they are not associated with their host of isolation (Preston, 2004). Genome sequence data suggest the presence of genes, for example, for the synthesis of toxins, which may not be of use against the host of isolation but might contribute to bacterial survival outside the niche of its host (Buell et al., 2003; Waterfield et al., 2003).

The research outlined in this paper involves the molecular characterisation of disease-causing *Pseudomonas* isolated from mushroom farms. *Pseudomonas* from this environment have received comparatively little study and only one mechanism by which they can cause disease has previously been characterised in detail, namely the role of the lipodepsipeptide biosurfactant and toxin tolaasin (Rainey et al., 1992; Soler-Rivas et al., 1999c). The mushroom environment is interesting as, unlike plant leaves or human lungs, it contains other organisms that *Pseudomonas* may interact with, either to gain nutrition, or to defend against bacteriovores to which the *Pseudomonas* constitute a food source (Fletcher et al., 1989; Grewal, 1991c). Mushroom pathogenic *Pseudomonas* are therefore of interest for the study of the mechanisms that allow them to cause disease on commercial mushroom farms but may also contribute towards our understanding of the molecular basis of opportun-istic pathogenesis and bacterial survival in nature. This paper first gives a short introduction to mushroom pathogenic *Pseudomonas* and then discusses interactions between *Pseudomonas* and the model organisms *Caenorhabditis elegans* and *Dictyostelium discoidium*. We then briefly describe the experimental approach we have adopted to identify factors which a group of mushroom pathogenic *Pseudomonas*, termed NZ strains, may use to be successful colonisers of the mycosphere.

2   Mushroom-Associated *Pseudomonas*

Both pathogenic and non-pathogenic *Pseudomonas* strains are commonly found on farms cultivating the edible button mushroom *Agaricus bisporus*. *P. putida* and *P. tolaasii* are the best characterised and show a dose and time-dependent chemotaxis towards mycelial exudates, although the identities of all the important factors have not been determined (Grewal and Rainey, 1991). Saprophytic *P. putida* strains in the casing layer are believed to be essential in triggering *A. bisporus* basidiodome initiation and in promoting the rate of radial hyphal extension (Grewal and Rainey, 1991; Rainey et al., 1990). The reported mechanism is the removal of 'self-inhibitory' substances and requires direct contact, i.e. it is not a diffusible or volatile factor (Rainey, 1991a). Thus control of bacterial diseases of mushrooms is challenging, since sterile soil systems are not suitable for mushroom fruiting body formation.

The best characterised mushroom disease caused by mushroom pathogenic *Pseudomonas* is blotch disease, which occurs on mushroom beds and post harvest. Blotch disease is typically characterised by discoloured, sunken lesions on mushroom caps. The best characterised blotch pathogen is *P. tolaasii* which causes brown blotch disease on sporophores of the cultivated mushroom *A. bisporus* (Tolaas, 1915; Wong and Preece, 1979; Rainey et al., 1992). Pathogenic bacteria multiply within the mycelium but disease symptoms are only apparent once the sporophore develops (Fletcher et al., 1989). Browning is due to melanin and quinine production, polyphenol oxidase (tyrosinase) activity catalysing the reaction between oxygen and phenolics released from vacuoles due to membrane disruption (Mamoun et al., 1999; Soler-Rivas et al., 1999c). Tyrosinase activity is induced by the *P. tolaasii* secreted toxin tolaasin, production of which is identifiable by the white-line-in-agar (WLA) reaction with another diffusible lipodepsipeptide from *P. reactans* (Wong and Preece, 1979; Soler-Rivas et al., 1999a; Munsch and Alatossava, 2002b). Tolaasin is an important factor in symptom development; indeed, non-Pseudomonad tolaasin detoxifying bacteria are being sought with a view to controlling disease symptoms (Moquet et al., 1996; Soler-Rivas et al., 1999b; Tsukamoto et al., 2002). *P. tolaasii* has been reported to have two morphologies, smooth and rough, only the former being pathogenic with the latter impaired in attachment but more mobile (Rainey, 1991b; Soler-Rivas et al., 1999c). Phenotype conversion from smooth to rough is dependent upon loss of the pheN gene, homologous to other bacterial two-component regulatory proteins (Soler-Rivas et al., 1999c).

In addition to *P. tolaasii* other *Pseudomonas* have been isolated with the ability to cause blotch disease (Bessette et al., 1985; Munsch et al., 2002). Godfrey et al. (2001a) isolated a range of blotch-causing Pseudomonads from mushroom farms in New Zealand and characterised these strains by mushroom bioassay, API biotyping and by testing for tolaasin production (Godfrey et al., 2001a). This study demonstrated that phylogenetically diverse *Pseudomonas* could cause similar disease symptoms, and found that browning could be caused by strains lacking tolaasin production, possibly due to protease activity. Interestingly one strain, NZI7, differed from *P. tolaasii* in multi-locus enzyme electrophoresis profile and was phylogenetically closer to *P. syringae* but was WLA positive (Godfrey et al., 2001b). A PCR-generated fragment from NZI7 showed much higher similarity to *P. tolaasii* tolaasin-encoding genes than to corresponding *P. syringae* syringomycin-encoding ones, suggesting acquisition by horizontal gene transfer of tolaasin-encoding genes (Godfrey et al., 2001b).

3   Interactions of *Pseudomonas* with Bacteriovores from the Mycosphere

3.1   Interactions with Caenorhabditis elegans

Caenorhabditis elegans is a bacteriovorous/saprophytic hermaphrodite rhabditid nematode commonly associated with mushroom cultivation, from which bacteria can be isolated from both the body surface and gut (Grewal and Hand, 1992). Nematodes are prevalent in the biosphere where associated bacteria can be found as food, symbionts, passengers or parasites (Poinar and Hansen, 1986; Grewal, 1991a, b). The standard laboratory N2 strain on which extensive genetics has been performed was originally isolated from a mushroom farm in near Bristol, UK (Chen et al., 2006). At low inoculums, *C. elegans* can reduce brown blotch symptoms and promote flushing, possibly by disseminating the *P. tolaasii* antagonist *P. reactans* and the sporophore-inducing *P. putida* (Grewal, 1991a, c; Grewal and Richardson, 1991).

Nematodes are affected by both nutritional conditions and bacterial species in monoxenic cultures (Poinar and Hansen, 1986) and selective feeding upon certain bacteria is sometimes observed (Poinar and Hansen, 1986; Grewal, 1991b). Feeding upon different strains can produce variation in reproductive capacity while products of bacterial metabolism can be nematicidal (Johnston and Titus, 1957; Grewal, 1991b). Diet also influences the ability of both *C. elegans* and wild relatives to locate and sense food; preconditioning on one bacterial strain can affect subsequent chemotaxis and utilisation of the same strain (Rodger et al., 2004). This substrate legacy effect may result in the use of a greater range of bacteria as food but has not been observed in 3D microcosms (Rodger et al., 2004).

Knowledge of how *C. elegans* perceives and combats pathogenic microorganisms remains in its infancy (Ruiz-Diez et al., 2003; Nicholas and Hodgkin, 2004). Anti-microbial defences include the cuticle and the pharyngeal grinder; the latter mechanically disrupts bacteria prior to entering the intestine, although some microbes can pass through and cause infection (Millet and Ewbank, 2004; Nicholas and Hodgkin, 2004) *C. elegans* has been developed as a model host system for *P. aeruginosa* pathogenesis (Pradel and Ewbank, 2004; Sifri et al., 2005). Three different modes of killing exist depending upon the culture conditions and strain used: fast killing, slow killing and paralysis (Ruiz-Diez et al., 2003). In strain PA14 rich, high salt media promotes fast killing due to a diffusible phenazine toxin; low salt media promotes slow killing, an infection-like process in the nematode gut regulated by quorum sensing and involving virulence factors such as MucD (Mahajan-Miklos et al., 1999; Tan et al., 1999; Tan, 2002). A lethal paralysis of nematodes occurs when *P. aeruginosa* PAO1 is grown on brain-heart infusion medium to high cell density and it is believed that cyanide is the key factor mediating this effect, which the Egl-9 *C. elegans* mutant is resistant to (Darby et al., 1999; Steinert et al., 2003). Given its importance to plant pathogenesis and also animals

(Huang et al., 1988; Ha and Jin, 2001) type III secretion may be expected to influence *C. elegans* interactions with *Pseudomonas* as well but this has not been demonstrated for *P. aeruginosa* (Wareham et al., 2005).

3.2 *Dictyostelium discoidium*

The social amoeba *D. discoidium* is a model organism that alternates between a unicellular and multicellular form, starvation causing individual cells to aggregate and produce a multicellular organism (Pozos and Ramakrishan, 2004; Eichinger et al., 2005). *D. discoidium* is a natural soil inhabitant, feeding on bacteria using a phagocytic mechanism akin to that of macrophages and producing antimicrobial molecules to combat bacterial growth in the phagosome (Steinert and Heuner, 2005). *Dictyostelium* is thought to be a natural reservoir for important human pathogens such as *Legionella pneumonophilia* which can live within the phagosome (Pradel and Ewbank, 2004).

Plating of *Dictyostelium* onto lawns of some environmental and pathogenic bacteria, such as *P. aeruginosa*, prevents *Dictyostelium* from feeding and multiplying (Cosson et al., 2002; Pukatzki et al., 2002). A *P. aeruginosa* mutant in the T3SS structural gene PscJ and the secreted protein ExoU lose virulence in this assay, as does a mutant of the quorum sensing transcription factor LasR (Pukatzki et al., 2002).

4 Characterization of Mushroom Pathogenic *Pseudomonas*

4.1 *Phenotype Survey of NZ Strains*

We are investigating the molecular basis of pathogenesis in mushroom-pathogenic *Pseudomonas* (NZ strains) and also evaluating NZ interactions with other organisms found in the mushroom environment, notably the nematode *C. elegans* and the amoeba *Dictyostelium discoidium*. NZ strains were originally isolated from mushroom farms in New Zealand owing to their differing abilities to cause varying degrees of discolouration on *Agaricus* tissue (Godfrey et al., 2001a, b). We have surveyed NZ strains for their ability to produce molecules that might mediate these interactions, including toxins, hydrogen cyanide, exoenzymes, quorum sensing molecules and type III secretion systems. We have also examined their ability to affect the growth of several other fungi and to inhibit other micro-organisms that colonise the mycosphere. We are currently undertaking MLST analysis of NZ strains to refine the existing phylogeny which we hope will provide a better understanding of their evolutionary history and may aid identification of factors common to particular lineages.

4.2 Genetic Characterisation of NZI7

We have performed a detailed genetic characterisation of one NZ strain called *Pseudomonas* sp. NZI7. This strain displays a number of interesting phenotypes, including mushroom pathogenesis, inhibition of the growth of Gram-positive and Gram-negative bacteria, inhibition of mycelial growth, inhibition of *D. discoidium* development and two types of nematode interaction (anti-predation and fast-killing), which are regulated in response to environmental factors. We have generated a collection of ~10,000 miniTn5::*gfp*::*lux* transposon mutants of strain NZI7, and screened these to identify NZI7 genes and gene clusters involved in the production of extracellular factors and in interactions with other organisms. NZI7 produces the toxin tolaasin which is a key factor in mushroom pathogenesis and antibacterial activity but not in inhibition of nematode or *Dictyostelium* predation. We have also identified genes whose disruption results in loss of resistance to nematode predation, some of which are located in a novel gene cluster. The two-component sensor kinase GacS plays a central role in regulating expression of both tolaasin and nematode resistance genes. We are currently characterizing tolaasin synthesis and nematode resistance genes in terms of their function, expression and regulation, both biochemically and also using *lux*-based gene reporters.

5 Conclusions

Pseudomonas are ubiquitous and found in a panoply of different niches. Plant-associated species may be pathogenic, neutral or growth promoting in their effects. Many factors, including production of toxins, exoenzymes, T3SSs and EPS have been shown to contribute towards successful bacterial colonisation of host tissues (Preston, 2000). In particular type III secretion has been established as a significant contributor to pathogenesis in *Pseudomonas* interactions with eukaryotes, from plants to Man. T3SSs have also been discovered in 'non-pathogenic' *Pseudomonas* but the ecological role(s) of these systems remains to be proven (Preston et al., 2001; Rezzonico et al., 2004). Evidence for a role in interactions with fungi is slowly beginning to emerge which may indicate that it plays a role in the myco-sphere as well as the phyllosphere (Rezzonico et al., 2005). To assess the significance of such factors it will be important to consider not only their interaction with the organism they were isolated from but also with other potential hosts in their environment. Mushroom pathogenic *Pseudomonas* represent a good resource to carry out such analyses and have not been examined in the context of T3SSs previously. The wide range of phenotypes we have observed in both *C. elegans* and *D. discoidium* interactions with NZ strains are also of considerable interest and may indicate that multiple mechanisms exist to allow bacteria to survive predation.

Acknowledgements Peter Burlinson is supported by a BBSRC graduate studentship. Gail Preston is a Royal Society University Research Fellow.

References

Bessette, A. E., Kerrigan, R. W. & Jordan, D. C. (1985) Yellow blotch of *Pleurotus ostreatus*. *Applied and Environmental Microbiology*, 50, 1535–1537.

Buell, C., Joardar, V., Lindeberg, M., Selengut, J., Paulsen, I., Gwinn, M., Dodson, R., Deboy, R., Durkin, A., Kolonay, J., Madupu, R., Daugherty, S., Brinkac, L., Beanan, M., Haft, D., Nelson, W., Davidsen, T., Zafar, N., Zhou, L., Liu, J., Yuan, Q., Khouri, H., Fedorova, N., Tran, B., Russell, D., Berry, K., Utterback, T., Van Aken, S., Feldblyum, T., D'ascenzo, M., Deng, W., Ramos, A., Alfano, J., Cartinhour, S., Chatterjee, A., Delaney, T., Lazarowitz, S., Martin, G., Schneider, D., Tang, X., Bender, C., White, O., Fraser, C. & Collmer, A. (2003) The complete genome sequence of the *Arabidopsis* and tomato pathogen *Pseudomonas syringae* pv. tomato DC3000. *Proceedings of the National Academy of Sciences of the United States of America*, 100, 10181–10186.

Chen, J., Lewis, E. E., Carey, J. R., Caswell, H. & Caswell-Chen, E. P. (2006) The ecology and biodemography of *Caenorhabditis elegans. Experimental Gerontology*, 41, 1059–1065.

Cosson, P., Zulianello, L., Join-Lambert, O., Faurisson, F., Gebbie, L., Benghezal, M., Van Delden, C., Curty, L. & Kohler, T. (2002) *Pseudomonas aeruginosa* virulence analyzed in a *Dictyostelium discoideum* host system. *Journal of Bacteriology*, 184, 3027–3033.

Darby, C., Cosma, C. L., Thomas, J. H. & Manoil, C. (1999) Lethal paralysis of *Caenorhabditis elegans* by *Pseudomonas aeruginosa. Proceedings of the National Academy of Sciences of the United States of America*, 96, 15202–15207.

Eichinger, L., Pachebat, J. A., Glockner, G., Rajandream, M. A., Sucgang, R., Berriman, M., Song, J., Olsen, R., Szafranski, K., Xu, Q., Tunggal, B., Kummerfeld, S., Madera, M., Konfortov, B. A., Rivero, F., Bankier, A. T., Lehmann, R., Hamlin, N., Davies, R., Gaudet, P., Fey, P., Pilcher, K., Chen, G., Saunders, D., Sodergren, E., Davis, P., Kerhornou, A., Nie, X., Hall, N., Anjard, C., Hemphill, L., Bason, N., Farbrother, P., Desany, B., Just, E., Morio, T., Rost, R., Churcher, C., Cooper, J., Haydock, S., Van Driessche, N., Cronin, A., Goodhead, I., Muzny, D., Mourier, T., Pain, A., Lu, M., Harper, D., Lindsay, R., Hauser, H., James, K., Quiles, M., Madan Babu, M., Saito, T., Buchrieser, C., Wardroper, A., Felder, M., Thangavelu, M., Johnson, D., Knights, A., Loulseged, H., Mungall, K., Oliver, K., Price, C., Quail, M. A., Urushihara, H., Hernandez, J., Rabbinowitsch, E., Steffen, D., Sanders, M., Ma, J., Kohara, Y., Sharp, S., Simmonds, M., Spiegler, S., Tivey, A., Sugano, S., White, B., Walker, D., Woodward, J., Winckler, T., Tanaka, Y., Shaulsky, G., Schleicher, M., Weinstock, G., Rosenthal, A., Cox, E. C., Chisholm, R. L., Gibbs, R., Loomis, W. F., Platzer, M., Kay, R. R., Williams, J., Dear, P. H., Noegel, A. A., Barrell, B. & Kuspa, A. (2005) The genome of the social amoeba *Dictyostelium discoideum. Nature*, 435, 43–57.

Feil, H., Feil, W. S., Chain, P., Larimer, F., Dibartolo, G., Copeland, A., Lykidis, A., Trong, S., Nolan, M., Goltsman, E., Thiel, J., Malfatti, S., Loper, J. E., Lapidus, A., Detter, J. C., Land, M., Richardson, P. M., Kyrpides, N. C., Ivanova, N. & Lindow, S. E. (2005) Comparison of the complete genome sequences of *Pseudomonas syringae* pv. syringae B728a and pv. tomato DC3000. *Proceedings of the National Academy of Science of the United States of America*, 102, 11064–11069.

Fletcher, J. T., White, P. F. & Gaze, R. H. (1989) *Mushrooms: Pest and Disease Control*, Andover, Intercept, UK.

Godfrey, S., Harrow, S., Marshall, J. & Klena, J. (2001a) Characterization by 16S rRNA sequence analysis of pseudomonads causing blotch disease of cultivated *Agaricus bisporus. Applied and Environmental Microbiology*, 67, 4316–4323.

Godfrey, S., Marshall, J. & Klena, J. (2001b) Genetic characterization of *Pseudomonas* 'NZ17 – a novel pathogen that results in a brown blotch disease of *Agaricus bisporus. Journal of Applied Microbiology*, 91, 412–420.

Grewal, P. (1991a) Effects of *Caenorhabditis elegans* (nematoda, rhabditidae) on the spread of the bacterium *Pseudomonas tolaasii* in mushrooms (*Agaricus bisporus*). *Annals of Applied Biology*, 118, 47–55.

Grewal, P. (1991b) Influence of bacteria and temperature on the reproduction of *Caenorhabditis elegans* (nematoda, rhabditidae) infesting mushrooms (*Agaricus bisporus*). *Nematologica*, 37, 72–82.

Grewal, P. (1991c) Relative contribution of nematodes (*Caenorhabditis elegans*) and bacteria towards the disruption of flushing patterns and losses in yield and quality of mushrooms (*Agaricus bisporus*). *Annals of Applied Biology*, 119, 483–499.

Grewal, P. & Hand, P. (1992) Effects of bacteria isolated from a saprophagous rhabditid nematode *Caenorhabditis elegans* on the mycelial growth of *Agaricus bisporus*. *Journal of Applied Bacteriology*, 72, 173–179.

Grewal, P. & Richardson, P. (1991) Effects of *Caenorhabditis elegans* (nematoda, rhabditidae) on yield and quality of the cultivated mushroom *Agaricus bisporus*. *Annals of Applied Biology*, 118, 381–394.

Grewal, S. & Rainey, P. (1991) Phenotypic variation of *Pseudomonas putida* and *P. tolaasii* affects the chemotactic response to *Agaricus bisporus* mycelial exudate. *Journal of General Microbiology*, 137, 2761–2768.

Ha, U. & Jin, S. (2001) Growth phase-dependent invasion of *Pseudomonas aeruginosa* and its survival within HeLa cells. *Infection and Immunity*, 69, 4398–4406.

Huang, H. C., Schuurink, R., Denny, T. P., Atkinson, M. M., Baker, C. J., Yucel, I., Hutcheson, S. W. & Collmer, A. (1988) Molecular cloning of a *Pseudomonas syringae* pv. syringae gene cluster that enables *Pseudomonas fluorescens* to elicit the hypersensitive response in tobacco plants. *Journal of Bacteriology*, 170, 4748–4756.

Johnston & Titus (1957) Further studies on microbiological reduction of nematode population in water-saturated soils. *Phytopathology*, 47, 525–526.

Lee, D. G., Urbach, J. M., Wu, G., Liberati, N. T., Feinbaum, R. L., Miyata, S., Diggins, L. T., He, J., Saucier, M., Deziel, E., Friedman, L., Li, L., Grills, G., Montgomery, K., Kucherlapati, R., Rahme, L. G. & Ausubel, F. M. (2006) Genomic analysis reveals that *Pseudomonas aeruginosa* virulence is combinatorial. *Genome Biology*, 7, R90.

Mahajan-Miklos, S., Tan, M., Rahme, L. & Ausubel, F. (1999) Molecular mechanisms of bacterial virulence elucidated using a *Pseudomonas aeruginosa Caenorhabditis elegans* pathogenesis model. *Cell*, 96, 47–56.

Mamoun, M., Moquet, F., Savoie, J., Devesse, C., Ramos-Guedes-Lafargue, M., Olivier, J. & Arpin, N. (1999) *Agaricus bisporus* susceptibility to bacterial blotch in relation to environment: biochemical studies. *FEMS Microbiology Letters*, 181, 131–136.

Margulies, M., Egholm, M., Altman, W. E., Attiya, S., Bader, J. S., Bemben, L. A., Berka, J., Braverman, M. S., Chen, Y. J., Chen, Z., Dewell, S. B., Du, L., Fierro, J. M., Gomes, X. V., Godwin, B. C., He, W., Helgesen, S., Ho, C. H., Irzyk, G. P., Jando, S. C., Alenquer, M. L., Jarvie, T. P., Jirage, K. B., Kim, J. B., Knight, J. R., Lanza, J. R., Leamon, J. H., Lefkowitz, S. M., Lei, M., Li, J., Lohman, K. L., Lu, H., Makhijani, V. B., Mcdade, K. E., Mckenna, M. P., Myers, E. W., Nickerson, E., Nobile, J. R., Plant, R., Puc, B. P., Ronan, M. T., Roth, G. T., Sarkis, G. J., Simons, J. F., Simpson, J. W., Srinivasan, M., Tartaro, K. R., Tomasz, A., Vogt, K. A., Volkmer, G. A., Wang, S. H., Wang, Y., Weiner, M. P., Yu, P., Begley, R. F. & Rothberg, J. M. (2005) Genome sequencing in microfabricated high-density picolitre reactors. *Nature*, 437, 376–380.

Millet, A. C. & Ewbank, J. J. (2004) Immunity in *Caenorhabditis elegans*. *Current Opinion in Immunology*, 16, 4–9.

Moquet, F., Mamoun, M. & Olivier, J. (1996) *Pseudomonas tolaasii* and tolaasin: comparison of symptom induction on a wide range of *Agaricus bisporus* strains. *FEMS Microbiology Letters*, 142, 99–103.

Munsch, P. & Alatossava, T. (2002b) The white-line-in-agar test is not specific for the two cultivated mushroom associated pseudomonads, *Pseudomonas tolaasii* and *Pseudomonas "reactans"*. *Microbiological Research*, 157, 7–11.

Munsch, P., Alatossava, T., Marttinen, N., Meyer, J., Christen, R. & Gardan, L. (2002) *Pseudomonas costantinii* sp nov., another causal agent of brown blotch disease, isolated from cultivated mushroom sporophores in Finland. *International Journal of Systematic and Evolutionary Microbiology*, 52, 1973–1983.

Nicholas, H. R. & Hodgkin, J. (2004) Responses to infection and possible recognition strategies in the innate immune system of *Caenorhabditis elegans*. *Molecular Immunology*, 41, 479–493.

Poinar, G. O. & Hansen, E. L. (1986) Associations between nematodes and bacteria. *Helmintological Abstracts (Series B)*, 55, 61–81.

Pozos, T. & Ramakrishan, L. (2004) New models for the study of mycobacterium-host interactions. *Current Opinion in Immunology*, 16, 499–505.

Pradel, E. & Ewbank, J. (2004) Genetic models in pathogenesis. *Annual Review of Genetics*, 38, 347–363.

Preston, G. (2004) Plant perceptions of plant growth-promoting *Pseudomonas*. *Philosophical Transactions of the Royal Society of London Series B-Biological Sciences*, 359, 907–918.

Preston, G., Bertrand, N. & Rainey, P. (2001) Type III secretion in plant growth-promoting *Pseudomonas fluorescens* SBW25. *Molecular Microbiology*, 41, 999–1014.

Preston, G. M. (2000) *Pseudomonas syringae* pv. tomato: the right pathogen, of the right plant, at the right time. *Molecular Plant Pathology*, 1, 263–275.

Pukatzki, S., Kessin, R. & Mekalanos, J. (2002) The human pathogen *Pseudomonas aeruginosa* utilizes conserved virulence pathways to infect the social amoeba *Dictyostelium discoideum*. *Proceedings of the National Academy of Sciences of the United States of America*, 99, 3159–3164.

Rainey, P. (1991a) Effect of *Pseudomonas putida* on hyphal growth of *Agaricus bisporus*. *Mycological Research*, 95, 699–704.

Rainey, P. (1991b) Phenotypic variation of *Pseudomonas putida* and *P.tolaasii* affects attachment to *Agaricus bisporus* mycelium. *Journal of General Microbiology*, 137, 2769–2779.

Rainey, P., Cole, A., Fermor, T. & Wood, D. (1990) A model system for examining involvement of bacteria in basidiome initiation of *Agaricus bisporus*. *Mycological Research*, 94, 191–195.

Rainey, P. B., Brodey, C. L. & Johnstone, K. (1992) Biology of *Pseudomonas tolaasii*, cause of brown blotch disease of the cultivated mushroom. *Advances in Plant Pathology*, 8, 95–117.

Rezzonico, F., Defago, G. & Moenne-Loccoz, Y. (2004) Comparison of ATPase-encoding type III secretion system *hrcN* genes in biocontrol fluorescent pseudomonads and in phytopathogenic proteobacteria. *Applied and Environmental Microbiology*, 70, 5119–5131.

Rezzonico, F., Binder, C., Defago, G. & Moenne-Loccoz, Y. (2005) The type III secretion system of biocontrol *Pseudomonas fluorescens* KD targets the phytopathogenic Chromista *Pythium ultimum* and promotes cucumber protection. *Molecular Plant-Microbe Interactions*, 18, 991–1001.

Rodger, S., Griffiths, B., Mcnicol, J., Wheatley, R. & Young, I. (2004) The impact of bacterial diet on the migration and navigation of *Caenorhabditis elegans*. *Microbial Ecology*, 48, 358–365.

Ruiz-Diez, B., Sanchez, P., Baquero, F., Martinez, J. & Navas, A. (2003) Differential interactions within the *Caenorhabditis elegans – Pseudomonas aeruginosa* pathogenesis model. *Journal of Theoretical Biology*, 225, 469–476.

Sifri, C., Begun, J. & Ausubel, F. (2005) The worm has turned – microbial virulence modeled in *Caenorhabditis elegans*. *Trends in Microbiology*, 13, 119–127.

Soler-Rivas, C., Arpin, N., Olivier, J. & Wichers, H. (1999a) The effects of tolaasin, the toxin produced by *Pseudomonas tolaasii* on tyrosinase activities and the induction of browning in *Agaricus bisporus* fruiting bodies. *Physiological and Molecular Plant Pathology*, 55, 21–28.

Soler-Rivas, C., Arpin, N., Olivier, J. & Wichers, H. (1999b) WLIP, a lipodepsipeptide of *Pseudomonas 'reactans'*, as inhibitor of the symptoms of the brown blotch disease of *Agaricus bisporus*. *Journal of Applied Microbiology*, 86, 635–641.

Soler-Rivas, C., Jolivet, S., Arpin, N., Olivier, J. & Wichers, H. (1999c) Biochemical and physiological aspects of brown blotch disease of *Agaricus bisporus*. *FEMS Microbiology Reviews*, 23, 591–614.

Steinert, M. & Heuner, K. (2005) *Dictyostelium* as host model for pathogenesis. *Cellular Microbiology*, 7, 307–314.

Steinert, M., Leippe, M. & Roeder, T. (2003) Surrogate hosts: protozoa and invertebrates as models for studying pathogen-host interactions. *International Journal of Medical Microbiology*, 293, 321–332.

Stover, C., Pham, X., Erwin, A., Mizoguchi, S., Warrener, P., Hickey, M., Brinkman, F., Hufnagle, W., Kowalik, D., Lagrou, M., Garber, R., Goltry, L., Tolentino, E., Westbrock-Wadman, S., Yuan, Y.,

Brody, L., Coulter, S., Folger, K., Kas, A., Larbig, K., Lim, R., Smith, K., Spencer, D., Wong, G., Wu, Z., Paulsen, I., Reizer, J., Saier, M., Hancock, R., Lory, S. & Olson, M. (2000) Complete genome sequence of *Pseudomonas aeruginosa* PAO1, an opportunistic pathogen. *Nature*, 406, 959–964.

Tan, M. (2002) Cross-species infections and their analysis. *Annual Review of Microbiology*, 56, 539–565.

Tan, M., Rahme, L., Sternberg, J., Tompkins, R. & Ausubel, F. (1999) *Pseudomonas aeruginosa* killing of *Caenorhabditis elegans* used to identify *P. aeruginosa* virulence factors. *Proceedings of the National Academy of Sciences of the United States of America*, 96, 2408–2413.

Tolaas (1915) A bacterial disease of cultivated mushrooms. *Phytopathology*, 5, 51–54.

Tsukamoto, T., Murata, H. & Shirata, A. (2002) Identification of non-pseudomonad bacteria from fruit bodies of wild Agaricales fungi that detoxify tolaasin produced by *Pseudomonas tolaasii*. *Bioscience Biotechnology and Biochemistry*, 66, 2201–2208.

Vodovar, N., Vallenet, D., Cruveiller, S., Rouy, Z., Barbe, V., Acosta, C., Cattolico, L., Jubin, C., Lajus, A., Segurens, B., Vacherie, B., Wincker, P., Weissenbach, J., Lemaitre, B., Medigue, C. & Boccard, F. (2006) Complete genome sequence of the entomopathogenic and metabolically versatile soil bacterium *Pseudomonas entomophila*. *Nature Biotechnology*, 24, 673–679.

Wareham, D., Papakonstantinopoulou, A. & Curtis, M. (2005) The *Pseudomonas aeruginosa* PA14 type III secretion system is expressed but not essential to virulence in the *Caenorhabditis elegans-P.aeruginosa* pathogenicity model. *FEMS Microbiology Letters*, 242, 209–216.

Waterfield, N., Daborn, P., Dowling, A., Yang, G., Hares, M. & Ffrench-Constant, R. (2003) The insecticidal toxin makes caterpillars floppy 2 (Mcf2) shows similarity to HrmA, an avirulence protein from a plant pathogen. *FEMS Microbiology Letters*, 229, 265–270.

Wong, W. & Preece, T. (1979) Identification of *Pseudomonas tolaasi* – white line in agar and mushroom tissue block rapid pitting tests. *Journal of Applied Bacteriology*, 47, 401–407.

Part V
Taxonomy and Evolution

The Evolution of the Pseudomonads

D.S. Guttman, R.L. Morgan, and P.W. Wang

Abstract A preliminary multilocus sequence typing (MLST) analysis of 367 strains within the genus *Pseudomonas* identified major clades corresponding to *P. syringae*, *P. viridiflava*, *P. fluorescens*, *P. putida*, *P. aeruginosa*, and *P. stutzeri*. Split decomposition revealed three primary radiations in the genus, with *P. aeruginosa* being quite distinct, significant gene flow between *P. syringae* and *P. viridiflava*, and moderate gene flow between *P. fluorescens* and *P. putida*. Both phylogenetic and population genetic tests indicate that *P. aeruginosa* has undergone a very recent selective sweep, which purged nearly all of the standing genetic variation in the species, consistent with epidemic population dynamics. *P. fluorescens* has also undergone a strong selective sweep, although this event appears to be quite ancient. Finally, as reported previously, *P. syringae* is an endemic pathogen that has maintained a relatively stable population among its plant hosts for a very long time.

Keywords *Pseudomonas*, MLST, evolutionary history, pathogen, epidemic, endemic, recombination

1   Introduction

The Pseudomonads are arguably one of the most fascinating and ecologically significant bacterial genera. The intense medical, agricultural, and biotechnological interest in this group has prompted the sequencing of 9 complete genomes, with another 15 in the pipeline as of October 2006 (www.genomesonline.org). This remarkably versatile group of organisms includes species such as the opportunistic human pathogen *P. aeruginosa*, which was identified by the Antimicrobial Availability Task Force of the Infectious Disease Society of America as one of the top six infectious disease threats (Talbot et al., 2006). It is the leading cause of

Centre for the Analysis of Genome Evolution and Function, University of Toronto, Toronto, Canada M5S3B2

Author for correspondence: David S. Guttman; e-mail: david.guttman@utoronto.ca

M'B. Fatmi et al. (eds.), *Pseudomonas syringae Pathovars and Related Pathogens.*   307
© Springer Science+Business Media B.V. 2008

mortality among Cystic Fibrosis (CF) patients, and one of the most common causes of hospital-associated pneumonia. This persistent pathogen attacks wounds, burns, eyes, the urinary tract, as well as causing general systemic infections (Rowland, 1999; Office of Laboratory Security, 2001; Anonymous, 2002). It is also pathogenic on a wide range of other organisms, including mammals, insects (Jander et al., 2000), worms (Mahajan-Miklos et al., 1999), amoeba (Pukatzki et al., 2002), fungi (Hogan and Kolter, 2002), and even plants (Rahme et al., 1995, 1997; Plotnikova et al., 2000).

Pseudomonas syringae is a foliar plant pathogen that causes a variety of blights, speck, and spot diseases in many important agricultural crops, including tomato, soybeans, rice, and tobacco, to name just a few. Well over 50 different pathogenic varieties (pathovars) have been named within this complex. In addition to being a significant agricultural pathogen, this species is also one of the most important model systems for the study of secreted virulence proteins and their role in pathogen-host interactions. *Pseudomonas viridiflava* is a very close relative to *P. syringae*, and perhaps should be considered to be part of the *P. syringae* species complex due to genetic and ecological commonalities. This plant pathogen has been singled out because it is a natural pathogen of the model plant species *Arabidopsis thaliana* (Jakob et al., 2002; Goss et al., 2004).

Pseudomonas fluorescens is a common environmental bacterium frequently recovered from soil, water and plant surfaces. Many strains of this species are used as biocontrol and plant-growth promoting agents, although strains associated with plant disease are also frequently found within this species complex.

Pseudomonas putida is an important bio-degradative species that is capable of eliminating some of the most deadly and challenging environmental toxins (Wackett, 2003). More strains of *P. putida* with significant degradative abilities have been isolated from the environment than any other bacterial species (Wackett, 2003). Recently, a strain of *P. putida* was isolated that is even capable of degrading polystyrene foam (Ward et al., 2006).

Pseudomonas stutzeri is a soil bacterium with important biodegradative and bioremediation capabilities. It is the only a bacterium capable of degrading the dry-cleaning solvent carbon tetrachloride without producing chloroform.

The extensive genetic diversity and wide variety of niches inhabited by members of this genus has prompted a number of taxonomic treatments. Whole genome DNA–DNA hybridization is still the recommended standard for assessing the relationship within and among bacterial species (Wayne et al., 1987). Nevertheless, this approach, and any other approach that relies on whole genome comparisons, must now be questioned based on our current understanding of the dynamic nature of bacterial genomes. Recent studies have shown that large fractions of bacterial genomes are not transmitted vertically (from mother to daughter cell) over evolutionary time (Hacker and Kaper, 2000; Dobrindt and Hacker, 2001), but rather acquired and lost in a horizontal manner. This so-called flexible or accessory genome can account for as much as 60% of the genetic complement of a species (Welch et al., 2002). In *P. syringae*, only half of all putative coding sequences are found in all three sequenced strains, indicating that as much as half of the genomic complement of this species should be considered to be part of the flexible genome (Guttman, unpublished data). Genetic material transferred horizontally will not have the same evolutionary history as the clonal, or vertically transmitted, backbone of the species. Consequently, the clonal evolutionary history of a species is best studied by restricting the analysis to those genes that are most likely to be transmitted in a vertical manner, such as essential housekeeping genes, or the so-called core genome.

Multilocus sequence typing (MLST) has been developed specifically to characterize the clonal evolutionary relationships among bacterial isolates (Maiden et al., 1998; Enright and Spratt, 1999). MLST is a highly discriminating, rapid, and portable DNA-based strain typing method in which regions from several housekeeping loci are sequenced from each strain. Since housekeeping loci are often under different levels of constraint, the use of many loci ensures that there will be enough variability to discriminate between the most closely related clones while still being able to track global clonal dynamics. The use of housekeeping genes also focuses the dataset on the core genome, and consequently, is more likely to represent the clonal history of the species. MLST is rapidly becoming the gold standard for strain typing. It has been shown to have the highest discriminatory power of any of the major typing methods (Nallapareddy et al., 2002; Peacock et al., 2002; Kotetishvili et al., 2003; Cooper and Feil, 2004; Revazishvili et al., 2004; Harbottle et al., 2006), and is the only method that permits precise phylogenetic analyses.

Molecular phylogenetic approaches have been used to study the evolution of *P. syringae* (Sawada et al., 1999; Sarkar and Guttman, 2004; Hwang et al., 2005), *P. viridiflava* (Goss et al., 2004), and *P. aeruginosa* (Curran et al., 2004; Vernez et al., 2005), and the genus *Pseudomonas* (Yamamoto et al., 2000). Many of these studies addressed strictly phylogenetic-based questions, and did not address more fundamental questions related to the biology of the organisms in question. Yamamoto et al. (2000) performed a very nice phylogenetic analysis of the Pseudomonads using two housekeeping genes, which should rightfully be considered the first MLST analysis of this group.

In this work, we describe a preliminary four-locus MLST analysis of all Pseudomonads. We expand upon the work done by Yamamoto et al. (2000) by addressing questions related to recombination, selection, and demographic history of each of the major species complexes. Finally, we address questions related to the meaning of bacterial species in the context of the Pseudomonads.

2   Materials and Methods

2.1   Bacterial and Yeast Strains

Three hundred and sixty-seven *Pseudomonas* strains and two outgroup species were used in this study (a detailed strain list is available from the authors upon request). Our definition of strain is any bacterial isolate that was collected at a unique time and place. Efforts were made to avoid using isolates collected within a year of each other by the same individual. All were grown in King's B medium (King et al., 1954) at 30°C.

2.2 Multilocus Sequence Typing

The four housekeeping genes sequenced were *rpoD*, encoding sigma factor 70, *gyrB*, encoding DNA gyrase B, *gltA* (also known as *cts*), encoding citrate synthase, and *gapA*, encoding glyceraldehyde-3-phosphate dehydrogenase. These loci are a subset of the seven used in the *P. syringae* MLST paper (Sarkar and Guttman, 2004), and were chosen because they consistently provide robust data, and their combined level of polymorphism is sufficient to reliably resolve evolutionary relationships.

The MLST primers used for DNA amplification and sequencing of the four loci are presented in Table 1 (Sarkar and Guttman, 2004; Hwang et al., 2005). These degenerate primers were designed based on global multiple sequence alignments from all available Pseudomonads. PCR amplification was performed on 50 ng of genomic DNA using MBI-Fermentas Taq polymerase, 200 μM nucleotides, 5% DMSO, and primer concentrations of 1 μM at an annealing temperature of 58°C for 30 cycles with Hybaid PCR Express thermal cyclers. DNA sequencing was performed with the CEQ-DTCS Quick Start kit on a Beckman-Coulter CEQ 8000 DNA Sequencer at 55°C with the addition of 1 M Betaine (Sigma). Forward and reverse sequences were obtained either using the PCR or internal sequencing primers for each locus (Table 2). The sequences were edited using BioEdit (www.mbio.ncsu. edu/BioEdit), aligned using ClustalW (Thompson et al., 1997) and GeneDoc (www.

Table 1 MLST primers

Name	Primer sequence	Use[a]
gapA + 264	CCGGCSGARCTGCCSTGG	PCR
gapA-931	ASSCCCAYTCGTTGTCRTACCA	PCR
gapA + 312	TCGARTGCACSGGBCTSTTCACC	SEQ
gapA-874	GTGTGRTTGGCRTCGAARATCGA	SEQ
gltA + 174	GCCTCBTGCGAGTCGAAGATCACC	PCR only
gltA-1192	CTTGTAVGGRCYGGAGAGCATTTC	PCR
gltA + 513	CCTGRTBGCCAAGATGCCKAC	SEQ
gltA-1130	CGAAGATCACGGTGAACATGCTGG	SEQ
gyrB + 133	CTGCACCAYATGGTSTTCGAGG	PCR
gyrB-1124	CGNGCDGCRTCGAKCATCTTGC	PCR
gyrB + 271	TCBGCRGCVGARGTSATCATGAC	SEQ
gyrB-1022	TTGTCYTTGGTCTGSGAGCTGAA	SEQ
rpoD + 147	CAGGTGGAAGACATCATCCGCATG	PCR only
rpoD-1222	CCGATGTTGCCTTCCTGGATCAG	PCR
rpoD + 377	GRAATCGCCAARCGYATYGA	SEQ
rpoD-1078	CGGTTGATKTCCTTGATCTCGGC	SEQ

[a] Any primer labeled as 'PCR' can also be used for sequencing. Those labeled as 'PCR only' cannot be used for sequencing due to their distance from the region of interest

Table 2 Population genetic analysis of the Pseudomonads

Species	N[a]	hap.[b]	π	θ	TajD[c]	P[d]	%GC	ρ	θ/ρ
P. aeruginosa	142	100	0.0080	0.0403	−2.6382	<0.001	0.64	0	–
P. fluorescens	69	65	0.0937	0.1183	−0.7346	>0.1	0.597	0.0857	1.38
P. putida	13	13	0.0901	0.1089	−0.7566	>0.1	0.609	0.0265	4.11
P. syringae	134	106	0.0632	0.0627	0.0274	>0.1	0.582	0.0111	5.65
P. viridiflava	8	8	0.0523	0.0578	−0.5182	>0.1	0.587	0.0046	12.56

[a] Number of strains, [b] Number of haplotypes, [c] Tajima's D, [d] Statistical significance of Tajima's D

psc.edu/biomed/genedoc), and trimmed to their minimal shared length in GeneDoc. Neighbor-joining (NJ) and Unweighted Pair Group Method with Arithmetic mean (UPGMA) phylogenetic analyses were performed on the combined datasets using MEGA ver3.1 (Kumar et al., 2004), and Phylip version 3.6.2 (Felsenstein, 1993). The trees were rooted with orthologous sequences from *Xanthomonas campestris* pv. vesicatoria 85–10 and *Erwinia carotovora* subsp. atroseptica SCRI1043. Split decomposition analysis was performed with SplitsTree ver3.2 (Huson, 1998). Analyses were performed as described in Sarkar and Guttman (Sarkar and Guttman, 2004). Population genetic analyses were performed with DnaSP ver4.1 (Rozas et al., 2003). Classic skyline plot analyses were performed with Genie ver3.0 (Pybus and Rambaut, 2002) using rooted UPGMA trees.

3   Results and Discussion

3.1   *Phylogenetics*

We report here on the multilocus sequence typing (MLST) of the genus *Pseudomonas*. Four loci totaling 2046 bp of sequence were obtained from 367 Pseudomonads. This strain set is heavily biased to towards *Pseudomonas syringae* and *Pseudomonas aeruginosa* and should be viewed strictly as a preliminary analysis.

Previous analyses (Sarkar and Guttman, 2004; Hwang et al., 2005) have revealed a minimal amount of recombination between the MLST loci; therefore, as in previous studies, we have concatenated the sequences for a combined analysis. A combined neighbor-joining (NJ) phylogenetic analysis reveals well-resolved and bootstrap-supported clades (Fig. 1), which roughly correspond to the *P. syringae*, *P. viridiflava*, *P. fluorescens*, *P. putida*, *P. aeruginosa*, and *P. stutzeri* species complexes. The fully expanded phylogenetic tree is available from the authors upon request.

As previously described (Hwang et al., 2005), the *P. syringae* species complex is comprised of five distinct clades or phylogroups. Phylogroup 1 is largely comprised of pathogens of tomato and Brassicaceous crops. Phylogroup 2 has the highest genetic diversity and broadest host diversity of any *P. syringae* phylogroup.

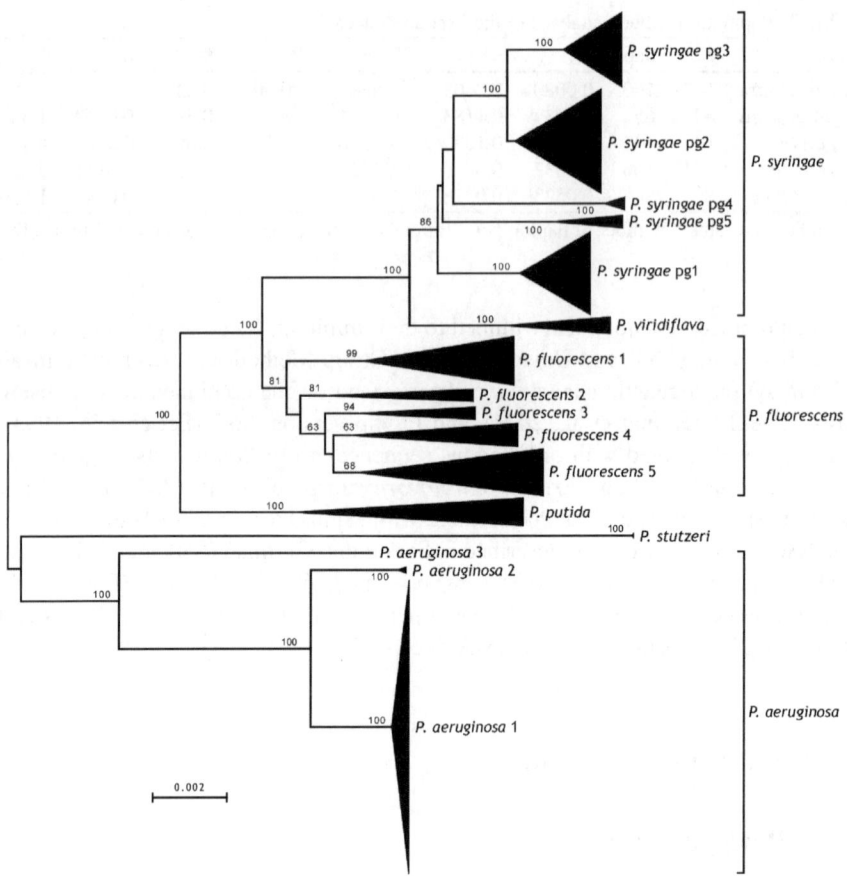

Fig. 1 Neighbor-joining phylogenetic analysis of the Pseudomonads. The back triangles represent clades in the phylogenetic tree. The height of each triangle is proportional to the number of strains. The depth of each triangle is proportional to the genetic diversity found in that clade. The numbers above the nodes are bootstrap scores using 500 pseudoreplicates. The scale bar along the bottom is genetic distance based on a Kimura 2-parameter model. *P. syringae* strains are clustered based on phylogroups previously described (Sarkar and Guttman, 2004; Hwang et al., 2005). *P. fluorescens* strains are clustered based solely on visually obvious clades and are numbered arbitrarily. The full phylogenetic tree is available from the author upon request

The only major host exclusively housed in this clade is for the pea pathogens. Phylogroup 3 is dominated by two closely related clades composed of soybean and kidney bean pathogens. There was one significant recombination event that occurred at or near the *gyrB* locus between phylogroups 2 and 3, which attracts two subclades in these two phylogroups. Phylogroup 4 is composed almost exclusively of monocot pathogens. Phylogroup 5 is largely comprised of radish pathogens.

The *P. syringae* and *P. viridiflava* complexes are very closely related, and form a very strongly supported clade within the Pseudomonads. The strains in both of these groups also have a great deal in common both ecologically and genetically.

Foremost of these similarities is the fact that both of these groups are pathogens that attack the aerial surfaces of plants. Furthermore, both use nearly indistinguishable type III secretion systems to inject virulence-associated effector proteins into their hosts (Goss et al., 2004; Araki et al., 2006). No study has yet investigated the extent of genetic exchange among strains of *P. syringae* and *P. viridiflava*. If extensive exchange is observed, particularly with respect to niche-specific (e.g. virulence-associated) genes, and given the genetic and ecological similarity, then a strong argument could be made for considering *P. viridiflava* as part of the *P. syringae* species complex.

The large *P. aeruginosa* clade stands out for it extreme paucity of genetic diversity. This group of 142 strains includes both clinical and environmental isolates collected from Europe, North and South America, and a lesser extent Asia and Africa. The clinical isolates were taken from a wide range of pathologies, including CF and non-CF pulmonary infections, septicemia and eye infection. It also includes strains carrying metallo-beta-lactamase resistance genes, and even environmental and clinical strains isolated in the late 1800s and early 1900, prior to the widespread use of antibiotics. There are very few clades within this group that significantly resolve (by bootstrap analysis) in this four-gene analysis, and no obvious relationships with respect to geography, clinical or environmental source. The addition of three more MLST loci does provide substantially more resolution (data not shown), but the phylogenetic structure still is minimal compared to the *P. syringae* complex. Four strains diverge from the main body of *P. aeruginosa* isolates. All of these strains have been confirmed to be *P. aeruginosa* via a high temperature – acetaminde assay. The most divergent of these strains was isolated by the Guttman laboratory from an old field site North of Toronto, Canada. The other three strains include both environmental and clinical isolates. One of these (*P. aeruginosa* PA7) is a highly resistant "superbug" from Buenos Aires, Argentina currently being sequenced by The Institute for Genomic Research (TIGR).

The *P. fluorescens* clade is extremely diverse and includes a very large number of named species including *Pseudomonas agarici*, *Pseudomonas marginalis* and *Pseudomonas chlororaphis* among others. In Fig. 1, we have clustered *P. fluorescens* strain into groups based simply on visually obvious divisions. The numbering of these clades is completely arbitrary, and not meant to imply relationships to previously described taxonomic groupings. Although there is a number of significant clades within this complex, the evolutionary relationships of taxonomically described strains provides few clear-cut guidelines for further partitioning this complex. For example, the small clade of six strains labeled *P. fluorescens* 2 contains strains previously described as *P. fluorescens*, *P. corrugata*, and *P. putida*.

The last large clade comprises the *P. putida* complex. We believe that this clade is very under-sampled in this current preliminary analysis. In addition to strains previously described as *P. putida*, there are also *P. fluorescens* strains and even the sequenced *Drosophila* pathogen *Pseudomonas entomophila* L48. Many of the remaining strains in this clade were isolated off plant leaves in the Toronto area.

Finally, we have only two strains in the *Pseudomonas stutzeri* group. Further sampling will be required before we can draw any conclusions about this group.

Split decomposition is a phylogenetic method that does not impose a bifurcating branching or tree-like structure on the dataset. It is useful for identifying reticulations or network structure that may be indicative of past recombination events. A split decomposition analysis of the full strain set reveals three major points of strain diversification (Fig. 2A). The *P. aeruginosa* complex is quite distinct from the rest of the Pseudomonads. Nearly all of the strains radiate out from a central point in a classic 'star phylogeny' pattern, although these branches are extremely short. The *P. syringae* complex is most distant from the *P. aeruginosa* complex, and shows the most complex structure both in terms of reticulations and varying branch length. The *P. viridiflava* strains clearly radiate from the same point as the five major *P. syringae* phylogroups, further supporting their inclusion in the *P. syringae* complex. The *P. fluorescens* complex falls between *P. syringae* and *P. aeruginosa*. This group radiates out from a central point in a 'star phylogeny' form similar to *P. aeruginosa*, except in this case the branch lengths are extremely long. The *P. putida* clade radiates from the same point as the *P. fluorescens* strain. A further examination of the *P. fluorescens* – *P. putida* radiation reveals some network interconnections, suggesting that there has been a small degree of recombination between these two groups (Fig. 2B).

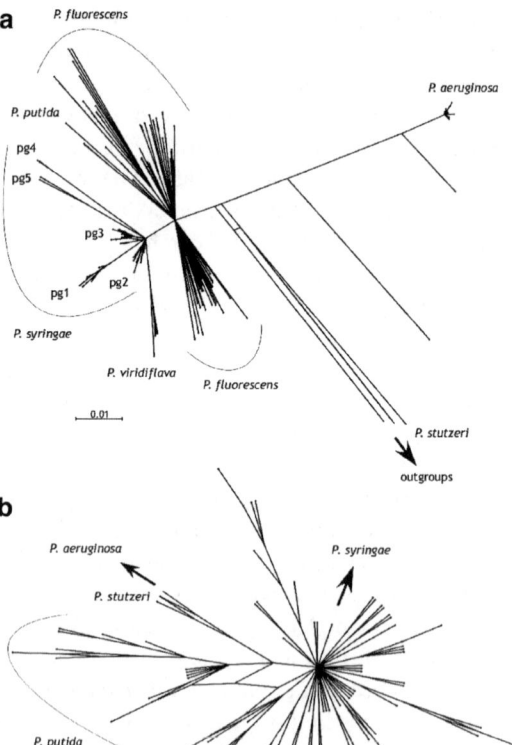

Fig. 2 Split decomposition analysis of the Pseudomonads.
a. Split decomposition analysis using the full dataset with branches proportional to the Hamming genetic distance.
b. Analysis of the *P. fluorescens* and *P. putida* clade with equal branch lengths showing network structure between *P. putida* and *P. fluorescens* strains

3.2   Population Genetics

A population genetic analysis of the major *Pseudomonas* species complexes reveals some remarkable differences. π and θ are two measures of intraspecific genetic variation (polymorphism). The former is based on pairwise comparisons of nucleotide diversity, while the latter is based on the number of segregating sites in the population. Table 2 shows that while the 142 *P. aeruginosa* isolates have 100 haplotypes, both measures of polymorphism in this species are substantially lower than the corresponding measures from the other species. The greatest difference is with respect to π, which is better at taking into account the frequency of polymorphisms than θ. The Tajima's D statistic (Tajima, 1989) evaluates discrepancies between θ and π to identify genes under positive or diversifying selection. Significant positive Tajima's D values indicate an excess of polymorphisms at intermediate frequency, and is typically associated with diversifying selection, while significant negative Tajima's D values indicate an excess of rare variants and is associated with purifying selection or selective sweeps. A selective sweep is the reduction or elimination of neutral polymorphism in a population as a result of its linkage to a variant that is increasing in frequency due to strong positive selection. *P. aeruginosa* has a very strongly significant negative Tajima's D, supporting a relatively recent selective sweep. This conclusion is further supported by the star phylogeny observed for this species, which is also associated with selective sweeps. None of the other *Pseudomonas* species complexes shows similar patterns, although all but *P. syringae* are negative. The positive, albeit non-significant, Tajima's D value for *P. syringae* indicates a slight excess of intermediate frequency or long-lived polymorphisms.

A further striking difference between the species complexes is with respect to the rate of recombination observed among the housekeeping loci. We used a recombination metric normalized by the rate of neutral mutation, which permits us to determine the relative probability that any single nucleotide will be changed due to mutation versus recombination. This value is obtained by taking the ratio of polymorphism, θ ($=2N_e\mu$), to the population recombination rate ρ ($=2N_e r$), where N_e is the effective population size, μ is the mutation rate, and r is the recombination rate. The most dramatic difference observed among the species complexes is that there was effectively no recombination observed among the *P. aeruginosa* isolates. The θ/ρ ratio for the other species ranged from a low of 1.38 for *P. fluorescens* to a high of 12.56 for *P. viridiflava*, indicating over a ten-fold difference in the relative influence of recombination, but overall, a much greater role for mutation in generating diversity than recombination.

3.3   Demographic History

We estimated the demographic history of each of the *Pseudomonas* species complexes using skyline plots. Skyline plots are coalescence-based analyses of demographic history, which provide a visual method to determine if population sizes

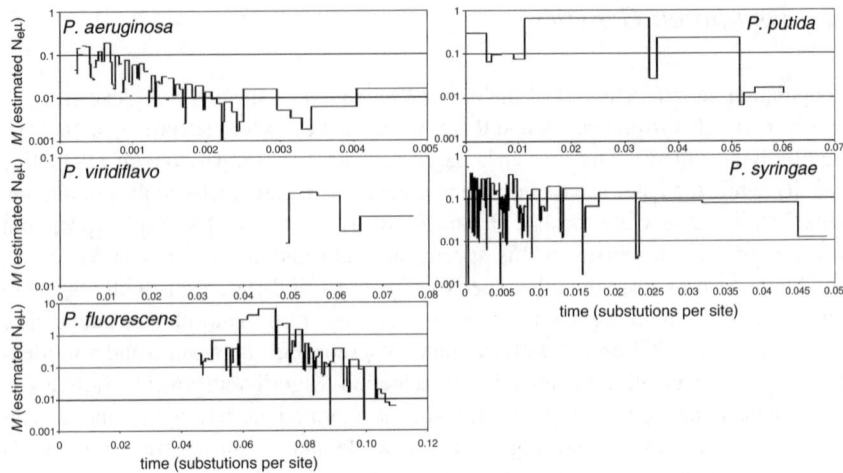

Fig. 3 Classic skyline plots. See text for details

have been constant or changed during recent evolutionary history. Skyline plots chart time as the number of substitutions per site along the X-axis, and an estimate of polymorphism ($M = N_e\mu$) along the Y-axis. Skyline plots that are roughly parallel to the X-axis indicate a constant population size, which is commonly associated with an endemic pathogen, while those that drop off as time increases indicate population expansion, which is commonly associated with epidemic pathogens. Sarkar and Guttman (2004) performed a skyline plot analysis of *P. syringae* that determined that the species had maintained roughly a constant population size through time, and the new analysis agrees with this assessment (Fig. 3). The skyline plot for *P. aeruginosa* shows a quite dramatic increase in population size all the way to the present, which is consistent with exponential population growth and a recent selective sweep. *P. viridiflava* appears to have a plot most consistent with roughly constant population size through time, although the sampling in this clade is currently not deep enough for a complete analysis. The skyline plot for *P. putida* is a bit difficult to interpret, but may indicate a rapid expansion in the distant past followed by roughly constant population size. Finally, *P. fluorescens* shows a period of extremely rapid exponential population growth in the distant past followed by a very long period of static population size.

4   Conclusions

There is currently a tremendous degree of debate as to whether bacterial species actually exist, or if the frequency of horizontal exchange is so high that it effectively eliminates species boundaries. We believe that the preponderance of MLST studies have fairly conclusively shown that for most bacteria, species boundaries do exist, although they are frequently porous. The next logical question is what accounts for the cohesion of these boundaries. MLST analyses can help tease apart the relative importance of genetics verses ecology in maintaining these boundaries.

This preliminary MLST analysis reveals some of fascinating complexities of the genus *Pseudomonas*, and provides insight into the nature of bacterial species. The phylogenetic analysis showed that although major clades corresponding to *P. syringae*, *P. viridiflava*, *P. fluorescens*, *P. putida*, *P. aeruginosa*, and *P. stutzeri*, could be identified, there may be significant gene flow between *P. syringae* and *P. viridiflava*, and perhaps between *P. fluorescens* and *P. putida*. Furthermore, the split decomposition analysis revealed three primary radiations in this genus, clustering *P. syringae* and *P. viridiflava*, and *P. fluorescens* and *P. putida*.

Both phylogenetic and population genetic analyses indicate that *P. aeruginosa* has undergone a very recent (in evolutionary terms) selective sweep, which purged nearly all of the standing genetic variation in the species. The cause of this sweep is not known, but it is intriguing to imagine that it may have to do with adaptation to the human niche. *P. fluorescens* also has undergone a strong selective sweep, although this event appears to be quite ancient. Finally, as reported by Sarkar and Guttman (2004), *P. syringae* appears to be an endemic pathogen, that has maintained a relatively stable population among its plant host for a very long time.

Much more extensive strain sampling will need to be performed in the *P. putida* and *P. stutzeri* complex before strong conclusions can be drawn. It will be particularly interesting to investigate the extent and evolutionary significance of gene flow between *P. fluorescens* and its sister taxa *P. syringae* and *P. putida*, with whom it shares numerous ecological similarities.

One feature common to all of the Pseudomonads is that there is very little horizontal genetic exchange in the core genome. This restricted recombination results in the Pseudomonads being one of the most highly clonal groups studied. In some species such as *P. aeruginosa*, this clonality could simply be due to the recent selective sweep, but this explanation can not account for the strong clonality observed in *P. syringae*. Given the consistently low rates of recombination observed among these species, it is quite possible that there is a more fundamental genetic basis for this observation, and perhaps this accounts for the species complexes observed today.

Acknowledgements D.S. Guttman is supported by funding from the Canada Research Chairs program, the Natural Sciences and Engineering Research Council of Canada, and Performance Plants Inc. of Kingston, Ontario.

References

Anonymous (2002). Chapter 267: cystic fibrosis. *The Merck Manual*. Whitehouse Station, NJ, Merck and Co.

Araki, H., D. Tian, E. M. Goss, K. Jakob, S. S. Halldorsdottir, M. Kreitman and J. Bergelson (2006). Presence/absence polymorphism for alternative pathogenicity islands in *Pseudomonas viridiflava*, a pathogen of Arabidopsis. *Proc. Natl. Acad. Sci. USA* **103**(15): 5887–5892.

Cooper, J. E. and E. J. Feil (2004). Multilocus sequence typing – what is resolved? *Trends Microbiol.* **12**(8): 373–377.

Curran, B., D. Jonas, H. Grundmann, T. Pitt and C. G. Dowson (2004). Development of a multilocus sequence typing scheme for the opportunistic pathogen *Pseudomonas aeruginosa. J. Clin. Microbiol.* **42**(12): 5644–5649.

Dobrindt, U. and J. Hacker (2001). Whole genome plasticity in pathogenic bacteria. *Curr. Opin. Microbiol.* **4**(5): 550–557.

Enright, M. C. and B. G. Spratt (1999). Multilocus sequence typing. *Trends Microbiol.* **7**(12): 482–487.

Felsenstein, J. (1993). *PHYLIP (Phylogeny Inference Package).* Seattle, WA, Distributed by author, Department of Genetics, University of Washington.

Goss, E. M., M. E. Kreitman and J. Bergelson (2004). Genetic diversity, recombination, and cryptic clades in *Pseudomonas viridiflava* infecting natural populations of *Arabidopsis thaliana. Genetics* **16**: 16.

Hacker, J. and J. B. Kaper (2000). Pathogenicity islands and the evolution of microbes. *Annu. Rev. Microbiol.* **54**: 641–679.

Harbottle, H., D. G. White, P. F. McDermott, R. D. Walker and S. Zhao (2006). Comparison of multilocus sequence typing, pulsed-field gel electrophoresis, and antimicrobial susceptibility typing for characterization of *Salmonella enterica* serotype Newport isolates. *J. Clin. Microbiol.* **44**(7): 2449–2457.

Hogan, D. A. and R. Kolter (2002). Pseudomonas-Candida interactions: an ecological role for virulence factors. *Science* **296**(5576): 2229–2232.

Huson, D. H. (1998). Splits tree: analyzing and visualizing evolutionary data. *Bioinformatics* **14**(1): 68–73.

Hwang, M. S. H., R. L. Morgan, S. F. Sarkar, P. W. Wang and D. S. Guttman (2005). Phylogenetic characterization of virulence and resistance phenotypes in *Pseudomonas syringae. Appl. Environ. Microbiol.* **71**(9): 5182–5191.

Jakob, K., E. M. Goss, H. Araki, T. Van, M. Kreitman and J. Bergelson (2002). *Pseudomonas viridiflava* and *P. syringae* – natural pathogens of *Arabidopsis thaliana. Mol. Plant-Microbe Interact.* **15**(12): 1195–1203.

Jander, G., L. G. Rahme and F. M. Ausubel (2000). Positive correlation between virulence of *Pseudomonas aeruginosa* mutants in mice and insects. *J. Bacteriol.* **182**(13): 3843–3845.

King, E. O., M. K. Ward and D. E. Raney (1954). Two simple media for the demonstration of phycocyanin and fluorescin. *J. Lab. Clin. Med.* **44**: 301–307.

Kotetishvili, M., O. C. Stine, Y. Chen, A. Kreger, A. Sulakvelidze, S. Sozhamannan and J. G. Morris, Jr. (2003). Multilocus sequence typing has better discriminatory ability for typing *Vibrio cholerae* than does pulsed-field gel electrophoresis and provides a measure of phylogenetic relatedness. *J. Clin. Microbiol.* **41**(5): 2191–2196.

Kumar, S., K. Tamura and M. Nei (2004). MEGA3: integrated software for molecular evolutionary genetics analysis and sequence alignment. *Brief. Bioinf.* **5**(2): 150–163.

Mahajan-Miklos, S., M. W. Tan, L. G. Rahme and F. M. Ausubel (1999). Molecular mechanisms of bacterial virulence elucidated using a *Pseudomonas aeruginosa Caenorhabditis elegans* pathogenesis model. *Cell* **96**(1): 47–56.

Maiden, M. C. J., J. A. Bygraves, E. Feil, G. Morelli, J. E. Russell, R. Urwin, Q. Zhang, J. J. Zhou, K. Zurth, D. A. Caugant, I. M. Feavers, M. Achtman and B. G. Spratt (1998). Multilocus sequence typing: a portable approach to the identification of clones within populations of pathogenic microorganisms. *Proc. Natl. Acad. Sci. USA* **95**(6): 3140–3145.

Nallapareddy, S. R., R. W. Duh, K. V. Singh and B. E. Murray (2002). Molecular typing of selected *Enterococcus faecalis* isolates: pilot study using multilocus sequence typing and pulsed-field gel electrophoresis. *J. Clin. Microbiol.* **40**(3): 868–876.

Office of Laboratory Security (2001). *Material Safety Data Sheet – Infectious Substances:* Pseudomonas *spp.* Ottawa, Health Canada.

Peacock, S. J., G. D. de Silva, A. Justice, A. Cowland, C. E. Moore, C. G. Winearls and N. P. Day (2002). Comparison of multilocus sequence typing and pulsed-field gel electrophoresis as tools for typing *Staphylococcus aureus* isolates in a microepidemiological setting. *J. Clin. Microbiol.* **40**(10): 3764–3770.

Plotnikova, J. M., L. G. Rahme and F. M. Ausubel (2000). Pathogenesis of the human opportunistic pathogen *Pseudomonas aeruginosa* PA14 in Arabidopsis. *Plant Physiol.* **124**(4): 1766–1774.

Pukatzki, S., R. H. Kessin and J. J. Mekalanos (2002). The human pathogen *Pseudomonas aeruginosa* utilizes conserved virulence pathways to infect the social amoeba *Dictyostelium discoideum*. *Proc. Natl. Acad. Sci. USA* **99**(5): 3159–3164.

Pybus, O. G. and A. Rambaut (2002). GENIE: estimating demographic history from molecular phylogenies. *Bioinformatics* **18**(10): 1404–1405.

Rahme, L. G., E. J. Stevens, S. F. Wolfort, J. Shao, R. G. Tompkins and F. M. Ausubel (1995). Common virulence factors for bacterial pathogenicity in plants and animals. *Science* **268**(5219): 1899–1902.

Rahme, L. G., M. W. Tan, L. Le, S. M. Wong, R. G. Tompkins, S. B. Calderwood and F. M. Ausubel (1997). Use of model plant hosts to identify *Pseudomonas aeruginosa* virulence factors. *Proc. Natl. Acad. Sci. USA* **94**(24): 13245–13250.

Revazishvili, T., M. Kotetishvili, O. C. Stine, A. S. Kreger, J. G. Morris, Jr. and A. Sulakvelidze (2004). Comparative analysis of multilocus sequence typing and pulsed-field gel electrophoresis for characterizing *Listeria monocytogenes* strains isolated from environmental and clinical sources. *J. Clin. Microbiol.* **42**(1): 276–285.

Rowland, B. M. (1999). Pseudomonas infections. *The PDR Encyclopedia of Medicine*. The Thomson Corporation.

Rozas, J., J. C. Sanchez-DelBarrio, X. Messeguer and R. Rozas (2003). DnaSP, DNA polymorphism analyses by the coalescent and other methods. *Bioinformatics* **19**(18): 2496–2497.

Sarkar, S. F. and D. S. Guttman (2004). The evolution of the core genome of *Pseudomonas syringae*, a highly clonal, endemic plant pathogen. *Appl. Environ. Microbiol.* **70**(4): 1999–2012.

Sawada, H., F. Suzuki, I. Matsuda and N. Saitou (1999). Phylogenetic analysis of *Pseudomonas syringae* pathovars suggests the horizontal gene transfer of *argK* and the evolutionary stability of *hrp* gene cluster. *J. Mol. Evol.* **49**(5): 627–644.

Tajima, F. (1989). Statistical methods for testing the neutral mutation hypothesis by DNA polymorphism. *Genetics* **123**: 585–595.

Talbot, G. H., J. Bradley, J. E. Edwards, D. Gilbert, M. Scheld and J. G. Bartlett (2006). Bad bugs need drugs: an update on the development pipeline from the Antimicrobial Availability Task Force of the Infectious Diseases Society of America. *Clin. Infect. Dis.* **42**(5): 657–668.

Thompson, J. D., T. J. Gibson, F. Plewniak, F. Jeanmougin and D. G. Higgins (1997). The CLUSTAL_X windows interface: flexible strategies for multiple sequence alignment aided by quality analysis tools. *Nucleic Acids Res.* **25**(24): 4876–4882.

Vernez, I., P. Hauser, M. V. Bernasconi and D. S. Blanc (2005). Population genetic analysis of Pseudomonas aeruginosa using multilocus sequence typing. *FEMS Immunol. Med. Microbiol.* **43**(1): 29–35.

Wackett, L. P. (2003). *Pseudomonas putida* – a versatile biocatalyst. *Nat. Biotechnol.* **21**(2): 136–138.

Ward, P. G., M. Goff, M. Donner, W. Kaminsky and K. E. O'Connor (2006). A two step chemo-biotechnological conversion of polystyrene to a biodegradable thermoplastic. *Environ. Sci. Technol.* **40**(7): 2433–2437.

Wayne, L. G., D. J. Brenner, R. R. Colwell, P. A. D. Grimont, O. Kandler, M. I. Krichevsky, L. H. Moore, W. E. C. Moore, R. G. E. Murray, E. Stackebrandt, M. P. Starr and H. G. Truper (1987). Report of the ad-hoc-committee on reconciliation of approaches to bacterial systematics. *Int. J. Syst. Evol. Microbiol.* **37**(4): 463–464.

Welch, R. A., V. Burland, G. Plunkett, 3rd, P. Redford, P. Roesch, D. Rasko, E. L. Buckles, S. R. Liou, A. Boutin, J. Hackett, D. Stroud, G. F. Mayhew, D. J. Rose, S. Zhou, D. C. Schwartz, N. T. Perna, H. L. Mobley, M. S. Donnenberg and F. R. Blattner (2002). Extensive mosaic structure revealed by the complete genome sequence of uropathogenic *Escherichia coli*. *Proc. Natl. Acad. Sci. USA* **99**(26): 17020–17024.

Yamamoto, S., H. Kasai, D. L. Arnold, R. W. Jackson, A. Vivian and S. Harayama (2000). Phylogeny of the genus *Pseudomonas*: intrageneric structure reconstructed from the nucleotide sequences of *gyrB* and *rpoD* genes. *Microbiology* **146**(Pt 10): 2385–2394.

Characterization of *Pseudomonas savastanoi* pv. *savastanoi* Strains Collected from Olive Trees in Different Countries

C. Moretti[1], P. Ferrante[1], T. Hosni[2], F. Valentini[1,2], A. D'Onghia[2], M. Fatmi[3], and R. Buonaurio[1]

Abstract To investigate the variability of a *Pseudomonas savastanoi* pv. *savastanoi* population, 53 isolates of the bacterium, isolated in Albania, Italy, Morocco, Portugal and Turkey from knots of several olive cultivars, were characterized at molecular levels by rep-PCR and f-AFLP. All bacterial strains were pathogenic on olive plants and produced fluorescent pigments on King's B medium. They were negative for levan, oxidase, arginine dihydrolase and potato soft rot and induced hypersensitive reaction in tobacco plants. Based on these results and on the amplicon generation of the *iaaL* and the *ptz* genes by PCR, all the isolates can be considered belonging to *P. savastanoi* pv. *savastanoi*. Rep-PCR analysis, performed using ERIC primers and the novel primers KRP2-KRP8 and KRPN2, demonstrated that fingerprints obtained with the novel primers were more polymorphic with respect to those generated with ERIC primers. UPMGA, performed combining data from the 3 primer sets, revealed 26 distinct fingerprints with an overall similarity of about 82%. Most of the isolates from Morocco and Umbria (Italy) as well as the pathovar reference strain of *P. savastanoi* pv. *savastanoi* are grouped in a cluster. In addition, all Turkish and Albanian isolates and most of the isolates from Apulia (Italy) form another cluster. F-AFLP analysis provided a significantly higher resolution than the rep-PCR and revealed high polymorphisms between the isolates. It has the potential to play a major role in characterizing *P. savastanoi* pv. *savastanoi* populations. Virulence tests were performed on the olive cv. Frantoio with 19 bacterial isolates, selected on the basis of rep-PCR fingerprints. Wide virulence variability among the isolates was found. Knot volume significantly correlates with knot weight, allowing the evaluation of disease

[1] Dipartimento di Scienze Agrarie e Ambientali, Università degli Studi di Perugia, via Borgo XX Giugno 74 - 06121 Perugia, Italy

[2] MAIB-CIHEAM The Mediterranean Agronomic Institute of Bari, via Ceglie 9 – 70010 Valenzano (Bari), Italy

[3] Institut Agronomique et Vétérinaire Hassan II, Complexe Horticole d'Agadir, BP. 18/S, Agadir, Morocco

Author correspondence: C. Moretti; e-mail: chiaraluce.moretti@unipg.it

M'B. Fatmi et al. (eds.), *Pseudomonas syringae Pathovars and Related Pathogens.*   321
© Springer Science+Business Media B.V. 2008

progress in a non-destructive manner. Sequencing of genetic traits of the bacterium likely involved in its pathogenicity and virulence is in progress.

Keywords Olive, *Pseudomonas savastanoi* pv. *savastanoi*, rep-PCR, f-AFLP, virulence

1 Introduction

Olive knot caused by *Pseudomonas savastanoi* pv. *savastanoi* represents a serious disease in many olive producing areas of the Mediterranean basin, which can significant affect olive yield and oil quality (Schroth et al., 1968, 1973). Disease symptoms are characterized by knots or galls on different parts of the plant, mainly on twigs and young branches (Young, 2004). Pathogenicity of the bacterium is *hrp/hrc*-dependent (Sisto et al., 2004), while its virulence is mainly due to the phytohormones indole-3-acetic acid (IAA) and cytokinins it produces (Comai and Kosuge, 1980; Nester and Kosuge, 1981; Surico et al., 1985). When examined at the genotypic level by rep-PCR or f-AFLP, populations of the bacterium appeared rather homogeneous (Scortichini et al., 2004; Sisto et al., 2006), while heterogeneity was found at the phenotypic level (Alvarez et al., 1998). Atypical levan-positive and no-fluorescent strains have been described (Marchi et al., 2005).

Our general goal is to determine the genetic variability of a wide population of *P. savastanoi* pv. *savastanoi* we are collecting from olive trees in countries of the Mediterranean basin, Portugal included. We here report the preliminary results obtained on 53 isolates of the bacterium collected in Albania, Italy, Morocco, Portugal and Turkey.

2 Materials and Methods

2.1 Sampling, Bacterial Isolation and Identification

Pseudomonas savastanoi pv. *savastanoi* was isolated from young knots collected from diseased olive trees of different cultivars, grown in several locations of Albania, Italy, Morocco, Portugal and Turkey. The internal water-soaked tissue of knot was excised and crushed in a few drops of sterile water. A loopful of the suspensions was streaked on nutrient agar (NA) and the plates incubated at 27 ± 2°C for 2 days. Bacterial colonies similar in appearance to *Pseudomonas savastanoi* were selected and purified on NA amended with 5% of sucrose (SNA). Fifty-three isolates were obtained and stored in 15% glycerol solution at −80°C (Table 1).

For bacterial identification, all isolates were submitted to the following phenotypic tests, according to the procedures described by Schaad et al. (2001): fluorescence

Table 1 Country, location, olive cultivar and number of isolates of *Pseudomonas savastanoi* pv. *savastanoi*

Country	Location	Olive cultivar	N° of isolates
Albania (AL)	Durazzo	Vayes Durres	3
Albania (AL)	Durazzo	Koker madhi i Durres	3
Albania (AL)	Durazzo	Ulliri i Durrsit	1
Albania (AL)	Berat	Syzeze Berat (K.M. Iberatit)	2
Albania (AL)	Lushnye	Zhàme Lushye	2
Italy (IU)	Umbria	Frantoio	4
Italy (IA)	Apulia	Oliva rossa	3
Italy (IA)	Apulia	Cellina di Nardò	3
Italy (IA)	Apulia	Ogliarola Garganica	7
Morocco (MO)	Marrakesh	Picholine marocaine	5
Morocco (MO)	Meknes	Picholine marocaine	8
Portugal (PO)	Castelo Branco	Cordovil	1
Turkey (TU)	Marmara	Kalamata	2
Turkey (TU)	Akdeniz	Gemlink	9

on King's B medium (KB); levan production, oxidase activity, potato soft rot, arginine dihydrolase activity and tobacco hypersensitivity (LOPAT tests). The strain LMG 2209[T] of *P. savastanoi* pv. *savastanoi*, obtained from the BCCM (The Belgian Coordinated Collections of Microorganisms, Gent, Belgium) was used as pathovar reference strain.

Identification of the bacterial isolates was also done by amplifying the *iaaL* and *ptz* genes, which are specific for *P. savastanoi* and code for an IAA-lysine synthetase and an isopentenyl transferase, respectively. DNA was extracted from bacterial cells, grown in Luria-Bertani broth for 16 h at $27 \pm 2°C$ in an orbital shaker at 200 rpm, with the GenElute Bacterial Genomic DNA Kit (Sigma-Aldrich, St. Louis, Missouri, USA), following the manufacturer's instructions. PCR was carried out using the primers iaaLFor and iaaLRev for the *iaaL* gene (Penyalver et al., 2000) and the primers ptzFor (5′-TTATTCTTGAGGGGGGGTC-3′) and ptzRev (5′-CGATATCCGTCAATATCTT-3′) for the *ptz* gene, which had been designed on the *ptz* gene sequence (NCBI = X03679) of strain LMG 2209[T] of *P. savastanoi* pv. *savastanoi*.

PCR amplifications were carried out in a final volume of 50 μl, using the following amplification mixture: 50 ng of total DNA; 1x PCR buffer (KCl 50 mM, Tris-HCl 10 mM, pH 8.3); 1.5 mM $MgCl_2$; 200 μM of each dNTP; 10 μM of each primer; 2U of Taq DNA Polymerase (Invitrogen, California, USA).

PCR was performed in the MyCycler (Bio-Rad) thermal cycler. Cycling conditions were: 5 min at 95°C, 35 cycles of 30 s at 94°C, 45 s at 69°C for the *iaaL* gene or 54°C for the *ptz* gene, and 50 s at 72°C, followed by 5 min at 72°C. Amplification products (495 and 251 bp for the *iaaL* and the *ptz* genes, respectively) were separated on a 1.2 agarose gel in TAE buffer (Tris 242 g l⁻¹; acetic acid 57.1 ml l⁻¹; Na_2EDTA 100 ml l⁻¹, pH 8.0) at 120 V for 2 h, stained with 0.75 mg l⁻¹ ethidium bromide, and visualized by UV transillumination.

2.2 Pathogenicity and Virulence Tests

Pathogenicity of all isolates was tested using 6-month-old olive (cv. Frantoio) plants. To prepare the inoculum, bacteria were grown on NA medium at $27 \pm 2°C$ for 24–48 h, suspended in sterile deionised water and spectrophotometrically adjusted to 10^8 cfu ml^{-1}. Ten microlitres of the bacterial suspensions or water (control plants) were placed in wounds (two per plant) made in the bark of olive stems with a sterile scalpel. Wounds in inoculated and control plants were protected with Parafilm M (American National Can, Chicago, Illinois, USA). Plants were maintained in transparent polycarbonate boxes to reach high RH values (90–100%) and kept in a growth chamber at 22–24°C, 70 μE m^{-2}s^{-1} illumination and 12 h light period. One–two months after the inoculation, disease symptoms were recorded.

The virulence test was preliminarily carried out only on 19 bacterial isolates, representative of each rep-PCR fingerprint (see Fig. 1), using 1-year-old olive (cv. Frantoio) plants (two plants per isolate). Inoculum preparation and inoculation were

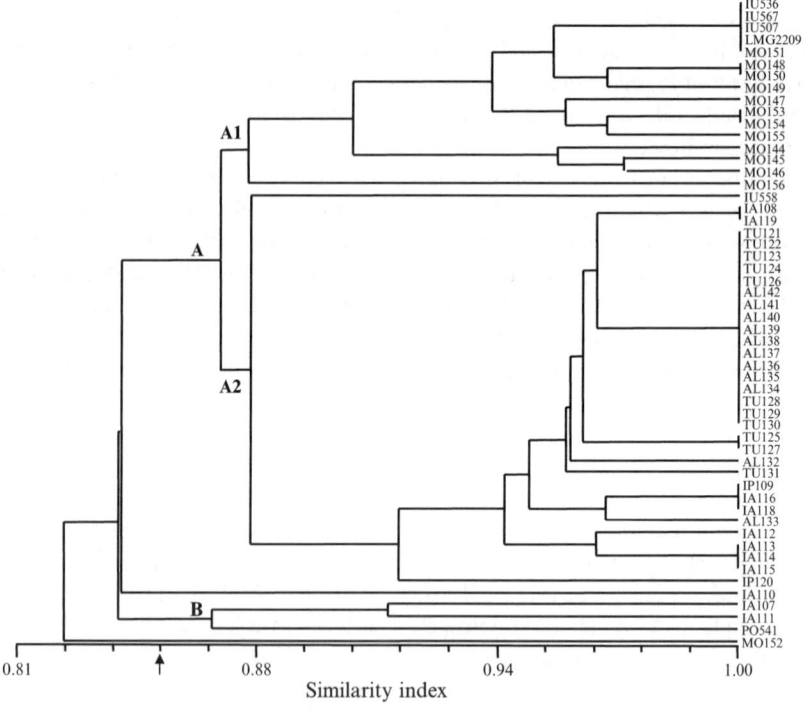

Fig. 1 UPGMA dendrogram and genetic relatedness of 53 isolates of *Pseudomonas savastanoi* pv. *savastanoi* obtained by rep-PCR analysis and ERIC, KRP2-KRP8 and KRPN2 primer sets. Bacterial isolates were collected from Albania (AL), Italy (I; Umbria region = IU; Apulia region = IA), Morocco (MO), Portugal (PO) and Turkey (TU). The arrow indicates the cutoff value (85%) for cluster analysis. LMG 2209^T = *Pseudomonas savastanoi* pv. *savastanoi* pathovar reference strain

carried out as above described except for three inoculation sites per plant were made. Three months after the inoculation, disease severity was recorded by determining the fresh weights of knots as well as knot volumes, as calculated by measuring length, width (subtracting the stem diameters measured above and under the knot) and height of the knot with a vernier caliper.

2.3   Rep-PCR and f-AFLP

All bacterial isolates and the strain LMG 2209[T] of *P. savastanoi* pv. *savastanoi* were subjected to rep-PCR analysis according to the procedure described by Rademaker and De Bruijn (1997), using the primers ERIC 1R-ERIC2 and the novel primers KRP2-KRP8 and KRPN2, designed by Tsygankova et al. (2004). DNA was extracted as reported in Section 2.1.

F-AFLP analysis was performed according to Vos et al. (1995) on all bacterial isolates and on the strain LMG 2209[T] of *P. savastanoi* pv. *savastanoi*. *Pseudomonas savastanoi* pv. *phaseolicola* (IPV-BO 1510 and 1676), *Pseudomonas syringae* pv. *syringae* (LMG 1247[T]), *Pseudomonas syringae* pv. *tomato* (NCPPB 1106[T]) and *Pseudomonas viridiflava* (LMG 2352[T]) were also analyzed for comparison. Total bacterial DNA (500 ng) was digested with *Eco*RI and *Mse*I restriction enzymes (New England Biolabs, Beverly, Massachusetts, USA) and the double stranded DNA adaptors were ligated to both ends of the restriction fragments (Table 2). For pre-amplifications, *Eco*RI and *Mse*I primers containing one selective nucleotide at the 3′-end were used (Table 2). The selective amplifications were performed using the FAM-*Eco*RI and *Mse*I primers with three and two selective nucleotides at the 3′-end, respectively (Table 2). The FAM-labelled fragments were separated by an ABI Prism 377 genetic analyzer.

DNA fingerprints obtained with rep-PCR were visually scored and the presence/absence of bands was collated into a binary data matrix, while the free-share software Genographer (version 1.6.0, J.J. Benham, Montana State University, USA)

Table 2 Adaptors and PCR primers used in f-AFLP analysis

Adaptors and primers	Sequences
Adaptors	
*Eco*RI	5′-CTCGTAGACTGCGTACC-3′
	3′-CTGACGCATGGTTAA-5′
*Mse*I	5′-GACGATGAGTCCTGAG-3′
	3′-TACTCAGGACTCAT-5′
Primers	
*Eco*RI + 1	5′-GACTGCGTACCATTCC-3′
FAM-*Eco*RI + 3	5′-GACTGCGTACCATTCC**CAG**-3′
FAM-*Eco*RI + 3	5′-GACTGCGTACCATTCC**CAA**-3′
*Mse*I + 1	5′-GATGAGTCCTGAGTAA**G**-3′
*Mse*I + 2	5′-GATGAGTCCTGAGTAA**GC**-3′
*Mse*I + 2	5′-GATGAGTCCTGAGTAA**GA**-3′

was used for scoring AFLP fingerprints. Cluster analysis was carried out on similarity matrices generated using the Dice's coefficient (Dice, 1945) and subjected to the unweighted pair-group method with arithmetic average (UPGMA) clustering algorithm, using the NTSYSpc software (Exeter Software, New York, USA), version 2.1.

3 Results and Discussion

All bacterial isolates produced fluorescent pigments when grown on KB medium. After 6 days of culture on SNA, they generated colonies white to pale yellow colour, circular (0.5–2.9 mm in diameter), and resembling a 'fried egg'. Indeed, they presented a slightly raised matt center and a flat transparent waved edge.

All isolates were negative for levan, oxidase, arginine dihydrolase and potato soft rot, and induced hypersensitive reaction when inoculated in tobacco plants (LOPAT tests).

PCR analysis revealed that all the bacterial isolates tested and the reference pathovar strain LMG 2209^T of *P. savastanoi* pv. *savastanoi* generated amplicons of the expected size for the *iaaL* and *ptz* genes (*data not shown*).

When inoculated in olive plants, all bacterial isolates induced knot formation in correspondence with the inoculation sites. On the basis of morphological, biochemical, physiological and pathogenicity tests as well as *iaaL* and *ptz* gene amplifications, all the isolates tested can be considered belonging to *Pseudomonas savastanoi* pv. *savastanoi*.

The genomic rep-PCR profiles consisted of bands ranging in size from 300 to 4000 bp (*data not shown*). Fingerprints obtained with this analysis and the KRP2-KRP8 and KRPN2 primer sets were more polymorphic with respect to those generated with the ERIC primers. UPGMA analysis was performed on a similarity matrix obtained combining data from the three primer sets and using Dice's coefficient. A cophenetic value of 0.87 was determined for this matrix, indicating a high goodness-of-fit for the cluster analysis. The dendrogram obtained revealed 26 distinct fingerprints with an overall similarity of about 82% among the *P. savastanoi* pv. *savastanoi* isolates tested, including the pathovar reference strain (Fig. 1). Very similar results were obtained by Scortichini et al. (2004), who examined a wide Italian population of *P. savastanoi* pv. *savastanoi* with rep-PCR. At an arbitrary 85% similarity cutoff level, it is possible to distinguish two main clusters (A and B) in the dendrogram (Fig. 1). Forty-nine isolates (92%) and the pathovar reference strain of *P. savastanoi* pv. *savastanoi* are in the cluster A, while the isolates IA107, IA111 and PO541 are in the cluster B. The isolate MO152 is not grouped in any cluster. Furthermore, two sub-clusters (A1 and A2) are evident into the cluster A. Most of the isolates from Morocco (12 of 13) and Umbria (Italy) (three of four) as well as the pathovar reference strain are grouped in the sub-cluster A1. All Turkish and Albanian isolates (22) and most of the isolates from Apulia (Italy) (10 of 13) are in the sub-cluster A2. Noteworthy is the high similarity (95–100%) of the isolates collected in Albania and Turkey. No obvious relationship was detected between isolates and the olive cultivars (*data not shown*).

F-AFLP analysis generated complex fingerprints, which consisted of 94 distinct fragments ranging in size from 100 to 500 bp. A cophenetic value of 0.91 was determined for the similarity matrix, indicating a high goodness-of-fit for the cluster analysis. Cluster analysis, performed with the Dice's coefficient, revealed high polymorphism between the isolates (24–92% of similarity) (Fig. 2). A big cluster (cluster A), including 81% of the total isolates assayed and the pathovar reference strain of *P. savastanoi* pv. *savastanoi*, and other minor clusters have been distinguished using an arbitrary 52% similarity cutoff level. Cluster A is divided in two sub-clusters (A1 and A2) with similarity indexes ranging from 59% to 92%. Sub-cluster A1 is, in turn, divided in two sub-sub-clusters (A1.1 and A1.2). Sub-sub-cluster A1.1 comprised 12 isolates, which were from Albania (three), Umbria (Italy) (three), Turkey (three), Apulia (Italy) (two) and Morocco (one). All the countries and olive cultivars (26 isolates) are represented in the sub-sub-cluster A1.2. Four isolates from Apulia (Italy) and one from Turkey are in the cluster A2. The pathovar reference strain LMG 2209^T of *P. savastanoi* pv. *savastanoi* generates a distinct fingerprint, which shows a similarity of 52% respect to the isolates of the

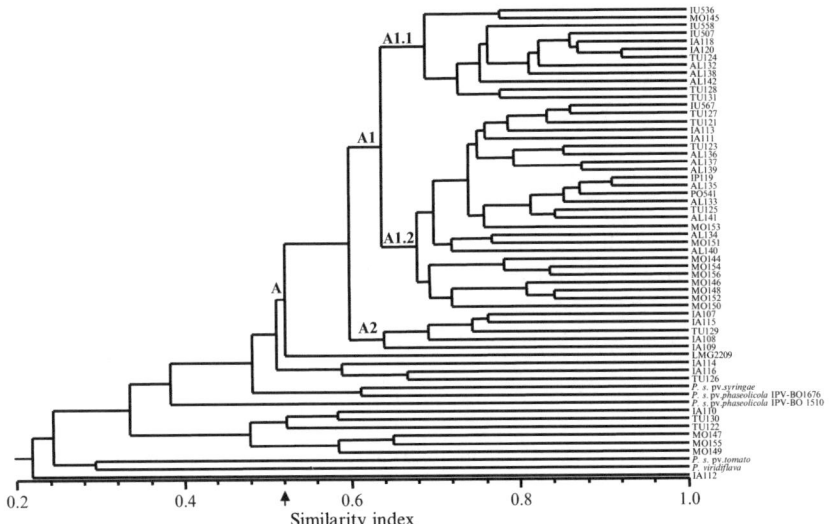

Fig. 2 UPGMA dendrogram and genetic relatedness of 53 isolates of *Pseudomonas savastanoi* pv. *savastanoi* obtained by f-AFLP analysis and *Eco*RI CAG-*Mse*I GA and *Eco*RI CAA-*Mse*I GC primer sets. Bacterial isolates were collected from Albania (AL), Italy (I; Umbria region = IU; Apulia region = IA), Morocco (MO), Portugal (PO) and Turkey (TU). The harrow indicates the cutoff value (52%) for cluster analysis. *Pseudomonas syringae* pv. *syringae* (LMG 1247^T); *Pseudomonas syringae* pv. *tomato* (NCPPB 1106^T); *Pseudomonas savastanoi* pv. *phaseolicola* (IPV-BO 1510, race 2); *Pseudomonas savastanoi* pv. *phaseolicola* (IPV-BO 1676, race 6) and *Pseudomonas viridiflava* (LMG 2352^T). LMG 2209^T = *Pseudomonas savastanoi* pv. *savastanoi* pathovar reference strain

cluster A. Ten isolates of *P. savastanoi* pv. *savastanoi* (four from Apulia-Italy, three from Morocco and three from Turkey) and the strains of *P. savastanoi* pv. *phaseolicola* (IPV-BO 1510 and 1676), *P. syringae* pv. *syringae* (LMG 1247[T]), *P. syringae* pv. *tomato* (NCPPB 1106[T]) and *P. viridiflava* (LMG 2352[T]) are not included in the cluster A.

We demonstrate that knot volume can be considered a suitable parameter for evaluating the virulence of *P. savastanoi* pv. *savastanoi* isolates. Indeed, it significantly correlates ($r^2 = 0.93$; $P \leq 0.01$) with the knot fresh weight, a parameter frequently used for this analysis. In addition, in comparison to knot weight, knot volume determination is a non-destructive assay, which permits evaluation of disease progress. Virulence test performed on 19 *P. savastanoi* pv. *savastanoi* isolates, representing the main fingerprints obtained by rep-PCR, shows a wide variability among the isolates (Table 3). Knot weight and volume of the most virulent isolate (IA111) are 40–45 times higher than those of the less virulent isolate (MO153). We did not find any correlation between degree of virulence and geographic origin of the isolates.

Further research is in progress to study the genetic heterogeneity demonstrated in the present work. Amplification and sequencing of some genetic traits likely involved in pathogenicity and virulence of *P. savastanoi* pv. *savastanoi* may be useful in understanding the diversity among *P. savastanoi* pv. *savastanoi* collected in different countries of the Mediterranean basin.

Table 3 Virulence of selected *Pseudomonas savastanoi* pv. *savastanoi* isolates collected in Albania (AL), Italy (Apulia-IA; Umbria-IU), Morocco (MO), Portugal (PO) and Turkey (TU), determined in olive plants (cv. Frantoio), 3 months after the inoculation

Isolates	Percentage of knots formed	Knot fresh weight (mg)		Knot volume[a] (mm³)	
MO153	16.7	18.33	a	24.18	a
IA120	100.0	31.67	ab	53.52	ab
IA108	100.0	173.33	abc	209.58	abc
MO152	66.7	175.00	abc	200.36	abc
MO156	33.4	193.33	abc	284.61	abc
IA107	100.0	216.67	abc	257.89	abc
IU507	100.0	360.00	abcd	522.80	abcde
AL135	100.0	361.67	abcd	392.55	abcd
TU125	100.0	373.33	abcd	479.19	abcde
IA112	100.0	393.33	bcd	500.61	abcde
TU131	100.0	395.00	bcd	524.64	abcde
AL133	100.0	460.00	cd	668.90	bcde
IA109	100.0	473.33	cd	586.83	abcde
MO154	100.0	476.67	cd	626.37	abcde
MO155	100.0	511.67	cd	765.03	cde
PO541	100.0	536.67	cd	786.96	cde
MO144	100.0	548.33	cd	815.02	cde
AL132	100.0	638.33	d	938.49	de
IA111	100.0	743.33	d	1090.83	e

[a] Volume was calculated measuring knot length, width and height with a vernier caliper. Fresh weigh and volume values are the means of 6 replicates. Data followed by the same letters are not statistically different at $p \leq 0.05$ (Duncan's multiple range test).

Acknowledgements This research has been supported in part by the FISR project: 'Miglioramento delle proprietà sensoriali e nutrizionali di prodotti alimentari di origine vegetale relative alla prima e seconda trasformazione'. The Authors thank Prof. Luigi Russi for helpful discussion on statistical analysis.

References

Alvarez F., Delosrios J., Jimenez P., Rojas A., Reche P., Troya M. T. (1998). Phenotypic variability in different strains of *Pseudomonas syringae* subsp. *savastanoi* isolated from different hosts. Eur. J. Plant Pathol. 104(6): 603–609.

Comai L., Kosuge T. (1980). Involvement of plasmid deoxyribonucleic acid in indoleacetic acid synthesis in *Pseudomonas savastanoi*. J. Bacteriol. 143: 950–957.

Dice L. (1945). Measurement of the amount of ecological association between species. Ecology 26: 297–302.

Marchi G., Viti C., Giovannetti L., Surico G. (2005). Spread of levan-positive populations of *Pseudomonas savastanoi* pv. *savastanoi*, the causal agent of olive knot, in central Italy. Eur. J. Plant Pathol. 112(2): 101–112.

Nester E. W., Kosuge T. (1981). Plasmids specifying plant hyperplasias. Annu. Rev. Microbiol. 35: 531–565.

Penyalver R., García A., Ferrer A., Bertolini E., López M. M. (2000). Detection of *Pseudomonas savastanoi* pv. *savastanoi* in olive plants by enrichment and PCR. Appl. Environ. Microbiol. 66(6): 2673–2677.

Rademaker J. L. W., De Bruijn F. J. (1997). Characterization and classification of microbes by rep-PCR genomic fingerprinting and computer-assisted pattern analysis. In: Caetano-Anolles G. and Gressfoff P. (eds.). Protocols, application and overviews, pp. 151–171. Wiley, New York.

Schaad N. W., Jones J. B., Chun W. (2001). Initial identification of common genera. In: Schaad N. W., Jones J. B. and Chun W. (eds.). Laboratory guide for identification of plant pathogenic bacteria, pp. 84–120. Third edition. APS, St. Paul, MN.

Schroth M. N., Hildebrand D. C., O'Reilly H. J. (1968). Off-flavor of olives from trees with olive knot tumors. Phytopathology 58: 524–525.

Schroth M. N., Osgood J. W., Miller T. D. (1973). Quantitative assessment of the effect of the olive knot disease on olive yield and quality. Phytopathology 63: 1064–1065.

Scortichini M., Rossi M. P., Salerno M. (2004). Relationship of genetic structure of *Pseudomonas savastanoi* pv. *savastanoi* populations from Italian olive trees and patterns of host genetic diversity. Plant Pathol. 53: 491–497.

Sisto A., Cipriani M. G., Morea M. (2004). Knot formation caused by *Pseudomonas savastanoi* subsp. *savastanoi* on olive plants is *hrp*-dependent. Phytopathology 94: 484–489.

Sisto A., Cipriani M. G., Cerboneschi M., Santilli E., Stea G., Tegli S. (2006). Characterisation of *Pseudomonas savastanoi* pv. *savastanoi* strains isolated from different host plants by f-AFLP. 13° Congresso Nazionale S.I.Pa.V., Foggia, Italy, 12–15 September 2006. Book of abstracts, p. 43.

Surico G., Iacobelis N. S., Sisto S. (1985). Studies on the role of indole-3-acetic acid and cytokinins in the formation of knots on olive and oleander plants by *Pseudomonas syringae* pv. *savastanoi*. Physiol. Plant Pathol. 26: 309–320.

Vos P., Hogers R., Bleeker M., Reijans M., van de Lee T., Hornes M., Friters A., Pot J., Paleman J., Kuiper M., Zabeau M. (1995). AFLP: a new technique for DNA fingerprinting. Nucleic Acids Res. 23(21): 4407–4414.

Tsygankova S. V., Ignatov A. N., Boulygina E. S., Kuznetsov B. B., Korotkov E. V. (2004). Genetic relationships among strains of *Xanthomonas campestris* pv. *campestris* revealed by novel rep-PCR primers. Eur. J. Plant Pathol. 110(8): 845–853.

Young J. M. (2004). Olive knot and its pathogens. Australas Plant Path. 33: 33–39.

Separate Origins and Pathogenic Convergence in *Pseudomonas avellanae* Lineages

M. Scortichini

Abstract *Pseudomonas avellanae* is the causal agent of hazelnut (*Corylus avellana* L.) decline both in northern Greece and central Italy, and two lineages related to the geographic origin of the pathogen have previously been pointed out. In order to verify the possible correlation between the genetic diversity and the geographic distance, forty representative strains, obtained from all the areas where the disease was so far observed, were analysed by combining the data obtained from repetitive-sequence PCR using ERIC and BOX primer sets and IS50-PCR, and statistical methods revealing the genetic diversity among the populations. The Mantel test performed with ERIC-PCR, BOX-PCR and IS50-PCR data revealed that the *P. avellanae* populations that are spatially distant from each other are also genetically dissimilar. The gene flow estimates confirm such data. Virulence tests were carried out on hazelnut trees trained in soils with different pHs. The highest virulence of *P. avellanae* strains was observed when they were inoculated on twigs of trees grown in subacidic soils (pH 4.8). Such data are in agreement with the more severe occurrence of the hazelnut decline both in Greece and Italy in orchards characterized by subacidic soils. The sudden occurrence of hazelnut decline in different geographic areas could represent a case of pathogenic convergence in locally adapted phytopathogenic pseudomonads.

Keywords *Corylus avellana*, acidic soils, Mantel test

1 Introduction

Pseudomonas avellanae (i.e., *P. syringae* pv. *avellanae*) is the causal agent of hazelnut (*Corylus avellana* L.) decline both in northern Greece and central Italy (Psallidas, 1987; Scortichini, 2002). Disease symptoms include the rapid wilting of

C.R.A. – Centro di Ricerca Sperimentale per la Frutticoltura, Via di Fioranello, 52, 00134 Roma, Italy

Author for correspondence: Marco Scortichini; e-mail: mscortichini@yahoo.it

M'B. Fatmi et al. (eds.), *Pseudomonas syringae Pathovars and Related Pathogens.*   331
© Springer Science+Business Media B.V. 2008

the branches and trees that can be observed from spring to autumn. In some circumstances, longitudinal cankers are noticed along the trunk. Interestingly, hazelnut decline occurs in both countries since the end of 1970s, and in the two areas the epidemics are more severe in subacid soils (Scortichini, 2002).

By contrast, the hazelnut cultivars utilized in the two countries are different. In northern Greece, the cultivar "Palaz" was introduced from Turkey at the end of 1960s, and, remarkably, in Turkey *P. avellanae* has not been recorded despite very extensive hazelnut cultivation. In central Italy, the cultivars "Tonda Gentile Romana" and "Nocchione" are locally adapted and cultivated since millennia, and the cultivar "Palaz" is completely unknown. In addition, exchange of hazelnut propagative materials between the two countries did not occur.

These findings prompted investigation of whether some relationships exist between the sudden and contemporary occurrence of the hazelnut decline in the two countries. This paper reports the possible separate origin of the two *P. avellanae* lineages inferred by the molecular typing of representative strains and appropriate statistical tests as well as the different virulence shown by the strains when they were inoculated on hazelnut twigs of trees grown in soils of different pHs.

2 Materials and Methods

2.1 Origin of the Lineages

Forty representative *Pseudomonas avellanae* strains isolated in Greece and Italy in different years and obtained from all the areas where the disease has been observed and representing the two lineages of the pathogen, were molecularly typed using repetitive-sequence PCR and ERIC, BOX and REP primer sets and insertion sequence PCR using the insertion IS50 of the transposon Tn5. Six populations of the pathogen were assessed, each representing the area from which the pathogen was isolated: four populations are from northern Greece and two from central Italy. The data sets were compiled as a matrix of strains and molecular fragments. Subsequently, the resulting data were analysed using statistical methods enabling verification of any correlation between the genetic diversity of the strains and the geographic areas from where the strains were originally isolated. Mean genetic diversity estimates within and among the populations of the two lineages were calculated using: (a) Nei's mean genetic diversity distance; (b) Nei's original measure of genetic identity and genetic distance; (c) Nei's unbiased measures of genetic identity and distance; (d) gene-flow estimates independent of population size; (e) Mantel test to compute the linear correlation between two proximity matrices to reveal whether environmental variables are intercorrelated between themselves. More details are reported by Scortichini et al. (2006).

2.2   Virulence Tests

In order to ascertain the virulence of *P. avellanae* strains, artificial inoculations were performed on twigs of hazelnut plants grown in soils characterized by different pH by using well established techniques (Scortichini et al., 2002). Soils of pH values of 4.8, 5.4, and 6.8 were collected from hazelnut orchards located in the province of Viterbo, central Italy, showing symptoms of bacterial decline. Isolations confirmed the presence of *P. avellanae* in all orchards. Subsequently, healthy hazelnut suckers of "Tonda Gentile Romana" were taken from an old hazelnut orchard of Rome province which never showed symptoms of decline. The suckers were planted in pots, each containing one soil of a certain pH. For each soil, 12 suckers were trained without adding any fertilizer for three years before the inoculation. During this period, the pH value was checked every year. The value never exceeded ±0.2 compared to the original one. *P. avellanae* type-strain BPIC 631, isolated in Greece, and ISPaVe 011, isolated in Italy, were tested. For both strains, inoculum grown for 48 h onto nutrient sucrose agar (NSA) at $25 \pm 1°C$, tenfold dilutions in sterile saline (0.85% of NaCl in distilled water) from $1–2 \times 10^6$ to $1–2 \times 10^4$ CFU ml^{-1} were prepared. For each strain, 10 µl of each suspension was placed with a micropipette on the surface of a leaf scare immediately after removal of the leaf at the base of the petiole. For each strains, each dose and each pH, 20 shoots longer than 25 cm were inoculated in early October. Symptoms were checked in the following spring when the length of twig necrosis was measured. The mean length of necrosis and the standard deviation were calculated.

3   Results

3.1   Origin of the Lineages

A total of 18 ERIC-PCR, 14 BOX-PCR and 28 IS50-PCR products were scored for all 40 *P. avellanae* strains. Both repetitive sequence and IS50 PCR, by showing some fragments unique to particular populations, proved to have a high level of discrimination between the populations of strains isolated from northern Greece and central Italy. The mean values of Nei's genetic diversity, h, for ERIC and BOX repetitive PCR and IS50-PCR data found for the six *P. avellanae* populations are reported in Table 1. Collectively, the genetic diversity values were greater for the strains isolated in central Italy. The measure of genetic identity among the four populations from northern Greece was high and varied from 0.891 to 0.984 with ERIC and from 0.808 to 0.975 with BOX. The genetic distances between the populations from northern Greece and central Italy ranged from 0.226 to 0.452 for ERIC-PCR data and from 0.129 to 0.387 for BOX-PCR data. IS50-PCR data revealed a similar but lower genetic identity among the different populations.

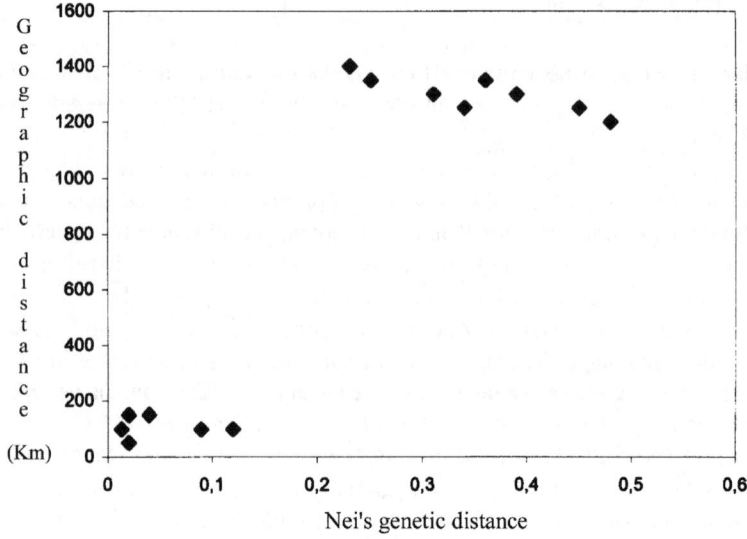

Fig. 1 Plot of geographic distances and Nei's genetic distance derived from ERIC-PCR for the four *Pseudomonas avellanae* populations obtained from Greece and from the two isolated in Italy using the Mantel test. The distribution of correlations from 10,000 permuted samples indicates a significant isolation by distance between the two *P. avellanae* lineages (P = 0.05). The strains from Italy are represented in the upper part of the plots. The points represent the linear correlation observed within the two *P. avellanae* lineages (from Scortichini et al. 2006)

The gene flow estimate, Nm was always <1: 0.007 for ERIC-PCR, 0.26 for BOX-PCR, and 0.10 for IS50-PCR, thus indicating a separate differentiation of the two *P. avellanae* lineages. The value of the Mantel test product–moment correlation coefficient was always positive: r = 0.89 for ERIC-PCR; 0.73 for BOX-PCR and = 0.74 for IS50-PCR, and significantly different from 0 at P = 0.05 (Fig. 1). Collectively, these data indicate that the two *P. avellanae* lineages from northern Greece and central Italy are spatially distant from each other are also genetically dissimilar in terms of genetic differentiation (Scortichini et al., 2006).

3.2 Virulence Tests

A clear relationships between a higher virulence of *P. avellanae* strains and soils with pH lower than 6.0 was observed upon the artificial inoculations (Fig. 2). In fact, by using the same doses, a more extensive twig necrosis was observed in the hazelnut twigs of plants grown in soil with pH values of 5.4 and 4.8. At the highest dose (i.e., 10,000 cells per leaf scar), with both strains, a twig necrosis greater than

20 cm was observed in the twigs of plant grown in soil with pH of 4.8. The virulence of both strains resulted similar and the same trend was observed.

4 Discussion

Data from genetic analysis points out that the geographically distant populations of *P. avellanae* are also dissimilar in terms of genetic diversity: Nei's genetic diversity values as well as gene flow estimates and Mantel tests all indicated such a possibility (Scortichini et al., 2006). This means, in our case, that although all pseudomonads inciting the hazelnut decline belong to *P. avellanae*, the genetic differences found in the two lineages using the neutral markers are, probably, deeply related to the geographic origin of the strains. If these data are considered jointly to the different histories of hazelnut cultivation in Greece and Italy, such findings would support the hypothesis of separate origins for the two lineages of the pathogen. The sole common link, is the sudden appearance of the disease in areas characterized by the presence of subacidic soils (pH < 5.0). Field observations carried out in the two countries are consistent in showing a higher incidence of hazelnut decline in areas characterized by subacidic soils. A similar environment such as the hazelnut trees growing in subacidic soils both in Greece and Italy might have locally selected similar pseudomonads showing pathogenic convergence to *C. avellana*. Data obtained in the present study would indicate a higher virulence of *P. avellanae* strains when inoculated in twigs of hazelnut grown in soil with pH lower than 6.0.

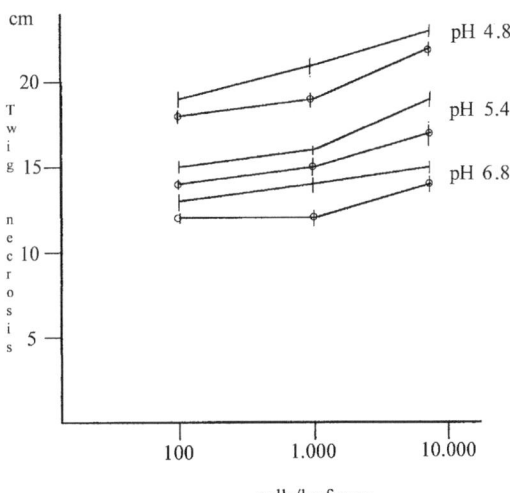

Fig. 2 Virulence of *Pseudomonas avellanae* strains expressed as mean length necrosis of twigs. The plants used for the artificial inoculations were grown for three years in pots containing soils of different pH. I: *P. avellanae* ISPaVe 011, isolated in Italy; °: *P. avellanae* BPIC 631, isolated in Greece. Vertical bar represent standard error

References

Psallidas P.G., 1987, The problem caused by *Pseudomonas syringae* pv. *avellanae* in Greece. Bull. PEPP/EPPO Bull. 17: 257–261.

Scortichini M., 2002, Bacterial canker and decline of Eurpoean hazelnut. Plant Dis. 86: 704–709.

Scortichini M., Marchesi U., Rossi M.P., Di Prospero P., 2002, Bacteria associated with hazelnut (*Corylus avellana* L.) decline are of two groups: *Pseudomonas avellanae* and strains resembling *P. syringae* pv. *syringae*. Appl. Environ. Microbiol. 68: 476–484.

Scortichini M., Natalini E., Marchesi U., 2006, Evidence for separate origins of the two *Pseudomonas avellanae* lineages. Plant Pathol. 55: 451–457.

Genetic Diversity Among Pseudomonad Strains Associated with Cereal Diseases in Russian Federation

E.V. Matveeva[1], A.N. Ignatov[2], V.K. Bobrova[3], I.A. Milyutina[3], A.V. Troitsky[3,2], V.A. Polityko[1], and N.W. Schaad[4]

Abstract Basal glume rot, caused by *Pseudomonas syringae* pv. *atrofaciens* (PSA), has emerged as a major bacterial disease of wheat, barley, and rye in Russia. Forty-nine suspect strains of PSA isolated from diseased cereal plants from different regions of Russia were tentatively identified by production of a fluorescent pigment and confirmed by pathogenicity tests on the host of origin. Each strain was then grouped according to LOPAT (*l*evan, *o*xidase, *p*otato rot, *a*rginine dihydrolase, and *t*obacco hypersensitivity) assays. Ten strains were assigned to LOPAT group 1a, 13 to group 1b, 5 to group 2, 4 to group 3, and 16 to group 5. Strains of each group were then characterized by 16S-23S rRNA Intergenic Transcribed Region (ITR) sequencing and fingerprinted by restriction fragment length polymorphism of ITR, and repetitive PCR using REP, ERIC, and BOX primers. A phylogenetic tree constructed from ITR sequence data revealed two discrete clusters, designated "syringae" and "fluorescens". ERIC-PCR did not work well. However, BOX PCR produced very useful differential genomic fingerprints. There was a high correlation between LOPAT group 1a and BOX PCR patterns. The remaining groups showed a low correlation to BOX PCR patterns and a high level of genetic diversity.

Keywords *Pseudomonas syringae* pv. *atrofaciens*, wheat, barley, rye

[1] Russian Research Institute of Phytopathology, Moscow region, 143080

[2] Centre "Bioengineering", Russian Academy of Sciences, Moscow, 117312

[3] Moscow State University, Moscow, 119992; Russia

[4] USDA-ARS, Foreign Disease-Weed Science Research Unit, Ft. Detrick, MD 21702, USA

Author for correspondence: Alexander Ignatov; e-mail: ignatov@biengi.ac.ru

M'B. Fatmi et al. (eds.), *Pseudomonas syringae Pathovars and Related Pathogens.*　　337
© Springer Science + Business Media B.V. 2008

1 Introduction

Several major bacterial diseases of cereals are emerging as wide-spread problems in Russia. Perhaps most important among them are bacterial leaf blight, caused by *P. syringae* pv. *syringae* (van Hall, 1987), and basal glume rot caused by *Pseudomonas syringae* pv. *atrofaciens* (Young et al., 1978; Dye Wilke 1978). Basal glume-like symptoms have recently been found on cereals (wheat, barley, rye) in every surveyed region of the European part of Russia (Matveeva et al., 2003). Although the Russian Federation occupies a large geographic region with contrasting soil, flora, and climate conditions, little is known of the genetic diversity among plant pathogenic pseudomonads in Russia. For the first time we show considerable genetic diversity exists among strains of *P. syringae* pv. *atrofaciens* in Russia.

2 Materials and Methods

2.1 Bacterial Strains and Phenotypic Identification

Diseased plant samples were collected in several European and Asian regions of Russia and mailed to Moscow (Table 1). Bacteria were isolated and purified on either King et al.'s medium B (KB) or KBC (Braun-Kiewnick and Sands, 2000) agar. After presumptive species identification based on fluorescent colonies on KB agar, each strain was tested for pathogenicity on its corresponding host plant by seed inoculation techniques, as described (Matveeva et al., 2003) Only virulent bacteria were maintained for further study. Each strain was grouped based on LOPAT (Levan production on sucrose, Oxidase reaction, Pectate activity on potato, Arginine dihydrolase activity, and hypersensitivity reaction on Tobacco (*Nicotiana tabacum* L.) and geranium (*Pelargonium zonale* L.) leaves) tests, as described (Lelliott et al., 1966; Braun-Kiewnick and Sands, 2000). All strains were maintained on KB agar slants and stored at −78°C.

2.2 Molecular Characterization

DNA was isolated from bacterial cultures following the method of Ausubel et al. (1995). PCR-amplifications were performed in a Perkin-Elmer 9700 thermo-cycler according to the recommended protocols for each primer combination. For restriction fragment length polymorphism (RFLP) analysis of PCR-amplified 16S-23S ITR DNA fragments, DNA was digested with six restriction endonucleases, *HaeIII*, *TaqI*, *MspI*, *Sau 96I*, *Bst KTI* and *MseI* as recommended by the manufacturer (MBI Fermentas, Lithuania). For 16S-23S ITR sequencing, DNA was amplified by PCR

Table 1 List of strains used in this study

Strain	Plant host	Cultivar	Region
W7, W9	Wheat (*T. aestivum*)	Spartanka	North-Caucasus
W905	–	Demira	–
W912	–	Ofelia	–
W901	–	Pobeda	–
W923	–	Demira	Central Chernozemyi
W918	–	Moskovskaya 5	–
W914, W920	–	Mironovskaya-808	–
W217-4	–	Mironovskaya-61	–
W930	–	Gonor	–
W6005	–	Moskovskaya 5	Nizhne-Volzhski
W427-8	–	Saratovskaya-8	–
W27-10	–	Saratovskaya-55	Sredne-Volzhski
W20-7	–	Saratovskaya-90	–
W922	–	Kazakhskaya-84	–
W237-5	–	Mironovskaya-808	–
W238-4; W238-9	–	Samsar	–
W-194-3	–	Niva-2	Ural
R915	Rye (*S. cereale*)	Orlovskaya-9	Central Chernozemyi
R216-4,7	–	Krona	–
R1/2, R-88	–	Purga	–
R102	–	Chulpan	–
R923	–	Bilina	Volgo-Vyatski
R72	–	Raduga	–
R56, R85	–	Vjatka-2	–
R218	–	Estafeta	–
R200	–	Chulpan	–
R111	–	L-503	–
R1/9	–	–	Sredne-Volzhski
B224-5	Barley (*H.vulgare*)	Odesski-100	Central Chernozemyi
B909	–	Kroshka	North-Caucasus
B92-2	–	Ekolog	–
Bt-5, Bt-6	–	Zazerski-85	–
Bt-8, Bt–21	–	–	Central Chernozemyi
B908	–	–	North-Caucasus
S10	Corn (*Zea mays.*)	Na	Central Chernozemyi
147–15	Oats (*Avena sativa*)	Na	West Siberia

Na, not available

with primers D21 and D22 (Manceau and Horvais, 1997). Additionally, primers B1 and B2 were used to amplify a fragment of *syrB* gene, involved in biosynthesis of syringomycin (Quigley and Gross, 1994). Genomic PCR fingerprinting with repetitive PCR (REP) using ERIC-, BOX-PCR and AP-PCR was performed as described (Louws et al., 1999). Amplified products were electrophoresed and purified from agarose gels with QiaQuick PCR purification kit, as described by the manufacturer (Qiagen, USA). DNA sequencing was carried out on ABI Prism 3100-Avant Genetic Analyzer DNA sequencer (Applied Biosystems, Foster City, CA). Sequences were aligned using CLUSTAL program (Thompson et al., 1994). For the phylogenetic tree, construction was performed using TREECON 1.3b package

(Van de Peer and De Wachter, 1997). Phenetic distance (1-Pearson's r) matrix was calculated based on biochemical traits and BOX-PCR profiles. The raw data were treated by the Factor Analysis by STATISTICA 6.0 program (StatSoft, USA). Phenetic trees were built by UPGMA algorithm using TREECON 1.3b.

3   Results

3.1   Phenotypic Identification

Forty-nine fluorescent pseudomonads, presumptively identified as *P. syringae* pv. *atrofaciens*, were isolated from diseased wheat, rye, barley, corn, and oat plants from different regions of Russia (Table 1). All HR positive highly pathogenic strains typed into each of the five LOPAT groups (Table 2). Eleven strains were typical *P. syringae* (LOPAT group 1). Thirteen strains were levan negative, typical of group 1b. Five were oxidase-negative but pectolytic positive similar to *P. viridiflava* (LOPAT group 2). A large number of strains were oxidase-positive; four belonged to LOPAT group 3 (arginine-dihydrolase negative), and 16 to group 5.

Table 2 Grouping of the fluorescent plant pathogenic pseudomonads according to LOPAT tests[a]

Strain	No	HR	Oxidase	ADG	Levan	Pectin	Group
W7; W9; W913; W918; W920; R1/9; B92-2; R218-3	8	+[b]	–	–	+	–	1a
W914; B908	2	+	–	–	w	–	1a
W905; W427-8; W237-5; W217-4; R915; R216-7; R200; W930; W901	8	+	–	–	w	–	1b
R916; R917; R216-3; R936	5	+	–	–	–	–	1b
S10	5	+	–	+	w	+	2
B224-5		+	–	–	w	+	2
R1/2		+	–	–	w	w	2
W194-3		+	–	–	w	–	2
W912		+	–	–	w	+	2
R111, R102	4	+	+	–	+	–	3
Bt5		+	+	–	–	+	3
W238-9		+	+	–	+	w	3
W922; Bt8; Bt21; 147-15; W6005	16	+	+	+	–	–	5
W237-5; W913		+	+	+	–	+	5
W238-4		+	+	+	+	–	5
R88; R56		+	+	+	+	w	5
R72; W923; R101; R85; B909; Bt6		+	+	+	+	–	5

[a]LOPAT: Levan, oxidase, pectin, arginine dehydrolase (AGD), tobacco hypersensitivity (HR)
[b]Plus, positive; minus, negative; w, weak positive

3.2 Molecular Characterization

RFLP and REP-PCR resulted in multi-band reproducible fingerprint patterns. Most bands were polymorphic among the strains studied. REP- and ERIC-PCR failed to generate enough common bands to result in useful genomic fingerprints. Although BOX-PCR produced the clearest and most differential set of genomic fingerprints, the low number of common bands hindered the resolving of genetic relationships of strains by cluster analysis. Visual inspection of the BOX fingerprints did suggest a division of strains into groups of similar genotypes. Most strains of phenetic LOPAT group 1a represented similar BOX-patterns, other groups contained strains of less similar BOX-patterns. There were several strains of unique BOX-patterns that had no common bands with other strains. The results of these assays indicate a high level of genetic diversity among the *Pseudomonas spp.* infecting cereals in Russia. Using phenotype and molecular data, factor analysis revealed two major groups at 68.16% and 13.58%. Based on factor analysis (Fig. 1) or UPGMA clustering (data not shown) only *P. syringae* typed into a separate group.

RFLP analysis of 16S-23S ITR revealed banding patterns demonstrating the considerable polymorphism between strains (Table 3). All strains typed into two groups based on RFLP patterns for *HaeIII*, *TaqI*, *MspI*, *Sau96I*, and *Bst KTI* restriction endonucleases (Table 4).

Twelve strains had restriction patterns with these endonucleases and differed greatly from those of the other 33 strains (groups I and II, respectively). The nucleotide sequences of ITR of ten *Pseudomonas* strains were determined. D21 and D22

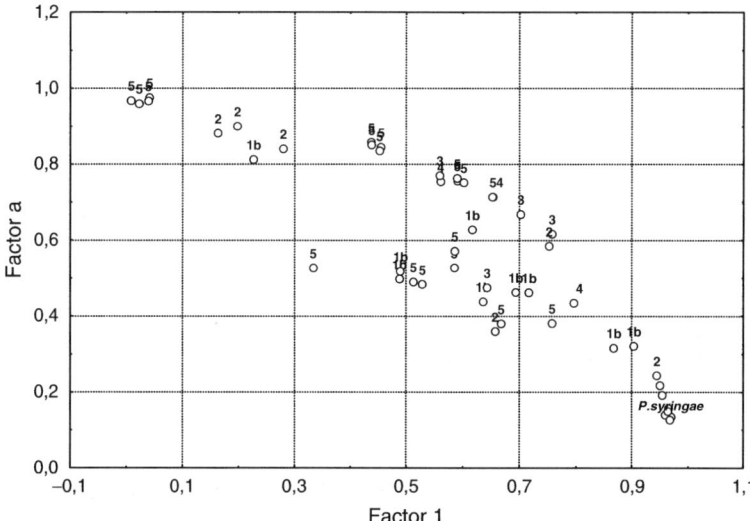

Fig. 1 LOPAT groups 1b, 2, 3, 4 and 5 of *Pseudomonas* strains placed in the co-coordinates of first (68.2%) and second (13.6%) main factors obtained by Factor Analysis of joint biochemical and PCR markers data. Strains of the group 1a (*P. syringae*) are in the right bottom corner

Table 3 Types of restriction patterns (A–G) obtained after digestion of PCR-amplified Intergenic Transcribed Region (ITR) with Mse I endonuclease in all the tested strains of pseudomonads

Fragment size, bp	Type of ITS							
	A1	A2	B	C	D	E	F	G
450				4 50				
400							40 0	
350	3502 0		35 0	3 50	35 0	3 50		
330								33 0
280		2 80						
200	20 0	20 0			20 0			
180					18 0	12 0	1 80	
120			1 20	12 0				
80			8 0	8 0				
70		7 0						7 0

Black field – presence of corresponding length band in ITR profile, blank field – absence

Table 4 Restriction Fragment Length Polymorphisms (RFLP)-PCR profiles of Intergenic Transcribed Region (ITR) in *Pseudomonas* strains obtained with six endonuclease enzymes and corresponding LOPAT grouping

RFLP Group	Strains	HaeIII	TaqI	MspI	AspS91	BstKtI	MseI	LOPAT
I-1	2	A[a]	A	A	D	H	E	1a
I-2	10	A	A	A	D	H	F	1a
II-3	2	B	B	–	B	B	C	1b
II-4	3	B	B	–	B	B	C	1b
II-5	1	B	B	B	A	C	B	5
II-6	2	B	B	–	A	C	D	5
II-7	1	B	B	–	A	E	D	5
II-8	1	B	B	–	E	C	D	4
II-9	2	B	B	–	E	I	D	1b
II-10	1	B	B	–	E	A	B	2
II-11	7	B	B	–	E	A	A1	2
II-12	2	B	B	B	E	I	A1	2
II-13	1	B	B	–	E	I	A1	5
II-14	2	B	B	–	E	D	B	5
II-15	1	B	B	–	E	D	D	5
II-16	3	B	B	–	E	E	D	5
II-17	1	B	B	–	E	G	E	1b
II-18	1	B	B	C	C	F	D	5
Out group	17	–	–	–	–	–	A2	5

[a]Letters A–F correspond to distinct RFLP-PCR profiles in Table 3 obtained with individual enzymes

primers amplified an ITR product of 525–555 bp in length; a single band was visible for most strains but two bands with a difference of about 80 bp were present in some strains (Tables 3 and 4). Based on ITR sequences available from GenBank two main clusters were revealed: "Syringae" and "Fluorescence" with bootstraps of 86% and 96% (Fig. 2).

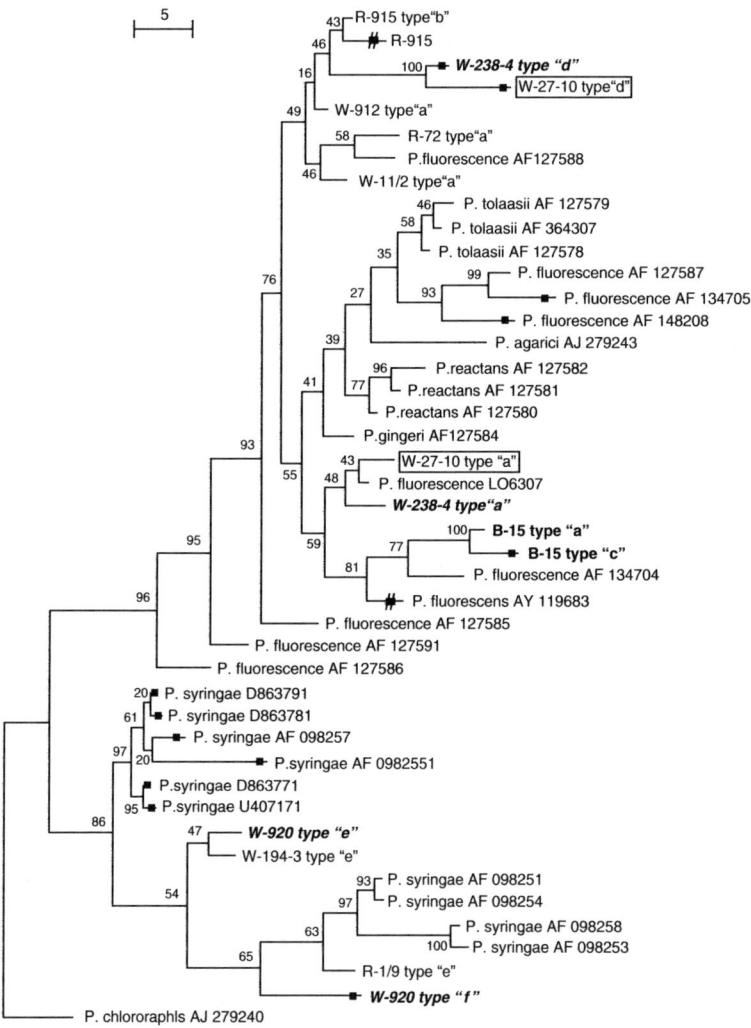

Fig. 2 Evolutionary tree based on 16S-23S Intergenic Transcribed Region (ITR) sequencing, shows 44 operons from 39 *Pseudomonas* strains. Posterior branch support values are indicated near internal nodes. Scale bar for branch length represents a number of expected substitutions per nucleotide site. Pairs of operons from the same strain are marked uniformly (by bold, italic bold, underlined, italic underlined or boxed)

The "Syringae" clade included all *P. syringae* accessions received from GenBank and three strains representative of RFLP pattern group I. The "Fluorescence" clade included all other sequenced strains of *Pseudomonas* isolated from diseased cereals and accessions listed in the GenBank as *Pseudomonas fluorescens, P. tolaasii, P. reactans, P. gingeri,* and *P. agarici*. Strains of *P. tolaasii* and *P. reactans* form subgroups, however, the relatedness was lower than 50%. Therefore, the exact structure of the "Fluorescence" cluster remains unresolved. The sequenced strains from cereals were most related to *P. fluorescens* accessions.

PCR amplification with primers B1 and B2 specific for the syringomycin gene *syrB* resulted in a product of 752 bp length and was amplified from all strains tested. Amplification products of R1/9 strain (RFLP pattern group I) and R1/2 strain (RFLP pattern group II) were cloned and sequenced. The similarity of nucleotide sequences of these products of W-1/9 and W1/2 strains to each other and the sequence of *syrB1* gene from *P. syringae* pv. *syringae* (the GenBank accession # U25130) were 95.0%, 96.5% and 93.5%, respectively.

The sequence data have been submitted to the DDBJ/EMBL/GenBank databases.

4 Discussion

The presence of two bands in the RFLP-PCR analysis of 16S-23S ITR indicated that different types of operons occurred in the genome of these strains. Similar conclusion has been made by Manceau and Horvais (1997) who found two very closely spaced bands after gel electrophoresis of amplified ITR from several strains of *P. syringae*. The results clearly revealed the high diversity among bacteria infecting cereal crops and potatoes in the Russian Federation. Despite a similarity in pathogenicity the strains exhibited a high level of diversity in both biochemical and genetic properties. Based on these results, the classification of pseudomonads pathogenic to wheat based on biochemical characteristics and genome data do not fully agree. Strains of *P. syringae* had some variation in levan production, strains of LOPAT group 2 (*P. viridiflava*) and group 3 differed in levan production and pectolytic activity, and strains of LOPAT group 5 varied in levan, pectin, and arginine dihydrolase tests.

Similar results were described before for atypical strains of *P. viridiflava* isolated from different host plants in Spain (Gonzales et al., 2003).

Acknowledgement   This work was supported in part by ISTC grant #1771p.

References

Ausubel, F.M., Brent, R., Kinston, R.E., Moore, D.D., Seidman, J.D., Smith, J.A., and Struhl, K. (1995) *Short Protocols in Molecular Biology*. Wiley, New York, pp. 2–15.

Braun-Kiewnick, A. and Sands, D.C. (2000) Pseudomonas. In: N.W. Schaad, J.B. Jones, and W. Chun (eds.) *Laboratory Guide for Identification of Plant Pathogenic Bacteria*, 3rd edn. APS, St. Paul, MN.

Gonzales, A.J., Rodicio, M.R., and Mendoza, M.C. (2003) Identification of an emergent and atypical *Pseudomonas viridiflava* lineage causing bacteriosis in plants of agronomic importance in a Spanish region. *Appl. Environ. Microbiol.* 69: 2936–2941.

Lelliott, R.A., Billing, E., and Hayward, A.C. (1966) A determinative scheme for the fluorescent plant pathogenic pseudomonads. *J. Appl. Bacteriol.* 29: 470–489.

Louws, F.J., Rademaker, I.L.W., and de Bruijn F.J. (1999) The three Ds of PCR-based genomic analysis of phytobacteria: diversity, detection and disease diagnosis. *Ann. Rev. Phytopathol.* 37: 81–125.

Manceau, C. and Horvais, A. (1997) Assessement of genetic diversity among strains of *Pseudomonas syringae* by PCR-restriction fragment length polymorphism analysis of rRNA operons with special emphasis on *P. syringae* pv. *tomato*. *Appl. Environ. Microbiol.* 63: 498–505.

Matveeva, E.V., Pekhtereva, E.Sh., Polityko, V.A., Ignatov, A.N., Nikolaeva, E.V., and Schaad, N.W. (2003) Distribution and virulence of *Pseudomonas syringae* pv. *atrofaciens*, causal agent of basal glume rot, in Russia. In: N. Jacobellis, et al. (eds.) *Pseudomonas syringae Pathovars and Related Pathogens*. Kluwer, pp. 97–105.

Quigley, N.B. and Gross, D.C. (1994) Syringomycin production among strains of *P. syringae* pv. *syringae*: conservation of the *syrB* and *syrD* genes and activation of phytotoxin production by plant signal molecules. *Mol. Plant-Microbe Interact.* 7: 78–90.

Thompson, J.D., Higgins, D.G., and Gibson, T.J. (1994) CLUSTAL W: improving the sensitivity of progressive multiple sequence alignment through sequence weighting, position specific gap penalties and weight matrix choice. *Nucleic Acids Res.* 22: 4673–4680.

Van de Peer, Y. and De Wachter, R. (1997) Construction of evolutionary distance trees with TREECON for Windows: accounting for variation in nucleotide substitution rate among sites. *Comp. Appl. Biosci.* 13: 227–230.

Characterization of Antimicrobial and Structural Metabolites from *Burkholderia gladioli* pv. *agaricicola*

A. Cimmino[1], P. Lo Cantore[2], G. Karapetyan[3], Z. Kaczynski[3], N.S. Iacobellis[2], O. Holst[3], and A. Evidente[1]

Abstract *Burkholderia gladioli* pv. *agaricicola*, the causal agent of the soft rot disease of the cultivated mushroom *Agaricus bitorquis*, was demonstrated to produce in culture metabolites with antimicrobial activity, which were extracted by acid water-saturated *n*-butanol and then purified by HPLC. Furthermore, this paper reports on the structural determination, based on the extensive use of chemical analyses and spectroscopic techniques (essentially NMR and MS), of the repeating unit of the capsular exopolysaccharide as well as that of the O-chain of lipopolysaccharide.

Keywords Cultivated mushrooms, *Agaricus bitorquis*, soft rot disease, exopolysaccharides, lipopolysaccharides

1   Introduction

Sporophores of the edible mushroom *Agaricus bitorquis* are attacked by *Burkholderia gladioli* pv. *agaricicola* which causes the so called soft rot disease characterized by rapid development of deep, oozing, rotted lesions of the pileus which renders the mushroom unmarketable. Several devastating cases of the disease have been observed in the UK (Atkey et al., 1991; Lincoln et al., 1991). Considering the importance of the disease, some studies were planned on the evaluation of the production, purification and characterization of bioactive secondary metabolites and

[1]Dipartimento di Scienze del Suolo, della Pianta, dell'Ambiente e delle Produzioni Animali, Università degli Studi di Napoli Federico II, Via Università, 100, 80055 Portici, Italy

[2]Dipartimento di Biologia, Difesa e Biotecnologie Agro-Forestali, Università degli Studi della Basilicata, Via dell'Ateneo Lucano 10, 85100 Potenza, Italy

[3]Division of Structural Biochemistry, Research Center Borstel, Leibniz-Center for Medicine and Biosciences, D-23845 Borstel, Germany

Author for correspondence: Antonio Evidente; e-mail: evidente@unina.it

M'B. Fatmi et al. (eds.), *Pseudomonas syringae Pathovars and Related Pathogens.*
© Springer Science+Business Media B.V. 2008

on the structural determination of exo- and lipo-polysaccharides (LPSs and EPSs, respectively) aimed at clarifying their role in the host–pathogen interaction.

2 Materials and Methods

2.1 Growth of Bacteria

Type strain ICMP11096 of *Burkholderia gladioli* pv. *agaricicola* was grown at 25°C under shaking (180 rpm). Erlenmeyer flasks (500 ml) were filled with 150 ml of minimal medium (MM) (Lavermicocca et al., 1997) and inoculated with 1.5 ml of a bacterial suspension containing 10^8 CFU ml^{-1}. After 48 h the cultures were centrifuged (20,000 g for 15 min) and the resulting supernatants were evaluated for antimicrobial activity toward strain ITM100 of *Bacillus megaterium* according to a previous established procedure (Lavermicocca et al., 1997), and then lyophilized and stored at −20°C. Cells were washed twice with saline and lyophilised.

2.2 Purification of Antimicrobial Metabolites

Several methods – including fractionated precipitation with salts [$CaCl_2$ or $(NH_4)_2SO_4$], acid (HCl), and extraction with different organic solvents with increasing polarity – were preliminary used to purify the antimicrobial metabolites from the culture filtrates of strain ICMP11096 of *Burkholderia. gladioli* pv. *agaricicola*. In particular, in the case of *n*-butanol extraction, lyophiles (corresponding to 500 ml culture filtrates) were suspended in water (50 ml), brought to pH 2 with 1N HCl and exhaustively extracted with water-saturated *n*-butanol (3 × 50 ml). The extracts were combined and partially dried under reduced pressure and finally lyophilised. The crude organic extracts, showing a high antimicrobial activity, was further purified by reversed-phase high performance chromatography (HPLC) on a Shimadzu instrument equipped with HPLC Pump LC-10 ADVP and UV-VIS Detector SPD-10AV. A semipreparative (25 cm × 10 mm) Altech C8 columns was used for the purification of the above mixture. A solvent system of water (0.1% TFA, v v^{-1}) and CH_3CN with flow rates of 5.0 ml min^{-1} was used for elution. Solvents were degassed before use. Seven homogeneous fractions were collected, partially dried under reduced pressure, lyophilised and then tested for the antimicrobial activity as reported below.

2.3 Isolation of the Polysaccharides

Lyophilized bacterial cells (4.57 g) were extracted with hot phenol/water (Westphal and Jann, 1965) and the resulting water phase was lyophilised (335 mg, 7.3% of the bacterial dry mass) and then an aliquot (284 mg) of the resulting lyophil was suspended in 10 ml of ultrapure Milli-Q water and centrifuged (4°C, 100,000 g, 5.5 h). The LPS-containing pellet was again washed with ultra pure Milli-Q water, ultra-centrifuged and freeze dried (LPS, 52.2 mg, 18.3% of the lyophilised water phase). Supernatants were combined, lyophilized and this material (181.9 mg, 63.9% of the lyophilised water phase) was further purified by gel-permeation chromatography on a Sephadex G-50 (Pharmacia) column as previously reported (Kaczynski et al., 2006) to obtain one fraction containing capsular EPS (145 mg). The LPS fraction (52.2 mg) was hydrolysed with aqueous 1% AcOH as previously reported (Karapetyan et al., 2006) and the lipid precipitate (lipid A) was removed by ultracentrifugation (100,000 g, 4°C, 4 h) and lyophilised (4.6 mg, 8.8% of LPS). The supernatant containing the carbohydrate portion was lyophilised (42.6 mg, 81.6% of LPS). An aliquot of this (18.7 mg) was purified by gel-chromatography on a Sephadex G-200 column as previously reported (Karapetyan et al., 2006) to give two main fractions consisting of EPS (4.1 mg) and of O-chain polysaccharide (OPS, 9.2 mg), respectively. An aliquot of the latter (5.4 mg) was treated with NaOCH$_3$, neutralized, dialyzed, and run successively on a TSK 55 (S) column as previously reported (Karapetyan et al., 2006) to obtain a high molecular mass polysaccharide (0.6 mg, 11.1% of the OPS).

2.4 Polysaccharide Chemical Analyses

EPS and OPS were acid hydrolysed and the resulting monosaccharide mixtures were analysed as alditol acetate by GLC using a Hewlett-Packard 5880 instrument equipped with a SPB-5 capillary column in the condition previously reported (Kaczynski et al., 2006). The absolute configuration of the sugars was determined by GLC of their acetylated (S)-2-butanol glycosides utilizing the above chromatographic conditions (Leontein and Lönngren, 1978; Gerwig et al., 1979). The methylation of the polysaccharide was carried out according to Ciucanu and Kerek (1984). The methylated samples were extracted, reduced, acetylated as previously reported (Kaczynski et al., 2006) and then the resulting products analysed by GLC-MS using a Hewlett-Packard 5989 instrument equipped with a HP-5MS capillary column (30 m × 0.25 mm × 0.25 μm) as previously reported (Kaczynski et al., 2006).

2.5 NMR Spectroscopy

NMR spectra were obtained for solutions in ^{2}H$_2$O with a Bruker DRX Avance and 600 MHz spectrometer (operating frequencies 600.31 MHz for ^{1}H NMR and 150.96 MHz for ^{13}C NMR) at 50°C. Chemical shifts were reported relative to internal acetone (^{1}H 2.225; ^{13}C 31.45). One-dimensional ^{1}H and ^{13}C NMR, and COSY, TOCSY, ROESY, DQFCOSY, as well as the ^{1}H,^{13}C-heteronuclear HMQC and HMBC experiments were recorded applying standard Bruker software.

2.6 Biological Assays

The antimicrobial activity of culture of strain ICMP 11096 of *Burkholderia gladioli* pv. *agaricicola* as well as of HPLC fractions was evaluated on the Gram-positive bacterium *Bacillus megaterium* strain ITM100 according to the procedure reported by Lavermicocca et al. (1997). The assays were performed twice with three replicates. The specific activity was expressed in active units (a.u.) per microgram.

3   Results and Discussion

3.1   Isolation, Purification and Characterisation of Toxic Metabolites

Among the methods used to purify the antimicrobial metabolites from the culture filtrates (500 ml) of strain ICMP11096 of *Burkholderia gladioli* pv. *agaricicola* only the extraction with *n*-butanol exhaustively recovered almost all the original antimicrobial activity present in the culture. The bioactive organic extract (500 mg), was analysed on a semipreparative reverse phase C8 column by HPLC yielding the chromatographic profile reported in Fig. 1. Seven groups of homogenous fraction were collected and tested for their antimicrobial activity on strain ITM100 of *Bacillus megaterium*. As showed by the histogram reported in Fig. 1, all the toxic

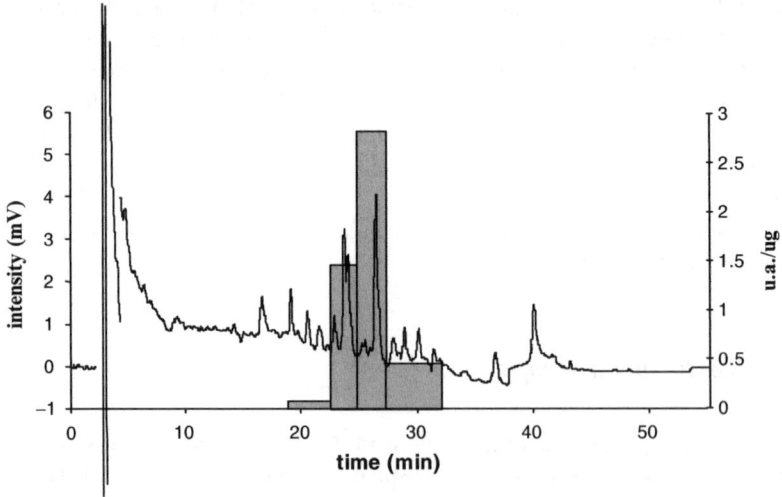

Fig. 1   HPLC profile of the *n*-butanol extracts of strain ICMP11096 of *Burkholderia gladioli* pv. *agaricicola* culture filtrates. The histogram reports the antimicrobial activity expressed in active units (a.u.) per microgram on strain ITM100 of *Bacillus megaterium*

metabolites were present in the fraction groups 2–5 collected between 18–32 min elution time.

Preliminary ¹H NMR analysis of the main active fraction indicated the probable peptide nature for the metabolite/s contained.

3.2 Polysaccharides

A putative capsular polysaccharide containing D-rhamnose was isolated from cells of strain ICMP1096 of *Burkholderia gladioli* pv. *agaricicola*. The structure of the exopolysaccharide was determined by chemical analyses and NMR spectroscopy (Kaczynski et al., 2006).

Sugar analyses of the polysaccharide identified D-rhamnose as the sole constituent. The ¹H NMR spectrum (Fig. 2) of the capsular polysaccharide showed four anomeric signals and four overlapping methyl signals characteristic for 6-deoxy-sugars. The sugar residues were labelled **A–D** in order to decreasing chemical shifts of the anomeric protons.

The ¹³C NMR spectrum contained 20 signals, however, since four of them possessed double intensity, a total of 24 carbon atoms were present, thus confirming a repeating unit comprising four hexoses. A ¹H,¹³C-heteronuclear multiple-quantum coherence (HMQC) experiment identified four anomeric carbon signals.

The anomeric configurations of all rhamnose residues were assigned by the coupling constants $^1J_{H-1,C-1}$ which were identified in another ¹H,¹³C HMQC experiment without decoupling. The complete structural characterization of the capsular polysaccharide

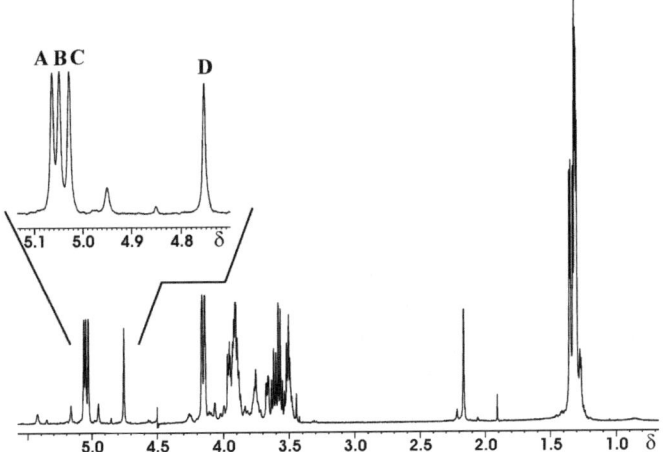

Fig. 2 The ¹H NMR spectrum of the capsular polysaccharide isolated from *Burkholderia gladioli* pv. *agaricicola*

was achieved by 1D and 2D ¹H and ¹³C NMR spectroscopy.¹H,¹H COSY, double-quantum-filtered COSY (DQFCOSY) and TOCSY, as well as ¹H,¹³C HMQC spectra allowed the complete assignment of all ¹H and ¹³C chemical shifts.

The 2D rotating frame nuclear Overhauser spectroscopy (ROESY) and ¹H,¹³C-heteronuclear multibond correlation (HMBC) experiments revealed the sequence of the sugar residues in the repeating unit.

In summary, the data identified the structure of a putative capsular polysaccharide from *Burkholderia gladioli* pv. *agaricicola* as:

A   **B**   **C**   **D**

→4)-α-D-Rha*p*-(1→3)-α-D-Rha*p*-(1→3)-α-D-Rha*p*-(1→3)-β-D-Rha*p*-(1→

A neutral O-specific polysaccharide (OPS) was obtained by mild acid hydrolysis of the lipopolysaccharide (LPS) of the same bacterium.

Sugar analysis of the polysaccharide identified rhamnose, mannose and galactose in an approximate molar ratio of 5:2:2. Methylation analysis revealed the polymer to be composed of 2-substituted rhamnopyranose and two 3- and 4-substituted hexopyranose residues.

The absolute configurations of the sugars were identified as D-mannose, D-galactose and D-rhamnose. The ¹H NMR spectrum (Fig. 3) of the O-specific polysaccharide contained four main signals in the anomeric region, a signal characteristic for a methyl group of rhamnose and one signal characteristic for an *O*-acetyl group. The sugar residues were labeled **A–C** according to decreasing chemical shifts of the anomeric protons. The couplings of the anomeric signals of residues **A** and **B** identified the *manno* and that of residue **C** the *galacto* configured residues. A COSY experiment allowed to locate the acetoxy group at C-2 of the mannose residue.

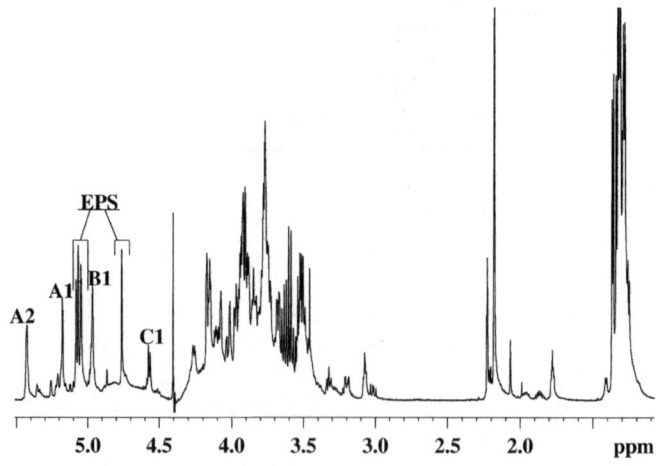

Fig. 3 The ¹H NMR spectrum of the OPS of the LPS from *Burkholderia gladioli* pv. *agaricicola*

Data from sugar analysis and from NMR spectroscopy indicated that the OPS consisted of a trisaccharidic repeating unit, containing one residue each of D-mannose, D-rhamnose and D-galactose.

The complete structural characterization of the polysaccharide was achieved by 1D and 2D ^{1}H and ^{13}C NMR spectroscopy.^{1}H,^{1}H COSY, double-quantum-filtered COSY (DQFCOSY), ROESY and TOCSY, as well as ^{1}H,^{13}C HMQC and HMBC spectra allowed the complete assignment of all ^{1}H and ^{13}C chemical shifts

The anomeric configurations of all residues were assigned by the coupling constants $^1J_{H-1,C-1}$ which were identified in another ^{1}H,^{13}C HMQC experiment recorded without decoupling. The $^1J_{H-1,C-1}$ revealed that the mannosyl and rhamnosyl residues had the α- and the galactosyl residue the β-anomeric configuration.

The ROESY and ^{1}H,^{13}C HMBC experiments revealed the sequence of the sugar residues in the repeating unit.

In summary, the data identified the structure of the O-specific polysaccharide from *Burkholderia gladioli* pv. *agaricicola* as:

$$Ac$$
$$\downarrow$$
$$2$$
$$\rightarrow3)\text{-}\alpha\text{-}\text{D-Man}p\text{-}(1\rightarrow2)\text{-}\alpha\text{-}\text{D-Rha}p\text{-}(1\rightarrow4)\text{-}\beta\text{-}\text{D-Gal}p\text{-}(1\rightarrow$$

$$\quad\quad A \quad\quad\quad\quad\quad B \quad\quad\quad\quad\quad C$$

This structure was also confirmed carrying out similar chemical and spectroscopic studies on the deacetylated polysaccharide obtained by alkaline hydrolysis of the OPS.

4 Conclusions

Antimicrobial metabolites produced by strain ICMP 1096 of *Burkholderia gladioli* pv. *agaricicola* appear to be low-medium molecular weight compound with a hydrophilic peptide nature. Further work is necessary for the structure determination of the above metabolites. The role, if any, of these in the virulence of the producing bacterium remain also to be determined. The structure of the repeating unit of the EPS and the O-chain of the LPS were determined (Kaczynski et al., 2006; Karapetyan et al., 2006).

The role of the exopolysaccharides (EPSs) produced by plant pathogenic bacteria in the host–pathogen interaction has not been completely clarified. They are considered either molecules able to avoid or delay the activation of plant defence, acting as signals in plant–pathogen cross-talk as well as having a significant role in the virulence of the pathogen of interest. In the last decades the chemical composition and structures of EPSs of several plant pathogenic bacteria have been characterised. Some EPSs have a very complex structure, being heterogeneous branched heteropolymers. The differences in the chemical composition and structures of the EPSs purified from different plant pathogens suggest that these molecules may play specific and different roles in plant–pathogen interaction (Corsaro et al., 2001; Evidente and Motta, 2002).

LPSs appear to play an important role in the interaction between plant pathogenic bacteria and the respective host plants (e.g., in the recognition process representing the first phase of such interaction) as well as in the virulence of the pathogens (Dow et al., 2000).

Although, the EPS and LPS structures of *Burkholderia gladioli* pv. *agaricicola* as well as of other bacterial pathogen of mushrooms (i.e. *Pseudomonas tolaasii, P. reactans*) have been determined (Molinaro et al., 2002, 2003; Karapetyan et al., 2006; Zbigniew et al., 2006) nothing is known on the role of these structural substances in the mushroom–pathogen interaction and/or in the pathogen virulence.

Acknowledgements This work was supported in part by grants from the Italian Ministry of University and Research (MIUR), Italian National Research Council (CNR). Contribution DISSPAPA-BOOK-1.

References

Atkey, P.T., Fermor, T.R., and Lincoln, S.P. (1991) Bacterial soft rot of *Agaricus bitorquis*. In *Proceedings of the 13th International Congress on The Science and Cultivation of Edible Fungi*, Rotterdam, Dublin, Irish Republic, 1–6 September 1991. A.A. Balkema, Rotterdam, The Netherlands, 431–435.

Ciucanu, I. and Kerek, F. (1984), A simple and rapid method for the permethylation of carbohydrates, *Carbohydrate Research* **131**: 209–217.

Corsaro, M. M., De Castro, C., Molinaro, A., and Parrilli, M. (2001) Structure of lipopolysaccharides from phytopathogenic Gram-negative bacteria. In *Recent Research Development in Phytochemistry* (Pandalai, S.G. ed.), Research Signpost, Trivandrum, India, Vol. 5, pp. 119–138.

Dow J.M., Newman M.A, and von Roepenack E. (2000) The induction and modulation of plant defense responses by bacterial lipopolysaccharides, *Annual Review Phytopathology* **38**: 241–261.

Evidente, A. and Motta, A. (2002) Bioactive metabolites from phytopathogenic bacteria and plants. In *Studies in Natural Products Chemistry* (Atta-ur-Rahman, ed.), Elsevier, Amsterdam, The Netherlands, Vol. 26, pp. 581–629.

Gerwig, G.J., Kamerling, J.P., and Vliegenthart, J.F.G. (1979) Determination of the absolute configuration of neutral monosaccharidein complex structure of neutral monosaccharide in complex carbohydrates by capillary GLC, *Carbohydrate Research* **77**: 1–7.

Kaczynski, Z., Karapetayn, G., Evidente, A., Iacobellis, N.S., and Holst, O. (2006) The structure of a putative exopolysaccharide of *Burkholderia gladioli* pv. *agaricicola, Carbohydrate Research* **341**: 285–288.

Karapetyan, G., Kaczynski, Z., Evidente, A., Iacobellis N.S., and Holst, O. (2006) The structure of the O-specific polysaccharide of the lipopolysaccharide from *Burkholderia gladioli* pv. *agaricicola, Carbohydrate Research* **341**: 930–934.

Lavermicocca, P., Iacobellis, N.S., Simmaco, M., and Graniti, A. (1997) Biological properties and spectrum of activity of *Pseudomonas syringae* pv. *syringae* toxins, *Physiological Molecular Plant Pathology* **50**: 129–140.

Leontein, K. and Lönngren, J. (1978) Assignment of absolute configuration of sugars by GLC of their acetylated glicosides formed from chiral alcohols, *Journal of Methods Carbohydrate Chemistry* **62**: 359–362.

Lincoln, S.P. Fermor, T.R. Stead, D.E., and Sellwood, J.E. (1991) Bacterial soft rot of *Agaricus bitorquis, Plant Pathology* **40**: 136–144.

Molinaro, A., Evidente, A., Iacobellis, N.S., Lanzetta, R., Lo Cantore, P., Mancino, A., and Parrilli, M. (2002) O-Specific chain structure from the lipopolysaccharide fraction of *Pseudomonas reactans*: a pathogen of cultivated mushrooms, *Carbohydrate Research* **337**: 467–471.

Molinaro, A., Bedini, E., Ferrara, R., Lanzetta, R., Parrilli, M., Evidente, A., Lo Cantore, P., and Iacobellis, N.S. (2003) Structural determination of the O-specific chain structure of the lipopolysaccharide from the mushroom pathogenic bacterium *Pseudomonas tolaasii, Carbohydrate Research* **338**: 1251–1257.

Westphal, O. and Jann, K. (1965) Bacterial lipopolysaccharide extraction with phenol–water and further application of the procedure, *Methods in Carbohydrate Chemistry* **5**: 83–91.

Tomato Pith Necrosis Disease Caused by *Pseudomonas* Species in Turkey

H. Saygili[1], Y. Aysan[2], N. Ustun[3], M. Mirik[4], and F. Sahin[5]

Abstract Pith necrosis disease of tomato caused by *Pseudomonas* species has been first time observed in Turkey since early 1990s. Since then it has spread out and caused severe damage on greenhouse grown-tomato in Mediterranean and Aegean regions of Turkey. Several surveys were conducted to determine the distribution of the disease and the identity of the bacterial pathogens causing pith necrosis on tomato in the regions in 2003 and 2004. A total of 62 *Pseudomonas* strains isolated were characterized by morphological, physiological, biochemical, pathogenicity test and fatty acid methyl ester (FAME) analysis. The strains in six different Pseudomonas species including *P. viridiflava*, *P. cichorii*, *P. corrugata*, *P. mediterraneae*, *P. fluorescens*, and unidentified *Pseudomonas* sp. were proven to be the causal agents of pith necrosis. Based on FAME analysis, *Pseudomonas* strains had 35 different fatty acids, respectively. The *Pseudomonas* strains were divided into three different groups according to cluster analysis of fatty acids. Fatty acid compositions of *Pseudomonas* strains were qualitatively similar. The strains were not separated by FAME analysis and the identification should be confirmed by LOPAT tests or PCR. Identification of *P. corrugata* and *P. mediterraneae* strains were confirmed by amplification of 1100 and 600 bp DNA fragment in PCR studies, respectively. Profiles of the all *Pseudomonas* strains were compared in BOX-PCR experiments. The study is also demonstrating that the causal agents of tomato pith necrosis caused by *Pseudomonas* species can be distinguished based on their symptom differences in the greenhouses.

[1] Ege University, Faculty of Agriculture, Department of Plant Protection, Izmir 35100, Turkey

[2] Cukurova University, Faculty of Agriculture, Department of Plant Protection, Adana 01330, Turkey

[3] Plant Protection Research Institute, 35040 Izmir, Turkey

[4] Namik Kemal University, Faculty of Agriculture, Department of Plant Protection, Tekirdag 59030, Turkey

[5] Yeditepe University, Faculty of Engineering and Architecture, Department of Genetic and Bioengineering, Istanbul 34755, Turkey

Author for correspondence: Hikmet Saygili; e-mail: hikmet.saygili@ege.edu.tr

Keywords Tomato, pith necrosis, FAME, PCR

1 Introduction

Tomato is an important greenhouse and field-grown vegetable with a production of 9,750,000 metric tons in 2003 in Turkey (FAO). Several yield-limiting bacterial diseases on tomato have caused considerable concern among commercial tomato growers in Turkey. One of them is stem and pith necrosis, caused by *Pseudomonas* species. Several names such as tomato stem necrosis, and pith necrosis have been used to describe the general pathological symptoms of the disease exhibited on all above ground parts of plants including stems, petiole, leaves and fruits. Spots on stem, lesions on petiole and fruit stalk, hollowing and the browning of the pith, discoloration of the vessels are defined as typical symptoms of the disease. The disease induces important damages between October and June in the greenhouses of Mediterranean and Aegean regions of Turkey.

Pith necrosis is a serious disease of tomato with 30 years of history. *Pseudomonas corrugata* is one of the most widespread pathogen known in Europe (Scarlett et al., 1978; Catara et al., 2002) and USA (Lai et al., 1983). But in Turkey, pith necrosis disease of tomato caused by *P. viridiflava, P. cichorii* and *P. corrugata* has been determined in many commercial greenhouses since early 1990s (Aysan, 2001; Ustun and Saygili, 2001; Sahin et al., 2004; Saygili et al., 2004, 2005; Basim et al., 2005). So far *P. viridiflava* (Aysan et al., 2004), has been known to be the most widespread species associated with stem necrosis in Turkey. However, there have been no detailed studies to differentiate all potential bacterial species causing stem and pith necrosis on tomatoes based on symptom development in the fields. The present study is providing new data to distinguish different *Pseudomonas* species causing pith necrosis of tomato according to disease symptoms. The aim of this study was also to identity the bacterial pathogens causing pith necrosis in the different regions of Turkey using morphological, physiological, and biochemical characters as well as pathogenicity tests, fatty acid methyl ester (FAME) analysis, and PCR.

2 Materials and Methods

Several surveys was conducted to determine the disease incidence and the identity of the bacterial pathogens causing pith necrosis on tomato grown in Antakya, Adana, Mersin, Antalya, Mugla and Izmir provinces in the Mediterranean and Aegean regions of Turkey during the late winter and spring of 2002, 2003, and 2004. During these surveys, all diseased tomato plants with typical symptoms of stem and pith necrosis were collected and average disease incidence for each location was determined. Symptom differences of stem and pith necrosis pathogens of tomato were recorded in the surveys. Small portions of infected stem showing lesions from dark green to brown were macerated and streaked onto Petri dishes of King's medium B. The plates were incubated at 25°C for 48 h, and the grown colonies were sub-cultured and purified.

In pathogenicity tests, tomato seedlings were inoculated by injecting 0.1 ml of bacterial suspension (10^7 cfu/ml) of the each isolates through the stem with a 1 ml syringae. The following five strains were used as positive references for their symptom development: CFPB 2101 *P. cichorii*, CFPB 2431 *P. corrugata*, CFPB 5447 *P. mediterranea*, CFPB 2102 *P. fluorescens* (from Dr. L. Sutra, INRA, France), and GSPB 1685 *P. viridiflava* (from Drs. H. Batur-Michaelis and E. Mavridis, Georg-August University Göttingen, Germany) kindly provided. The plants, injected only sterile distilled water, were used as negative control. After inoculation all plants were placed in a growth room at 25°C, 16/8 h day/night and 70% relative humidity and checked daily for the symptom development.

Bacteria were re-isolated from inoculating plants showing disease symptoms. The isolates were primarily tested according to the LOPAT tests (L: production of levan, O: oxidase reaction, P: soft rot in potato, A: arginin dihidrolase activity, T: hypersensitive reaction on tobacco) and ability to produce acid from carbon compounds as described by Leliott and Stead (1987). The identification results were confirmed by ELISA for *P. viridiflava* as described by Aysan et al. (2004), fatty acid methyl ester analysis (FAME) in Sherlock Microbial Identification System (MIS Sherlock 4.0 MIDI, Inc., Newark, DE) with database of prokaryotes (De Boer and Sasser, 1986; Janse et al., 1992), and PCR for *P. corrugata* (PC5/1: 5´-CCA CAG GAC AAC ATC TCC AC-3´ and PC5/2: 5´-CAGGCGCTTTCTGGAACATG-3´), and *P. mediterranea* (PC1/1: 5´-GGA TAT GAG CCA GGT CTT CG-3´ ve PC1/2: 5´-CGC TCA AGC GCG ACT TCAG-3´) (Catara et al., 2002). A total of 62 *Pseudomonas* spp were identified at the species level, and then used for further studies.

3   Results and Discussion

Based on external disease symptoms, the average disease incidence caused by *Pseudomonas* was estimated as 15–20%. Pith necrosis, caused by *Pseudomonas* species, is characterized by yellowing and wilting of the lower leaves, various sizes of irregular dark brown-black lesions on stem, peduncle and leaf stalks, adventitious root on the cracked stem, brown discoloration of the pith, sometimes no external symptom. *Pseudomonas viridiflava* is characterized as general wilting and yellowing on tomato plants, brown-black spot limited on the pruning sites of the stem and canker of the petioles. Internal symptoms of the disease are known to be browning and collapse of the pith and vascular discoloration. Fruit symptoms are induced by only *P. viridiflava* but not other *Pseudomonas* species associated pith necrosis. Fruit symptoms are initially round black spots. Then soft rotting was observed around spots. The disease incidence was approximately

Table 1 Symptom differences of *Pseudomonas* species causal agent of pith necrosis

Pathogen	General symptoms	External symptom on stem	Internal symptom	Fruit lesions
P. viridiflava	Yellowing of the leaf stalks and wilting of the lower leaves	Small, brown-black spots limited on the pruning sites of the stem	Pith necrosis, browning and collapse of the pith and vascular discoloration	Initially round black spots, soft rotting around spot
P. cichorii	Yellowing and wilting of the lower leaves which progress upwards	Large, irregular dark brown-black lesions on stem. They cover large areas on the lower parts of the stem	Pith necrosis	No fruit lesions
P. corrugata	Yellowing of the upper leaves	Darkbrown-black lesions and cracks on stem. Adventitious root on the cracked stem	Pith necrosis	No fruit lesions
P. mediterraneae	Symptoms are very similar to *P. corrugata*			
P. fluorescens	Yellowing and wilting of the lower leaves which progress upwards	Irregular, dark brown-black areas on stem	Yellow, brown discoloration of the pith	No fruit lesions
Unidentified *Pseudomonas* sp.1	Symptoms are very similar to *P. corrugata* and *P. mediterraneae*. LOPAT character is different			
Unidentified *Pseudomonas* sp.2	Symptoms are very similar to *P. cichorii*. Colonies of the strains were centre blue on King's medium B, 7 days after incubation			

100% in one of the commercial greenhouses of Antakya in 2002. Our previous studies showed that *P. viridiflava* and *P. cichorii* are dominant pathogens among *Pseudomonas* species for Turkey. *P. cichorii* is characterized as general wilting, chlorosis of the young leaves, and enlarged brown-black spots on the stem nodes and branches coming from them, initially yellow discoloration and then dark brown discoloration of the pith, vascular discoloration, but no fruit symptoms. Symptoms caused by *P. fluorescens* were yellowing and wilting of the lower leaves which progressed upwards, small spots on stem, brown discoloration of the pith. Symptoms caused by *P. corrugata* and *P. mediterraneae* are characterized as yellowing of the upper leaves, gray to brown lesions on the stem, cracks of the stem and adventitious root from the cracks and no fruit symptoms. Internal symptoms of the disease are yellowing, browning, hollowing and collapse of the pith and vascular discoloration. The study is also demonstrating that the causal agents of tomato pith necrosis caused by *Pseudomonas* species can be distinguished based on their symptom differences in the greenhouse (Table 1).

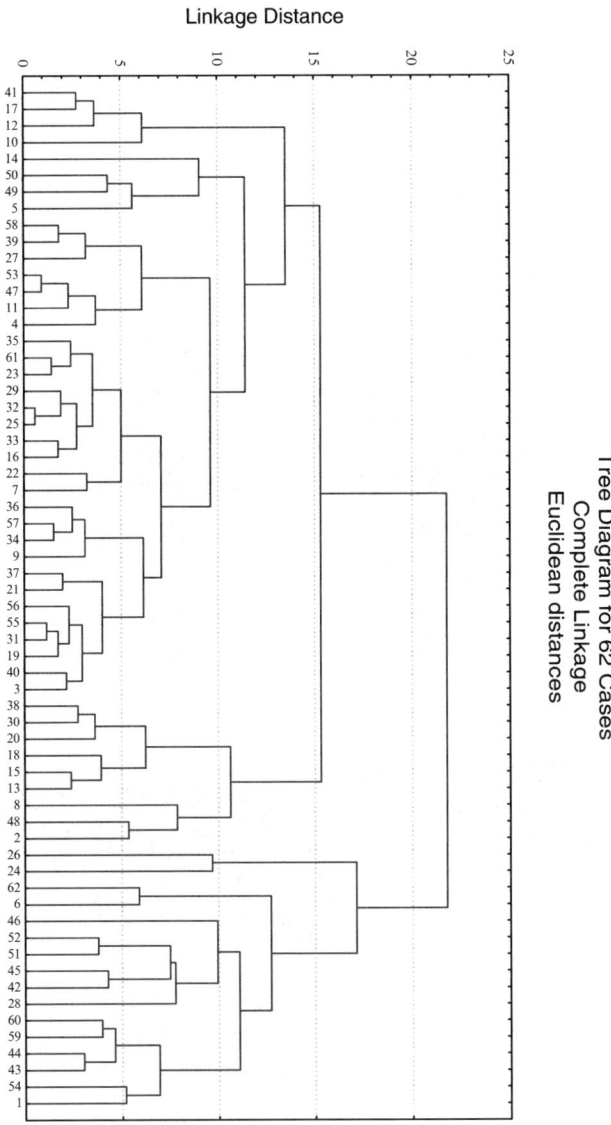

Fig. 1 Cluster analysis of fatty acids of pith necrosis strains

Of the 62 strains of *Pseudomonas* species, 14, 13, 12, 8, 8, and 7 strains were identified as *P. viridiflava*, *P. mediterraneae*, unidentified *Pseudomonas* spp., *P. cichorii*, *P. fluorescens*, and *P. corrugata* respectively (Table 2). Based on FAME analysis, *Pseudomonas* strains had 35 different fatty acids, respectively. Identification of *Pseudomonas* strains was confirmed as with similarity indices of 59% and 92% in FAME analysis. The *Pseudomonas* strains were divided into three different groups according to cluster analysis of fatty acids. Fatty acid compositions of *Pseudomonas* strains were qualitatively similar (Table 3, Fig. 1). The strains were not separated by FAME analysis and the identification should be confirmed by LOPAT tests or PCR. *P. viridiflava* strains strongly reacted with the polyclonal antisera specific *P. viridiflava* in indirect-ELISA tests with 0.9640–1.1960 at A_{405} wavelength of mean absorbance values. In PCR assays, specific bands at 600, and 1100 bp were observed for strains of *P. mediterraneae* and *P. corrugata* (Fig. 2), respectively. In BOX-PCR studies, bands profiles of the strains were confirmed as identical to the reference strains (Fig. 3). In pathogenicity tests, the bacteria were re-isolated from the inoculated plants and all strains were pathogenic on tomato.

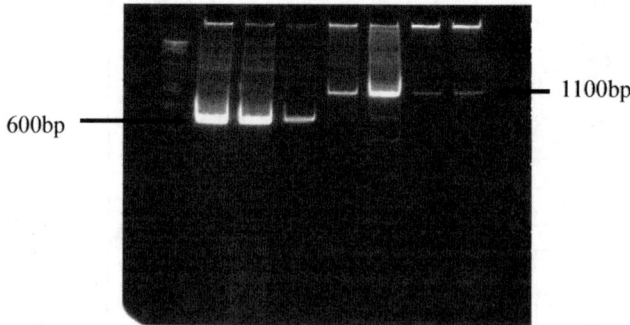

Fig. 2   M: 100 bp DNA marker; 1–3: *P. mediterraneae* strains; 4–7: *P. corrugate* strains

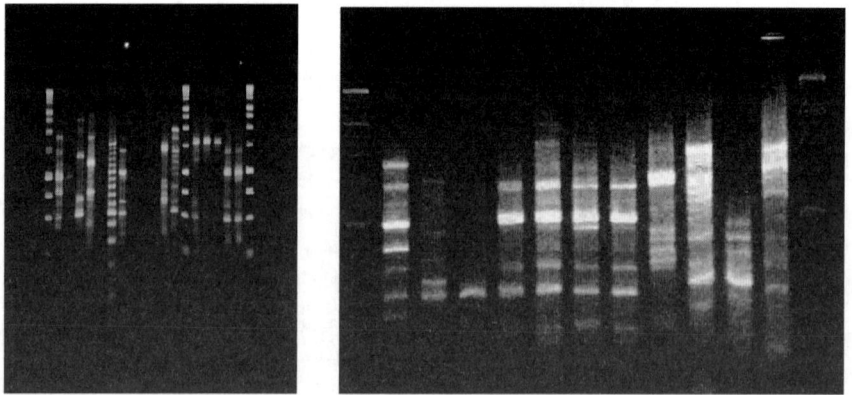

Fig. 3   BOX-PCR profiles of *Pseudomonas* strains

Table 2 Morphological, biochemical and physiological characteristics of *Pseudomonas* strains

Comperative tests	Fourteen strains (Pvir)	Thirteen strains (Pmed)	Twelve strains (unidentified)	Eight strains (Pcic)	Eight strains (Pfluo)	Seven strains (Pcor)
Gram reaction	–	–	–	–	–	–
Fluorescent pigment	+	–	–	+	+	+	–
Gelatin liquefaction	Nd	Nd	Nd	Nd	Nd	+	Nd
Levan production	–	–	+	+	–	–	–
Oxidase reaction	–	+	+	+	+	+	+
Potato rot	+	–	–	–	–	–	–
Arginine dihydrolase activity	–	+	+	–	–	+	+
HR test on tobacco	+	+	+	+	+	–	+
Sorbitol	+	–	–	–	–	Nd	–
Raffinose	–	–	–	–	–	Nd	–
Maltose	–	–	–	–	–	–	–
Myo-Inositol	+	+	+	+	+	+	+
D(−) Fruktose	+	+	+	+	+	+	+
Sucrose	–	+	+	–	–	+	+
L(+) Arabinose	+	+	+	+	+	+	+
D(−) Mannitol	+	+	+	+	+	+	+
D(+) Xylose	+	+	+	+	+	+	+
L(+) Rhamnose	–	–	–	–	–	–	–
D(+) Galaktose	+	+	+	+	+	+	+
D(−) Arabinose	–	–	–	–	–	–	–
Trehalose	Nd	+	Nd	Nd	–	+	+
Lactose	–	–	–	–	–	–	–
D(+) Mannose	+	+	+	+	+	+	+
L(+) Tartaric Acid	Nd	–	Nd	Nd	+	–	–
Histamine	Nd	+	Nd	Nd	–	–	–
Benzoic acid	Nd	–	Nd	Nd	–	–	–
Keto-D-Gluconic Acid	Nd	+	Nd	Nd	–	–	–
D(−) Tartarate	Nd	–	Nd	Nd	+	–	–
Meso Tartaric Acid	Nd	+	Nd	Nd	+	–	–

–: negative, +: positive, Nd: not determined

Table 3 Fatty acid compositions of pith necrosis strains

Fatty acids	Strains	Range	Mean	Standard derivation
8:0 3OH	6	0.31–1.01	0.55	0.25
10:0	8	0.13–0.59	0.32	0.17
10:0 2OH	1	0.13	0.13	–
10:0 3OH	62	2.42–7.98	3.98	1.39
12:0	62	1.43–7.68	4.37	1.18
11:0 ISO 3OH	2	1.17–1.48	1.33	0.1
Unknown 12.484	3	0.20–1.04	0.51	0.46
12:0 2OH	62	1.58–5.60	3.25	0.94
12:1 3OH	14	0.29–2.72	1.06	0.71
12:0 3OH	62	3.35–5.69	4.10	0.41
Unknown 13.957	4	0.24–0.91	0.42	0.33
14:0	27	0.19–0.61	0.35	0.14
15:0 Iso	1	5.3	5.30	–
15:0 Antestio	1	3.31	3.31	–
15:0	8	0.13–0.60	0.29	0.17
Unknown 14.502	2	0.23–0.45	0.34	0.14
Sum in feature 2	11	0.39–1.14	0.75	0.25
Sum in feature 3	62	25.60–42.56	34.86	4.19
16:0 Iso	1	0.09	0.09	–
16:0	62	18.50–31.75	25.22	3.05
17:0 Antestio	1	0.33	0.33	–
17:0 Iso	13	0.07–1.17	0.54	0.26
Iso 17:1 w5c	2	0.19–2.39	1.29	1.56
17:1 w8c	6	0.16–0.39	0.26	0.1
17:0 Cyclo	41	0.33–8.32	2.90	2.47
17:0	9	0.27–0.51	0.43	0.1
Sum in feature 5	2	0.31–0.35	0.33	0.02
16:0 3OH	9	0.31–0.55	0.46	0.1
18:1 w9c	2	0.31–4,59	2.45	3.03
18:1 w7c	62	12.47–27.98	20.08	3.62
18:0	47	0.31–1.49	0.89	0.35
11 methyl 18:1 w7c	12	0.25–1.08	0.58	0.22
19:0 10 methyl	1	0.54	0.54	–
19:0 Cyclo w8c	7	0.30–2.37	0.73	0.73
Sum in feature 7	3	0.44–0.71	0.56	0.14
Summed feature 2	12:0 Alde			
Summed feature 3	16:1 w7c/15iso 2OH			
Summed feature 5	18:2 w6,9c/18:0 Ante			
Summed feature 7	Un 18.846/19:1 w6c			

The present study provided the evidence that seven different *Pseudomonas* species including *P. fluorescens*, *P. mediterraneae*, *P. viridiflava*, *P. cichorii*, *P. corrugata*, two unidentified *Pseudomonas* species are the causal agents of pith necrosis on tomato greenhouses in Turkey.

The presence of some unusual strains in two unidentified *Pseudomonas* species has been first time determined to be causal agents of pith necrosis on tomato. A further study is necessary to identify these strains at species level. This study also shown that tomato pith necrosis disease caused by *Pseudomonas* species may be distinguished based on their symptom differences in the field/greenhouse conditions.

Acknowledgement The research was supported by grant no. TOGTAG 3122 from the Scientific and Technical Research Council of Turkey and grant no. EU 04-DPT-03 from Prime Ministry State Planning Organization of Turkey.

References

Aysan, Y., 2001. Bacterial stem necrosis of tomato in the greenhouse in the eastern Mediterranean region of Turkey. 11th Congress of the Mediterranean Phytopathological Union and 3rd Congress of the Sociedade Portuguesa de Fitopatologia. Evora, Portugal, 17–20 September. pp. 301–303.

Aysan, Y., N. Yildiz, and F. Yucel, 2004. Identification of *Pseudomonas viridiflava* on tomato by traditional methods and enzyme-linked immunosorbent assay. Phytoparasitica 32(2):146–153.

Basim, H., E. Basim, S. Yilmaz, and M. Ilkucan, 2005. First report of pith necrosis of tomato caused by Pseudomonas mediterraneae in Turkey. Plant Pathology 54:240.

Catara, V., L. Sutra, A. Morineau, W. Achouak, R. Christen, and L. Gardan, 2002. Phenotypic and genomic evidence for the revision of *Pseudomonas corrugata* and proposal of *Pseudomonas mediterranea* sp. *nov*. International Journal of Systematic and Evolutionary Microbiology 52:1749–1758.

De Boer, S.H. and M. Sasser, 1986. Differentiation of *Erwinia carotovora* ssp. *carotovora* and *E. carotovora* ssp. *atroceptica* on the basis of cellular fatty acid composition. Canadian Journal of Microbiology 32:796–800.

Janse, J.D., J.H.J. Derks, B.E. Spit, and W.R. van der Tuin, 1992. Classification of fluorescent soft rot *Pseudomonas* bacteria, including *P. marginalis* strains, using whole cell fatty acid analysis. Systemic and Applied Microbiology 15:538–553.

Lai, M., D.C. Opgennorth, and J.B. White, 1983. Occurrence of *Pseudomonas corrugata* on tomato in California. Plant Disease 67:110–112.

Leliott, R.A. and D.E., Stead, 1987. Diagnostic Procedures for Bacterial Plant Diseases in Methods for the Diagnosis of Bacterial Diseases of plants 58–59. Blackwell Scientific Publications, 216 p.

Sahin, F., Y. Aysan, and H. Saygili, 2004. The first observation of pith necrosis on tomato caused by some *Pseudomonas* species in Turkey. Acta Horticulturae Nr 695:291–293.

Saygili, H., Y. Aysan, F. Sahin, N. Ustun, and M. Mirik, 2004. Occurence of pith necrosis caused by *Pseudomonas flourescens* on tomato plants in Turkey. Plant Pathology 53(6):803.

Saygili, H., F. Sahin, Y. Aysan, and M. Mirik, 2005. New symptoms of tomato soft rot diseases in Turkey. Acta Horticulturae Nr 695:93–95.

Scarlett, C.A., J.T. Fletcher, P. Roberts, and R.A. Leliott, 1978. Tomato pith necrosis caused by *Pseudomonas corrugata* n. sp. Annals of Applied Biology 88:105–114.
Ustun, N. and H. Saygili, 2001. Pith necrosis of greenhouse tomatoes in Aegean region of Turkey. 11th Congress of the Mediterranean Phytopathological Union and 3rd Congress of the Sociedade Portuguesa de Fitopatologia. Evora, Portugal, 17–20 September, pp. 70–73.

Part VI
New Emerging Pathogens

Emerging Plant Pathogenic Bacteria and Global Warming

N.W. Schaad

Abstract Several bacteria, previously classified as non-fluorescent, oxidase positive pseudomonads, *Ralstonia*, *Acidovorax*, and *Burkholderia* have emerged as serious problems worldwide. Perhaps the most destructive is *R. solanacearum* (RS), a soilborne pathogen with a very wide host range. RS race 3, biovar 2 infects potato and geranium during cooler weather making it an additional threat. *Acidovorax avenae* subsp. *avenae* has emerged as a disease of upland rice in Southern Europe during periods of high temperatures and *B. gladioli* has emerged as a serious pathogen of orchids in Thailand. *B. andropogonis* has been identified for the first time on jojoba in eastern Australia; the plant grows under very high temperatures and the disease occurs in nursery stock grown under overhead watering. *Burkholderia glumae* is emerging as a problem on rice in the southern United States and the pathogen has become much more prevalent in southern South Korea during periods of high temperatures. Why the increase in these bacteria? One possible explanation is global warming. A common trait among them is an optimum growth temperature of 32–36°C; most grow well up to 41°C, whereas most other plant pathogenicbacteria grow best at lower temperatures. An increase in extreme weather conditions, including extended heat waves, long periods of rain, and storms such as hurricanes appear to favor these high-temperature bacteria. There has been a gradual increase in the mean global temperature over the past century. Recent summer temperatures in the southern USA. have been 1–2°C higher than that which is optimum for many crops. Evidence is growing that the increase in the occurrence of extreme weather, including heat waves, continual rains, and hurricanes is caused by rising CO_2 levels from expanding world economies. Several heat waves have occurred in Europe and the US in recent years. The year 1995 was the warmest since global records began in 1856, and the 2003 heat wave killed hundreds in Europe. An increase in violent and extreme summer storms has occurred the past several years. Ice core studies, photographic recordings of retreating glaciers, and recorded temperatures provide solid evidence that global warming is occurring at

USDA/ARS/FD-WSRU, Ft. Detrick, MD 21702, USA

Author for correspondence: N.W. Schaad; e-mail: norman.schaad@ars.usda.gov

M'B. Fatmi et al. (eds.), *Pseudomonas syringae Pathovars and Related Pathogens.*   369
© Springer Science+Business Media B.V. 2008

an alarming rate. If such conditions continue, damage caused by these emerging heat-loving plant pathogenic bacteria should be expected to increase.

Keywords Crop diseases, global warming

1 Introduction

This is a review of the emergence of several heat-loving bacteria and the possible role of global warming in their emergence. Several bacteria previously classified as non-fluorescent, oxidase positive pseudomonads, *Ralstonia*, *Acidovorax*, and *Burkholderia* species have emerged recently as serious problems worldwide (Stead et al., 2003).

2 Ralstonia

Potentially, the most damaging bacterium is the high-threat, APHIS Select agent bacterium *Ralstonia solanacearum* race 3, biovar 2. The common *R. solanacearum* race 3 biovar 1 strain is found in solanaceous crops throughout the tropics and subtropics (Kelman, 1953; Titatarn, 1986) and is a limiting factor in commercial production of tomatoes and potatoes. Simplot Corporation (Boise, ID), a major contractor for providing potatoes for McDonald's restaurants, attempted for several years to grow Idaho cultivar 'Netted Gem' potatoes in higher elevations in several southeast Asian countries but failed due to *R. solanacearum*. Whereas *R. solanacearum* race 3 biovar 1 causes damage in warmer tropical climates, biovar 2 also infects during relatively cool weather making it an additional threat to the major potato production areas in temperate climates. More recently the biovar 2 was identified on geraniums growing in commercial greenhouses in Pennsylvania (Kim et al., 2003). Because the infected geraniums posed a direct threat to the US potato industry, thousands of plants were destroyed resulting in a loss of several million dollars to the flower nursery industry.

3 Acidovorax

Acidovorax avenae subsp. *citrulli* has emerged over the past 20 years as a major seedborne pathogen on several cucurbits including watermelon, melons, and pumpkins (Latin and Hopkins, 1995; Schaad et al., 2003). Presence of the disease has threatened the existence of the watermelon seed industry due to possible legal action. The seedling stage of the disease was reported for the first time in 1965 and occurred only in certain Plant Introduction lines growing for increase at the USDA Plant Introduction Station in Griffin, Georgia. (Webb and Goth, 1965). The causal organism was not described until 13 years later (Schaad et al., 1978). The fruit rotting stage of the disease (watermelon fruit blotch) was described in 1988 by Wall et al. (1990) in the Mariana Islands. Unusually heavy rains occurred in 1987 during fruit maturation and widespread epidemics resulted with many growers losing their entire crop (Wall and Santos, 1988). Since that time, watermelon fruit blotch (WFB) has emerged as a serious threat to the estimated $450 million USA whole-sale watermelon crop and $75 million seed and transplant industry (Schaad et al., 2003). WFB has become widespread during periods of increased hurricane activity and warmer, wetter seasons and is a serious problem with crops grown under centre pivot sprinkler irrigation in Texas (T. Isakeit, 1997; personal communication). The disease has become widespread in China since being identified in 2000 (Zhao, 2001), and has become an important disease in watermelons in Thailand (Kawicha et al., 2002).

A. avenae subsp. *avenae* (*Pseudomonas avenae*) was described for the first time in oats (Manns, 1909). Additionally, the organism has been described causing losses in both field and sweet corn in the USA. (Johnson et al., 1949; Sumner and Schaad, 1977) but only sporadic losses have occurred in field corn. Rice strains of *A. avenae* subsp. *avenae* have caused severe losses in rice throughout Southeast Asia, China, and Africa (Kihupi et al., 1999). Most recently, *A. avenae* subsp. *avenae* has emerged as the cause of sterile panicles in rice in Italy (Cortesi et al., 2005). The condition was diagnosed initially as a physiological condition but recent studies have shown the condition to be caused by *A. avenae* subsp. *avenae* (Cortesi et al., 2005). Surveys showed the disease to be present in late season dur-ing periods of abnormally warm weather; currently crop loss estimates are only 1–2% but that should increase with increasing temperatures. The occurrence of higher than normal temperature in Europe and the rice growing area of Italy during the past few years favors *A. avenae* subsp. *avenae*. If such a pattern continues, the disease will most likely continue to increase. Spikelet induced sterility occurs in rice in Asia when the temperature exceeds 35°C at flowering (Matsui et al., 2000). Seed sterility has been reported to occur in the second and third crop in Japan dur-ing periods of unusually high temperatures and rain (Yoshida et al., 1981). It is possible that high temperature stress and rice sterility, which occurs sporadically throughout Asia (Yoshida et al., 1981) may be in part caused by *A. avenae* subsp. *avenae* as shown in Italy (Cortesi et al., 2005) and not solely environmental factors as described (Matsui et al., 2000). A new species of *Acidovorax*, *A. anthurii*, recently has been described causing severe losses in *Anthurium* spp. in tropical West Indies (Gardan et al., 2000). This disease should be monitored closely. All plant pathogenic *Acidovorax* species except a recently proposed species, *A. vale-rianellae* (Gardan et al., 2002), are normally found during hot, wet weather and in the warmer parts of the world. In contrast, this newly proposed species has been reported from lamb's lettuce (*Valerianella locusta*), a cool season crop in southern France (Gardan et al., 2002). Either the disease occurs late in the sea-son during heat waves or the organism is only distantly related to *A. avenae*. In fact, preliminary 16S rDNA and ITS sequencing showed *A. valerianellae* formed a separate branch more closely related to *A. facilis* than to *A. avenae* (Schaad, unpublished data).

4   Burkholderia

Burkholderia includes several important pathogens: *B. glumae* on rice; *B. gladioli* on onion, garlic, iris, tulip, orchid, and gladiolus; *B. caryophylli* on carnation; *B. andropogonis* on sorghum (Bradbury, 1986), and *B. agaricicola* on mushrooms (Lincoln et al., 1991). Bacterial seedling rot and panicle blight caused by *B. glumae* is perhaps the most serious bacterial disease of rice in Japan (Uematsu et al., 1997). Panicle blight has recently emerged as a problem in certain rice cultivars in the southern United States (Sayler et al., 2006). The occurrence of rice seedling rot has become much more prevalent in southern South Korea due to recent increased summer temperatures (Jeong et al., 2003). *B. gladioli* was described in Thailand on *Dendrobium* orchid in 1983 (Chuenchitt et al., 1983) and has emerged as a serious, widespread disease in many different orchids in Thailand (P. Thammakijjawat, Thailand Department of Agriculture, personal communications). Bacterial leaf stripe of sorghum, caused by *B. andropogonis*, is common in sorghum and often causes yield loss in warm, humid growing areas (Moffet et al., 1986). Recently, *B. andropogonis* has been identified on jojoba (*Simmondsia chinesis*), an important plant grown in eastern Australia for lubricants, pharmaceuticals, and cosmetics (Cother et al., 2004). The plant grows under very high temperatures and the disease occurs in nursery stock grown under overhead watering. A common link to the emergence of all the above bacteria is an optimum growth temperature of 32–36°C; most grow well at 41–42°C (Schaad et al., 1975, 1978; Bradbury, 1986). In contrast, few other plant bacteria grow well above 36°C (Bradbury, 1986).

5   Global Warming

Why the increase in these heat-loving bacteria? The most likely explanation is the influence of global warming on the World's climate. The evidence for global warming is quite broad, including, major shifts in recorded temperature and precipitation, melting glaciers and reduced snow cover, and more frequent and severe storms and droughts. Studies have shown global warming is caused primarily by heat-trapping greenhouse gas (GHG) emissions (IPCC, 2001). Industrial activity, from the burning of fossil fuels such as coal, oil, and gas, generates CO_2 and other gases which trap the sun's rays in the atmosphere and enhance the natural "greenhouse effect" (Gore, 2006). Although automobiles and industry are considered the major contributors of GHG, the increase in air traffic is emerging as a major factor. "Concentrations of CO_2, the prime GHG, have increased 30% during the past 100 years" (Fallon, 1997). Before the early 1800s the atmosphere contained about 290 ppm CO_2; by 1995 the figure was 360 ppm (IPCC, 2001). According to World Meteorological Organization, CO_2 reached 379 ppm in 2006 (Anonymous, 2006). It is estimated the CO_2 level will increase to 800 ppm by the end of the century and bring serious consequences to plants (IPCC, 2001).

Strong evidence for global warming is provided by ice core studies and photographic recordings of glaciers (Thompson, 2004; Thompson et al., 2006). Changes in $d^{18}O$ levels in glacial ice cores reveal "... that within another century human activities may have nudged global-scale climate conditions closer to those which prevailed before 5,000 years ago, during the early Holocene" (Thompson et al., 2006). Most glaciers outside the polar regions have retreated significantly (Fig. 1, Thompson et al., 2006). Photos taken of Qori Kalis, the largest glacier in the Peruvian Andes, from 1978 to 2005 show the glacier retreating and the subsequent formation of the proglacier lake (Fig. 2, Thompson et al., 2006). Of the three remaining ice fields on Mt. Kilimanjaro only 18% remain (Thomspon et al., 2006). Ice core evidence indicating that rising temperature serves as the primary variable driving the retreat of most tropical ice fields (Thompson et al., 2006) is supported by satellite-derived lower troposphere temperatures (Mears and Wentz, 2005). Satellite imagery shows the Artic ice pack has shrunk nearly 30% over the past 25 years and during the winter of 2004 it shrunk by 28,000 square miles (Struck, 2006b). Based upon $d^{18}O$ measurements of the Huascaran glacier ice cap cores (Fig. 1), temperatures in the tropical Pacific have risen 1.5–2.0°C between the last 11,000–8,000 years ago (Thompson et al., 2006). There has been a gradual increase in the mean temperature of 0.5–1°C over the past century and a projected increase of 1.4–5.8°C for this century (IPCC, 2001). If this trend is not reversed, the earth could warm by 1.0–3.5°C within 50–100 years and greatly alter the Earth's climate (IPCC, 2001).

There is growing evidence that rising GHG emissions from growing world economies will likely alter the mean seasonal temperature and increase the occurrence of extreme climate, including heat waves, continual rains, and hurricanes (Arnell, 2002; Diffenbaugh et al., 2005). Computer models predict an increase of 3°C over the next century which will increase risks of extreme weather such as flooding and wildfires (Scholze et al., 2006). In fact, the recent August–September 2006 'Day' fire near Ojai, California was one of the largest fires ever recorded in the US. The fire, covering over 245 square miles, was fuelled by strong Santa Anna winds and increased temperatures over the period. Global warming is predicted to impact human health (Patz et al., 2005), water (Barnett et al., 2005), and crops (White et al., 2006).

Extreme upward heat shifts can reduce the quality of many crops and higher night temperatures are known to reduce rice yields (Peng et al., 2004). High quality wine grape production, which is dependent upon a very narrow range of temperatures and rain during berry maturation, will be especially susceptible to climate change (White et al., 2006). It is predicted the premium wine production area in the US could decline by up to 81% by the late twenty-first century if temperatures continue to rise at current rates (White et al., 2006). In the wine regions of California, Oregon, and Washington, the mean growing season temperatures increased by 0.89°C from 1948 to 2002 (Jones, 2005). Using historical climate records and a regional climate model, high quality wine grape production areas are predicted to decline by 50% (White, 2006). These are highly significant effects since suitable soil for growing wine grapes is limited.

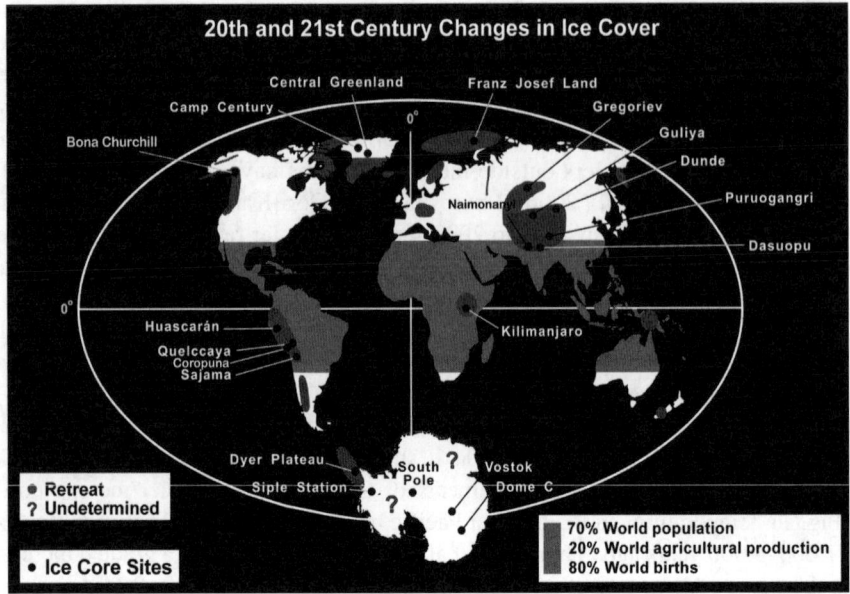

Fig. 1 Changes in twentieth- and twenty-first-century ice cover. Taken from: Thompson, L.G., et al. 2006. Abrupt tropical climate change: past and present. Proc. Nat. Acad. Sci. 103:10536–10543

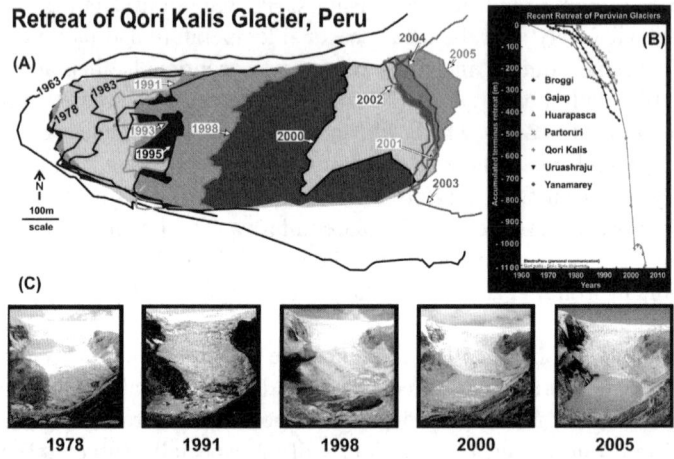

Fig. 2 Retreat of Qori Kalis glacier in Peru. Taken from: Thompson, L.G., et al. 2006. Abrupt tropical climate change: past and present. Proc. Nat. Acad. Sci. 103:10536–10543

Extreme weather conditions, including extended heat waves, continual periods of rain, and storms such as hurricanes favor diseases caused by these heat-loving bacteria. Several heat waves have occurred in Europe and the USA over the past several years. The year 1995 was the warmest year in Europe since global records began in 1856 (Beniston, 2004) and 2003's heat wave killed hundreds (Scott et al.,

2004; International Federation of Red Cross, 2004). Grape ripening records show that temperatures in 2003 were higher than any year since 1370 (Chuine et al., 2004). The year 2005 was the warmest recorded in the USA and that year was surpassed by 2006 (Heath, 2006); Sacramento, California recorded a record 10 consecutive days over 43°C in late July.

Changes in the frequency of extreme temperatures may have a greater effect on biological and agricultural systems than changes in mean climate. Projected changes of periods of higher humidity/precipitation will increase plant diseases (Carroll and Wilcox, 2003; Pardo et al., 2005) and result in the need for extensive disease control. Bacterial blight of corn and wheat, caused by *A. avenae* subsp. *avenae*, is much more severe at higher temperatures (Schaad and Sumner, 1980). Under greenhouse conditions, good infection occurred with daytime temperatures of 30–35°C whereas very poor infection occurred at 21°C or less (Johnson and Robert, 1949). Under controlled growth chamber conditions, an increase in temperatures from 30°C day/22°C night to 30°C day/26°C night resulted in an increase in disease of 40% and 44% for corn (Table 1) and wheat (Table 2), respectively (Schaad and Sumner, 1980). Growth studies in liquid medium showed the mean doubling time of four strains of *A. avenae* subsp. *avenae* decreased from 77 to 57 min when the temperature increased from 29°C to 36°C (Table 3) (Schaad and Sumner, 1980). Besides promoting bacterial diseases favored by high temperatures, increased temperatures and accompanying rise in CO_2 have a negative effect on the seed development of many crops (Allen and Vu, 2006). Seed yield generally decreases about 6% for every 0.5°C above a plant's optimal mean daily temperature (OMDT) (Allen and Vu, 2006). According to controlled growth chamber studies (Allen and Vu, 2006), "as carbon dioxide rates rise, crop plants may grow larger, but seed development and yield is another matter". Milder winters may reduce frost and/or freezing damage to subtropical crops such as citrus and winter vegetables growing in transition areas, but insect vectors such as sharpshooters and aphids will flourish and the transmission of plant diseases are expected to increase as well. Studies have shown that increased temperatures have resulted in increased human vector-based diseases (Kovats et al., 2001). The milder winters in the USA have increased the numbers and range of the glassy winged sharpshooter in California and have resulted in a significant increase in the spread of *Xylella fastidiosa subsp. fastidiosa*, the causal agent of Pierce's disease of grape (Hopkins and Purcell, 2002). Similarly, leaf scorch in oak, maple, and other shade trees has increased in the South and East Coast. Canadian Forest Service scientists report that the average winter temperature in Canada has risen over 2°C over the last century. The resulting warmer winters have resulted in an increase in the over wintering of lodgepole pine beetles. Large-scale infestations of beetles are destroying large numbers of pine trees in British Columbia (Struck, 2006a). The recent increase in summer temperatures over the southern US has been 1–2°C higher than the optimum for most crops (Allen and Vu, 2006). Increased temperatures favor these heat-loving pathogens because crop plants infected by these pathogens often suffer at high temperatures. Even rice, an important food plant in the tropics, does not set seed well at temperatures above 36°C (Matsui et al., 2000).

Table 1 Effect of temperature on severity of bacterial blight of corn[a]

Day/night temperature, °C[b]	Mean lesion area[c]
30/26	6.3
30/22	4.5
22/18	4.2
22/14	3.9
18/14	2.7

[a]Each of eight plants for each temperature combination inoculated at 3–4 leaf stage using 0.1 ml of inoculum containing 10^6 colony forming units *Acidovorax avenae* subsp. *avenae* ATCC 19860
[b]Plants grown in a phytotron with a 12 h day/night
[c]Recorded after 7 days, mm², log^{10}

Table 2 Effect of temperature on severity of bacterial blight of wheat[a]

Day/night temperature, °C[b]	Mean lesion area[c]
30/26	1.3
30/22	0.9
22/18	0.2
22/14	0.4
18/14	0.3

[a]Each of eight plants for each temperature combination inoculated at 3–4 leaf stage using 0.1 ml of inoculum containing 10^6 colony forming units *Acidovorax avenae* subsp. *avenae* ATCC 19860, C-21, C-138, and C-139
[b]Plants grown in a phytotron with a 12 h day/night
[c]Recorded after 7 days, mm², log^{10}

Table 3 Effect of temperature on growth of *Acidovorax avenae* subsp. *avenae*[a]

Temperature, °C	Doubling time (mim)[b]
36	46–67
29	64–91
21	100–212
15	172–310

[a]Determined in liquid medium 523 on a rotary shaker
[b]Mean range of four strains, ATCC 19860, C-10, C-12, and C-13

6   Conclusions

The projected increase in global temperatures will likely favor continual emergence of bacteria that prefer high temperatures. Furthermore, we should expect to see the emergence of additional heat-loving bacteria on additional crops in the not so far future.

Acknowledgements A special thanks to Lonie Thompson for allowing the use of Figs. 1 and 2.

References

Allen, L. H. Jr. and Vu, J.C., 2006, Rising temperatures and plant productivity. In Jim Core (ed). Agriculture Research, August 2006, p. 21., U.S. Department of Agriculture, Baltsville, MD.

Anonymous, 2006, Greenhouse gases hit high. Washington Post. Nov. 4, National News, p. A11.

Arnell, N.W., 2002, The consequences of CO_2 stabilization for the impacts of climate change. Climate Change **53**:413–466.

Barnett, T.P., Pierce, D.W., AchutaRoa, K.M., Santer, B.D. Gregory, J.M., and Washington, W.M., 2005, Penetration of human-induced warming into the world's oceans. Science **309**:284–287.

Beniston, M., 2004, The 2003 heatwave in Europe: a shape of things to come? An analysis based on Swiss climatological data and model simulations. Geophys. Res. Lett. **31**:2022–2026.

Bradbury, J.F., 1986, Guide to plant pathogenic bacteria. Kew, Surrey, UK: CAB International Mycological Institute.

Carroll, J.E. and Wilcox, W.F., 2003, Effects of humidity on the development of grape vine powdery mildew. Phytopathology **93**:1137–1144.

Chuenchitt, S., Dhirabhava, W.l., Karnjanarat, S., Buangsuwon, D., and Uematsu, T., 1983, A new bacterial disease on orchids *Dendrobium* sp. caused by *Pseudomonas gladioli*. Kasetsart J. **17**:26–36.

Chuine, L., Yiou, P., Viovy, N., Seguin, B., Daux, V., and Ladurie, E.L., 2004, Grape ripening as a past climate indicator. Nature **432**:289–290.

Cortesi, P., Bartoli, F., Song, W.Y., and Schaad, N.W., 2005, First report of *Acidovorax avenae* ssp. *avenae* on rice panicle in Italy. J. Plant Pathol. **87**:75–76.

Cother, E.J., Noble, D., Peters, B.J., Albiston, A., and Ash, G.J., 2004, A new bacterial disease of jojoba (*Simmondsia chinensis*) caused by *Burkholderia andropogonis*. Plant Pathol. **53**:129–135.

Diffenbaugh, N.S., Pal, J.S., Trapp, R.J., and Giorgi, F., 2005, Fine scale processes regulate the response of extreme events to global climate change. Proc. Acad. Sci. USA **102**:15774–15778.

Fallon, S., 1997, A few degrees makes a world of difference. The Planet Newsletter **4**:1–4.

Gardan, L., Dauga, C., Prior, P., Gillis, M., and Saddler, G.S., 2000, *Acidovorax anthurii sp.nov.*, a new phytopathogenic bacterium which causes bacterial leaf spot of anthurium. Int. J. Syst. Evol. Microbiol. **53**:253–246.

Gardan, L., Stead, D.E., Dauga, C., and Gillis, M., 2002, *Acidovorax valerianellae sp. nov.*, anovel pathogen of lamb's lettuce (*Valerianella locusta* (L.) Laterr.) Int. J. Syst. Evol. Microbiol. **53**:795–800.

Gore, A., 2006, An Inconvenient Truth: The Planety Emergency of Global Warming and What We Can Do About It. Rodale, Emmaus, PA.

Heath, B., 2006. This was hottest summer since 1936, report says., September 15, 2006. USA Today, p. A04.

Hopkins, D.L. and Purcell, S., 2002, *Xylella fastidiosa*: cause of Pierce's disease of grape vine and other emergent diseases. Plant Dis. **86**:1056–1066.

International Federation of Red Cross and Red Cresent Societies, 2004, World Disaster Report 2004, Ch. 2.

IPCC, 2001, Climate change 2001: the scientific basis. Contribution of Working Group I to the Third Assessment Report of the Intergovernmental Panel on Climate Change. Cambridge University Press, Cambridge, UK, New York.

Isakeit, T., 1997, First report of infection of honeydew with *Acidovorax avenae* subsp. *citrulli*. Plant Dis. **78**:831.

Jeong, Y., Kim, J., Kim, S., Kang, Y., Nagamatsu, T., and Hwang, I., 2003, Toxoflavin produced by *Burkholderia glumae* causing rice grain rot is responsible for inducing bacterial wilt in many field crops. Plant Dis. **87**:890–895.

Johnson, A.G., Robert, A.L., and Cash, L., 1949, Bacterial leaf blight and stalk rot of corn. J. Agric. Res. **78**:719–732.

Jones, G.V., 2005, Climate changes in the western United States grape growing regions. Acta Horticulturae (ISHI). **689**:41–60.

Kawicha, P., Thummabenjapone, P., and Sirithorn, P., 2002, Genetic diversity within *Acidovorax avenae* subsp. *citrulli* in cucurbit production areas of Northeast Thailand. Summary First International Conference on Tropical and Subtropical Plant Diseases. November 5–8, 2000, Chieng Mai, Thailand, p. 125.

Kelman, A., 1953, The bacterial wilt caused by *Pseudmonas solanacearum*. North Carolina Agr. Exp. Sta. Tech. Bull. 99.

Kihupi, A.L., Mabagala, R.B., and Mortensen, C.N., 1999, Occurrence of *Acidovorax avenae* subsp. *avenae* in rice seed in Tanzania. African Pl. Protect. **5**:55–58.

Kim, S.H., Olson, T.N., and Schaad, N.W., 2003, *Ralstonia solanacearum* race 3 biovar 2, the causal agent of brown rot of potato, identified in geraniums in Pennsylvania, Delaware, and Connecticut. Plant Dis. **87**:450.

Kovats, R.S, Campbell-Lendrum, D.H., McMichael, A.J., Woodward, A., and Cox, J.S., 2001, Early effects of climate change: do they include changes in vectorborne-disease. Philos. Trans. R. Soc. B **356**:1057–1068.

Latin, R. and Hopkins, D.L., 1995, Bacterial fruit blotch of watermelon. Plant Dis. **79**:761–765.

Lincoln, S.P., Fermor, T.R, Stead, D.E., and Sellwood, J.E., 1991, Bacterial soft rot of *Agaricus bitorquis*. Plant Pathol. **40**:136–144.

Manns, T.F., 1909, The blade blight of oats-a bacterial disease. Ohio Agr. Exp. Sta. Res. Bull. **210**:91–167.

Matsui, T., Omasa, K., and Horie, T., 2000, High temperature at flowering inhibits swelling of pollen grains, a driving force for the dehiscence in rice (*Oryza sativa* L.). Plant Prod. Sci. **3**:430–434.

Mears, C.A. and Wentz, F.J., 2005, The effect of diurnal correction on satellite-derived lower tropospheric temperature. Science **309**:1548–1551

Moffet, M.L., Hayward, A.C., and Fahy, P.C., 1986, Five new hosts of *Pseudomonas andropogonis* occurring in eastern Australia: host range characterization of isolates. Plant Pathol. **35**:34–43.

Pardo, E., Martin, S., Sanchis, V., and Ramos, A.J., 2005, Impact of relative humidity and temperature on visible growth and OTA production of ochratoxigenic *Aspergillus ochracenus* isolates on grape. Food Microbiol. **22**:383–389.

Patz, J.A., Campbell-Lendrum, D., Holloway, T., and Foley, J.A., 2005, Impact of regional climate change on human health. Nature **438**:310–317.

Peng, S., Huang, J.L., Sheehy, J.E., Laza, R.C., Visperas, R.M., Zhong, X.H., Centeno, G.S., Khush, G.S., and Cassman, K.G., 2004, Rice yields decline with higher night temperature from global warming. Proc. Natl. Acad. Sci. USA **101**:9971–9975.

Sayler, R.J., Cartwright, R.D., and Yang, Y., 2006, Genetic characterization and real-time PCR detection of *Burkholderia glumae*, a newly emerging bacterial pathogen of rice in the United States. Plant Dis. **90**:603–606.

Schaad, N.W. and Sumner, D.R., 1980, Influence of temperature and light on severity of bacterial blight of corn, oats, and wheat. Plant Dis. **64**:481–483.

Schaad, N.W., Kado, C.I., and Sumner, D.R., 1975, Synonmy of *Pseudomonas avenae* Manns 1905 and *Pseudomonas alboprecipitans* Rosen 1922. Int. J. Syst. Bacteriol. **25**:133–137.

Schaad, N.W., Sowell, G. Jr., Goth, R.W., Colwell, R.R., and Webb, R.E., 1978, *Pseudomonas pseudoalcaligenes* subsp. *citrulli* subsp. nov. Int. J. Syst. Bacteriol. **28**:117–125.

Schaad, N.W., Postnikova, E., and Randhawa, P., 2003, Emergence of *Acidovorax avenae* subsp. *citrulli* as a crop threatening disease of watermelon and melon. In N.S. Iacobellis et al. (eds). *Pseudomonas syringae* and Related Pathogens. Kluwer, Dordrecht, The Netherlands, pp. 573–581.

Scholze, M., Knoor, W., Arnell, N.W., and Prentice, I.C., 2006, A climate-change risk analysis for world ecosystems. Proc. Natl. Acad. Sci. USA **103**:13116–13120.

Scott, P.A., Stone, D.A., and Allen, M.R., 2004, Human contribution to the European heat wave of 2003. Nature **432**:610–614.

Stead, D.E., Stanford, Aaspin, A., and Heeney, J., 2003, Current status of some new and some old plant pathogenic pseudomonads. In N.S. Iacobellis et al. (eds). *Pseudomonas syringae* and Related Pathogens. Kluwer, Dordrecht, The Netherlands, pp. 561–572.

Struck, D., 2006a, 'Rapid Warming' Spreads Havoc in Canada's Forests. Tiny beetles destroying pines. Washington Post, March 1, p. A01.

Struck, D., 2006b, Melting Artic makes way for man. Washington Post, November 5, p. A01.

Sumner, D.R. and Schaad, N.W., 1977, Epidemiology and control of bacterial blight of corn. Phytopathology. **67**:1113–1118.

Thompson, L.G., 2004, High altitude, mid and low-latitude ice core records: implications for our future. In L.D. Cecil, J.R. Green, and L.G. Thompson (eds). Earth Paleo-environments: Records Preserved in Mid-and Low-Latitude Glaciers (Developments in Paleoenvironmental Research, Vol. 9), Kluwer, Dordrecht, The Netherlands, pp. 3–15.

Thompson, L.G., Mosley-Thompson, E., Davis, M.E., Henderson, K.A., Brecher, H.H., Zagorodnov, V.S. et al., 2002, Kilimanjaro ice core records: evidence of Holocene climate change in tropical Africa. Science **298**:589–593.

Thompson, L.G., Mosley-Thompson, E., Brecher, H., Davis, M., Leon, B., Les, D., Lin, P., Mashiotta, T., and Mountain, K., 2006, Abrupt tropical climate change: past and present. Proc. Nat. Acad. Sci. USA **103**:10536–10543.

Titatarn, V., 1986, "Bacterial Wilt in Thailand". In G.J. Persley (ed). Bacterial Wilt Disease in Asia and the South Pacific: Proceedings of an International Workshop held at PCARRD, LasBanos, Philippines, 8–10 October, 1985. ACIAR Proceedings No.13,145 pp.

Uematsu, T., Yoshimura, D., Nishiyama, K, Ibaraki, T., and Fuji, H., 1997, Occurrence of bacterial seedling rot in nursery flat, caused by grain rot bacterium *Pseudomonas glumae*. Ann. Phytpathol. Soc. Jpn. **42**:310–312.

Wall, G.C. and Santos V.M., 1988, A new bacterial disease on watermelon in the Mariana islands (Abstr.). Phytopathology **78**:1605.

Wall, G.C., Santos, V.M., Cruz, F.J., and Nelson, D.A., 1990, Outbreak of watermelon fruit blotch in the Mariana Islands. Plant Dis. **74**:80.

Webb, R.E. and Goth, R.W., 1965, A seed-borne bacterium isolated from watermelon. Plant Dis. Rep. **49**:818–821.

White, M.A., Diffenbaugh, N.S., Jones, G.V., Pal, J.S., and Giorgi, F, 2006, Exteme heat reduces and shifts United States premium wine production in the 21st century. Proc. Nat. Acad. Sci. USA **103**:11217–11222.

Yoshida, S., Satake, T., and Mackill D.J., 1981, High-temperature stress in rice. IRRI Research Paper Series, No. 67, International Rice Research Institute (IRRI), Los Baños, Philippines.

Zhao, T., 2001, Pathogen identification of Hami melon bacterial fruit blotch. Acta Phytpathologica Sinica **4**:357–364.

Angular Leaf Spot of Cucurbits: A Bacterial Disease in Expansion in Morocco

M. Fatmi, M. Bougsiba, and T. Hosni

Abstract Angular leaf spot was first reported in Morocco in 1996 on melon grown under plastic houses in the region of Souss-Massa, Agadir. The disease was then encountered on melon in other melon producing areas during the period from late fall to spring and was associated with high humidity. In 2002, similar symptoms of the disease were observed on cucumber grown under plastic houses. In the fall 2003, the disease was observed for the first time on zucchini grown in both plastic house and open field. In spring 2006, the disease was observed for the first time on cantaloupe grown under small plastic tunnels. The losses induced by this disease varied with the region and depended on the agricultural practices used.

The isolates were identified as *Pseudomonas syringae* pv *lachrymans* on the basis of morphological, physiological, and biochemical tests. Serologically all the isolates recovered from cucumber, melon and zucchini reacted positively by immunofluorescent staining method (IF) using a polyclonal antibody. However there were some differences as compared to reference strains regarding production of fluorescent pigment on KB medium and levan production. Only 65% of the isolates were fluorescent and levan positive.

From pathogenicity studies, all the obtained isolates induced similar symptoms upon inoculation to their original plant host. But some differences were obtained when the isolates were cross inoculated. Some cucumber isolates were not pathogenic to melon or to zucchini. Likewise some melon isolates were not pathogenic to cucumber or to zucchini. Some zucchini isolates were not pathogenic to cucumber or to melon.

The relatedness among Moroccan isolates of *Pseudomonas syringae* pv. *lachrymans* was analyzed by rep-PCR using REP consensus primers. Two reference strains (UMG5070T and CCUGU49342) were also included in this study. Results obtained revealed that 18 out of 20 isolates (90%) generated a distinct group with high similarity superior to 70%. Low level of similarity was recorded between this group and reference strains with less than 35% of similarity. Correlation between fingerprint clustering and host origin was not found.

Département de Protection des Plantes, Institut Agronomique et Vétérinaire Hassan II, Complexe Horticole d'Agadir, BP. 18/S, Agadir, Morocco

Author for correspondence: M'barek Fatmi; e-mail: fatmi@iavcha.ac.ma

M'B. Fatmi et al. (eds.), *Pseudomonas syringae Pathovars and Related Pathogens.*   381
© Springer Science+Business Media B.V. 2008

Keywords Cucurbits, *Pseudomonas syringae* pv. *lachrymans*, characterisation, rep-PCR

1 Introduction

Cucurbits play an important role in the economy of Morocco. The cultivated species include melon, cantaloupe, cucumber, squash, watermelon, zucchini and pumpkin and are grown under both plastic houses and open field. In the last decade, cucurbits have known a great extension reaching acreage of about 50,000 ha in 2005 and a total production of about 160,000 tons. More than 50% of the production is exported towards Europe providing a source of hard currency for the country. However, the development of these crop species was faced to several problems and particularly plants diseases. Among theses diseases, bacterial angular leaf spot, caused by *Pseudomonas syringae* pv. *lachrymans* (Bradbury, 1986), constitutes a major problem to the production of the cucurbits and particularly in winter and spring.

Bacterial angular leaf spot is a serious disease worldwide and has been reported on several species of the *Cucurbitaceae* (Van Gundy and Walker, 1957; Hopkins and Schenk, 1972; Scortichini and Tropiano, 1991; El-Sadek et al., 1992; Harighi, 2007). The causal agent, *Pseudomonas syringae* pv. *lachrymans* has a worldwide distribution and is responsible for several outbreaks resulting in severe losses of the crop production (Ark, 1954; Van Gundy and Walker, 1957; Hopkins and Schenk, 1972; Scortichini and Tropiano, 1991; El-Sadek et al., 1992). The bacterium induces necrotic spots on leaves, stems, and fruits. The pathogen is transmitted through both contaminated and infected seeds (Leben, 1981) and persists in soil within the plant debris (Kritzman and Zutra, 1983).

Symptoms similar to those of angular leaf spot were first encountered on melon in 1996, then on cucumber in 2002, on zucchini in 2003 and on cantaloupe in 2006 in the valley of Souss-Massa region of Agadir, Morocco (Fatmi, unpublished). The objective of this work is to identify the causal agent and investigate the diversity of the isolates from different host plants.

2 Materials and Methods

2.1 Isolation and Bacterial Isolate Collection

Isolations were made from diseased cucurbit plants (melon, cucumber, zucchini and cantaloupe) originating from the Souss-Massa Valley, the main cucurbit producing region in Morocco. The collection of the studied isolates was undertaken from 1996 to 2006. Samples of the diseased leaves were collected and isolation was performed as described by Scortichini and Tropiano (1991). Briefly, the samples were washed using tap water to remove soil particles, then surface sterilized in 0.5% sodium hypochlorite for 30 s, rinsed twice using sterile distilled water, (SDW), and placed in sterile glass Petri dishes. Tissues in between necrotic and chlorotic areas were selected and crushed in sterile mortars containing (SDW). After 15–20 min, serial tenfold dilutions in tubes were carried out and from each tube 0.1 ml of the suspension was spread onto Petri dishes containing medium B of King (KB) (King et al., 1954). The plates were then incubated at 24°C for 48 h. Representative colonies were purified by repetitive streaking onto KB medium. The purified isolates were stored in 20% glycerol at −20°C for further characterization.

2.2   Biochemical and Nutritional Tests

Gram staining, biochemical, physiological and nutritional tests were performed according to the methods reported by Schaad et al. (2001). These tests include glucose fermentation/oxidation, production of fluorescent pigment, levan production, oxidase reaction, potato rot, arginine dihydrolase production, nitrate reduction, gelatine and starch hydrolysis, catalase reaction, growth at 37°C and 40°C, utilization of adonitol, inositol, D-mannitol, D-sorbitol, L-lactate, D(−) tartrate and L(+) tartrate. Hypersentive reaction on tobacco leaves (Var. *Xanthi*) was performed as described (Klement, 1963). Two reference strains of *Pseudomonas syringae* pv. *lachrymans* (UNMG5070T and CCUG49342) were used in this study.

2.3   Pathogenicity Test

Bacterial inocula were made from a 48 h-old culture grown on KB medium from all tested isolates in SDW. The bacterial suspensions were adjusted to about 10^8 CFU/ml using a spectrophometer (optical density of 0.1 at 640 nm). Each isolate was inoculated to five potted seedlings at the second true leaf stage of cucumber (cultivar *Admirable F1*), of melon (cultivar *Lavi*) and zucchini (cultivar *Blitz F1*). The inoculation was performed by dusting the cotyledons and the leaves with rubbing with a piece of cheesecloth previously dipped in the inoculum. Control seedlings were inoculated with sterile distilled water. After inoculation the plants were covered separately by plastic bags for 24 h. The plants were kept in a greenhouse under favorable temperature (25°C) and relatively high humidity conditions. The plants were watered and fertilized as required. The plants were observed regularly for disease symptoms development. Plants that showed dark green water soaked spots were recorded positive (El-Sadek et al., 1992).

2.4   Serological Characterization

A reference strain of *Pseudomonas syringae* pv. *lachrymans*, isolated from cucumber
(UNMG5070T), was used for antiserum production as described (Rowhani et al.,
1994). Briefly, from a 48 h old culture previously grown on KB agar medium at
25°C a bacterial suspension was prepared in sterile phosphate-buffered saline
(PBS) and washed twice by centrifugation at 10,000 g for 10 min. Washed cells
were suspended in PBS, adjusted to 10^8 CFU/ml (OD: 0.1 at 640 nm), and heated to
80°C for 10 min. The cell suspension was stored at 4°C until use. For immunization
a local rabbit was used for antibody production. The rabbit was initially immunized
with 1.0 ml of cell suspension emulsified with equal volume of complete Freund's
adjuvant followed by a second injection with 1.0 ml emulsified with incomplete adjuvant 14 days after the first immunization. The injections were administered subcutane-
ously with 0.2 ml of emulsion in ten locations along the back of the rabbit. Blood
was collected 28 and 34 days after the second injection and antibody titer was
evaluated by indirect immunofluorescent method (De Boer, 1990).

Bacterial suspensions from all the tested isolates were prepared from a 48 h
old culture previously grown on KB agar medium for 48 h at 25°C in sterile PBS
and adjusted to about 10^8 CFU/ml as previously described. The reactions of the
tested isolates to the antiserum were determined using indirect immunofluorescent
method (De Boer, 1990).

2.5   Molecular Characterization

A subset of 20 Moroccan isolates of *Pseudomonas syringae* pv. *lachrymans* including seven, seven, and six isolates obtained, respectively, from cucumber, melon,
and zucchini, and the two reference strains (UMG5070T and CCUGU49342) were
used in this study.

2.6   Total Genomic DNA Extraction

From a 48 h-old pure culture grown on KB medium plates, cells were removed with
a sterile loop (taking one loop full), and resuspended in 6 ml of Luria-Bertani broth
(LB) (10 g tryptone; 5 g yeast extract; 10 g NaCl in 1 l of SDW). Tubes were shaken
to homogenize the sample and placed on a rotary shaker at 200 rpm at 27°C for
24 h, according to Ausubel et al. (1988). The total bacterial DNA was extracted
with GenElute Bacterial Genomic DNA Kit, Mini (Sigma) following manufacturer's instructions and purified DNA was stored at 4°C. In order to check the quality
of the DNA extracted and RNA contamination, 10 µl from each sample was mixed
with 2 µl of the loading dye 6X (containing 40% of sucrose and 0.25% of bromophenol

blue), and electrophoresed in a 0.7% agarose gel. The gel was run in TAE 0.5X buffer at 100 V and 25 mA for 1 h; and visualized by staining with ethidium bromide (0.3 µg/ml) and UV illumination. Also, DNA concentration and purity were assessed by absorbency at 260 and 280 nm.

2.7 PCR Amplification

The subset of isolates were subjected to rep-PCR analysis according to the procedure described by Rademaker and De Bruijn (1997), using the REP1R-REP2l primers. Rep-PCR was performed with the repetitive extragenic palindromic (REP)-PCR primers 1R (5′-I(Inosine)IIICGICGICATCIGGC-3′) and 2I (5′-ICGICTTATCIGGCCTAC-3′). Amplification was performed in a 25-µl volume containing 1 µl (0.3 µg/µl) of each of the two opposing primers, 1 µl (~20 ng) of genomic DNA, 1.25 µl (10 mM) of dNTPs, 0.4 µl (2 U/µl) of Taq polymerase (DynaZyme), 0.2 µl (20 mg/ml) of BSA, 2.5 µl of DMSO, and 5 µl of 5× Gitschier buffer (83 mM (NH$_4$)$_2$ SO$_4$; 335 mM tris-HCl pH 8.8; 33.5 mM MgCl$_2$; 33.5 EDTA pH 8.8; 150 mM β-mercapto-ethanol). The reaction mixture was initially denatured for 2 min at 95°C, and then subjected to 35 cycles of 3 min at 94°C, 30 s at 92°C, 1 min at 40°C, and 8 min at 65°C; and a final extension for 8 min at 65°C with a model 2720 thermocycler (Applied Biosystems).

2.8 Visualization

The amplified fragments in 10 µl were size fractionated through a 1.5% agarose gel matrix in 0.5X Tris-acetate-EDTA (TAE; 40 mM Tris-acetate and 1 mM EDTA, pH 8.0) buffer For 14 h at 4°C and a constant voltage of 50 V to yield fingerprint patterns. After electrophoresis, the fragments of DNA separated in the gel were coloured for 3 h in a 0.75 mg/l solution of ethidium bromide in 0.5X TAE buffer. Then, the ethidium bromide-stained agarose gel was visualized in a light UV transilluminator and a digital picture was captured through digital television camera (EuroClone). The molecular sizes of fragments generated were estimated by comparison with simultaneously run Ladder Mix (MBI Fermentas, Burlington, ON, Canada).

The contrast and the brightness of the gel digital picture have been improved with Adobe Photoshop 7 software. Differently sized DNA bands were identified by comparing their migration distance with that of the marker, normalized, and a similarity matrix was generated (0 = absence of the band; 1 = presence of the band). Cluster analysis was performed on the similarity matrix produced using Dice's coefficient (Dice, 1945) and subjected to the Unweighted Pair Group Method with Arithmetic Mean (UPGMA) clustering algorithm, using NTSYSpc software (Exeter Software, New York, NY, USA) version 2.1.

3   Results

3.1   *Identification*

Forty-three bacterial isolates were obtained from melon, cucumber and zucchini
(Table 1). All of them were aerobes, do not grow at 40°C and induce hypersensitivity when infiltrated to tobacco leaves. All the isolates were negative in oxidase
reaction, potato rot, arginine dihydrolase production, in the reduction of nitrate,
starch hydrolysis. The reference strains gave the same results. Regarding levan and
fluorescent pigment production, the isolates reacted differently. Only 37.3% of the
isolates produced fluorescent pigment on KB medium and 76.7% of the isolates
were positive in the production of levan. 18.6% of the isolates were fluorescent and
levan production positive. The reference strains were fluorescent and levan production positive. All the isolates assimilated D-mannitol, inositol, D-sorbitol and L(+)
tartrate but did not assimilate adonitol, L-lactate and D(−) tartrate as did the reference strains.

The titer of the obtained antiserum was 1/400. All the isolates reacted positively
with the antiserum at the dilution of 1/100. Seventy-nine percent of the isolates

Table 1 Pathogenicity of isolates *Pseudomonas syringae* pv. *lachrymans* on melon,
cucumber and zucchini

Host origin	Isolates	Pathogenicity on		
		Melon	Cucumber	Zucchini
Melon	P03-19, P03-20, P03-24, P04-44, P04-45, P04-50, P04-51	+	−	−
	P03-25, P03-26, P04-48, P04-52, P04-53, P04-54, P04-56	+	+	+
	P03-17, P04-47	+	+	−
	P96-01, P96-02, P03-14, P03-18, P03-21, P03-22, P03-27, P04-49, P04-55	+	−	+
Cucumber	[a]UNMG5070T, [a]CCUG49342	−	+	−
	P02-04, P02-05, P02-06, P02-07, P02-09, P02-10	+	+	+
	P02-12	+	+	−
Zucchini	P03-31, P03-32, P03-34, P03-38, P03-41,	+	+	+
	P03-30, P03-33, P03-37, P03-39	+	−	+
	P03-42, P03-43	−	+	+

[a] References strains

reacted positively with the antiserum at the dilution 1/200. The cells presented similar morphology as compared to the reference strains.

All the isolates proved to be pathogenic when inoculated to their respective plant host. Specific symptoms were observed 1 week after inoculation. Reisolations gave the same results both in morphological characters, biochemical tests and hypersensitivity on tobacco. The control plants (inoculated with SDW) did not show any symptoms of the disease. Some differences in symptom development were obtained when the isolates were cross inoculated. Some cucumber isolates were not pathogenic to melon or to zucchini. In the same way, some melon isolates were not pathogenic to cucumber or to zucchini. Some zucchini isolates were not pathogenic to cucumber or to melon (Table 1).

3.2 Relatedness Among Isolates of Pseudomonas syringae pv. lachrymans Isolated from Morocco

Rep-PCR using REP consensus primers generated complex fingerprint patterns consisting of 20 or more distinct bands ranging in size from 0.2 kb to approximately 4 kb (Fig. 1). The dendrogram obtained revealed two distinct groups arbitrarily designated types A and B (Fig. 2). Eighteen isolates (90%) and the pathovar reference strain UMG5070T are in the group A, while the isolate P03-31, P02-06, and the pathovar reference strain CCUGU49342 are in the group B. Group A isolates are 35% similar while isolates in group B have similarity of 45%. The similarity between groups A and B was 19%. Further subdivision concern group A into two distinct subgroups arbitrarily designated types A1 and A2 (Fig. 2). Subgroup A1 isolates had more than 70% similarity and contained three different clusters presenting identical fingerprinting patterns (100% similarity), whereas the smaller subgroup A2 contained only the pathovar reference strain UMG5070T.

Cluster analysis based on the fingerprints obtained by rep-PCR analysis showed no clear relationship between the host and geographic origins of the isolates and their grouping. Isolates of *P. syringae* pv. *lachrymans* with similar host ranges are thought to retain a subset of translocated effectors that are necessary for virulence in a specific host. Therefore, to understand the pathogenicity and host range of a given *P. syringae* pv. *lachrymans* isolate, it is important to identify the effectors that are expressed and translocated by that isolate.

4 Discussion

Pseudomonas syringe pv. *lachrymans* proved to be a serious problem to melon, cucumber, and zucchini plants grown under plastic houses during winter and spring seasons in Morocco. With the extension of cucurbits in Morocco, this pathogen will

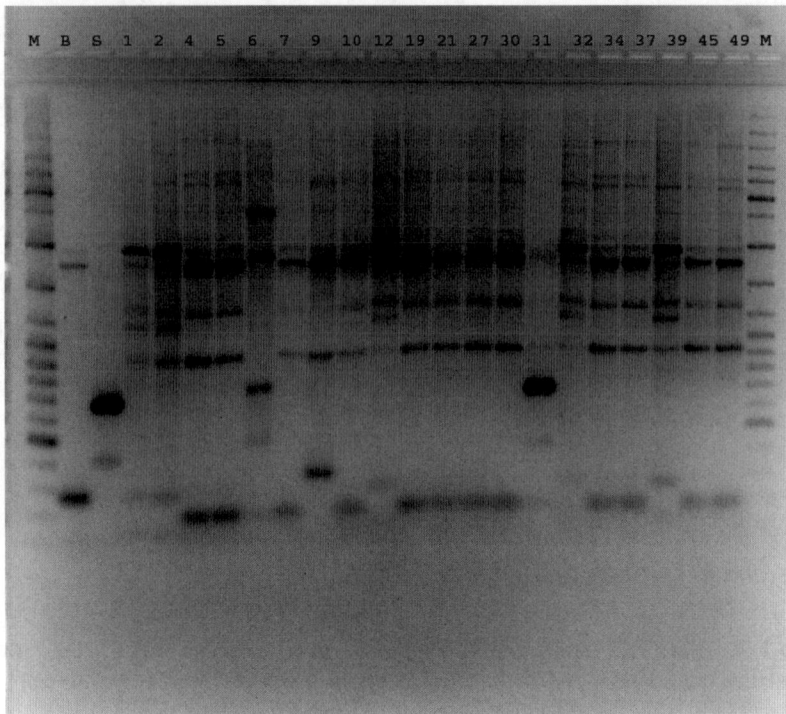

Fig. 1 Agarose gel electrophoresis of polymerase chain reaction (PCR) fingerprint patterns obtained from total genomic DNA of *P. syringae* pv. *lachrymans* isolates using primers corresponding to repetitive extragenic palindromic sequences (rep-PCR). Lane M: Ladder mix (MBI Fermentas). Lane 1: Reference strain UMG5070T (B); lane 2: Reference strain CCUGU49342 (S); lanes 3–20: *P. syringae* pv. *lachrymans* isolates from Morocco (referred here by the serial number of the isolates and not the hole number because of space)

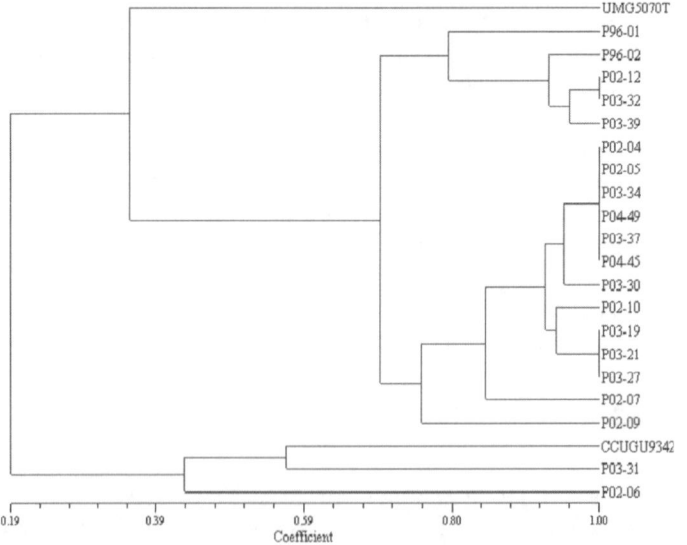

Fig. 2 Dendrogram of relatedness derived by cluster analysis of similarities between *P. syringae* pv. *lachrymans* isolates derived from rep-PCR DNA fingerprints generated from the REP consensus primers. Analysis was performed using the similarity coefficient of Dice and UPGMA clustering

constitute a limiting factor for the production of these crops. It has been reported that this disease is capable of causing serious damages to cucurbits cultivated in temperate, subtropical and dry climate regions and in winter, spring, or summer (Ark, 1954; Hopkins and Schenk, 1972; Umekawa and Watanabe, 1982). The aggressiveness of this disease is essentially related to high relative humidity content of the air and temperatures ranging from 15°C to 28°C (Umekawa and Watanabe, 1982). Such conditions are met in crops grown under plastic houses in both winter and spring seasons. The question that was raised by the growers is how this pathogen was introduced to Morocco since this pathogen has never been observed until 1996 in areas where cucurbits have been cultivated for many years. *Pseudomonas syringe* pv. *lachrymans* is a seedborne (Leben, 1981) and can survive in infected plant debris (Wiles and Walker, 1951; Van Gundy and Walker, 1957; Kritzman and Zutra, 1983; Leben, 1983). Consequently, it would be useful to ascertain the presence of the pathogen in commercial seeds to prevent primary source of inoculum.

Regarding the diversity of the studied isolates, all of them reacted similarly to the reference strains. However there were some differences as compared to reference strains regarding production of fluorescent pigment on KB medium and levan production. Thirty-five percent of the isolates did not produce a fluorescent pigment on KB medium and were levan production negative. This might be due to some differences among the isolates. Differences were also obtained when the isolates were cross inoculated. Some cucumber isolates were not pathogenic to melon or to zucchini. Likewise some melon isolates were not pathogenic to cucumber or to zucchini. Some zucchini isolates were not pathogenic to cucumber or to melon. Further studies are necessary to asses the variability of the isolates in levan and fluorescent pigment production and eventually their parasitic specialization on different species of *Cucurbitaceae*.

The rep-PCR results showed that 90% of the isolates generated a distinct group with high similarity superior to 70%. Low level of similarity was recorded between this group and reference strains with less than 35% of similarity. Correlation between fingerprint clustering and host origin was not found.

References

Ark, P.A., 1954. Angular leaf spot of squash. Plant Dis. Rep. 38: 201–203.

Ausubel, F.M., Brent, R., Kingston, R.E., Moore, D.D., Seidman, J.G., Smith, J.A., Struhl, K., 1988. Current Protocols of Molecular Biology. Wiley, New York.

Bradbury, J.F., 1986. Guide to Plant Pathogenic Bacteria. CAB International Mycological Institute, Ferry Lane, Kew, UK.

De Boer, S.H., 1990. Immunofluorescence for bacteria. In: Hampton, R., Ball, E., De Boer, S. (eds.), Serological Methods for Detection and Identification of Viral and Bacterial Plant Pathogens. A Laboratory Manual. APS, 389 pp.

Dice, L., 1945. Measurement of the amount of ecological association between species. Ecology 26: 297–302.

El-Sadek, S.A.M., Abdel-Latif, M.R., Abdel-Gawad, T.I., and Hussein, N.A., 1992. Occurrence of angular leaf spot disease in greenhouse cucumbers in Egypt. Egypt. J. Microbiol. 27(2): 157–175.

Harighi, B., 2007. Angular leaf spot of cucumber caused by *Pseudomonas syringae* pv. *lachrymans* in Kurdistan. Plant Dis. 91(6):769. Abstract.

Hopkins, D.L. and Schenk, N.C., 1972. Bacterial leaf spot of watermelon caused by *Pseudomonas syringae* pv *lachrymans*. Phytopathology 62: 542–545.

King, E.O., Ward, M.K., and Raney, D.E., 1954. Two simple media for the demonstration of pyocyanin and fluorescein. J. Clin. Med. 44: 301–307.

Klement, Z., 1963. Rapid detection of the pathogenicity of phytopathogenic Pseudomonads. Nature 199: 299–300.

Kritzman, G. and Zutra, D., 1983. Systemic movement of *Pseudomonas syringae* pv. *lachrymans* in the stem, leaves, fruits and seeds of cucumber. Can. J. Plant Pathol. 5: 273–278.

Leben, C., 1981. Survival of *Pseudomonas syringae* pv. *lachrymans* with cucumber seed. Can. J. Plant Pathol. 3: 247–249.

Leben, C. 1983. Association of *Pseudomonas syringae* pv. *lachrymans* and other bacterial pathogens with roots. Phytopathology 73: 577–581.

Rademaker, J.L.W. and De Bruijn, F.J., 1997. Characterization and classification of microbes by rep-PCR genomic fingerprinting and computer-assisted pattern analysis. In: Caetano-Anolles, G. and Gressfoff, P. (eds.), Protocols, Application and Overviews, pp. 151–171. Wiley, New York.

Rowhani, A., Feliciano, A.J., Lips, T., and Gubler, W.D., 1994. Rapid identification of Xanthomonas fragariae in infected strawberry leaves by enzyme-linked immunosorbent assay. Plant Dis. 78: 248–250.

Schaad, N.W., Jones, J.B., and Chun, W., 2001. Laboratory Guide for Identification of Plant Pathogenic Bacteria, 3rd Edition. APS, 373 pp.

Scortichini, M. and Tropiano, F.G., 1991. Occurrence of *Pseudomonas syringae* pv. *lachrymans* in black zucchini squash. Petria 1: 117–182.

Van Gundy, S.D. and Walker, J.C., 1957. Seed transmission, overwintering, and host range of the cucrbit angular leaf spot pathogen. Plant Dis. Rep. 41: 137–140.

Umekawa, M. and Watanabe, Y., 1982. Temperature and humidity to the occurrence of angular leaf spot of cucumber grown under plastic house. Ann. Phytopathol. Soc. Japan 48: 301–307.

Wiles, A.B. and Walker, J.C., 1951. The relation of *Pseudomonas lachrymans* to cucumber fruits and seeds. Phytopathology 41: 1059–1064.

Panicle Sterility and Grain Discolouration: New and Emerging Bacterial Diseases of Rice in Italy

P. Cortesi[1], F. Bartoli[1], C. Pizzatti[1], D. Bertocchi[1], and N.W. Schaad[2]

Abstract In Italy, panicle sterility and grain discolouration have emerged as serious problems throughout the rice growing areas. Surveys showed that the incidence of panicle sterility was highly variable, varying from low, barely observable in some fields, to more than 50% in other ones. In the Lombardy region the estimated average crop loss due to sterility was 1.1% and 0.1% in 2003 and 2004, respectively. Grain discolouration reduces rice quality. At harvest, susceptible cultivars can have more than 30% discoloured grains. The etiology for both diseases is uncertain and early diagnosis difficult because diseased plants are symptomless. We have demonstrated for the first time that both diseases have a bacterial etiology. Bacterial colonies typical of *Acidovorax avenae* subsp. *avenae* were purified from sterile panicles and discoloured rice samples. All recovered strains were confirmed as *Aaa* by classical PCR using *Aaa*-specific primers. Isolation of *Aaa* on a semi-selective agar medium did not always work well. However, an improved BIO-PCR assay was used to test 113 sterile panicle samples and 74% were positive for the presence of *Aaa*, but negative for *Burkholderia glumae* (*Bg*). The bacteria associated with discoloured grains were mostly composed of *Pseudomonas* and *Pantoea* species. Based on the 16S rDNA sequences, 85% of the pseudomonads population was classified as *P. straminea*, *P. fulva*, *P. putida*, *P. psychrotolerans*, *P. stutzeri* and *P. fluorescens*. *Aaa* and *Pseudomonas* strains were tested for pathogenicity on seedlings in the greenhouse and on plants at the booting stage in the field. Strains of *Aaa* caused soft rotting of seedlings, panicle sterility and grain discolouration and they showed significant intraspecific variation for virulence. Although most *Pseudomonas* spp. were not pathogenic, strains of *P. stutzeri* and strains presumptively identified as *P. fulva* induced grain discolouration incidence significantly higher than the control. In addition, we have found, for the first time, *P. ananatis* on rice in Italy.

[1]Institute of Plant Pathology, State University of Milan, via Celoria, 2 – 20133 Milan, Italy

[2]USDA/ARS-FDWSRU, Ft. Detrick, MD, USA

Author for correspondence: Paolo Cortesi, e-mail: paolo.cortesi@unimi.it

Keywords Etiology, diagnosis, *Acidovorax avenae* subsp. *avenae, Burkholderia glumae, Pseudomonas stutzeri, Pantoea ananatis*

1 Introduction

Panicle sterility has recently emerged as a problem in some fields of rice (*Oryza sativa* subsp. *japonica*) in Italy. The sterile panicles are clearly visible during late ripening because they remain green and straight, standing out above healthy ones.

Panicle sterility can be caused by adverse physiological conditions such as low (Satake, 1976) and high (Ziska and Manalo, 1996) temperature at flowering, drought (Sheoran and Saini, 1996), nutrient deficiency (Zhou et al., 1994; Gunawardena et al., 1998) or toxic compounds (Song et al., 1990; Kimura et al., 1994) and by diseases, including barley yellow dwarf virus (BYDV), sheath rot, blast and brown spot of rice caused by fungi such as *Sarocladium oryzae, Magnaporthe oryzae* and *Cochliobolus miyabeanus*, respectively (Osler and Moletti, 1982; Marchetti and Petersen, 1984; Ou, 1985; Webster and Gunnel, 1992; Singh and Dodan, 1995), and sheath brown rot, grain rot and bacterial brown stripe caused by *Burkholderia glumae* (*Bg*), *Pseudomonas fuscovaginae* and *Acidovorax avenae* subsp. *avenae* (*Aaa*), respectively (Zeigler and Alvarez, 1987, 1990; Cottyn et al., 1996; Jaunet et al., 1996).

However, it is doubtful that in Italy fungi or viruses are the causal factors of panicle sterility since these pathogens always cause specific symptoms, while the random or random circular spatial distribution of the disease is consistent with a seed-borne pathogen. *Bg, P. fuscovaginae* and *Aaa* are known as seed-borne bacteria of rice (Uematsu et al., 1976; Shakya et al., 1986; Zeigler and Alvarez, 1987). *Bg* has recently emerged as serious problem in Arkansas (Shahjahan et al., 2000; Sayler et al., 2006) and *Aaa*, a destructive seed-borne bacterium in tropical and sub-tropical environments (Shakya et al., 1985; Cottyn et al., 1996), was recently found on rice in Italy (Cortesi et al., 2005). Based on these reports, *Bg* and *Aaa* were therefore investigated as possible causal agents of panicle sterility in Italy.

Rice grain discolouration occurs in all major rice-growing areas in Italy with incidence ranging from 1% to 30%, with medium and short-grain rice cultivars more susceptible than long-grain ones (Pizzatti et al., 2004; Pizzatti, 2005). The disease is of great importance in tropical and sub-tropical countries (Zeigler and Alvarez, 1987; Duveiller et al., 1988, 1990; Rott et al., 1989; Cottyn et al., 1996) and can only be diagnosed after milling as brownish-black irregular lesions, chalkiness and brittleness of kernels.

Grain discolouration has a complex etiology. In tropical and sub-tropical countries it is associated with fungal and bacterial diseases such as sheath rot, blast and brown spot, and sheath brown rot, grain rot and bacterial brown stripe, respectively. In contrast, in Italy, the disease was attributed first to the phytopathogenic fungi *C. miyabeanus* and *M. oryzae*, which were found frequently on discoloured rice

(Cofelice et al., 2002; Haegi et al., 2002), and more recently, to the bug *Trigonotylus caelestialium* (Giudici and Villa, 2006). However, recent findings failed to support such conclusions. First, although fungicide treatments controlled brown spot on foliage, they were ineffective against grain discolouration (Cortesi and Giuditta, 2003). Second, brown spot incidence was significantly, but negatively, correlated with incidence of discoloured grains (Pizzatti et al., 2004; Pizzatti, 2005). Third, Giudici and Villa (2006) reproduced the symptoms of grain discolouration by placing adults of *T. caelestialium* on panicles in a greenhouse experiment, but failed to present any data about the life cycle and populations consistency of this leaf bug in Italy. Preliminary evidence suggests a bacterial etiology for both panicle sterility and grain discolouration (Pizzatti et al., 2004; Cortesi et al., 2005; Pizzatti, 2005), however additional investigations are needed to confirm the evidence.

The specific objectives of this study were the following: (I) to assess the spatial distribution and incidence of panicle sterility, (II) to search for *Bg* and *Aaa* in sterile panicles, (III) to isolate the bacteria associated with discoloured grains, and (IV) to assess the pathogenicity of bacteria associated with the two diseases.

2 Materials and Methods

2.1 Panicle Sterility

2.1.1 Surveys

The spatial distribution of panicle sterility in rice fields in the regions of Piedmont and Lombardy, the most important rice-growing areas in Italy, was studied for 2 years. In 2003, 104 farms distributed in 88 localities (55 farms in 48 localities in Piedmont and 49 farms in 40 localities in Lombardy) and consisting of about 10,000 ha (about 5% of Italian rice-growing area) were surveyed. In 2004, the survey was limited to 42 farms in 34 localities in Lombardy; in this region 36 farms in 30 localities were surveyed in both 2003 and 2004. On each farm, 2–3 weeks before harvest, symptoms of sterility were visually checked along a diagonal transect in all the fields. Disease incidence and severity were recorded for almost all farms in Lombardy. In each field where sterility was observed, a ring, 60 cm diameter, was randomly thrown five times along the transect and panicles and sterile panicles within the ring were counted, and percent of sterile panicles/hectare calculated. Severity was estimated on random samples of sterile panicles as percent of spikelets devoid of grains/panicle. An estimate of crop loss was calculated as the product of incidence and severity.

The etiology of panicle sterility was investigated by isolating from sterile panicles for both *Bg* and *Aaa*. A total of 113 samples of sterile panicles were collected from 21 rice cultivars. Each samples consisted of 20–30 sterile panicles collected from the second node from the top of the plant. Additionally, 13 and 2 symptomless samples were collected in 2003 and 2004, respectively, in fields where panicle sterility was observed. As control three samples were collected in fields were the diseases was never observed.

2.1.2 Detection and Identification of *Bg* and *Aaa*

In each sample, the presence of *Bg* and *Aaa* was determined using two methods: (1) isolation onto the semi-selective media CCNT (Kawaradani et al., 2000) and SP (Song et al., 2004) for *Bg* and *Aaa*, respectively; (2) molecular-based BIO-PCR (Schaad et al., 2003). Briefly, 20 g of spikelets and rachis from each sample were soaked in sterile water for 20 min, surface sterilized in 1% sodium hypochloride for 1 min and then rinsed twice in sterile water. Spikelets were broken and incubated overnight in 5 ml/g of phosphate buffer solution (PBS), containing 0.025% Tween 20 (pH 7.4). After incubation, the suspension was centrifuged 10 min at 10,000 × g. The precipitate was retained and bacteria were enriched by adding 1.5 ml/g of the liquid medium CCNT for *Bg* or SP for *Aaa*. The suspension was incubated at 30°C for 18 h on a rotary shaker at 150 rpm, then let sit 45 min at 5°C. For direct isolation, the suspension was serially diluted and 500 μl of the dilution was plated onto the respective agar solid medium. Plates were incubated for 4 days at 30°C and after incubation were observed for colonies typical of the two bacteria. Suspected *Aaa* colonies were transferred to YDC medium, which confers a typical tan brown colour to *Aaa*, and tested for poly-β-hydroxybutyrate (PHB) accumulation, hypersensitivity reaction on tobacco leaves, pathogenicity on corn seedlings and growth at 41°C (Sumner and Schaad, 1977; Jones et al., 2001). Molecular identification of *Acidovorax* sp. was obtained with the primers WFB (Walcott and Gitaitis, 2000), while *Aaa* was identified with AVA10, AVA11 and AVA12 primers in a semi-nested PCR (Song et al., 2004).

BIO-PCR was carried out according to Song et al. (2004) and with an additional enrichment step as follows: 500 μl of the two suspensions from liquid CCNT and SP were plated onto triplicate plates of each respective agar medium. After incubation for 5 days at 36°C, the surface of each plate was washed with 5 ml of sterile distilled water. One milliliter of the suspension was transferred to a 1.5 ml microcentrifuge tube and centrifuged 10 min at 10,000 × g, at 5°C. The pellet was retained, resuspended in 300 μl of sterile distilled water and boiled for 10 min. After cooling the tube on ice and centrifuging 10 min at 20,000 × g at 5°C, the supernatant containing the DNA was stored at 20°C. To detect *Bg* and *Aaa*, 5 μl of the DNA suspension from each of the two media was used as template for PCR amplification using *B. glumae*-specific primers GL12 and GL13 (Takeuchi et al., 1997) and *Aaa*-specific primers AVA10, AVA11 and AVA12 (Song et al., 2004). Reactions were performed in a total volume of 25 μl, containing 1× reaction buffer (Promega; Milan, Italy), 200 μM of each nucleotide triphosphate, 1.5 mM MgCl$_2$, 0.16 μM each AVA primer, or GL primer, 5 μl DNA suspension and 1 U Taq-DNA polymerase (Promega; Milan, Italy). PCR reactions were performed as described previously (Takeuchi et al., 1997; Song et al., 2004), using an i-Cycler (Bio-Rad Laboratories

Richmond, CA, USA). PCR products were analyzed by electrophoresis in 2% agarose gels in TBE buffer and stained with ethidium bromide.

2.1.3 Pathogenicity Tests

Pathogenicity tests were carried out in a greenhouse with 3–4-week-old seedlings grown in plastic boxes and in the field on plants at booting stage. Bacterial strains were incubated overnight in TSB at 30°C on a rotary shaker at 150 rpm. After centrifugation (3500 rpm), the cells were suspended in PBS to a final concentration of about 10^8 cfu/ml. Each of ten seedlings per box was inoculated with about 0.1 ml of each bacterial suspension, using a 1 ml syringe and small gauge needle. For each strain, 30 seedlings were inoculated in the greenhouse and five plants were inoculated in the field. In greenhouse, *Aaa* strain CAa4 (Song et al., 2004) was used as a positive control and plants inoculated with sterile water were used as negative controls. Frequencies of infected seedling, sterile spikelets and discoloured grains were recorded and the data, transformed as arcsin v%, were submitted to ANOVA and Duncan's multiple range tests.

To satisfy Koch's postulates, bacteria were isolated from infected seedlings and plants from the greenhouse and field experiments. Strains identity was determined by comparing the BOXA1R fingerprints of inoculated and isolated strains (Louws et al., 1999; Louws and Cuppels, 2001).

2.2 Grain Discolouration

To identify bacteria associated with discoloured grains, seeds of 35 samples of the susceptible short grain cultivar "Selenio" with an incidence of discoloured grains exceeding 4% were used. For each sample, 10 g of discoloured grains were washed with sterile water, partially crushed in a mortar, suspended in 100 ml of PBS containing 0.025% Tween 20 (pH 7.4), and incubated at 4°C (Cottyn et al., 2001). After 2 h incubation, 100 μl suspension was serially diluted to 10^{-5} and 500 μl plated onto King et al.'s medium B (KMB) and Nutrient Agar (NA) and the plates incubated at 28°C for 4 days. Representative colonies were purified by streaking onto NA and stored at −80°C in TSB with 20% glycerol. *Aaa* was specifically searched for 12 rice samples exceeding 4% of discoloured grains through direct isolation on semi-selective medium as described above.

2.2.1 Identification of Bacteria

Gram staining, catalase activity, and glucose fermentation/oxidation were performed, as described (Schaad et al., 2001). Suspect *Pseudomonas* strains were tested for levan production on sucrose, oxidase reaction, pectolytic and arginine dihydrolase activities, production of pigments on Pseudomonas agar P, fluorescence on KMB, and starch and gelatin hydrolysis (Braun-Kiewnick and Sands, 2001). API 20 NE tests (Biomériuex, Rome, Italy) were used for further characterization of Gram-negative aerobic strains. All fermentative strains on Triple Sugar Iron Agar (TSI) medium were further characterized using API 20 E tests (Biomériuex). Cocci Gram-positive strains were identified using API Staph system (Biomériuex) and tested for Lisostaphin resistance and growth on Mannitol Phenol-red Agar medium (Geary and Stevens, 1986).

2.2.2 Molecular Identification

Molecular identification was restricted to representative strains of each biochemical group. Representative strains were grown 24 h in liquid culture on a rotary shaker (150 rpm) at 30°C and total DNA extracted and purified, as described (Wilson, 1987). Presumptive identification of pseudomonads was obtained with the genus-specific 16S-based PCR primer PsMg (Braun-Howland et al., 1993; Johnsen et al., 1999). Positive strains were further characterized through Amplified Ribosomal DNA Restriction Analysis (ARDRA) of the ITS1 region, digested with the TaqI endonuclease (Braun-Howland et al., 1993) and 16S rDNA sequencing. Representative strains of each ARDRA group, sharing the same restriction fragment banding pattern, were identified based on 16S rDNA partial sequences (about 1200 nucleotides) of the 1500 bp fragment obtained with the 16SF and 16SR primers (Ibrahim et al., 1993). The sequences were compared with those available in the NCBI Genbank using the BLAST program and subsequently aligned with the ClustalX program (version 1.81). A phylogenetic tree was then constructed by the Neighbor Joining method and the support for branching was assessed with 1000 replication of a Bootstrap analysis (Felsenstein, 2002).

Identification of suspected Aaa strains was confirmed using AVA PCR primers (Song et al., 2004).

Suspect strains of the rice pathogen P. ananatis were identified using the species-specific PCR primers As2b-Sn2c (R. Gitaitis University of Georgia, Tifton, USA). PCR products were analyzed by electrophoresis in 1% agarose gels in TAE buffer and stained with ethidium bromide.

2.2.3 Pathogenicity tests

Pathogenicity tests for representative strains of Pseudomonas and Pantoea spp. were done in the field using late boot-stage plants of susceptible rice cultivar "Selenio". Approximately 1 ml of each bacterial suspension, prepared as described above, was injected into the boot of each plant, using a 5 ml syringe with standard gauge needle. Five plants for each strain were inoculated and five control plants were inoculated with sterile water. At maturity, the grains were harvested and the frequencies of discoloured grains were recorded and data analysed, as described above.

3 Results

3.1 Panicle Sterility

3.1.1 Surveys

Panicle sterility of rice was found in the Piedmont and Lombardy regions in both 2003 and 2003 and 2004 (Table 1). In 2003, the frequencies of localities and farms with panicle sterility were roughly the same, with values between 46% and 51% in the two regions. In Lombardy, the frequency of localities with the disease increased slightly to 50% in 2004, whereas the frequency of farms with panicle sterility decreased to 41%. Within regions, the frequencies of localities and farms with panicle sterility varied among provinces: values between 50% and 67% were recorded for Novara in Piedmont, and Milano in Lombardy, respectively. In contrast, in Lodi province, Lombardy, no panicle sterility was observed. Looking at the geographical distribution of the localities of the two regions where panicle sterility was found did not reveal any spatial pattern in either year (data not shown).

Panicle sterility was found in 22 out of 38 rice cultivars surveyed (Table 1). Twenty-nine cultivars were surveyed both in Piedmont and in Lombardy, however the number of cultivars surveyed in each province varied between 22 in Vercelli and 12 in Novara provinces of Piedmont, and between 25 in Pavia and 9 in Lodi provinces in Lombardy. On each farm, the average number of rice cultivars grown was three, while the average number of infected cultivars varied between 0, in Lodi province, and 0.82 in Pavia province, in 2003. However, the average number of infected cultivars did not change significantly between provinces and regions, or between years in Lombardy ($P = 0.05$). Covariance analysis showed independence between the number of infected cultivars and the number of cultivars grown in each farm, and the two factors were not significantly correlated ($P > 0.05$).

In Lombardy, the incidence of panicle sterility, as percent of panicles showing some sterility, was 1.06% in 2003, whereas in 2004 it was 0.12%, significantly less than in 2003 ($P < 0.05$). Within Lombardy, incidence in Pavia was significantly higher than in Milano in both years ($P < 0.05$) (Table 2). In each year, disease incidence on each cultivar varied from 0% to 22%, in 2003 and from 0% to 2%, in 2004. The cultivars with disease incidence significantly higher ($P < 0.05$) than the mean incidence calculated for the province were, in 2003, the cultivars "Arborio" and "Baldo" in Milano province, 1.26% ± 0.163 ($N = 5$) and 0.52% ± 0.175 ($N = 3$), respectively, and the cultivars "S.Andrea", "Ariete", and "Thaibonnet" in Pavia provinces with 22.1% ± 0.542 ($N = 4$), 7.58% ± 0.373 ($N = 9$), 3.85% ± 1.28 ($N = 3$), respectively. In 2004, the cultivars with highest disease incidence were "Loto" in Milano province, 0.29% ± 0.005 ($N = 4$), and "Ariete", "Ambra" and "Gladio" in Pavia province, 2.22% ± 0.739 ($N = 3$), 0.61% ± 0.097 ($N = 3$), 0.29% ± 0.024 ($N = 12$), respectively.

Table 1 Panicle sterility distribution in Piedmont and in Lombardy

Year	Region	Province	Hectare surveyed	Locality surveyed	Farm surveyed	Cultivar surveyed	Locality with panicle sterility (%)	Farm with panicle sterility (%)	Cultivar with panicle sterility (%)	Cultivar grown per farm (x) (Ē[a] ± s.e.)	Cultivar with panicle sterility per farm (y) (Ē ± s.e.)	Covariance between x and y	Correlation between x and y	T[b]	P[c]
2003	Piedmont	Alessandria and Biella	578	7	7	18	42.9	42.9	27.8	3.29 ± 0.892	0.57 ± 0.299	0.48	0.256	0.593	0.575
		Novara	1185	10	12	12	50.0	58.4	41.7	3.00 ± 0.300	0.75 ± 0.251	0.36	0.402	1.388	0.192
		Vercelli	3765	31	36	22	45.2	50	36.4	3.00 ± 0.287	0.61 ± 0.107	−0.03	−0.026	0.150	0.882
		All provinces pooled	5528	48	55	29	45.8	50.9	41.4	3.04 ± 0.225	0.64 ± 0.009	0.11	0.090	0.661	0.511
	Lombardy	Lodi	332	3	3	9	0	0	0	4.67 ± 1.230	0	n.d.	n.d.	n.d.	n.d.
		Milano	813	9	13	10	66.7	61.5	50.0	2.54 ± 0.313	0.69 ± 0.175	−0.15	−0.217	0.736	0.476
		Pavia	3247	28	33	25	46.4	51.5	56.0	2.94 ± 0.310	0.82 ± 0.160	0.52	0.318	1.866	0.071
		All provinces pooled	4393	40	49	29	47.5	51.0	58.6	2.94 ± 0.243	0.73 ± 0.120	0.23	0.164	1.141	0.260
2004	Lombardy	Lodi	312	3	3	7	0	0	0	4.00 ± 0.999	0	n.d.	n.d.	n.d.	n.d.
		Milano	1083	10	13	12	50	38.5	41.7	2.92 ± 0.350	0.62 ± 0.241	0.55	0.505	1.939	0.076
		Pavia	3099	21	26	19	42.8	46.2	57.9	3.04 ± 0.316	0.58 ± 0.137	0.26	0.227	1.141	0.265
		All provinces pooled	4494	34	42	22	50.0	40.5	59.1	3.07 ± 0.231	0.55 ± 0.114	0.28	0.249	1.627	0.111

[a]Average ± standard error

[b]Student's T value for the correlation null hypothesis

[c]Significance of the T-test

Table 2 Incidence of panicle sterility in Lombardy

Year of survey	Province	N[a]	Area surveyed (ha)	Incidence (%) ± s.e.[b]
2003	Lodi	14	332.1	0
	Milano	33	813.4	0.27 ± 0.042
	Pavia	63	1842	1.59 ± 0.008
	All provinces pooled	110	2987.5	1.06 ± 0.003
2004	Lodi	12	312.5	0
	Milano	38	1082.9	0.004 ± 0.0001
	Pavia	78	3098.9	0.169 ± 0.0014
	All provinces pooled	128	4494.3	0.118 ± 0.0006

[a]N = Sample size
[b]Standard error

3.1.2 Detection and Identification of *Bg* and *Aaa*

The disease survey allowed to collect 113 samples from 45 farms: 83 samples from 45 farms in 2003 and 30 samples from 17 farms in 2004 (Table 3).

Bg was not isolated on CCNT agar medium and the results of the BIO-PCR assay were negative for all samples.

Isolation of suspected *Aaa* on SP agar medium was successful and a total of 274 strains were obtained: 239 from 30 samples collected in 2003 and 35 from 4 samples collected in 2004. Molecular identification of *Aaa* with PCR was positive for 24 out of 274 strains, which were named Iaa, followed by a number (Fig. 1). These strains were obtained from six samples collected in 2003: five were samples constituted of sterile panicles and one sample was constituted of asymptomatic panicle collected in a field that had panicle sterility. Five strains were obtained from two farms and two cultivars, "Balilla" and "Thaibonnet", in Piedmont, while 19 strains were from three farms and three cultivars, "Ambra", "Pegaso" and "S.Andrea", in Lombardy. The bacterium was isolated mainly from samples collected in fields where disease incidence was high. All other strains belonged to the genus *Acidovorax*, because all gave the expected PCR product using the primers WFB (Walcott and Gitaitis, 2000), but not the expected PCR product with the primers OAF (Song et al., 2004). These strains were named as A, followed by a number.

Detection of *Aaa* by BIO-PCR was much more efficient. The bacterium was found in 87% of farms and in 88% of cultivars, and in 74% and 77% of samples collected in 2003 and in 2004, respectively. *Aaa* was found both in diseased and in symptomless samples collected in diseased fields with frequencies of 74% and 85% in 2003 and 78% and 100% in 2004, respectively (Table 3). The number of *Aaa*-positive samples was not different for the two regions in 2003 ($\chi^2 = 2.73$, $P = 0.435$), and it was not different for the 2 years in Lombardy ($\chi^2 = 0.67$: $P = 0.880$). In 2003,

Table 3 Molecular detection of *Burkholderia glumae* (*Bg*) and *Acidovorax avenae* subsp. *avenae* (*Aaa*) in rice panicles collected in Italy over 2 years

Year	Region	Province	Sampling (N[a])		Bg[b] Samples (N)				Aaa[c]			Farm positive for Aaa	Cultivar positive for Aaa
			Farm	Cultivar	Control	Diseased	Symptomless from field with panicle sterility	All samples pooled	Control	Diseased	Symptomless from field with panicle sterility		
2003	Peidmont	All provinces pooled	19	11	1	30	8	0	0	22	7	17	10
		All provinces pooled	26	17	1	38	5	0	0	28	4	22	14
2004	Lombardy	All provinces pooled	17	13	1	27	2	0	0	21	2	15	12

[a]Sample size

[b]Number of samples positive for presence of *Bg* using BIO-PCR

[c]Number of samples positive for presence of *Aaa* using BIO-PCR

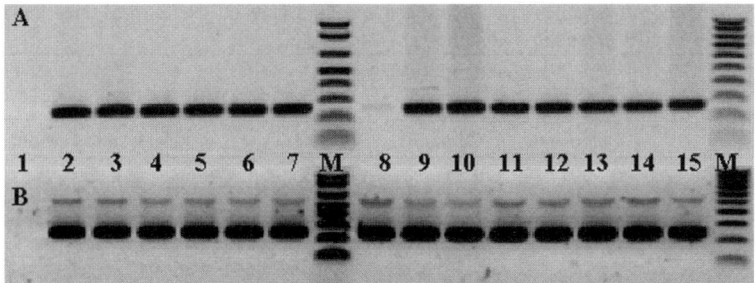

Fig. 1 *Acidovorax avenae* subsp. *avenae* (*Aaa*) amplicons obtained with AVA primers (Song et al., 2004) in semi-nested PCR. (A) 262 bp first round amplicons. (B) 241 bp second round amplicons. Lane 1 = negative control; Lanes 2–3 = CAa4 and CAa6, *Aaa* reference strains; Lanes 4–15 = *Aaa* Italian strains; M = 1 Kb Plus DNA Ladder and 100 pb ladder (Gibco, Milan, Italy)

Table 4 ANOVA for incidence of seedling rot on three rice cultivars experimentally infected with *Acidovorax* sp. and *A. avenae* subsp. *avenae* strains

Source of variation	Degree of freedom	Mean square	F value	P
Replication	2	0.041	0.9699	>0.1
Strain	7	1.861	43.753	<0.001
Rice cultivar	2	0.012	0.2754	>0.1
Strain × cultivar	14	0.034	0.8053	>0.1
Error	46	0.043		

the number of farms where the bacterium was found was not different between Piedmont and Lombardy ($\chi^2 = 0.09$: $P = 0.924$) and was not different in Lombardy between 2003 and 2004 ($\chi^2 = 0.02$: $P = 0.891$). *Acidovorax avenae* subsp. *avenae* was never found in healthy samples collected on farms where the disease was not observed (Table 3).

3.1.3 Pathogenicity Tests

Analysis of variance for incidence of rotted seedlings of three rice cultivars infected with strains of *Acidovorax* sp. and *Aaa* showed significant differences of virulence among strains, but no differences of susceptibility among cultivars and no significant interaction between strains virulence and cultivar susceptibility (Table 4). All strains identified as *Aaa* caused severe seedling rot within 2 weeks after inoculation. Virulence of the Italian strains of *Aaa* was similar to reference strain Caa4. The mean frequency of rotted seedling for the three rice cultivars varied from 56% for Iaa182 to 87% for IAa210, which was the most virulent strain tested. The strains identified as *Acidovorax* sp. were significantly less virulent than the *Aaa* strains and their mean frequency of rotted seedling for the three rice cultivars was 7.8%, value not significantly different than the control (Table 5).

Table 5 Virulence of *Acidovorax avenae* subsp. *avenae* strains on seedlings of three Italian rice cultivars grown under greenhouse conditions

| | Rotted seedling (%) | | | |
| | Rice cultivar | | | |
Strain	Baldo	Carnaroli	Selenio	Mean rotted seedling (%)
Control	0	0	3.33	2.22 c[a]
A30[b]	10	3.33	10	7.78 c
A208[b]	6.67	13.33	3.33	7.78 c
IAa93[c]	70	56.67	73.33	66.67 ab
IAa182[c]	60	60	46.67	55.56 b
IAa210[c]	80	86.67	90.33	86.67 a
IAa258[c]	73.33	83.33	70	75.56 ab
CAa4[d]	56.67	43.33	46.67	48.89 b

[a]Duncan's multiple range test $P = 0.05$. Mean with the same letter are not significantly different
[b]*Acidovorax* sp. strains
[c]Italian *Aaa* strains
[d]*Aaa* reference strain (Song et al., 2004)

Table 6 ANOVA for incidence of panicle sterility on rice cultivar "Volano" experimentally infected with *Acidovorax* sp. and *A. avenae* subsp. *avenae* strains

Variable	Source of variation	Degree of freedom	Mean square	F value	P
Panicle sterility	Replication	4	0.075	0.3672	>0.1
	Strain	6	0.063	2.3112	0.046
	Error	24	0.027		
Grain discolouration	Replication	4	0.002	0.1862	>0.1
	Strain	6	0.011	0.9994	>0.1
	Error	24	0.011		

Analysis of variance for field pathogenicity tests on rice cultivar "Volano", infected at the booting stage, showed significant differences for incidence of panicle sterility, but not for incidence of grain discolouration (Table 6). Virulence of the strains, as percent of sterile spikelets/panicle, varied. The *Aaa* strains IAa93 and IAa182 and the *Acidovorax* sp. strain A30 were the most virulent and caused about 25% of sterile spikelets/panicle, significantly higher than 10% observed on the control. The strains IAa210 and IAa258 and the *Acidovorax* sp. strain A208 were less virulent and caused about 16% of sterile spikelets/panicle, not significantly different than the control (Table 7).

The BOXA1R fingerprints of *Acidovorax* sp. and *Aaa* strains used for pathogenicity test on seedlings were identical to strains re-isolated from the inoculated seedlings (Fig. 2A). In contrast, in field experiment the BOXA1R fingerprints of *Aaa* strains were identical to strains isolated from sterile panicles of cultivar "Volano" inoculated into the stem, but, interestingly, the *Aaa* strain IAa182 was

Table 7 Virulence of *Acidovorax* sp. and *Acidovorax avenae* subsp. *avenae* (*Aaa*) strains on rice cultivar "Volano" experimentally infected at booting stage in the field

Strain	No. of spikelets/ panicle	No. of grains/ panicle	Sterile spikelets/ panicle (%)	Discoloured grains (%)
Control	64.6	59.0	9.42 b[a]	2.53
A30[b]	54.6	41.6	23.16 a	1.45
A208[b]	54.4	46.4	16.24 ab	0.98
IAa93[c]	64.4	49.4	24.56 a	1.27
IAa182[c]	56.2	40.2	25.70 a	2.09
IAa210[c]	47.0	39.4	15.80 ab	2.84
IAa258[c]	40.4	34.2	16.68 ab	5.29

[a]Duncan's multiple range test $P = 0.1$. Mean with the same letter are not significantly different
[b]*Acidovorax* sp. strains
[c]Italian *Aaa* strains

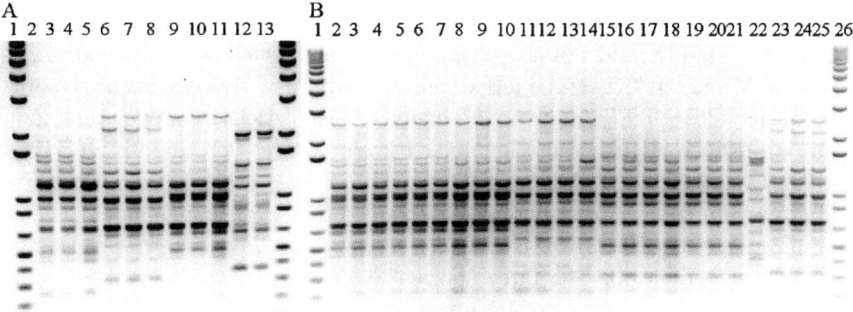

Fig. 2 BOXA1R fingerprints of *Acidovorax* sp. and *Acidovorax avenae* subsp. *avenae* (*Aaa*) strains. (A) Infected rice seedlings. Lanes 1 and 13: 1 Kb Plus DNA Ladder (Gibco, Milan, Italy); Lanes 2–4: *Aaa* strain CAa4 inoculated (lane 2) and CAa4 re-isolated; Lanes 5–7: *Aaa* strain IAa258 inoculated (lane 5) and IAa258 re-isolated; Lanes 8–10: *Aaa* strain IAa210 inoculated (lane 8) and IAa210 re-isolated; Lane11: *Acidovorax* sp. strain A208 inoculated; Lane12: A208 re-isolated. (B) Infected rice plants in the field. Lanes 1 and 26: 1 Kb Plus DNA Ladder; Lanes 2–10: *Aaa* strain IAa 210 inoculated (lane 2) and re-isolated from three plants; Lanes 11–14: *Aaa* strain IAa93 inoculated (lane 11) and re-isolated from two plants; Lanes 15–20: *Aaa* strain IAa182 inoculated (lane 15) and re-isolated from three plants; Lane 21: *Aaa* strain IAa182 isolated from one plant inoculated with *Acidovorax* sp. strain A30; Lane 22: *Acidovorax* sp. strain A30 inoculated; Lane 23–25: *Aaa* strain IAa258 inoculated (lane 23) and re-isolated from two plants

isolated from one panicle of cultivar "Volano" inoculated into the stem with *Acidovorax* sp. strain A30 (Fig. 2B). This findings can justify the higher virulence observed in field experiment for *Acidovorax* sp. strain A30 resulted non-virulent on rice seedlings.

3.2 Grain Discolouration

3.2.1 Identification of Bacteria

From 35 discoloured grain samples 409 bacterial strains were obtained: 79% of them were Gram-negative, rod-shaped and catalase-positive, whereas the remaining 21% were Gram-positive. The majority of Gram-positive strains were cocci (about 60%), rarely arranged in tetrads and all catalase-positive. One hundred thirty-five (42%) out of 324 Gram-negative strains were aerobe, whereas the remaining 189 strains (58%) were anaerobe, and identified as *Enterobacteriaceae*. Eighty-two aerobe strains were identified as pseudomonads based upon production of a 450 bp amplicon following PCR amplification with the PsMg primer.

Amplification of the ITS1 region yielded one 1500 bp amplicon for 67 strains and two to three amplicons, from 1000 to 1500 bp, for 15 strains, which were Taq I digested. Twenty-eight polymorphic restriction fragments were obtained, resulting in 24 ARDRA patterns, thus resolving additional diversity within the biochemical groups. More than 50% of the pseudomonads population belonged to five ARDRA groups (II, III, IV, VII and VIII). All fluorescent pseudomonads were in ARDRA groups V, VI and IX; five non-fluorescent yellow-pigmented pseudomonad strains shared the V and VI ARDRA pattern (Table 8). Partial 16S rDNA sequences (about 1200 nucleotides) of 17 strains representative of each ARDRA group were compared with those available in Genbank. Eleven strains were identified with a similarity value of 99%: six strains were identified as *P. psychrotolerans* and five as *P. fulva*. Similarity values ≥98% were obtained for the remaining strains, which were presumptively identified as *P. straminea, P. fulva, P. putida, P. stutzeri* and *P. mendocina* (Table 8). The three species *P. straminea, P. fulva* and *P. putida* grouped together in a phylogenetic tree based on 16S rDNA sequences and this group was genetically distant from *P. psychrotolerans, P. stutzeri* and *P. fluorescens* strains. *Pseudomonas fulva* strains, assigned to five different ARDRA groups, were the most diverse, whereas the *P. psychrotolerans* group showed little genotypic diversity (data not shown).

The non-pseudomonad strains ($N = 37$) were identified as *Chryseomonas luteola* (5 strains), *Sphingomonas paucimobilis* (15 strains) and *Agrobacterium radiobacter* (2 strains) according to API 20E tests; 15 strains were not identified.

Fifteen strains suspected *Acidovorax* spp. were found in 5 out of 12 rice samples with grain discolouration exceeding 4%. All strains had the typical tan-brown colour on YDC medium, accumulated PHB and grew at 41°C. However, only four of them could be identified as *A. avenae* subsp. *avenae* based on the species-specific AVA primers.

The *Enterobacteriaceae* population was mainly composed of *Pantoea* spp. (about 67%). Biochemical characterization of the *Pantoea* spp. group ($N = 104$) using the API 20 E system revealed the presence of three major phenetic groups; 23 strains did not group. Strains belonging to phenetic groups A and B differed only in citrate utilization, whereas those in group C failed to reduce nitrate, but were positive for citrate utilisation, indole production, acid production from melibiose

Table 8 Molecular characterization of the pseudomonads isolated from rice discoloured grains

ARDRA group[a]	16S rDNA identification (Blast similarity value[b])	No. of strains
II	*P. psychrotolerans* (99%)	10
III	*P. psychrotolerans* (99%)	6
IV	*P. psychrotolerans* (99%)	14
XIV	n.i.[c]	2
XV	n.i.	2
I	*P. fulva – P. straminea* (99%)	3
VIII	*P. psychrotolerans* (99%)	7
X	*P. fulva – P. putida* (98%)	2
XI	n.i.	1
XII	*P. psychrotolerans* (99%)	3
XVIII	n.i.	1
XIX	*P. mendocina* (98%)	2
XX	*P. fulva–P. straminea* (99%)	3
XXI	n.i.	1
V	P. fulva (99%)	5
VI	*P. fulva* (99%)	4
VII	*P. psychrotolerans* (99%)	6
XIII	n.i.	1
IX	*P. putida* (98%)	3
XVI	*P. fulva* (99%)	1
XVII	*P. fulva* (99%)	1
XXII	*P. fulva* (99%)	2
XXIII	*P. fluorescens* (99%)	1
XXIV	*P. stutzeri* (98%)	1

[a]Strains within the same group shared identical TaqI Amplified Ribosomal DNA Restriction Analysis (ARDRA) pattern
[b]Blast similarity value obtained aligning 16S rDNA partial sequences
[c]Not identified

and amylose, but not from rhamnose. None of the *Pantoea* strains in groups A, B and C were identified as *P. ananatis*. However, 4 of the 23 strains that did not group were identified as *P. ananatis*. The *P. ananatis* strains were positive for indole and they did not reduce nitrate, as expected for this species.

3.2.2 Pathogenicity Tests

Among the 12 pseudomonads and 20 *Pantoea* strains, only one *P. stutzeri* strain induced discoloured grains significantly higher than the control. *P. fulva* strains showed variable pathogenicity: only the strain belonging to the ARDRA group V induced disease incidence significantly higher than the control. *P. psychrotolerans*, *P. fluorescent*, *P. putida* and *Pantoea* spp. did not cause grain discolouration (data not shown for *Pantoea* spp.) (Table 9). However, *P. ananatis* strains induced symptoms of palea browning, a new disease of rice in Italy.

Table 9 Pathogenicity on rice cultivar "Selenio" of *Pseudomonas* spp. strains representative of the major Amplified Ribosomal DNA Restriction Analysis (ARDRA) groups

Strain	ARDRA group	Discoloured grains (%)
Control		1,99 c[a]
P. stutzeri	XXIV	8,88 a
P. fluorescens	XXIII	5,17 abc
P. fulva	V	6,58 ab
P. fulva	VI	6,55 abc
P. fulva	XVI	2,66 bc
P. fulva	XXII	2,58 bc
P. fulva- P. straminea	I	3,97 abc
P. putida	IX	4,21 abc
P. psychrotolerans	II	1,82 c
P. psychrotolerans	III	3,59 bc
P. psychrotolerans	IV	3,12 bc
P. psychrotolerans	VII	1,56 c

[a]Duncan's multiple range test $P = 0.05$. Mean with the same letter are not significantly different

[b]Strains within the same group shared identical TaqI Amplified Ribosomal DNA Restriction Analysis pattern

4 Discussion

Panicle sterility and grain discolouration are widespread in Italy and both diseases can be found in all of the most important provinces where rice is grown. A common feature of both diseases is the lack of specific diagnostic symptoms. In fact, discoloured grains become visible only after milling as brownish-black irregular lesions on kernels and sterile panicles remain upright during ripening because the spikelets are devoid of grains and are greenish in colour.

We demonstrated for the first time that panicle sterility in Italy is a disease caused by a bacterium, and it is not due to physiological disorders caused by abiotic factors (Satake, 1976; Song et al., 1990; Kimura et al., 1994; Zhou et al., 1994; Sheoran and Saini, 1996; Ziska and Manalo, 1996; Gunawardena et al., 1998). A 2-year survey showed that the spatial distribution of sterile panicle within fields is random. Plants boring sterile panicles can be clustered in patches, no bigger than one or a few square meters or scattered throughout the field. Panicle sterility was found in about 50% of localities and farms surveyed and on more than 50% of cultivars, both in Piedmont and in Lombardy. This disease does not have a clearly discernible spatial distribution within farms or within localities. Disease incidence can be severe; in some fields 50% of the plants showed sterile panicles, whereas in others the percentage of sterile panicles was near or below the threshold of detection. In Lombardy in 2003, the mean incidence was about 1%, whereas in 2004 the incidence decreased to 0.1%. In both years disease incidence in Pavia province (1.59% and 0.16%) was significantly higher than in Milano province (0.27 and 0.004). In contrast, we did not find the disease in Lodi province in both years.

The tenfold increase in panicle sterility incidence could be the results of higher temperatures occurred in 2003 than in 2004, and could be an indication of the negative effect of global warming on crops production (Schaad, this book). Virulence of *Aaa* is sensitive to temperature. Blight symptoms on wheat were generally more severe at higher temperatures, as well as the growth of infected plants, whereas on corn lesion development was not affected by temperature, but the growth of infected plant was significantly reduced with temperature ranging from 22/14°C to 30/22°C (Schaad et al., 1980). Based on these evidences we are currently investigating the effect of temperature on virulence of *Aaa* on rice.

Although *Bg* has been reported to cause severe epidemics of panicle blight in a wide area from Louisiana to Texas, USA (Shahjahan et al., 2000), the organism was not found in Italy. In contrast, the seed-borne bacterium *Aaa* found in 2005 for the first time on rice in Italy (Cortesi et al., 2005) was showed in this paper to be responsible for panicle sterility. Direct isolation of *Aaa* from sterile panicles using the semi-selective medium SP (Song et al., 2004) gave poor results. Still the bacterium was isolated in 3.6% of diseased samples, proving the presence of the pathogen. Attempts to diagnose the disease through molecular means were much more successful. By using BIO-PCR *Aaa* was found in 75% of sterile panicle samples and in 89% of samples constituted of asymptomatic panicles collected in fields where the disease was visible, but it was never found in samples collected in fields where the disease was not detected. The BIO-PCR revealed the presence of *Aaa* both on sterile panicle and, with higher frequency, on asymptomatic panicles collected in fields where the disease was present. This fact suggests that the bacterium is wide spread in the field where the disease is present and it can be possible that it survives better into the kernels than into the sterile spikelets. The bacterium was repeatedly found in 83% of the farms were the disease had been observed for two consecutive years. In contrast, *Aaa* was not detected in asymptomatic samples collected from farms where the disease was never observed. BIO-PCR can be a highly sensitive diagnostic tool (Schaad et al., 2003; Song et al., 2004), and our results confirmed that it was 20 times more efficient in detecting the bacterium than direct isolation on semi-selective medium.

Aaa was detected in 88% of the Italian cultivars surveyed. Such result indicates a widespread contamination of rice seed lots. Similar results were recently obtained in Burkina Faso, where all the nine seed lots of foreign rice cultivars resulted in various levels of seeds discoloration, low germination, and up to 20% of seedlings with brown stripe symptoms (Somda et al., 2005). Despite the worldwide distribution of *Aaa* (Cottyn et al., 1996; Cortesi et al., 2005; Somda et al., 2005), our finding showed, for the first time, that *Aaa* can cause panicle sterility in the absence of brown stripe or other visible symptoms. Moreover, *Aaa* can contaminate seeds of symptomless plants in fields with panicle sterility. A low level of seed contamination, which does not affect germination, could be easily overseen and result in considerable spread of the disease.

In Italy, the bacterium was found repeatedly in 83% of farms, but not always on the same cultivar. Therefore, we suggest that *Aaa* may be repeatedly introduced in the same farm through contaminated seed lots or the bacterium may overwinter in diseased fields on rice or on other hosts, thus increasing the inoculum over the time. We will thoroughly test these hypothesis in the future.

Grain discolouration is a complex disease in tropical and sub-tropical countries attributed to several fungi and bacteria, but no conclusive demonstration of its etiology is available (Zeigler and Alvarez, 1987; Duveiller et al., 1988, 1990; Rott et al., 1989; Cottyn et al., 1996). To date, in Italy the role of fungi has been emphasized (Caufin and Moletti, 1988; Cofelice et al., 2002; Haegi et al., 2002), however we demonstrated that symptoms of brown spot, induced by *C. myabeanus*, are not associated with the incidence of discoloured grains, as it was found in USA (Marchetti and Petersen, 1984; Pizzatti, 2005). Moreover, effective control of brown spot was not associated with lower incidence of discoloured grains (Cortesi and Giuditta, 2003; Pizzatti, 2005).

In this study, we have clearly shown that grain discolouration in Italy has a bacterial etiology. The disease is caused by several bacteria including species in the families *Pseudomonadaceae* and *Enterobacteriaceae*, with the genera *Pseudomonas* and *Pantoea* being the most common. The bacteria in Italy were primarily non-fluorescent, yellow-pigmented strains. Within the pseudomonads the saprophytic *P. fluorescens*, *P. putida* and *P. psychrotolerans* were the most common, but failed to cause grain discolouration. On the contrary, *P. stutzeri*, two strains presumptively identified as *P. fulva*, and the strain of *Aaa* isolated from discoloured grains, significantly increased the incidence of discoloured grains. Within the genus *Pantoea* we found, for the first time, *P. ananatis* on rice in Italy. This pathogen is known to cause palea browning (Webster and Gunnel, 1992), but as not been described as a casual agent of grain discolouration (Zeigler and Alvarez, 1990; Cottyn et al., 2001). Unlike the report of Cottyn et al. (2001) *P. fuscovaginae*, *P. syringae* and *B. glumae* were not found associated to grain discolouration in Italy.

In conclusion we have demonstrated that grain discolouration and panicle sterility in Italy have a bacterial etiology, and, among others, we found the seed-borne bacteria *Aaa* and *P. ananatis*. Epidemics of bacterial diseases must be prevented, particularly in the absence of effective bactericides. Therefore, in order to prevent the spread of the diseases it is of crucial importance to implement strategies to obtain bacteria-free certified seed lots.

Acknowledgements Authors thank all farmers and people who have made this research possible. This research was funded by Regione Lombardia (Regional Research Program in Agriculture 2001–2003 and 2004–2006), State University of Milan (FIRST Research Grant) and Ente Nazionale Risi.

References

Braun-Howland E.B., Vescio P.A. and Nierzwicki-Bauer S.A. (1993) Use of simplified cell blot technique and 16S rRNA-directed probes for identification of common enviromental isolates. Applied and Environmental Microbiology, 59: 3219–3224.

Braun-Kiewnick A. and Sands D.C. (2001) Gram negative bacteria: *Pseudomonas*. In: Schaad N.W., Jones J.B. and Chun W. (Eds.). Laboratory Guide for Identification of Plant Pathogenic Bacteria, pp. 84–120. The American Phytopathological Society, St. Paul, MN.

Caufin A. and Moletti M. (1988) Indagine riguardante i funghi presenti sul risone alla raccolta nelle diverse zone risicole italiane. In: 10° Convegno Internazionale sulla Risicoltura, Vercelli 1988: 271–278.

Cofelice G., Conca G., Infantino A., Riccioni L., Russo S. and Porta-Puglia A. (2002) Osservazioni sulla microflora e le alterazioni cromatiche del risone e del riso lavorato. Informatore Fitopatologico, 52: 58–63.

Cortesi P. and Giuditta L. (2003) Epidemiologia dell'elmintosporiosi e del brusone e difesa del riso. Informatore Fitopatologico, 53: 41–51.

Cortesi P., Bartoli F., Pizzatti C., Song W.Y. and Schaad N.W. (2005) First report of *Acidovorax avenae* subsp. *avenae* on rice in Italy. Journal of Plant Pathology, 87: 76.

Cottyn B., Cerez M.T., Van Outryve M.F., Barroga J., Swings J. and Mew T.W. (1996) Bacterial diseases of rice. I. Pathogenic bacteria associated with sheath rot complex and grain discoloration of rice in the Philippines. Plant Disease, 80: 429–437.

Cottyn B., Regalado E., Lanoot B., De Cleene M., Mew T.W. and Swings J. (2001) Bacterial populations associated with rice seed in the tropical environment. Phytopathology, 91: 282–292.

Cottyn B., Van Outryve M.F., Cerez M.T., De Cleene M., Swings J. and Mew T.W. (1996) Bacterial diseases of rice. II. Characterization of pathogenic bacteria associated with sheath rot complex and grain discoloration of rice in the Philippines. Plant Disease, 80: 438–445.

Duveiller E., Miyajima K., Snacken F., Autrique A. and Maraite H. (1988) Characterization of *Pseudomonas fuscovaginae* and differentiation from other fluorescent pseudomonads occurring on rice in Burundi. Journal of Phytopathology, 122: 97–107.

Duveiller E., Notteghem J.L., Rott P., Snacken F. and Maraite H. (1990) Bacterial sheath brown rot of rice caused by *Pseudomonas fuscovaginae* in Malagasy. Tropical Pest Management, 36: 151–153.

Felsenstein J. (2002) Phylip 3.6 (alpha3). Department of Genome Sciences.

Geary C. and Stevens M. (1986) Rapid lysostaphin test to differentiate Staphylococcus and Micrococcus species. Journal of Clinical Microbiology, 23: 1044–1045.

Giudici M.L. and Villa B. (2006) *Trigonotylus caelestialium* kirkaldy (*Heteroptera, Miridae, Mirinae, Stenodemini*) su riso in Italia. Informatore Fitopatologico, 56: 18–23.

Gunawardena T.A., Fukai S. and Blamey F.P.C. (1998) Spikelet sterility induced by nitrogen firtilization and low temperature in rice. In: Proceedings of the 9th Australian Agronomy Conference, Wagga Wagga, Australia 1998: 1–3.

Haegi A., Infantino A. and Porta-Puglia A. (2002) Ruolo della tossina di *Bipolaris oryzae* nelle alterazioni cromatiche del riso. Informatore Fitopatologico – La Difesa delle Piante, 2: 52–57.

Ibrahim A., Goebel B.M., Liesack W., Griffiths M. and Stackebrandt E. (1993) The phylogeny of the genus *Yersinia* based on 16S rDNA senquences. FEBS Microbiology Letters, 14: 173–177.

Jaunet T., Notteghem J.L. and Rapilly F. (1996) Pathogenicity process of *Pseudomonas fuscovaginae*, the causal agent of sheath brown rot of rice. Journal of Phytopathology, 144: 425–430.

Johnsen K., Enger O., Jacobsen C.S., Thirup L. and Torsvik V. (1999) Quantitative selective PCR of 16S ribosomal DNA correlates well with selective agar plating in describing population dynamics of indigenous *Pseudomonas* spp. in soil hot spots. Applied and Environmental Microbiology, 65: 1786–1788.

Jones J.B., Gitaitis R.D. and Schaad N.W. (2001) Gram negative bacteria: *Acidovorax* and *Xylophilus*. In: Schaad N.W., Jones J.B. and Chun W. (Eds.). Laboratory Guide for Identification of Plant Pathogenic Bacteria, pp. 121–138. The American Phytopathological Society, St. Paul, MN.

Kawaradani M., Okada K. and Kusakari S. (2000) New selective medium for isolation of *Burkholderia glumae* from rice seeds. Journal of General Plant Pathology, 66: 234–237.

Kimura Y., Shimada A. and Nagai T. (1994) Effect of glutamine synthetase inhibitors on rice sterility. Bioscience Biotechnology and Biochemistry, 58: 669–673.

Louws F.J. and Cuppels D.A. (2001) Molecular techniques. In: Schaad N.W., Jones J.B. and Chun W. (Eds.). Plant Pathogenic Bacteria: Laboratory Guide for Identification, pp. 321–332. APS, St. Paul, MN.

Louws F.J., Rademaker J.L. and de Bruijn F.J. (1999) The three ds of PCR-based genomic analysis of phytobacteria: diversity, detection, and disease diagnosis. Annual Review of Phytopathology, 37: 81–125.

Marchetti M.A. and Petersen H.D. (1984) The role of *Bipolaris oryzae* in floral abortion and kernel discoloration in rice. Plant Disease, 68: 288–291.

Osler R. and Moletti M. (1982) Response of four cultivars of *Oryza sativa* inoculated with different numbers of the aphid *Rhopalosiphum padi* infected with the rice "giallume" strain of Barley Yellow Dwarf Virus. Phytopathologische Zeitschrift, 105: 51–60.

Ou S.H. (1985) Rice Diseases. Commonwealth Mycological Institute, Kew, UK, pp. 380.

Pizzatti C. (2005) Maculatura delle cariossidi del riso: eziologia della malattia, caratterizzazione della microflora batterica e difesa. Tesi di Dottorato, Università degli Studi di Milano, Milano,. pp. 143.

Pizzatti C., Pedrali D. and Cortesi P. (2004) Lack of correlation between incidence of rice grain discolouration and panicle brown spot. Journal of Plant Pathology, 86: 329–330.

Pizzatti C., Pedrali D., Scarpellini M., Franzetti L. and Cortesi P. (2004) Isolation and characterization of bacteria associated with rice grain discolouration. Journal of Plant Pathology, 86: 330.

Rott P., Notteghem J.L. and Frossard P. (1989) Identification and characterization of *Pseudomonas fuscovaginae*, the causal agent of bacterial sheath brown rot of rice, from Madagascar and other countries. Plant Disease, 73: 133–137.

Satake T. (1976) Sterile-type cool injury in paddy rice plants. In: Proceedings of the Symposium on Climate and Rice, Los Banos, Philippines 1976: 281–300.

Sayler R.J., Cartwright R.D. and Yang Y. (2006) Genetic characterization and Real-time PCR detection of *Burkholderia glumae*, a newly emerging bacterial pathogen of rice in the United States. Plant Disease, 90: 603–610.

Schaad N.W., Frederick R.D., Shaw J., Schneider W.L., Hickson R., Petrillo M.D. and Luster D.G. (2003) Advances in molecular-based diagnostic in meeting crop biosecurity and phytosanitary issues. Annual Review of Phytopathology, 41: 305–324.

Schaad N.W., Jones J.B. and Chun W. (2001) Laboratory Guide for Identification of Plant Pathogenic Bacteria. APS, St. Paul, MN, pp. 373.

Schaad N.W., Sumner D.R. and Ware G.O. (1980) Influence of temperature and light on severity of bacterial blight of corn, oats, and wheat. Plant Disease, 64: 481–483.

Shahjahan A.K.M., Rush M.C., Groth D. and Clark C. (2000) Panicle blight. Rice Journal, 104: 26–28.

Shakya D.D., Chung H.S. and Vinther F. (1986) Transmission of *Pseudomonas avenae*, the cause of bacterial stripe of rice. Journal of Phytopathology, 116: 92–96.

Shakya D.D., Vinther F. and Mathur S.B. (1985) Worldwide distribution of bacterial stripe pathogen of rice identified as *Pseudomonas avenae*. Phytopatholologische Zeitschrift, 114: 256–259.

Sheoran I.S. and Saini H.S. (1996) Drought-induced male sterility in rice: changes in carbohydrate levels in enzyme activities associated with the inibition of starch accumulation in pollen. Sexual Plant Reproduction, 9: 161–169.

Singh R. and Dodan D.S. (1995) Sheath rot of rice. International Journal of Tropical Plant Disease, 13: 139–152.

Somda I., Ouedraogo S.L., Dakouo D. and Mortensen C.N. (2005) Prévalence de *Acidovorax avenae* subsp. *avenae*, agent des rayures bactériennes du riz dans les semences de base produites au Burkina Faso. Tropicoltura, 23: 85–90.

Song M.T., Kim J.K. and Choe Z.R. (1990) Effect of chemicals on inducing grain sterility of rice. Korean Journal of Crop Science, 35: 309–314.

Song W.Y., Kim H.M., Hwang C.Y. and Schaad N.W. (2004) Detection of *Acidovorax avenae* ssp. *avenae* in rice seeds using BIO-PCR. Journal of Phytopathology, 152: 667–676.

Sumner D.R. and Schaad N.W. (1977) Epidemiology and control of bacterial leaf blight of corn. Phytopathology, 67: 1113–1118.

Takeuchi T., Sawada H., Suzuki F. and Matsuda I. (1997) Specific detection of *Bukholderia plantarii* and *B. glumae* by PCR using primers selected from the 16S–23S rDNA spacer regions. Annals of Phytopathological Society of Japan, 63: 455–462.

Uematsu T., Yoshimura D., Nishiyama K., Ibaraki T. and Fujii H. (1976) Occurence of bacterial seedling rot in nursery flat, caused by grain rot bacterium *Pseudomonas glumae*. Annals of Phytopathological Society of Japan, 42: 310–312.

Walcott R.R. and Gitaitis R.D. (2000) Detection of *Acidovorax avenae* subsp. *citrulli* in watermelon seed using immunomagnetic separation and polymerase chain reaction. Plant Disease, 84: 470–474.

Webster R.K. and Gunnel P.S. (1992) Compendium of Rice Diseases. The American Phytopathological Society, St. Paul, MN, pp. 62.

Wilson K. (1987) Preparation of genomic DNA from bacteria. In: Ausubel F.M., Brent R., Kingston R.E., Moore D.D., Seidman J.G., Smith J.A. and Struhl K. (Eds.). Current Protocols in Molecular Biology, pp. 2.4.1–2.4.5. Wiley, New York.

Zeigler R.S. and Alvarez E. (1987) Bacterial sheath brown rot of rice caused by *Pseudomonas fuscovaginae* in Latin America. Plant Disease, 71: 592–597.

Zeigler R.S. and Alvarez E. (1990) Characteristics of *Pseudomonas* spp. causing grain discoloration and sheath rot of rice, and associated Pseudomonad epiphytes. Plant Disease, 74: 917–922.

Zhou Q.F., Lu S.N., Long G.X., Pan M.Q. and Zhang J.C. (1994) Sterility of rice plants with relation to copper deficiency under field conditions. Pedosphere, 4: 285–288.

Ziska L.H. and Manalo P.A. (1996) The effect of increasing night temperature can reduce seed set and potential yield of tropical rice. Australian Journal of Plant Physiology, 23: 791–794.

Pseudomonas Blight of Raspberry in Serbia

A. Obradović[1], V. Gavrilović[2], M. Ivanović[1], and K. Gašić[1]

Abstract Collapse of shoot tips and upper leaves, followed by wilting and dieback of entire shoots were observed on raspberry plants (cv. Willamette) grown in Serbia. Characteristic symptoms associated with oozing out from the diseased tissue indicated bacteria as a possible causal agent of the disease. Isolations resulted in bacterial isolates belonging to fluorescent pseudomonads. Results of LOPAT tests and studying of other biochemical and physiological characteristics identified the raspberry isolates as *Pseudomonas syringae*. Identification of the isolates by fatty acid analysis (MIDI Inc., Sherlock Version 4.5.) indicated that three of the isolates are closely related to the pathovar *syringae* (similarity index: 0.842–0.923), while the other six isolates clustered in a separate group with the pathovar *tomato* as a first choice but with rather low similarity index (0.515–0.778).

Keywords Bacterial disease, *Pseudomonas syringae*, raspberry, dieback, blight

1   Introduction

Being mostly export oriented, raspberry production plays an important role in fruit production in Serbia (Mišić et al., 2004). In spite of good experience in raspberry disease management practices, fruit growers were surprised by sporadic occurrence of blighting symptoms on shoots of young raspberry plants (cv. Willamette). Collapse of shoot tips and upper leaves, followed by wilting and dieback of entire shoots were observed in the end of May and beginning of June of 2002, 2003 and 2004 (Gavrilović et al., 2004) in the major raspberry growing region in Serbia. Small drops of pale yellow bacterial ooze were observed on the surface of the

[1] Department of Plant Pathology, Faculty of Agriculture, Nemanjina 6, 11080 Belgrade-Zemun, Serbia

[2] Institute for Plant Protection and Environment, T. Drajzera 9, 11000 Belgrade, Serbia

Author for correspondence: Aleksa Obradović; e-mail: aleksao@agrifaculty.bg.ac.yu

M'B. Fatmi et al. (eds.), *Pseudomonas syringae Pathovars and Related Pathogens.*   413
© Springer Science + Business Media B.V. 2008

diseased tissue. Cross sections of the diseased shoots revealed brown discoloration of vascular tissue. In some cases the symptoms were confused with a number of other diseases and disorders of raspberry. Therefore, we collected samples of the diseased plants in order to isolate and identify the causal agent.

2 Materials and Methods

Fragments of inner tissue, collected from diseased plants, were triturated in sterile tap water and the resulting suspensions were streaked onto nutrient agar sucrose (NAS) medium and King's medium B (KB) plates. The plates were incubated at 28°C for 48 h. Representative colonies were transferred to new KB plates and stored for further testing.

Hypersensitivity of the isolates was tested by infiltrating tobacco leaves (cv. Xanthi) with bacterial suspensions (10^8 CFU/ml). Inoculum was prepared from 24 h old cultures grown on KB medium and suspended in sterile tap water and concentration was adjusted to approximately 10^8 CFU/ml. Tobacco plants were kept at room temperature and monitored for confluent necrosis for 24 h. Pathogenicity of the isolates was investigated by inoculating young raspberry shoots and detached leaves, as well as unripe pear fruits (cv. William's). Shoots and pear fruits were inoculated by prickling with a bacteriological needle dipped into the suspension of bacteria, while stalks of detached leaves were dipped directly into the inoculum followed by wrapping the shoot prick points and stalk ends with wet cotton. Inoculated raspberry leaves and pear fruits were incubated in high humidity conditions for 5 days.

Bacteriological tests including Gram reaction, production of a fluorescent pigment on King's B medium, LOPAT tests, O/F metabolism of glucose, GATTa tests, growth at 37°C, nitrate reduction, utilization of sucrose, trehalose, mannitol, and ice nucleation activity (INA) were done using the standard procedures (Klement et al., 1990; Arsenijević, 1997; Schaad et al., 2001). Cellular fatty acid composition of 9 isolates was determined and used to identify the pathogen. Bacteria were grown, extracted, and analyzed according to the standard MIDI (Microbial Identification System, MIDI, Newark, DE) protocol (Technote #101-Identification of bacteria by GC analysis of fatty acids). Extracts were analyzed using the Sherlock System. Results were then used to compare unknown isolates to each other, and to data from the MIDI database, in order to identify the pathogen.

3 Results

3.1 Bacteriological Characteristics

Bacteria isolated from the diseased raspberry tissue formed round, shiny, grey–white colonies producing green fluorescent pigment on KB medium and typical levan-type growth on NAS plates. The isolates were Gram-negative, strict aerobes,

gelatin and aesculin hydrolysis positive, but negative for oxidase, pectate hydrolysis, arginine dihydrolase and tyrosinase activity. They utilized sucrose and mannitol
but not trehalose and L-tartrate, did not reduce nitrates and were variable in ability
to induce ice in INA test. Temperature of 37°C was lethal for the raspberry isolates.
According to biochemical and physiological test results, the causal agent of raspberry disease was identified as *Pseudomonas syringae*.

When the data generated by fatty acid analysis were processed by MIDI software, the isolates isolated from raspberry clustered into two groups (Euclidean distance ≤6) indicating a difference in fatty acid composition (Fig. 1). According to
the fatty acid profiles three isolates displayed greatest similarity with *Pseudomonas
syringae* pv. *syringae*. However, the other six raspberry isolates formed a separate
cluster from *P. s.* pv. *syringae*, with *P. s.* pv. *tomato* suggested as related organism
but with rather low similarity index (0.515–0.778).

3.2 Pathogenicity

Symptoms similar to those of natural infections were reproduced in raspberry
inoculation tests. Stalk end necrosis, progressing towards leaf veins, was observed
2 days after inoculation. Three days later, leaf tissue surrounding the vein became
dark brown or black and eventually necrotic. On inoculated raspberry shoots dis-
colored necrotic lesion developed at the point of prickling 2 days after inoculation, followed by wilting and necrosis of the upper part of the shoots in 5 days.
Raspberry isolates caused dark brown to black discoloration of inoculated pear
fruit tissue. Affected tissue became sunken and necrotic and unlike the fruits
inoculated with isolate of *Erwinia amylovora* no drops of bacterial exudate were
observed. All raspberry isolates induced hypersensitive reaction of tobacco leaves
in 24 h.

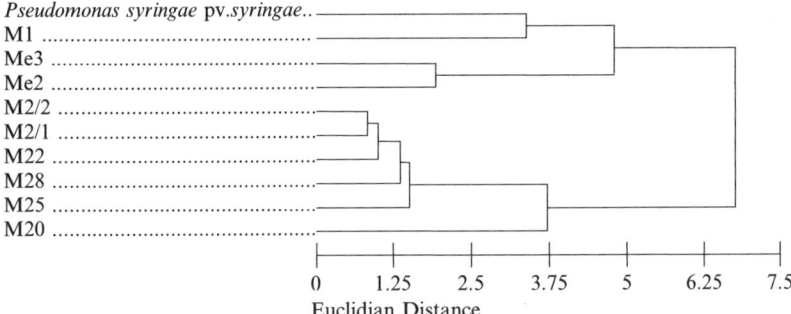

Fig. 1 Dendrogram representing cluster analysis based on fatty acid methyl ester profiles of
raspberry isolates (M1, Me2, Me3, M2/1, M2/2, M20, M22, M25, M28), and data for *Pseudomonas
syringae* pv. *syringae* from the software database

4 Discussion

Unlike pathogenic fungi, bacteria do not play an important role in pathology of raspberry in Serbia. *Pseudomonas syringae* was isolated from diseased raspberry plants in 1970 and 1971 (Ranković and Šutić, 1973). Since then, there were no other reports of the disease occurrence and severity, indicating that this is not an economically important disease in this country. Other than Serbia, *Pseudomonas* blight was recorded only in some parts of the United States and Canada, and also considered a relatively minor disease of red raspberry in that part of the world (Pepin, 1991). Although *P. syringae* is a wide spread pathogen of many plant species, *Pseudomonas* blight was not frequently observed on raspberry plants. The disease might have occurred more frequently but remained undetected due to indistinctive symptomatology contributing to confusion with a number of other diseases and disorders. Environmental factors play an important role in pathogenesis of the disease (Pepin, 1991). It is known that natural infections take place in cool weather conditions favoring formation of ice within plant tissue and causing damage, followed by more intensive bacterial colonization. This may be a part of an explanation for sporadic records of the disease in Serbia.

Ranković and Šutić (1973) reported about the problem in checking pathogenicity of the isolates indicating presence of pathogenic and non-pathogenic isolates within the population of isolates sharing the same bacteriological characteristics. They also observed loss of virulence after series of bacterial transfer on nutrient agar medium. In our experiments results of pathogenicity tests were not consistent for some isolates. Low temperature and high humidity or continuous presence of moisture contributed to the consistency of the results. The isolates also vary in their ability to cause ice nucleation. Variable pathogenicity to raspberry, inconsistent ice nucleation activity and toxin production of pseudomonads associated with diseased raspberries were already reported (Pepin, 1991).

The organism causing *Pseudomonas* blight of red raspberries has not been characterized to a pathovar level. Based on fatty acid profiles our raspberry isolates formed two groups, where three isolates clustered with *P. s.* pv. *syringae* and other six had no closely related bacteria found in MIDI database. This indicates that raspberry isolates belong to a less known part of the fluorescent pseudomonad population. Their opportunistic behavior keeps them undetected on asymptomatic plants representing potential threat for young and intensively growing plant tissue under conditions favorable for the infection and bacterial colonization. However, factors influencing sudden occurrence of *Pseudomonas* blight causing serious damage of raspberries should be further studied in order to prevent spreading of the disease and significant loss in the production.

Acknowledgements We thank E.R. Dickstein from Bacterial Identification and Fatty Acid Analysis Laboratory, Plant Pathology Department, University of Florida, for GC analysis of fatty acids of raspberry isolates.

References

Arsenijević M., 1997, Bakterioze biljaka (Bacterial Diseases of Plants), S Print, Novi Sad, Yugoslavia.

Gavrilović V., Milijašević S., Arsenijevic M., 2004, *Pseudomonas syringae* parazit maline u Srbiji (*Pseudomonas syringae* raspberry pathogen in Serbia). Jugosl. Vocar. 38(147–148): 183–190.

Klement Z., Rudolph K., Sands D.C., 1990, Methods in Phytobacteriology. Akadémiai Kiadó, Budapest.

Mišič P., Tešović Z., Stanisavljević M., Milutinović M., Nikolić M., Milenković S., 2004, Malina u Srbiji i Crnoj Gori (Raspberry in Serbia and Montenegro). Jugosl. Vocar. 38 (145–146): 5–22.

Pepin H.S., 1991, *Pseudomonas* blight. In: Ellis M.A., Converse R.H., Williams R.N. and Williamson B. (eds) Compendium of Raspberry and Blackberry Diseases and Pests. The American Phytopathological Society, St. Paul, MN.

Ranković M., Šutic D., 1973, Etioloska proucavanja oreolne pegavosti nekih sorti maline (The etiology of halo blight in certain raspberry cultivars). Zast. Bilja 126: 311–316.

Schaad N.W., Jones J.B., Chun W., 2001, Laboratory Guide for Identification of Plant Pathogenic Bacteria. The American Phytopathological Society, St. Paul, MN.

Studies on Plant Pathogenic Bacterium Causal Agent of Soybean Bacterial Spots (*Pseudomonas syringae* pv. *glycinea* (Coerper) Young et al.)

M. Ignjatov[1], J. Balaž[2], M. Milošević[1], M. Vidić[3], and T. Popović[2]

Abstract Bacterial blight is a disease of soybean appearing regularly with high intensity in Serbia. Isolation of pathogen was done using beef extract (MPA) and the medium enriched with sucrose (NSA) from infected soybean leaves of several varieties grown in Vojvodina province, during 2005.

Only the representative isolates were chosen for analyses. Pathogenicity of the isolates including the reference strain (National Collection of Plant Pathogenic Bacteria, United Kingdom-NCPPB 3318) was determined by inoculating soybean plants at cotyledon stage (cv. Balkan), by dipping and spraying with bacterial suspensions (conc. 10^8 CFU/ml). Pathogenicity of isolates was also confirmed on tobacco plants (HR positive reaction).

Cultural, biochemical and physiological characteristics were tested according to methods cited by Lelliot and Stead (1987), Klement et al. (1990), Schaad (1980). The colonies are chosen by fluorescence on KB agar, and by the LOPAT tests (levan production, oxidase reaction, potato rot, arginine dihydrolase production and tobacco hypersensitivity: +,−,−,−,+). The studied isolates of bacterium did not hydrolyze starch, gelatine and aesculin. They did not reduce nitrate but metabolized glucose oxidatively and were catalase positive. All the isolates produced acid from glucose, galactose, xylose, arabinose, manose, sucrose, rafinose and inositol.

Bacterium identification was confirmed by a serological method (agglutination tests by Express Kit – NEOGEN Europe Ltd., Scotland, UK), with appropriate antibodies of *Pseudomonas syringae* pv. *glycinea*. The studied isolates formed whitish sediment in the case of positive reaction.

Differential soybean sortiment: Acme, Chippewa, Flambeau, Harosoy, Lindarin, Merit and Norchief was used to determine which physiological race the isolates belonged to. Young plants were inoculated in two ways: by rubbing leaves with sterile cotton swab

[1] National Laboratory for Seed Testing, Maksima Gorkog 30, Novi Sad, Serbia

[2] Faculty of Agriculture, Department for Plant Protection, Trg Dositeja Obradovica 8, University of Novi Sad, Serbia

[3] Institute of field and vegetable crops, Maksima Gorkog 30, Novi Sad, Serbia

Author for correspondence: Jelica Balaz; e-mail: balazjel@polj.ns.ac.yu

M'B. Fatmi et al. (eds.), *Pseudomonas syringae Pathovars and Related Pathogens.*
© Springer Science+Business Media B.V. 2008

dipped in an aqueous bacterial suspension (Alvarez et al., 1995), and by spraying leaves under pressure (Cross et al., 1966; Balaž et al., 1990; Prom and Venette, 1997).

Based on the results of these studies we concluded that strains which cause blight of soybeans obtained from Vojvodina, belong to *Pseudomonas syringae* pv. *glycinea*, race 4.

Keywords Soybean, bacterial blight, *Pseudomonas syringae* pv. *glcyinea*, race 4, identification

1   Introduction

Bacterial blight (*Pseudomonas syringae* pv. *glycinea* (Coerper) Young et al.) is the most common bacterial disease of soybean. *P. s.* pv. *glycinea* has a worldwide distribution and occurs wherever soybeans are grown. Disease development is favored by cool, wet weather. Infected seed and plant debris are the main source of inoculum. Pathogen populations on seedlings, originating from infected seeds, are probably formed by cotyledons injured by abrasive soil particles. These populations then become a significant inoculum source for leaves that develop later (Daft and Leben, 1972). During years with a rainy spring bacterial blight appears with high intensity in Serbia (Balaž et al., 1995), and can cause great damage especially in early maturing varieties (Vidić and Balaž, 1997). Losses caused by the disease were estimated to approximately 11% annually (Sinclair and Backman, 1989). Because of the existence of more than one physiological race of this pathogen and favorable climatic conditions for development of diseases in Vojvodina region, detection of bacterial nature of disease has a great significance for crop protection, improvement and development of resistant soybean lines and varieties.

2   Materials and Methods

2.1   Isolation

Isolation of pathogens was done from diseased soybean leaves with characteristic spots, using standard methods of smearing macerated material on nutritive agar medium (NSA and MPA) (Shaad, 1980; Fahy and Hayward, 1983; Lelliot and Stead, 1987; Arsenijević, 1997). Samples were collected from several localities and different soybean varieties.

2.2   Pathogenicity Test

2.2.1   Hypersensitive Reaction (HR) on Tobacco

Pathogenicity of obtained bacterial isolates was tested on tobacco plant (*Nicotiana tabacum* L.). Bacterial suspensions (conc. 10^6 CFU/ml) were injected into the intercellular spaces of an intact tobacco leaf with a hypodermic needle (Klement et al., 1990). The reference strain (National Collection of Plant Pathogenic Bacteria, United Kingdom-NCPPB 3318) was included in this HR test.

2.2.2 Pathogenicity on Soybean

Soybean seeds (cv. Balkan) were sown in plastic boxes with wet and sterile sand covered with plastic lids. After 5 days, soybean seedlings (cotyledon stage) were used for pathogenicity test. Pathogenicity test was done on fully expanded soybean cotyledons by dipping and spraying with aqueous suspensions of the investigated isolates.

1. **Inoculation by dipping:** the method applied by Šutić (1951) for testing susceptibility of cotton varieties to *Xanthomonas campestris* pv. *malvacearum*, was used as a basis in this investigation. Wounded (punctured with sterile needle) and intact cotyledons of soybean were dipped into a bacterial suspension (conc. 10^8 CFU/ml) for 2 h. Plant roots were covered with wet paper in order to prevent drying. After 2 h, plants were transplanted to pots containing a 3:1 mixture of sterile substrate (Klasmann 2) and sand. After inoculation, the seedlings were kept at room temperature ($\approx 25°C$).
2. **Inoculation by spraying:** Wounded cotyledons were inoculated by spraying with an aqueous bacterial suspension (conc. 10^8 CFU/ml). The plants were incubated for 48 h in a mist chamber at 25°C, then transferred to room temperature and observed daily during 15 days. During the experiments plants were watered regularly and humidified with a hand sprayer. Control plants were inoculated with distilled water.

2.3 Morphological and Growth Characteristics

Gram reaction was tested using KOH test (Suslow et. al., 1982). Growth characteristics were determined on the basis of appearance, colour, size and shape of colonies developed on nutritive media MPA and NSA (Arsenijević, 1997; Schaad et al., 2001). King's medium B (King et al., 1954) was used for green fluorescent pigment production.

2.4 Biochemical and Physiological Characteristics

Biochemical and physiological characteristics of obtained bacterial isolates were studied. The following tests were used: levan production, oxidase and catalase test, O/F, metabolism of arginine, potato tuber rot, gelatin liquefaction, aesculin and starch hydrolysis, nitrate reduction, carbon hydrates metabolism (Lelliot and Stead,1987; Arsenijević, 1988; Schaad, 2001).

2.5 Agglutination Test

Agglutination test (express agglutination test) was used for *P. s.* pv. *glycinea* isolates identification with appropriate antybodies (Express Kit, NEOGEN Europe Ltd., Scotland, UK). Small quantities of testing bacterial colonies were tranferred on a testing card with applicator sticks and mixed with one drop of test reagent. Positive reaction is clearly indicated as granular agglutination, after 60 s. In the case of negative reaction, the drop of test reagent remains transparent (no agglutination). The results were compared with positive and negative controls of Express Kit.

2.6 Identification of Physiological Races

Reactions of differential soybean cultivars (Acme, Chippewa, Flambeau, Harosoy, Lindarin, Merit and Norchief) were used for race separation of *P. s.* pv. *glycinea* according to Cross et al. (1966). Isolates were applied onto leaves (in stage of ⅓–½ of fully developed leaf area) of differential cultivars in two ways: by rubbing leaves with a sterile cotton swab soaked in an aqueous bacterial suspension (Alvarez et al., 1995), and by spraying leaves under pressure with the bacterial isolates (Cross et al., 1966; Balaž et al., 1990; Prom and Venette, 1997). Aqueuos suspensions of inoculum were prepared from cultures grown 24 h on KB – agar and were adjusted using McFarlands scale (conc. 10^6–10^7 CFU/ml). After inoculation plants were covered with wet polyethilene bags for 48 h. Three days after inoculation plants were examined daily and the final estimation was done after 14 days. Leaves with water – soaked lesions were considered susceptible and leaves with no hydrosis and browning were designated as resistant.

3 Results and Discussion

3.1 Isolation

Three days after isolation, colonies on nutrient agar were circular, smooth, shiny and white. Five isolates were chosen for further investigations (Table 1).

Table 1 Bacterial isolates from soybean

Isolate	Variety	Locality	Date of isolation
B2/4	Balkan	Sombor	20.06.2005
B13/2	Balkan	Begec	24.06.2005
R12/2	Rita	Subotica	24.06.2005
R9/3	Ravnica	Backa Topola	25.06.2005
P15/2	Proteinka	Bijeljina	12.07.2005

3.2 Pathogenicity Tests

3.2.1 Hypersensitive Reaction (HR) on Tobacco

Pathogenicity test on tobacco showed that all the tested isolates caused hypersensitive reaction after 24 h, as the consequence of incompatible relationship between pathogen and host.

3.2.2 Pathogenicity on Host Plants (Soybean)

Pathogenicity test showed that all the isolates caused greasy and individual spots in all applied methods (Fig. 1). It was also observed that pathogenicity depended on applied inoculation method and isolates (Table 2). The first symptoms appeared 3 days after inoculation using spraying method. Investigated isolates showed different degrees of virulence. In the case of inoculation by dipping in bacterial suspension without injury, only P15/2 isolate caused appearance of individual, greasy spots after the 5th day, which revealed its strong pathogenicity, while R9/3 and B13/2 isolates were characterized by later appearance of symptoms (8 and 10 days, respectively after inoculation). Later appearance of symptoms was characteristic for R9/3 isolate, inoculated by dipping wounded cotyledons (5 days after inoculation) indicating its lower pathogenicity. No differences were detected between isolates B2/5 and R12/2 in their pathogenicity.

3.3 Morphological and Growth Characteristics

Morphological characteristics showed that the tested isolates were Gram negative. Isolates formed characteristic water-soluble, green-fluorescent pigment on KB-agar and produced levan on nutrient agar enriched with sucrose.

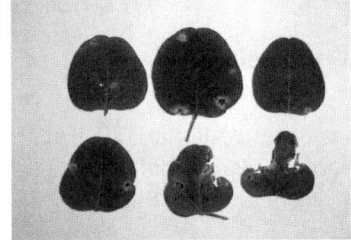

Fig. 1 Symptoms on soybean cotyledons and leaves

Table 2 Pathogenicity of bacterial strains on sobean cotyledons

Isolate	Inoculation method[a]	Symptom development[b] at different dates of observation following inoculation (days)					
		3 and 4	5	6	8	10	12
B2/5	I	–	–	–	+	++	+++
	II	+, +	++	+++	++++	Leaf	Leaf
	III	+, ++	+++	++++[c]	Leaf	Leaf	Leaf
B13/2	I	–	–	–	–	+	++
	II	+,+	++	+++	++++	Leaf	Leaf
	III	+,++	+++	++++[c]	Leaf	Leaf	Leaf
R9/3	I	–	–	–	+	++	+++
	II	–	+	++	++++	Leaf	Leaf
	III	+,++	+++	++++[c]	Leaf	Leaf	Leaf
R12/2	I	–	–	–	+	++	+++
	II	–,+	++	+++	++++	Leaf	Leaf
	III	+,++	+++	++++[c]	Leaf	Leaf	Leaf
P15/2	I	–	+	++	+++	+++	++++
	II	+,+	++	+++	++++[c]	Leaf	Leaf
	III	+,++	+++	++++[c]	Leaf	Leaf	Leaf

[a]Inoculation method I: inocultaion by dipping intact cotyledons; II: inoculation by dipping wounded cotyledons; III: inoculation by spraying wounded cotyledons
[b]+ 1–2 individual spots on cotyledons; ++ 3 individual spots on cotyledons; +++ spots fusion into dark, bent in surface; ++++ initiation of cotyledon necrosis
[c]Appearance of spontaneious infections (on uninjured parts of tissue); leaf – appearance of symptoms on leaves

3.4 Biochemical and Physiological Characteristics

Results of biochemical and physiological tests showed that the isolates were aerobes, oxidase and arginine dihydrolase negative. They were catalase positive, did not hydrolyse gelatine, aesculin and starch and did not reduce nitrate. All the isolates produced acid from glucose, galactose, xylose, arabinose, manose, sucrose, rafinose and inositol. Results of LOPAT tests showed that the isolates belonged to the group of fluorescent phytopathogenic *Pseudomonas*.

3.5 Agglutination Test

Antigenes of tested bacterial isolates were linked to antibodies of *P. s.* pv. *glycinea*, forming large molecular aggregates observed as granular sediments. Positive reaction was clearly indicated as granular agglutination with all the isolates.

3.6 Identification of Physiological Races

The investigated strains caused susceptible reactions in all inoculated soybean differential varietes. Specific symptoms were observed on plants inoculated by rubbing leaves, 3 days after inoculation. Symptoms on plants inoculated by spraying appeared after 48–72 h. Daily observation of symptoms showed an increase in number and size of spots on soybean plants. Based on the results of these studies, it was clear that all the isolates, obtained in Vojvodina province, caused appearance of wet, greasy spots and belong to *Pseudomonas syringae* pv. *glycinea*, race 4.

These results are in accordance with those obtained by Balaž et al. (1990). These authors also found the presence of race 4 in Serbia's agroecological conditions. First data about investigation of physiological strains of *P. s.* pv. *glycinea*, were obtained by Cross et al. (1966). According to studies of Fett and Sequeira (1981) carried out in Wisconsin (USA), most of the obtained isolates belonged to race 4. AboMoch et al. (1995), loc. cit. Balaž and Vidić (2001), concluded by testing populations of *P. s.* pv. *glycinea* in Europe (isolates collected in France, Germany, Hungary, Poland, Italy, Ukraine and former Yugoslavia), that race 4 was the most distributed. Prom and Venette (1997), testing 164 isolates originating from 170 most disseminated varieties from North Dakota pointed out that race 4 was predominating (63%) in relation to other recorded races (race 6, 22%; race 2, 7%; race 3, 0.3%, and race 5, 0.1%). The pathogenicity of 199 isolates of *P. s.* pv. *glycinea* obtained from five provinces in China was studied on seven soybean varieties, during 1990–1994. Eight separate pathogenic races (1, 2, 4, 5, 9, 10, 11, 12) were distinguished by the reactions of the seven different varieties. Race 4 was dominant (56%) in China (Gao Jie, 1998).

4 Conclusion

On the basis of pathogenic, morphological, biochemical, physiological and serological characteristics, we concluded that the studied isolates belong to *Pseudomonas syringae* pv. *glycinea*, race 4.

References

AboMoch F., Mavridis A., Rudolph K. (1995) Determination of races of *Pseudomonas syringae* pv. *glycinea* occuring in Europe. *Journal of Phytopathology*, 143(1):1–5.

Alvarez E., Braun E. J., McGee D. C. (1995) New assays for detection of *Pseudomonas syringae* pv. *glycinea* in soybean seed. *Plant Diseases*, 79(1).

Arsenijević, M. (1988) Bakterioze biljaka. Naučna knjiga. Beograd.

Arsenijević, M. (1997) Bakterioze biljaka. S print. Novi Sad.

Balaž J., Vidić, M. (2001) Iznalaženje otpornijih genotipova soje prema dominantnim rasama *Pseudomonas syringae* pv. *glycinea*. In: Bošković J., Bošković M. Veternik (eds) Primena sistema gen za gen, pp. 219–239.

Balaž F., Tošić M., Balaž J. (1995) Zaštita biljaka-bolesti ratarskih i povrtarskih biljaka. Novi Sad.

Balaž J., Arsenijević M., Vidić M. (1990) Bakteriološke karakteristike i fiziološke rase *Pseudomonas syringae* pv. *glycinea* (Coerper) Young, Dye et Wilkie parazita soje. *Zaštita bilja*, 41(4), br. 194: 423–429. Beograd.

Cross J. E., Kennedy B. W., Lambert J. W., Cooper R. L. (1966) Pathogenic races of the bacterial blight pathogen of soybeans, *Pseudomonas glycinea*. *Plant Disease Reporter*, 50:557–560.

Daft G.C. and Leben C. (1972) Bacterial blight of soybeans: seedling infection during and after emergence. *Phytopathology*, 62(10):1167–1170. View Abstract.

Fahy P. C., Hayward A. C. (1983) Media and methods for isolation and diagnostic test. In: Fahy, P. C., Persley, G. J (eds) Plant Bacteria Diseases. A Diagnostic Guide. Academic, Australia.

Fett W. F., Sequeira L. (1981) Further characterization of the physiological races of *Pseudomonas glycinea*. *Canadian Journal of Botany* 59:283–287.

Gao Jie (1998) Physiological specialization of the bacterial blight pathogen of soybeans *Pseudomonas syringae* pv. *glycinea*. *Journal of Jiling Agricultural University* 20:10–12.

King E. O., Ward M. K., Raney D. E. (1954) Two simple media for demonstration of pyocyanin and fluorescein. *Journal of Laboratory and Clinical Medicine* 44:301–307.

Klement Z., Rudolph K., Sands D. C. (1990) Methods in Phytobacteriology. Akademiai Kiado. Budapest.

Lelliott R. A., Stead D. E. (1987) Methods for the Diagnosis of bacterial Diseases of Plants. British Society for Plant Pathology. Blackwell, Oxford, London.

Prom L. K., Venette J. R. (1997) Races of *Pseudomonas syringae* pv. *glycinea* on commercial soybeans in eastern North Dakota. *Plant Disease* 81:541–544.

Schaad N. (1980) Laboratory guide for identification of plant pathogenic bacteria. APS, St. Paul, MN.

Schaad N., Jones J. B., Chun W. (2001) Laboratory guide for identification of plant pathogenic bacteria. APS Press, St. Paul, MN.

Sinclair J. B., Backman P. A. (1989) Compendium of Soybean Diseases. Third edition: 1–106. APS Press, St. Paul, MN.

Suslow T. V., Schroth M. N., Isaka M. (1982) Application of a rapid method for Gram differentination of plant pathogenic and saprophytic bacteria without staining. *Phytopathology*, 72:917–918. APS Press, St. Paul, MN.

Šutić D. (1951) Otpornost nekih sorata pamuka prema *Bacterium malvacearum* E. F. Smith. *Zaštita bilja*, br. 3. Beograd.

Vidić M., Balaž J. (1997) Reakcija genotipova soje prema *Pseudomonas syringae* pv. *glycinea*. *Zaštita bilja*, 48(2), br. 220: 119–125. Beograd.

Bacterial Diseases of *Agaricus bisporus* in Serbia

A. Obradović, K. Gašić, and M. Ivanović

Abstract An increase of mushroom production in Serbia contributed to the increased incidence and importance of problems related to mushroom pathology. Apart from the decline in mushroom quality due to a maturation or senescence process, mushroom caps are often colonized by mycopathogenic microorganisms causing growth deformation and decay. From samples of diseased white button mushrooms we isolated bacterial isolates associated with either small creamy or brown spots, with greasy appearance and oozing, or large, sunken and dark brown lesions spread over the diseased pileus. The isolates were selected for further investigation according to their pathogenicity to *A. bisporus* sporocarp slices. According to the pathogenic and bacteriological characteristics of the isolates, most of them were identified as *Pseudomonas agarici* Young (1970). The rest of the isolates remained unidentified.

Keywords Mushroom, *Pseudomonas agarici*, bacterial spot, decay

1 Introduction

The cultivation of button mushroom, or champignon (*Agaricus bisporus* (Lange) Imbach), is a fast growing small farm business in Serbia. However market demand for fresh champignons increases constantly setting high quality standards. The mushroom quality declines rapidly in the bed due to maturation or senescence processes, as well as microbial colonization of mushroom caps. The dry bubble (*Verticillium fungicola*) and wet bubble (*Mycogone perniciosa*) diseases were reported as the most frequently occurring diseases of cultivated mushrooms in

Department of Plant Pathology, Faculty of Agriculture, Nemanjina 6, 11080 Belgrade-Zemun, Serbia

Author for correspondence: Aleksa Obradovic; e-mail: aleksao@agrifaculty.bg.ac.yu

M'B. Fatmi et al. (eds.), *Pseudomonas syringae Pathovars and Related Pathogens.*   427
© Springer Science+Business Media B.V. 2008

Fig. 1 *Agaricus bisporus*. Samples of diseased pilei showing symptoms associated with complex population of bacteria (natural infection)

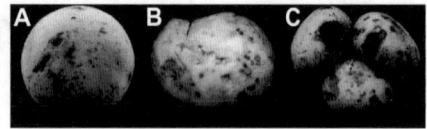

Serbia (Potočnik et al., 2004). However, occurrence of mushroom bacterial diseases in Serbia has never been reported to date. Last year, samples of diseased white button mushrooms were received from two mushroom farms in central Serbia. They exhibited symptoms similar to those described as "drippy gill" (Young, 1970; Gill and Cole, 2000) as well as other symptoms with no diagnostic value. Bacteria were associated with either small creamy or brown spots, with greasy appearance and oozing (Fig. 1A and B), or large, sunken and dark brown lesions spread over the diseased pileus (Fig. 1C). The objective of this study is to identify the causal agent of this disease.

2 Materials and Methods

2.1 Isolation and Maintenance of Bacterial Isolates

Bacteria were isolated either form pileus fragments or individual droplets oozing out from the diseased mycelial tissue. Symptomatic *Agaricus* sporocarps were rinsed in tap water and dipped into a solution of commercial bleach and water (1:5 v/v) for 1 min followed by rinsing in sterile distilled water. Small pieces were excised from the zone between symptomatic and asymptomatic tissue using sterile scalpel blade and placed into 100 µl of sterile tap water and triturated and the resulting suspension was streaked onto King's medium B (KB) plates (King et al., 1954). If oozing from the tissue was observed (Fig. 1B) bacteria were isolated by picking up individual droplets with a sterile loop and direct streaking onto KB plates. Following incubation at 27°C for 48 h, single colonies were transferred onto new KB plates and incubated for another 24 h after which the cultures were either stored at −80°C or maintained by subculturing for further testing.

2.2 Pathogenicity Test

Inoculum was prepared from 24 h old cultures by suspending bacterial growth in sterile tap water and adjusting the concentration to approximately 10⁸ CFU/ml. Caps of excised *A. bisporus* sporocarps were either cut longitudinally in 5 mm thick slices or sectioned in cube-shaped pieces. Pathogenicity was tested by placing a 10 µl drop of the inoculum on the central part of the slice or top of the cube, followed by incubation in a humid chamber at room temperature. Occurrence of symptoms was recorded daily.

2.3 Characterization of the Isolates

The isolates were characterized by performing standard bacteriological tests (Schaad et al., 2001). Known mushroom pathogenic bacteria reference isolates were used for comparison. Fatty acid composition of nine representative isolates was analyzed according to the standard MIDI (Microbial Identification System, MIDI, Newark, DE) protocol (Technote #101-Identification of bacteria by GC analysis of fatty acids). Extracts were analyzed using the Sherlock System and compared to data from the MIDI database.

3 Results and Discussion

An increase of mushroom production in Serbia intensified cropping and farming practices for maximizing yield and quality, raising the production value and consequently bringing in focus mushroom pathology. In order to respond to the farmers observations and search for answers we studied etiology of various symptoms mostly of white button mushrooms (*A. bisporus*). Many times isolations resulted in heterogeneous populations of bacteria, with unclear role of the majority of them. However, Gram-negative and weak-fluorescent bacteria were isolated from *A. bisporus* caps showing symptoms of small creamy or brown greasy spots with oozing (Fig. 1B). The isolates induced pale-brown discoloration of inoculated champignon cap slices or cubes. Similar symptoms were also caused by the reference isolate of *P. agarici* (ICMP 13775). The isolates were strict aerobes, LOPAT profile − + − − −, assimilated mannitol, but not trehalose and sucrose, and were negative for growth at 37°C, nitrate reduction, gelatin hydrolysis and ice nucleation. Based on results of pathogenicity and biochemical tests, the investigated isolates were identified as *P. agarici*. Close relatedness of nine mushroom isolates from Serbia and *P. agarici* was confirmed by the analysis of cellular fatty acid composition (Fig. 2). When the data were processed by MIDI software, the mushroom Isolates and *P. agarici* clustered into one group (Euclidean distance ≤ 6).

However, many bacterial Isolates associated with diseased *Agaricus* sporocarps differed in some characteristics from *P. agarici* reference isolate and therefore remained unidentified. This indicated that the population of bacteria present in *A. bisporus* culture is complex. Symptoms, if not associated with oozing, are not of diagnostic importance, contributing to the possible confusion with symptoms

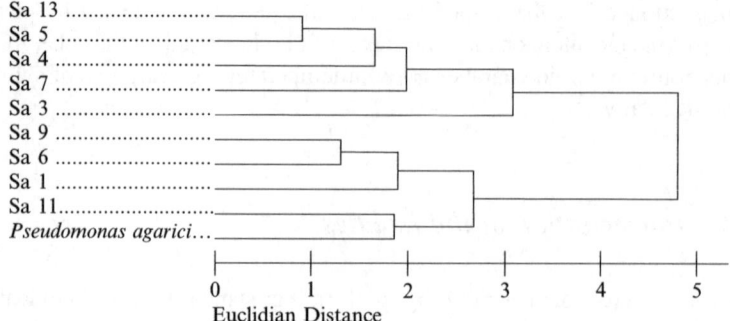

Sa 13
Sa 5
Sa 4
Sa 7
Sa 3
Sa 9
Sa 6
Sa 1
Sa 11
Pseudomonas agarici…

0   1   2   3   4   5

Euclidian Distance

Fig. 2 Dendrogram representing cluster analysis based on fatty acid methyl ester profiles of mushroom Isolates (Sa 1, Sa 3, Sa 4, Sa 5, Sa 6, Sa 7, Sa 9, Sa 11, Sa 13), and data for *Pseudomonas agarici* from the software database

caused by other biotic or abiotic factors and consequently to inadequate disease management. This is the first report of *P. agarici* as the causal agent of bacterial disease of cultivated mushrooms in Serbia.

Acknowledgements We thank E.R. Dickstein from Bacterial Identification and Fatty Acid Analysis Laboratory, Plant Pathology Department, University of Florida, for GC analysis of fatty acids.

References

Gill W. and Cole T. (2000) Aspects of the pathology and etiology of 'drippy gill' disease of the cultivated mushroom *Agaricus bisporus*. Can. J. Microbiol. 46: 246–258.

King E.O., Ward M.K., and Raney A.B. (1954) Two simple media for the demonstration of pyocyanin and fluorescin. J. Lab. Clin. Med. 44: 301–307.

Potočnik I., Tanović B., Vračarević M., Obradović A., and Todorović B. (2004) Prouzrokovači truleži šampinjona u Srbiji (Champignon rot causal agents in Serbia). Biljni lekar, 1: 40–44.

Schaad N.W., Jones J.B., and Chun W. (2001) Laboratory Guide for Identification of Plant Pathogenic Bacteria. The American Phytopathological Society, St. Paul, MN.

Young J.M. (1970) Drippy gill: a bacterial disease of cultivated mushrooms caused by *Pseudomonas agarici* n. sp. N. Z. J. Agric. Res. 13: 977–990.

Author Index